Earthquake Engineering and Structural Dynamics

Earthquake Engineering and Structural Dynamics

Edited by **Agnes Nolan**

NY RESEARCH PRESS

New York

Published by NY Research Press,
23 West, 55th Street, Suite 816,
New York, NY 10019, USA
www.nyresearchpress.com

Earthquake Engineering and Structural Dynamics
Edited by Agnes Nolan

International Standard Book Number: 978-1-63238-473-7 (Hardback)

Contents

Preface

This book provides comprehensive insights into the fields of earthquake engineering and structural dynamics. It comprises of research work contributed by various experts and researchers in the field of earthquake engineering. Both these disciplines focus on providing solutions to the problems created by damaging earthquakes, by planning, designing, constructing and managing earthquake-resistant structures and facilities. The significant studies included in this book, chart new and vital directions of research in the field of seismology. Several studies included discuss the probability of structural damages, others present approaches related to mitigating the risks of structural damage caused by earthquakes. Through this book, we attempt to further enlighten the readers about the new concepts of these disciplines. This book studies, analyses and upholds the pillar of earthquake engineering and structural dynamics. It is a ripe text for engineers, seismologists, geologists, researchers and students associated with these fields.

After months of intensive research and writing, this book is the end result of all who devoted their time and efforts in the initiation and progress of this book. It will surely be a source of reference in enhancing the required knowledge of the new developments in the area. During the course of developing this book, certain measures such as accuracy, authenticity and research focused analytical studies were given preference in order to produce a comprehensive book in the area of study.

This book would not have been possible without the efforts of the authors and the publisher. I extend my sincere thanks to them. Secondly, I express my gratitude to my family and well-wishers. And most importantly, I thank my students for constantly expressing their willingness and curiosity in enhancing their knowledge in the field, which encourages me to take up further research projects for the advancement of the area.

Editor

Application of story-wise shear building identification method to actual ambient vibration

Kohei Fujita , Ayumi Ikeda and Izuru Takewaki *

Department of Architecture and Architectural Engineering, Kyoto University, Kyoto, Japan

A sophisticated and smart story stiffness system identification (SI) method for a shear building model is applied to a full-scale building frame subjected to micro-tremors. The advantageous and novel feature is that not only the modal parameters, such as natural frequencies and damping ratios but also the physical model parameters, such as story stiffnesses and damping coefficients, can be identified using micro-tremors. While the building responses to earthquake ground motions are necessary in the previous SI method, it is shown in this paper that the micro-tremor measurements in a full-scale five-story building frame can be used for identification within the same framework. The SI using micro-tremor measurements leads to the enhanced usability of the previously proposed story-wise shear building identification method. The degree of auto-regressive eXogenous models and the cut-off frequencies of band-pass filter are determined to derive reliable results.

Keywords: system identification, ambient vibration, stiffness evaluation, identification function, multiple objective genetic algorithm

Edited by:
Tomaso Trombetti,
Università di Bologna, Italy

Reviewed by:
Bing Qu,
California Polytechnic
State University, USA
Stefano Silvestri,
Università di Bologna, Italy

*Correspondence:
Izuru Takewaki,
Department of Architecture and
Architectural Engineering,
Kyoto University,
Kyoto 615-8540, Japan
e-mail: takewaki@archi.kyoto-u.ac.jp

Introduction

Much research interest has been directed recently to system identification (SI) of civil, mechanical, and aerospace structures in response to the increasing need of enhancement of safety and upgrade of ability of damage detection (or damage diagnosis) of various kinds of structures (Hart and Yao, 1977; Beck and Jennings, 1980; Hoshiya and Saito, 1984; Kozin and Natke, 1986; Agbabian et al., 1991; Koh et al., 1991; Ghanem and Shinozuka, 1995; Hjelmstad et al., 1995; Shinozuka and Ghanem, 1995; Doebling et al., 1996; Hjelmstad, 1996; Masri et al., 1996; Housner et al., 1997; Kobori et al., 1998; Johnson and Smyth, 2006; Nagarajaiah and Basu, 2009; Fujino et al., 2010; Ji et al., 2011). Such need results from the accelerated demand of rapid assessment on material aging issues and of continuing use of buildings after earthquakes. The Building Continuity Plan (BCP) and resilience-oriented design framework also supports the increasing need of advancement of structural health monitoring technologies. It is also well-recognized that SI plays an important and principal role in reducing gaps between the constructed structural systems and their structural design models (model refinement).

Basically, two kinds of branches of the SI technique are known. One is called the modal parameter SI, which is well-established, and another is called the physical parameter SI. The modal parameter SI is well-established and various types of research have been conducted and accumulated so far [for example, see Hart and Yao (1977) and Beck and Jennings (1980)]. Since the modal parameters are global parameters and stable in principle, many useful techniques have been proposed. In the modal parameter SI, for evaluating the natural frequency and damping ratio, the observations at two places at least (usually the base and the top) are necessary. Furthermore, for evaluating modal shape identification, the simultaneous observations (or interpolation from fewer observations) at many places are usually required that is often a cumbersome task in the modal parameter SI method. When the observations at all the floors of a building are possible, all story stiffnesses of the shear building model with known floor masses may be identified within an acceptable accuracy (Hjelmstad et al., 1995).

In contrast to such modal parameter SI, the physical parameter SI has been developed for direct identification of physical parameters (stiffness and damping coefficients). For example, Takewaki and Nakamura (2000; 2005) introduced a method based on the work by Udwadia et al. (1978). In that method, a shear building model is used and stiffness and damping coefficients of a specific story are identified directly from the floor accelerations just above and below the specific story. However, the method by Takewaki and Nakamura (2000; 2005) has a difficulty resulting from the small signal/noise (SN) ratio in the low frequency range and cannot be applied to ambient vibration data, e.g., micro-tremors, and to high-rise buildings with large height-to-width aspect ratios. The former problem has been a major and most difficult problem in the field of SI and the latter problem has been tackled by Minami et al. (2013) and Fujita et al. (2013) by using a shear-bending model.

Hernandez-Garcia et al. (2010a;b) have developed an interesting method of damage detection using a floor-by-floor approach to enhance the efficiency and accuracy of the identification results. Xing and Mita (2012) have devised a time-domain substructure damage identification method for shear buildings by focusing on a substructure consisting of one story. Zhang and Johnson (2013) have developed another substructure identification method for shear buildings, which considers the noise effect of recorded data on identification accuracy and utilizes an iterative inductive procedure from the top story. Furthermore, a combined method of the modal parameter SI and the physical parameter SI is also well used (for example, Shinozuka and Ghanem, 1995; Barroso and Rodriguez, 2004). After modal parameters are identified, physical parameters are determined by solving inverse problems in which the existence and uniqueness of solutions are principal problems. Takewaki and Nakamura (2009) have developed a method for identifying the temporal variation of modal properties of a base-isolated building during an earthquake by using the relation between auto-regressive eXogenous (ARX) model parameters and modal parameters.

The difficulty arising in the limit manipulation in the method by Takewaki and Nakamura (2000; 2005) has been overcome by introducing an ARX model in the previous paper (Maeda et al., 2011). The weakness of a small signal-to-noise (SN) ratio in the low frequency range in the method (Takewaki and Nakamura, 2000; 2005) has been avoided by using the ARX model and introducing new constraints on the ARX parameters. Another difficulty due to small vibration levels of micro-tremor has been tackled by introducing a combination of the ARX model, filtering in the frequency domain (low and high-cut filter), and averaging in the time domain (sequential time-window shift for Fourier transformation and averaging). In this paper, for obtaining reliable and stable identification of story stiffnesses, a practical procedure to decide the combination of the degree of ARX models and cut-off frequencies of filtering is investigated by applying a multi-objective optimization algorithm. It is shown that the previously

proposed story-wise stiffness identification method is applicable in the new approaches to actual recorded data of an ambient vibration level with the help of this sophisticated combination.

System Identification Method for Physical Parameters of Shear Building Model

Governing Equations of Shear Building Model

Consider an N-story shear building model with viscous damping as shown in **Figure 1**. Let m_j and k_j denote the mass of the j-th floor and the story stiffness of the j-th story and let c_j be the viscous damping coefficient of the j-th story. Since the formulation in the frequency domain is appropriate in the present formulation, all the governing equations are expressed in the frequency domain. Let "i" denote the imaginary unit.

Keeping the relations $\dot{U}(\omega) = i\omega U(\omega)$, $\ddot{U}(\omega) = -\omega^2 U(\omega)$ in mind, the equations of motion in the frequency domain for this shear building model subjected to the horizontal ground acceleration $\ddot{u}_g$ is expressed as:

$$(-\omega^2 \mathbf{M} + i\omega \mathbf{C} + \mathbf{K})\mathbf{U}(\omega) = -\mathbf{M1}\ddot{U}_g(\omega) \qquad (1)$$

where $\mathbf{U}(\omega)$ and $\ddot{U}_g(\omega)$ are the Fourier transforms of the horizontal displacements $\mathbf{u}(t) = \{u_j(t)\}$ of floors relative to ground and the ground base acceleration $\ddot{u}_g$, respectively. The vector $\mathbf{1}$ in the right-hand side of Eq. (1) denotes $\mathbf{1} = \{1, 1, \ldots, 1\}^T$. The mass, stiffness, and damping matrices of shear building model are defined by:

$$\mathbf{M} = \text{diag}\,(m_1, m_2, \cdots, m_N), \qquad (2a)$$

$$\mathbf{K} = \begin{bmatrix} k_1 + k_2 & -k_2 & & \\ -k_2 & k_2 + k_3 & -k_3 & \\ & & \ddots & \\ & & -k_N & k_N \end{bmatrix}, \qquad (2b)$$

$$\mathbf{C} = \begin{bmatrix} c_1 + c_2 & -c_2 & & \\ -c_2 & c_2 + c_3 & -c_3 & \\ & & \ddots & \\ & & -c_N & c_N \end{bmatrix} \qquad (2c)$$

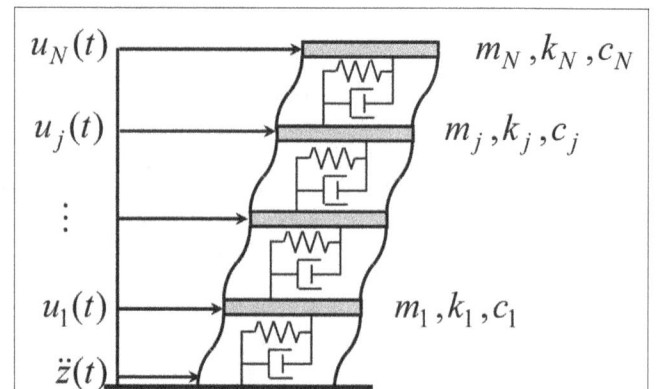

FIGURE 1 | N-story shear building model with viscous damping.

The transfer function $\mathbf{T}(\omega)$ can be defined from Eq. (1) as the ratio of the horizontal displacement to the ground acceleration.

$$\mathbf{T}(\omega) \equiv \mathbf{U}(\omega)/\ddot{U}_g(\omega) = -\mathbf{A}(\omega)^{-1}\mathbf{M1} \qquad (3)$$

where $\mathbf{A}(\omega) = -\omega^2\mathbf{M} + \mathrm{i}\,\omega\mathbf{C} + \mathbf{K}$.

Sub-system Identification Based on Identification Function

Based on the mathematical formulation for the SI of the physical parameters (Takewaki and Nakamura, 2000), the story stiffness and damping coefficient can be expressed by using the limit manipulation of the identification function $f_j(\omega)$ for $\omega \to 0$. The story stiffness and damping coefficient are derived as:

$$k_j = \lim_{\omega \to 0} \left(M_j \times \mathrm{Re}\left[-\omega^2 \frac{(\ddot{U}_g + \ddot{U}_j)}{(\ddot{U}_g + \ddot{U}_{j-1}) - (\ddot{U}_g + \ddot{U}_j)} \right] \right) \qquad (4)$$

$$c_j = \lim_{\omega \to 0} \left(M_j \times \frac{d}{d\omega}\left\{ \mathrm{Im}\left[-\omega^2 \frac{(\ddot{U}_g + \ddot{U}_j)}{(\ddot{U}_g + \ddot{U}_{j-1}) - (\ddot{U}_g + \ddot{U}_j)} \right] \right\} \right) \qquad (5)$$

where $M_j = \sum_{i=j}^{N} m_i$. Referring Takewaki and Nakamura (2000), the identification function can be defined by:

$$f_j(\omega) \equiv -\omega^2 M_j \frac{\ddot{U}_g + \ddot{U}_j}{\ddot{U}_{j-1} - \ddot{U}_j} \qquad (6)$$

In Eqs (4–6), $\ddot{U}_0 = 0$. From Eqs (4) and (5), it is concluded that the story stiffness and damping coefficient can be derived only from the accelerations just above and below floors of the object story in the previously proposed identification method (Takewaki and Nakamura, 2000).

Identification of Stiffness and Damping Coefficient using ARX Model

Taylor Series Expansion of Transfer Function using ARX Parameters

In the previous SI method (Takewaki and Nakamura, 2000; 2005) for shear building models as formulated in Eqs (4) and (5), the limit manipulation of the identification function $f_j(\omega)$ for $\omega \to 0$ was needed. However, when the identification functions are evaluated from the raw data such as actual vibration testing data, it is often the case that the identification functions become unstable and exhibits a large variability in the low frequency range. To overcome this difficulty, an ARX model is introduced, which is a time-domain model. The reliability of the ARX model in this direction has been confirmed and the applicability of the ARX model to shear building models and shear-bending models has been demonstrated in Maeda et al. (2011) and Minami et al. (2013).

The identification function $f_j(\omega)$ defined by Eq. (6) can be rewritten by introducing the transfer function $G_{j,j-1}(\omega)$ between j-th and $(j-1)$-th floors as

$$f_j(\omega) = -\frac{\omega^2 M_j}{\frac{1}{G_{j,j-1}(\omega)} - 1} \qquad (7)$$

where $G_{j,j-1}(\omega) \equiv (\ddot{U}_g + \ddot{U}_j)/(\ddot{U}_g + \ddot{U}_{j-1})$. The reason of the unstable phenomenon of the identification function in the low frequency range can be understood from Eq. (7). From the theoretical investigation, the limit value of the transfer function $G_{j,j-1}(\omega)$ at $\omega = 0$ should be $G_{j,j-1}(0) = 1$, because the j-th floor and $(j-1)$-th floor move identically at $\omega \to 0$. Therefore, the limit value of the denominator of the identification function $1/G_{j,j-1}(\omega) - 1$ for $\omega \to 0$ is zero. Furthermore, the limit value of the numerator of the identification function $\omega^2 M_j$ for $\omega \to 0$ is also zero. For these reasons, the limit value of the identification function $f_j(\omega)$ for $\omega \to 0$ is theoretically indefinite.

By using ARX parameters, the transfer function can also be expressed as (see Appendix)

$$G_{j,j-1}(\omega) = \frac{b_1 e^{-\mathrm{i}\omega T_0} + \cdots + b_n e^{-\mathrm{i}n\omega T_0}}{1 + a_1 e^{-\mathrm{i}\omega T_0} + \cdots + a_n e^{-\mathrm{i}n\omega T_0}} \qquad (8)$$

For evaluation of the limit value of the transfer function $G_{j,j-1}(\omega)$ for $\omega \to 0$, the formulation of the Taylor series expansion of the transfer function $G_{j,j-1}(\omega)$ is meaningful. The Taylor series expansion of the transfer function $G_{j,j-1}(\omega)$ in terms of ARX parameters can be defined as

$$G_{j,j-1}(\omega) \simeq A_0 + A_1\omega + A_2\omega^2 + \cdots \qquad (9)$$

Considering the relationship of ARX parameters in Eq. (8) and the coefficients of the Taylor series expansion in Eq. (9), the coefficients A_0, A_1, A_2 of the Taylor series expansion can be formulated in terms of the ARX parameters $\{a_k\}, \{b_k\}$ as follows.

$$A_0 = b_{\mathrm{sum}}/(1 + a_{\mathrm{sum}}) \qquad (10)$$

$$A_1 = \mathrm{i}T_0 \frac{(1 + a_{\mathrm{sum}})\sum_{k=1}^{n-1}(n-k)b_k - b_{\mathrm{sum}}\left\{ n + \sum_{k=1}^{n-1}(n-k)a_k \right\}}{(1 + a_{\mathrm{sum}})^2} \qquad (11)$$

$$A_2 = -\frac{T_0^2}{2}\left[\frac{\sum_{k=1}^{n-1}(n-k)^2 b_k}{1 + a_{\mathrm{sum}}} \right.$$
$$-\frac{b_{\mathrm{sum}}\left\{ n^2 + \sum_{k=1}^{n-1}(n-k)^2 a_k \right\}}{(1 + a_{\mathrm{sum}})^2}$$
$$-\frac{2\sum_{k=1}^{n-1}(n-k)b_k\left\{ n + \sum_{k=1}^{n-1}(n-k)a_k \right\}}{(1 + a_{\mathrm{sum}})^2}$$
$$\left. +\frac{2b_{\mathrm{sum}}\left\{ n + \sum_{k=1}^{n-1}(n-k)a_k \right\}^2}{(1 + a_{\mathrm{sum}})^3} \right] \qquad (12)$$

where $a_{\mathrm{sum}} \equiv \sum_{k=1}^{n} a_k, b_{\mathrm{sum}} \equiv \sum_{k=1}^{n} b_k$.

By introducing the real and imaginary parts of A_j as $A_j = A_j^R + \mathrm{i}A_j^I$ and substituting the properties of the real and imaginary parts of A_j into Eqs (10–12), the transfer function can be reduced to:

$$G_{j,j-1}(\omega) \simeq \left(A_0^R + A_2^R \omega^2 + \cdots \right) + \mathrm{i}\left(A_1^I \omega + A_3^I \omega^3 + \cdots \right) \qquad (13)$$

Constraints on ARX Parameters Derived from the Limit Value of Taylor Series Expansion of Transfer Function

As mentioned in the previous section, it is meaningful to note that Eqs (14a,b) can be derived from the mechanical interpretation, i.e., the j-th floor and $(j-1)$-th floor move identically at $\omega \to 0$ and $G_{j,j-1}(\omega)$ should not include linear terms of ω judging from Eqs (4–6).

$$\lim_{\omega \to 0} \mathrm{Re}\left\{G_{j,j-1}(\omega)\right\} = 1, \tag{14a}$$

$$\lim_{\omega \to 0} \frac{d}{d\omega} \mathrm{Im}\left\{G_{j,j-1}(\omega)\right\} = 0 \tag{14b}$$

From Eqs (13) and (14a, b), $A_0^R = 1, A_1^I = 0$. Furthermore, by substituting these equations into Eqs (10) and (11), the following equations can be derived.

$$\sum_{k=1}^{n} a_k + 1 = \sum_{k=1}^{n} b_k \tag{15}$$

$$\sum_{k=1}^{n-1} (n-k)b_k = n + \sum_{k=1}^{n-1} (n-k)a_k \tag{16}$$

For enhancing the reliability of the proposed SI method, these relations will be used as the constraints in the estimation of the ARX parameters. It is found that Eq. (15) is a linear equation for the ARX parameters. A batch processing least-squares estimation method (see Appendix) (Adachi, 2009) provides:

$$\mathbf{R}\boldsymbol{\theta} = \mathbf{f} \tag{17}$$

$\mathbf{R}$, $\boldsymbol{\theta}$, and $\mathbf{f}$ are defined in Appendix.

By applying the Lagrange multiplier method, the linear constraint can be incorporated into the batch processing least-squares estimation method as $\mathbf{p}^T\boldsymbol{\theta} = -1$ where $\mathbf{p} = \{1, \ldots, 1, -1, \ldots, -1\}^T$. Therefore, the present method is reduced to the problem for solving the following equations.

$$\begin{bmatrix} \mathbf{R} & \mathbf{p} \\ \mathbf{p}^T & 0 \end{bmatrix} \begin{Bmatrix} \boldsymbol{\theta} \\ \lambda \end{Bmatrix} = \begin{Bmatrix} \mathbf{f} \\ -1 \end{Bmatrix} \tag{18}$$

Objective Functions for Determination of Order of ARX Model

It is important to investigate how to determine the order of the ARX model. Let us introduce the following objective function in the time domain as the ensemble average of amplitude vector in time domain.

$$F_t = \sum_{i=1}^{N_t} f_t(t_i)/N_t \tag{19}$$

where $f_t(t_i)$ is defined by using an estimated time history u_{ARX} from ARX parameters and raw measurement data u_{raw} as:

$$f_t(t_i) = \left(\frac{u_{\mathrm{ARX}}(t_i) - u_{\mathrm{raw}}(t_i)}{\sigma_{\mathrm{raw}}} \times \frac{u_{\mathrm{raw}}(t_i) - \bar{u}_{\mathrm{raw}}}{\sigma_{\mathrm{raw}}} \right)^2 \tag{20}$$

In Eq. (20), σ_{raw} is the SD of u_{raw}, and $\bar{u}_{\mathrm{raw}}$ is the mean value of the time history. The first term $[u_{\mathrm{ARX}}(t_i) - u_{\mathrm{raw}}(t_i)]/\sigma_{\mathrm{raw}}$ of the

FIGURE 2 | Five-story steel building.

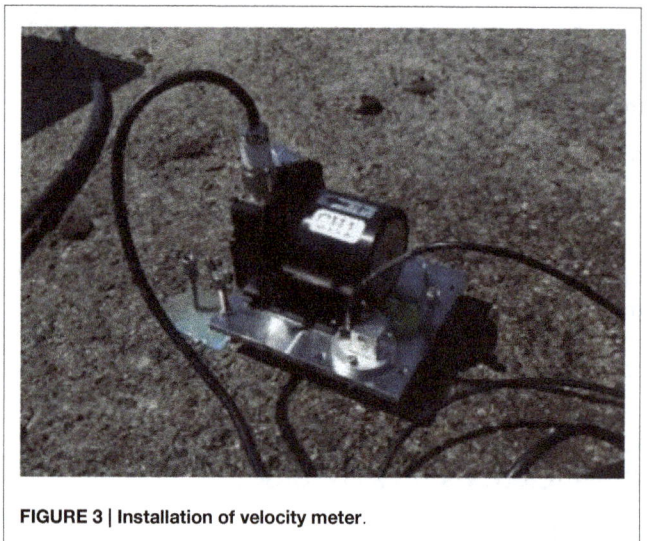

FIGURE 3 | Installation of velocity meter.

objective function in time domain indicates the accuracy of u_{ARX} compared with u_{raw}. The second term $[u_{\mathrm{raw}}(t_i) - \bar{u}_{\mathrm{raw}}]/\sigma_{\mathrm{raw}}$ can be regarded as the weighting function for the amplitude level of the time history, i.e., the influence of data on the objective function is small when $u_{\mathrm{raw}}(t_i) \approx \bar{u}_{\mathrm{raw}} \approx 0$. When the estimated time history u_{ARX}, which can be computed from the ARX parameters as $u_{\mathrm{ARX}} = \boldsymbol{\theta}^T \varphi(k)$, matches with the measured data in an acceptable accuracy, the objective function Eq. (19) is close to 0.

On the other hand, let us introduce another objective function in the frequency domain.

$$F_\omega = \sum_{i=1}^{\mathrm{NN}} \left(|G_{\mathrm{ARX}}(i)| - |G_{\mathrm{raw}}(i)|\right)^2 / \sum_{i=1}^{\mathrm{NN}} |G_{\mathrm{raw}}(i)|^2 \tag{21}$$

where G_{raw} denotes the transfer function computed from the records after appropriate post-data processing, e.g., filtering and ensemble averaging, and G_{ARX} is the transfer function evaluated by the ARX model. Note that $\omega_1 = \omega_l$ and $\omega_{\mathrm{NN}} = \omega_u$, where ω_l and

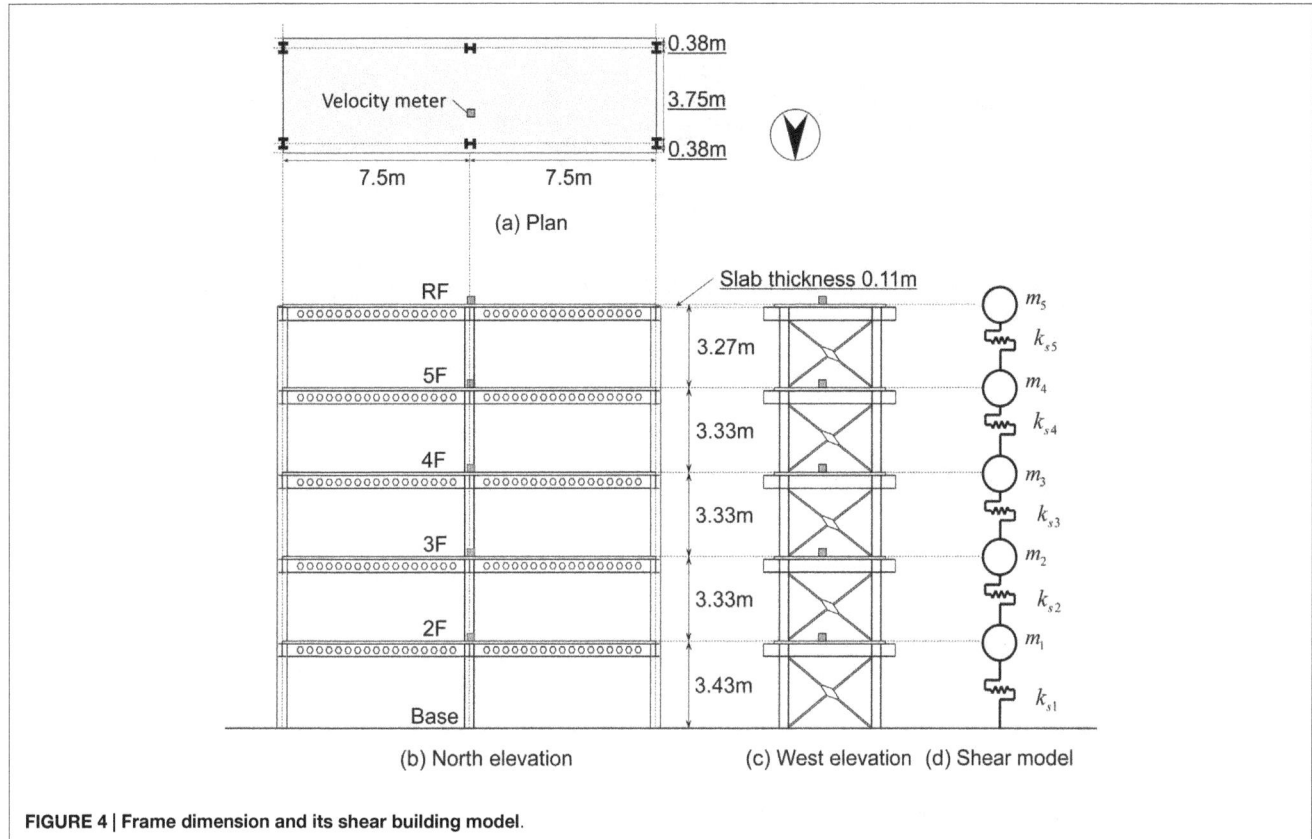

FIGURE 4 | Frame dimension and its shear building model.

ω_u are the lower and upper cut-off frequencies in the band-pass filter. It can be said that, if the objective function in the frequency domain is close to 0, the function in terms of the ARX model matches well with the record.

Application: Ambient Vibration Data in Steel Structure

Ambient Vibration Measurement in Five-Story Steel Frame Structure

Actual micro-tremor observations were conducted in an experimental real-size building at the Uji campus, Disaster Prevention Research Institute of Kyoto University, Japan. An overview photograph of the building is shown in **Figure 2**. The servo-type velocity meters (VSE-15D; *Tokyo Sokushin*) were installed in several stories. The velocity resolution of this velocity meter is 10^{-4} mm/s. **Figure 3** shows a photo of an installed velocity meter. The frame dimension and its shear building model are shown in **Figure 4**. The locations of velocity meters for three patterns of identification are presented in **Table 1**. In all measurement patterns, the velocity meters are fixed at the basement and roof floor so as to obtain the transfer function of the objective frame structure. Pattern A is aimed at identifying the story stiffnesses of the first and second stories and pattern B the fourth and fifth stories. On the other hand, pattern C is set for identifying the third story. The ambient measurements data were recorded in the long-span direction and short-span direction, respectively. The floor

TABLE 1 | Location of velocity meter and measurement pattern for stiffness identification.

	Pattern A; Object: 1st and 2nd story	Pattern B; Object: 4th and 5th story	Pattern C; Object: 3rd story
Roof	Located	Located	Located
5th story	–	Located	–
4th story	–	Located	Located
3rd story	Located	–	Located
2nd story	Located	–	–
1st story	Located	Located	Located

masses are estimated as $m_1 = 28.7 \times 10^3$ kg, $m_2 = 28.2 \times 10^3$ kg, $m_3 = m_4 = 27.6 \times 10^3$ kg, and $m_5 = 26.6 \times 10^3$ kg by preliminary investigation.

Ambient Data Measurement

Figure 5 shows some examples of recorded ambient data at the base, 3rd floor, 4th floor, and roof in the short-span direction in Pattern C. Durations of the measurement time were 5 and 10 min. **Figure 6** presents the transfer function of the roof velocity to the base for each pattern of the short-span direction measurements. In **Figure 6**, the transfer functions evaluated by raw data, i.e., no post-processing, are compared with those after smoothing and filtering. From these preliminary analyses of steel structure, natural frequencies can be estimated as shown in **Table 2**.

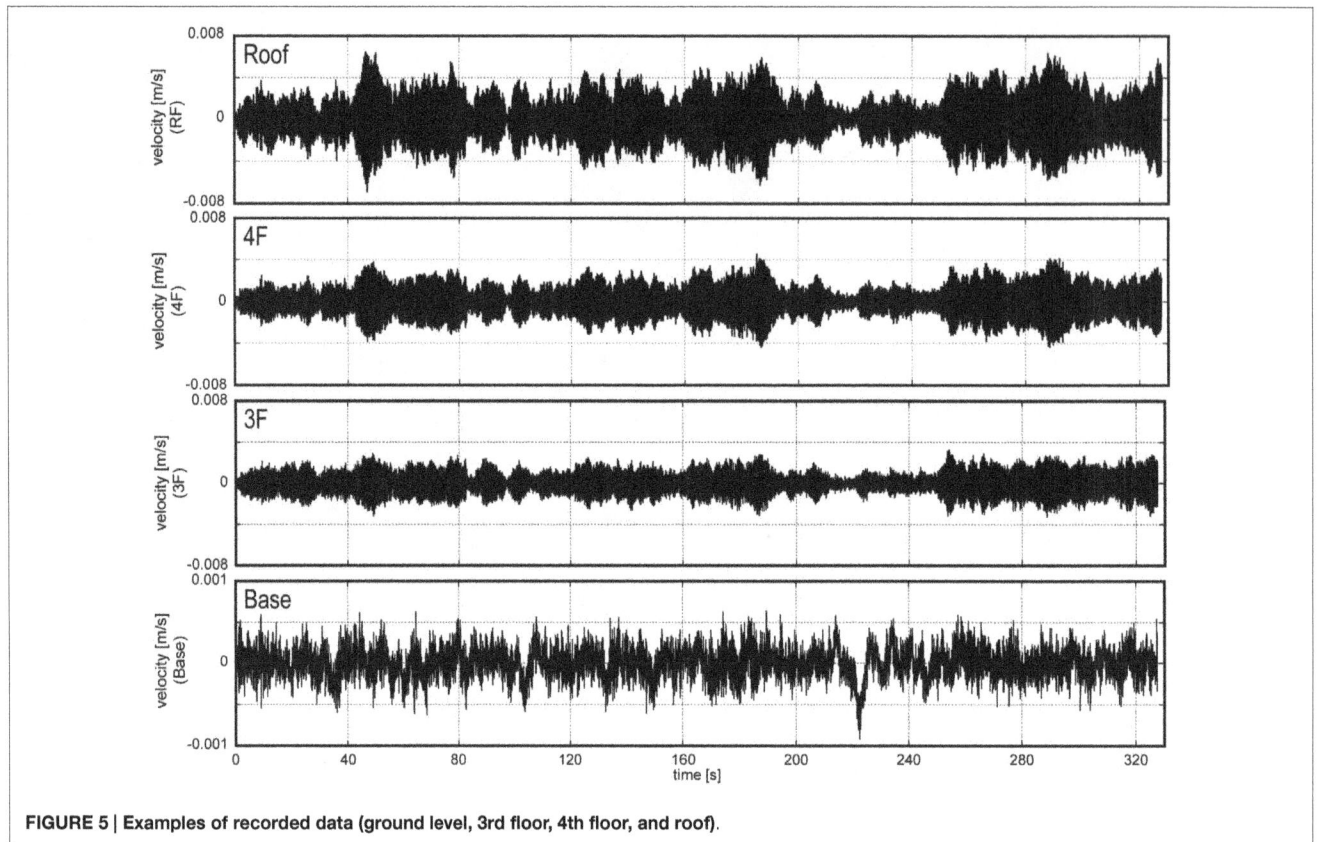

FIGURE 5 | Examples of recorded data (ground level, 3rd floor, 4th floor, and roof).

Optimal Combination of Order of ARX Model and Filtering Parameters

Determination of Identification Parameters by Iterative Manipulation

As seen in **Figure 6**, the transfer functions evaluated by raw data exhibit large variability in the low frequency range. For this reason, it is difficult to obtain the limit value of the identification function directly from the raw data. It is well known that the order of ARX models is important to obtain reliable identification results. Furthermore, it is also observed that the band-pass filter as post-data processing stabilizes the variability of the ARX modeling. Taking these unknown parameters for the structural identification into account, it is needed to investigate the influence of the combination of these unknown parameters on the stability of the identification results.

One of the solutions to this problem, i.e., the identification method combining the previous method (Hernandez-Garcia et al., 2010b; Xing and Mita, 2012) with the band-pass filtering and the ARX modeling is presented here. **Figure 7** shows the outline of the identification procedure including the band-pass filtering and the ARX modeling. In the flowchart (**Figure 7**), three procedures are needed; Procedure 1 is for determination of the order of the ARX model, which was derived by the parametrical analysis to minimize objective functions defined by Eqs (19) and (21); procedure 2 is for determination of the lower cut-off frequency; and procedure 3 is for that of the upper cut-off frequency.

Finally, the story stiffnesses are evaluated from the limit value of the identification function in Eq. (4) by selecting unknown identification parameters. If the stability of the limit value is not acceptable, it is needed to select the order of ARX models again in procedure 1. **Figure 8** presents examples of SI by using the measured ambient data of the real structure. The red lines, which can be derived by iterative manipulations, are the finally selected identification parameters for the order of the ARX model and cut-off frequencies. However, as seen in **Figure 8**, the objective functions in terms of these identification parameters vary drastically. Therefore, these procedures of iterative flows to determine several identification parameters may depend on the knowledge of structural engineers so as to derive reliable identification results.

Determination of Identification Parameters by Multi-Objective Optimization

It is important to select an appropriate combination of identification parameters for reliable identification of the story stiffnesses. In this section, this combination will be derived by applying a multi-objective optimization solver based on genetic algorithm. The design parameters for optimization are the order of the ARX model, lower and upper bounds of cut-off frequency. The objective functions are given by Eqs (18) and (20). In the application to the measured ambient data as shown in the following section, four velocity sensors are used simultaneously, which means that the transfer functions and identification functions can be evaluated

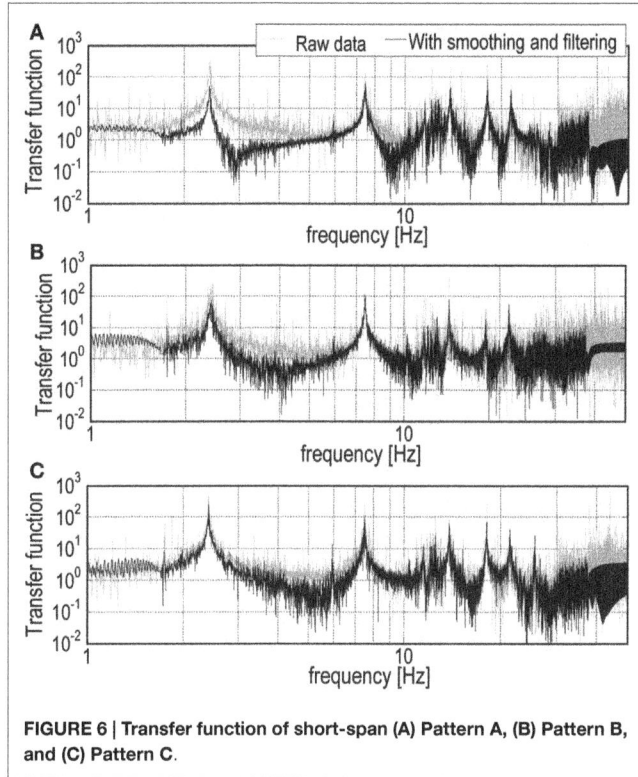

FIGURE 6 | Transfer function of short-span (A) Pattern A, (B) Pattern B, and (C) Pattern C.

TABLE 2 | Estimated natural frequencies in short-span direction from transfer function (roof/base).

1st	2nd	3rd	4th	5th
2.42 Hz	7.46 Hz	12.10 Hz	13.84 Hz	18.10 Hz

at three different consecutive stories. For the application to this experimental data, the number of the objective functions in the multi-objective optimization algorithm is six, i.e., the error rate in time domain and frequency domain at three consecutive stories.

The multi-objective optimization algorithm (MOGA-II) implemented in modeFRONTIER 4.5 (*ESTECO*) is used to investigate the optimal combination of the identification parameters. Initial design parameters are provided by the uniform random number method. **Figure 9** describes the multi-dimensional analysis in the post-data processing for all individuals in the multi-objective optimization (the number of designs is 8000). The points connected by lines represent a generated design and the corresponding error rates of objective functions. From this figure, an excellent combination of the identification parameters can be derived by determining an allowable range of objective functions. **Figure 9** shows two results (top and bottom illustrations) for different allowable ranges of the objective functions. In the bottom illustration, the error rate in time domain is widen, which causes the wide variability of design parameters. As for the order of ARX models, two different ranges of ARX order can

be seen as 26–36 and 58–60 by widening the error rate in time domain. From this multi-dimensional analysis, the evaluation of the objective functions influences the decision making on the identification parameters. **Figure 10** shows the comparison of the transfer function (roof/base) obtained from raw data with the transfer functions derived by the optimal combination of the order of ARX models and the cut-off frequencies in the multi-objective optimization (nine combinations are selected here). In **Figure 10**, no difference can be seen in these combinations, which demonstrates the reliability of the evaluation of error rate in the multi-objective optimization procedure.

Stability of Stiffness Identification by Limit Value of Identification Function

By applying the multi-objective optimization algorithm, several designs of the combination of identification parameters are obtained, which can make the objective functions in time and frequency-domain minimized in an acceptable manner. In this section, stability of the story stiffnesses is investigated by applying the SI method to the various (and optimized) identification functions derived by the multi-objective optimization. As explained before, the present SI method needs the evaluation of the limit value of identification functions. In the previously proposed SI method in the same framework, this limit value was just determined by selecting the raw value of the identification function at $\omega \approx 0$ (called *Method 1*). In this sense, *Method 1* is a simple and primitive one. **Figure 11** shows the cumulative frequency of the identified story stiffness k_i ($i = 1, 2, \ldots, 5$) for various combinations derived from the multi-dimensional analysis in **Figure 9** where the number of optimal designs is about 150. It can be observed that identified story stiffnesses by *Method 1* have large variability although the selected combination groups of the identification functions are similar. This means instability of the story stiffness identification.

For enhancing the stability of story stiffness identification, other approaches to limit value evaluation are proposed as:

Method 2: Using gradient sensitivity of identification function.

Method 3: Using model pole derived by ARX model.

A conceptual illustration of these methods for limit value evaluation is shown in **Figure 12**. In *Method 2*, the gradient sensitivity of the identification function is evaluated successively in the frequency domain. The limit of the identification function is determined by evaluating the mean value of the identification functions in a stable range where the gradient of the identification function is sufficiently small (see Figure12: *Method 2*). While, in *Method 3*, a raw value of the identification function at a particular frequency determined by the frequency for the model pole of ARX model is used as the limit value of the identification function. The particular frequency point is given by assuming an appropriate constant value, e.g., 0.5, as the ratio to the model pole frequency (see Figure12: *Method 3*). Let us see **Figure 11** again

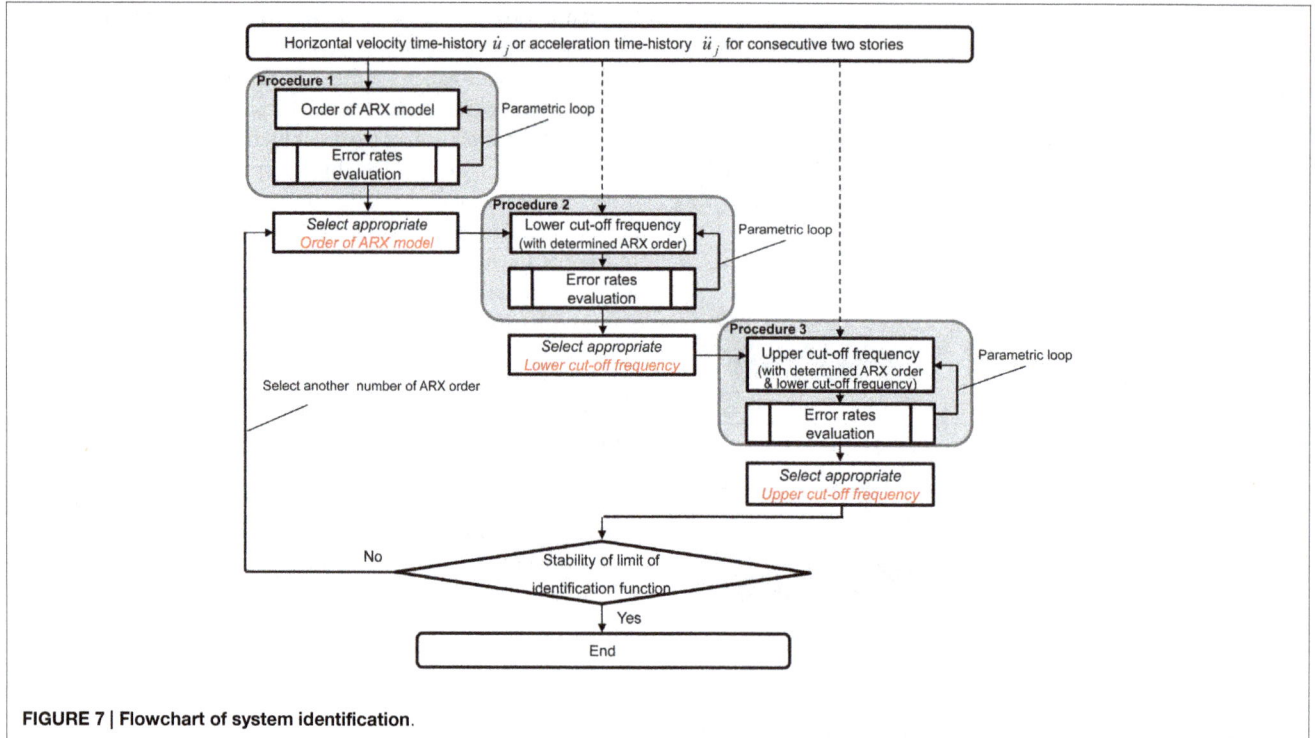

FIGURE 7 | Flowchart of system identification.

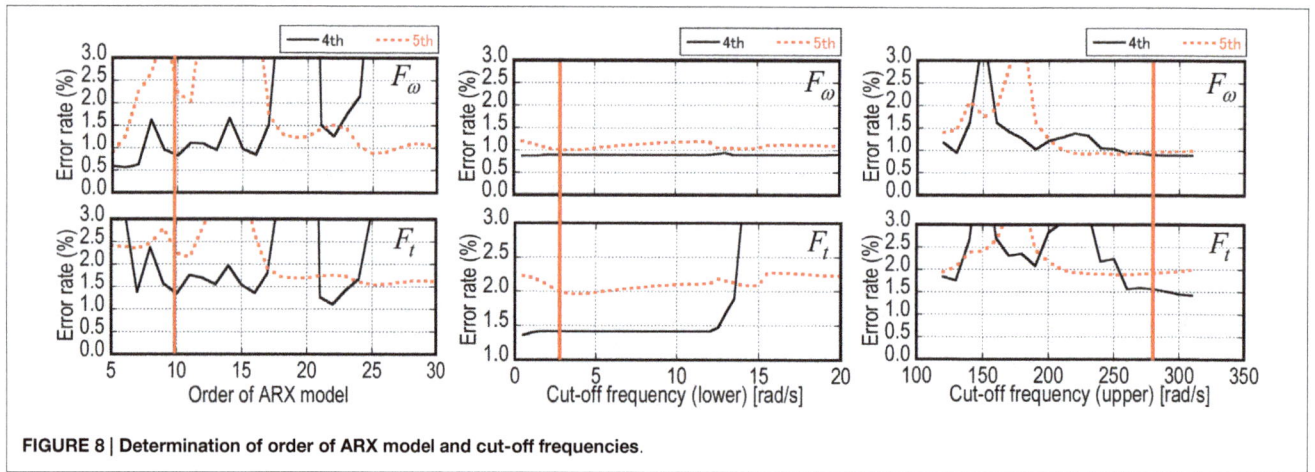

FIGURE 8 | Determination of order of ARX model and cut-off frequencies.

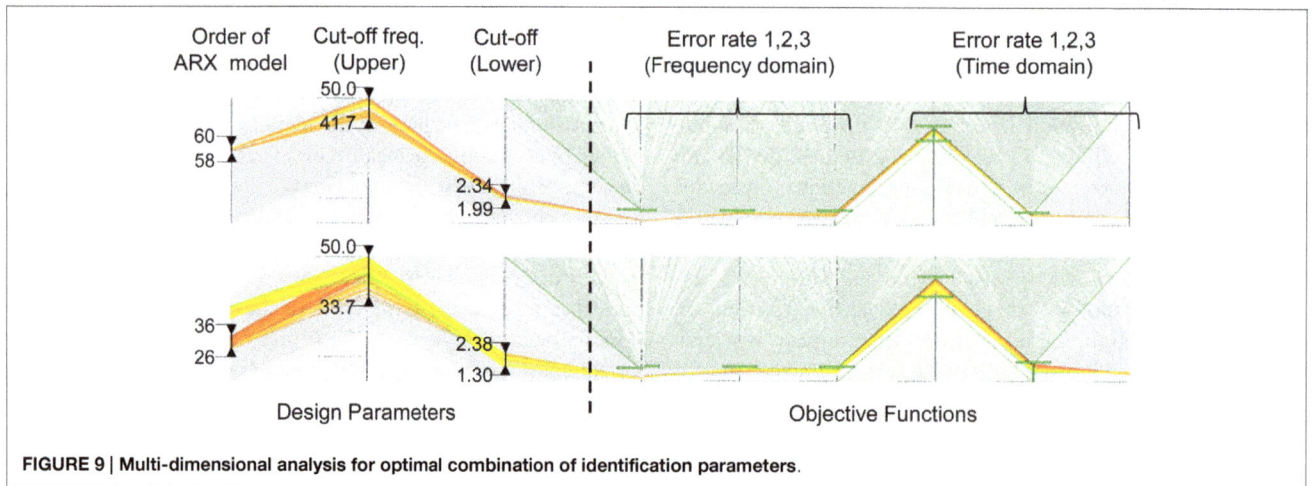

FIGURE 9 | Multi-dimensional analysis for optimal combination of identification parameters.

where the cumulative frequency of story stiffnesses is evaluated by the proposed methods. The variability of the identified story stiffnesses can be evaluated from the comparison of the gradient in the cumulative frequency shown in **Figure 11**. As for the first, second, and third stories, identified story stiffnesses have smaller variability compared with those by *Method 1*. However, improvement of stability for the fourth and fifth story stiffnesses cannot

be observed even in applying *Method 2 and 3*. This means that it is difficult to determine story stiffnesses in these cases. This instability needs further investigation on the relationship between the objective functions, Eqs (19) and (21), and the identification

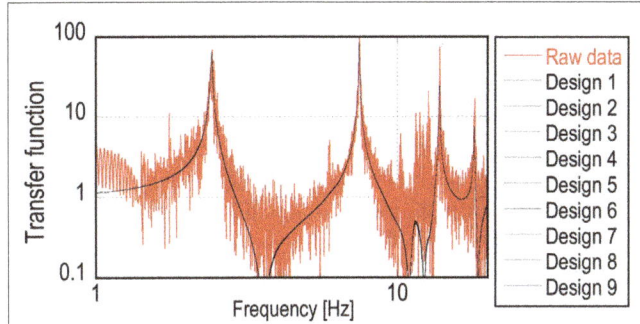

FIGURE 10 | Comparison of transfer functions (raw data VS ARX model).

TABLE 3 | Identified story stiffness.

$\times 10^4$ kN/m	1st story	2nd story	3rd story	4th story	5th story
Frame analysis	10.82	9.83	9.56	9.23	8.49
Method 1	9.10	9.13	8.13	6.25	6.48
Methods 2, 3	9.05	8.56	6.83	4.86	4.76

TABLE 4 | Natural frequencies of identified shear model.

Hz	1st mode	2nd mode	3rd mode	4th mode	5th mode
Reference (**Table 2**)	2.42	7.46	12.1	13.8	18.1
Method 1	2.47	6.77	11.0	13.7	16.6
Method 2, 3	2.37	6.20	9.91	12.4	15.5

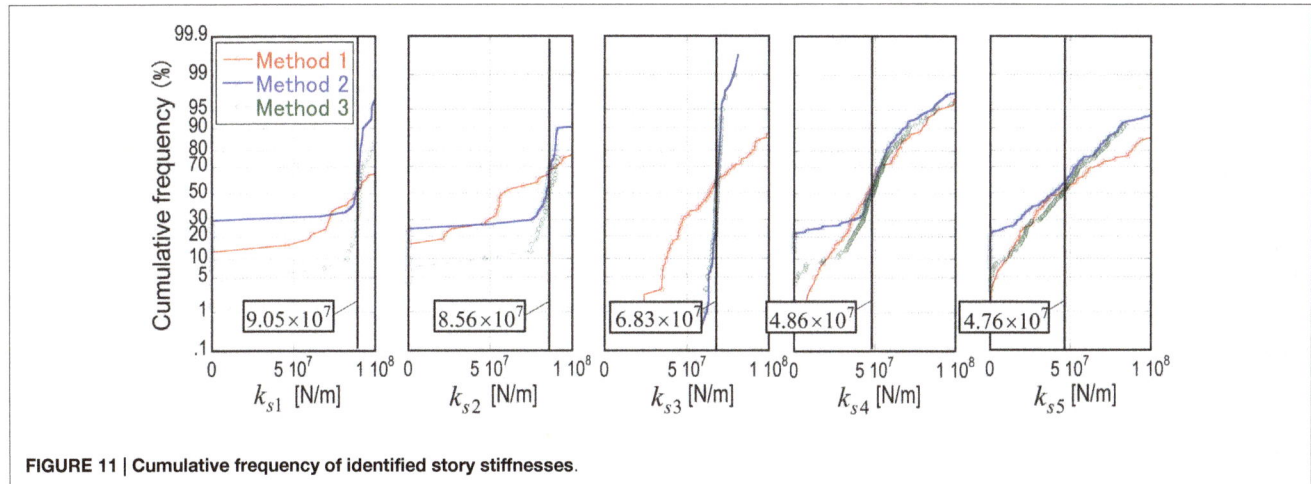

FIGURE 11 | Cumulative frequency of identified story stiffnesses.

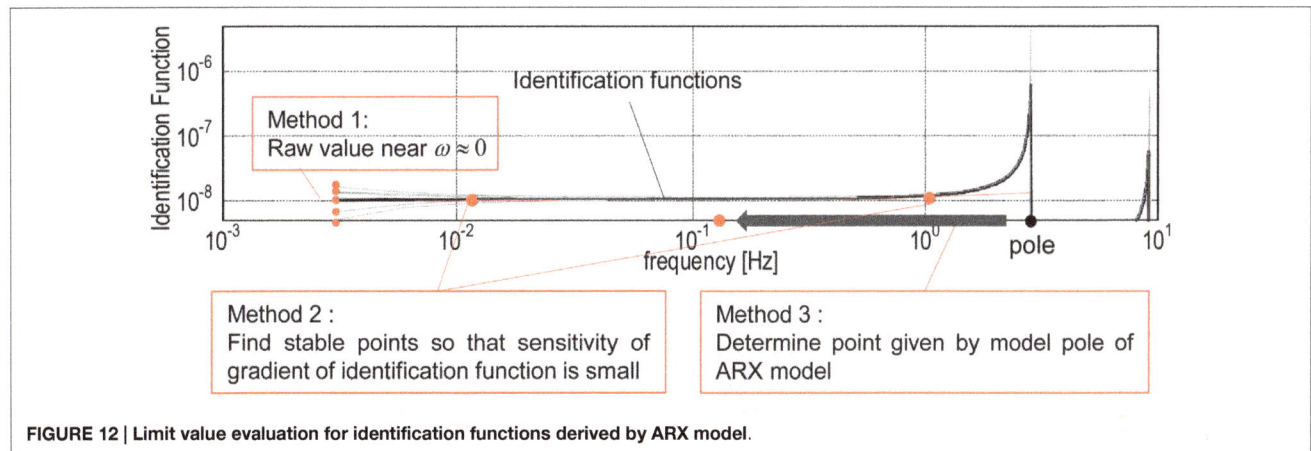

FIGURE 12 | Limit value evaluation for identification functions derived by ARX model.

function derived by the ARX model. **Tables 3** and **4** summarize the identified story stiffnesses and natural frequencies obtained by the shear model using the identified story stiffnesses. The identified story stiffnesses derived by *Method 2, 3* in **Table 3** are determined by the mean values of the selected optimal designs in **Figure 11**. In **Table 3**, a frame analysis was conducted by using a frame model including measured section size of columns and beams. It should be mentioned that the boundary condition of the numerical frame analysis may not correspond to the real structure. It is difficult to define "reference stiffness values" because a real building structure cannot be expressed by a shear building model exactly.

Conclusion

A previously proposed story-wise stiffness identification method using an ARX model for a shear building structure has been applied to the case where the shear building is subjected to micro-tremors measured in a full-scale steel frame structure. While earthquake ground motions and the building responses to such inputs are necessary in the previous method, it has been shown that micro-tremors can be used for identification within the same framework. This enhanced the usability of the previously proposed identification method. The difficulty in the selection of the combination of identification parameters, i.e., the order of the ARX model and the cut-off frequencies of band-pass filtering, has been tackled by a multi-objective optimization algorithm. Advantageous features are as follows:

(1) Micro-tremors can be used as an input for the previously proposed SI method for shear building models regardless of its small vibration level.

(2) In order to investigate identification parameters for reliable identification based on the ambient vibration, the objective functions in time and frequency domains have been proposed by applying filtering in the frequency range (low and high cuts) and averaging in time domain (sequential time-window shift for Fourier transformation and averaging).

(3) Stability in the story stiffness evaluation for optimal combinations of identification parameters derived by the multi-objective optimization has been compared among various limit-value estimation approaches. It has been observed that the limit value by selecting the raw value of the identification function at $\omega \approx 0$ is not necessarily appropriate. Proposed approaches such as using gradient sensitivity of the identification function or the model pole of the ARX model make it stable to derive story stiffnesses.

Acknowledgements

This research is partly supported by the Grant-in-Aid for Scientific Research (No. 24246095) in Japan. This support is gratefully acknowledged.

References

Adachi, S. (2009). *Fundamentals of System Identification*. Tokyo: Denki University Press.

Agbabian, M. S., Masri, S. F., Miller, R. K., and Caughey, T. K. (1991). System identification approach to detection of structural changes. *ASCE J. Eng. Mech.* 117, 370–390. doi:10.1061/(ASCE)0733-9399(1991)117:2(370)

Barroso, L. R., and Rodriguez, R. (2004). Damage detection utilizing the damage index method to a benchmark structure. *ASCE J. Eng. Mech.* 130, 142–151. doi:10.1061/(ASCE)0733-9399(2004)130:2(142)

Beck, J. L., and Jennings, P. C. (1980). Structural identification using linear models and earthquake records. *Earthquake Eng. Struct. Dyn.* 8, 145–160. doi:10.1002/eqe.4290080205

Doebling, S. W., Farrar, C. R., Prime, M. B., and Shevitz, D. W. (1996). *Damage Identification and Health Monitoring of Structural and Mechanical Systems from Changes in their Vibration Characteristics: a Literature Review*. Report LA-13070-MS. Los Alamos National Laboratory, Los Alamos.

Fujino, Y., Nishitani, A., and Mita, A. (2010). *Proceedings of 5th World Conference on Structural Control and Monitoring (5WCSCM)*. Tokyo.

Fujita, K., Ikeda, A., Shirono, M., and Takewaki, I. (2013). System identification of high-rise buildings using shear-bending model and ARX model: experimental investigation. *Proc. of ASEM13(ICEAS13)*, 2803–2815.

Ghanem, R., and Shinozuka, M. (1995). Structural-system identification I: theory. *ASCE J. Eng. Mech.* 121, 255–264. doi:10.1061/(ASCE)0733-9399(1995)121:2(255)

Hart, G. C., and Yao, J. T. P. (1977). System identification in structural dynamics. *ASCE J. Eng. Mech. Div.* 103, 1089–1104.

Hernandez-Garcia, M. R., Masri, S. F., Ghanem, R., Figueiredo, E., and Farrar, C. R. (2010a). An experimental investigation of change detection in uncertain chain-like systems. *J. Sound Vib.* 329, 2395–2409. doi:10.1016/j.jsv.2009.12.024

Hernandez-Garcia, M., Masri, S. F., Ghanem, R., Figueiredo, E., and Farrar, R. A. (2010b). A structural decomposition approach for detecting, locating, and quantifying nonlinearities in chain-like systems. *Struct. Control Health Monit.* 17, 761–777. doi:10.1002/stc.396

Hjelmstad, K. D. (1996). On the uniqueness of modal parameter estimation. *J. Sound Vib.* 192, 581–598. doi:10.1006/jsvi.1996.0205

Hjelmstad, K. D., Banan, Mo.R., and Banan, Ma.R. (1995). On building finite element models of structures from modal response. *Earthquake Eng. Struct. Dyn.* 24, 53–67. doi:10.1002/eqe.4290240105

Hoshiya, M., and Saito, E. (1984). Structural identification by extended Kalman filter. *ASCE J. Eng. Mech.* 110, 1757–1770. doi:10.1061/(ASCE)0733-9399(1984)110:12(1757)

Housner, G., et al. (1997). Special issue, structural control: past, present, and future. *ASCE J. Eng. Mech.* 123, 897–971. doi:10.1061/(ASCE)0733-9399(1997)123:9(897)

Ji, X., Fenves, G. L., Kajiwara, K., and Nakashima, M. (2011). Seismic damage detection of a full-scale shaking table test structure. *ASCE J. Struct. Eng.* 137, 14–21. doi:10.1061/(ASCE)ST.1943-541X.0000278

Johnson, E., and Smyth, A. (eds) (2006). *Proceedings of 4th World Conference on Structural Control and Monitoring (4WCSCM)*. San Diego, CA: IASC.

Kobori, T., Seto, Y. K., Iemura, H., and Nishitani, A. (eds) (1998). *Proceedings of 2nd World Conference on Structural Control*. Kyoto: John Wiley & Sons.

Koh, C. G., See, L. M., and Balendra, T. (1991). Estimation of structural parameters in time domain: a substructure approach. *Earthquake Eng. Struct. Dyn.* 20, 787–801. doi:10.1002/eqe.4290200806

Kozin, F., and Natke, H. G. (1986). System identification techniques. *Struct. Saf.* 3, 269–316. doi:10.1016/0167-4730(86)90006-8

Maeda, T., Yoshitomi, S., and Takewaki, I. (2011). Stiffness-damping identification of buildings using limited earthquake records and ARX model. *J. Struct. Construct. Eng.*, Architectural Inst. of Japan 666, 1415–1423. doi:10.3130/aijs.76.1415

Masri, S. F., Nakamura, M., Chassiakos, A. G., and Caughey, T. K. (1996). A neural network approach to the detection of changes in structural parameters. *ASCE. J. Eng. Mech.* 122, 350–360. doi:10.1061/(ASCE)0733-9399(1996)122:4(350)

Minami, Y., Yoshitomi, S., and Takewaki, I. (2013). System identification of super high-rise buildings using limited vibration data during the 2011 Tohoku (Japan) earthquake. *Struct. Control Health Monit.* 20, 1317–1338. doi:10.1002/stc.1537

Nagarajaiah, S., and Basu, B. (2009). Output only modal identification and structural damage detection using time frequency & wavelet techniques. *Earthquake Eng. Eng. Vib.* 8, 583–605. doi:10.1007/s11803-009-9120-6

Shinozuka, M., and Ghanem, R. (1995). Structural-system identification II: experimental verification. *ASCE J. Eng. Mech.* 121, 265–273. doi:10.1061/(ASCE)0733-9399(1995)121:2(265)

Takewaki, I., and Nakamura, M. (2000). Stiffness-damping simultaneous identification using limited earthquake records. *Earthquake Eng. Struct. Dyn.* 29, 1219–1238. doi:10.1002/1096-9845(200008)29:8<1219::AID-EQE968>3.0.CO;2-X

Takewaki, I., and Nakamura, M. (2005). Stiffness-damping simultaneous identification under limited observation. *ASCE J. Eng. Mech.* 131, 1027–1035. doi:10.1061/(ASCE)0733-9399(2005)131:10(1027)

Takewaki, I., and Nakamura, M. (2009). Temporal variation of modal properties of a base-isolated building during an earthquake. *J. Zhejiang Univ. Sci. A* 11, 1–8. doi:10.1631/jzus.A0900462

Udwadia, F. E., Sharma, D. K., and Shah, P. C. (1978). Uniqueness of damping and stiffness distributions in the identification of soil and structural systems. *ASME J. Appl. Mech.* 45, 181–187. doi:10.1115/1.3424224

Xing, Z., and Mita, A. (2012). A substructure approach to local damage detection of shear structure. *Struct. Control Health Monit.* 19, 309–318. doi:10.1002/stc.439

Zhang, D. Y., and Johnson, E. A. (2013). Substructure identification for shear structures with nonstationary structural responses. *ASCE J Eng Mech.* 139, 1769–1779. doi:10.1061/(ASCE)EM.1943-7889.0000626

Conflict of Interest Statement: The authors declare that the research was conducted in the absence of any commercial or financial relationships that could be construed as a potential conflict of interest.

Appendix: ARX model

Let k denote a discrete time step number (Adachi, 2009). When the output, the input, and white noise are denoted by $y(\cdot)$, $u(\cdot)$, and $w(\cdot)$, the ARX model is described by:

$$y(k) + a_1 y(k-1) + \cdots + a_{n_a} y(k-n_a) = b_1 u(k-1) + \cdots$$
$$+ b_{n_b} u(k-n_b) + w(k) \tag{A1}$$

The transfer function can be expressed in terms of the shift operator q.

$$G(q) = B(q)/A(q) \tag{A2}$$

where

$$A(q) = 1 + a_1 q^{-1} + \cdots + a_{n_a} q^{-n_a}, \tag{A3}$$
$$B(q) = b_1 q^{-1} + \cdots + b_{n_b} q^{-n_b} \tag{A4}$$

The relation between the Z transformation and the Fourier transformation is given by Eq. (A5) and the transfer function in terms of the variable q in Eq. (A2) can be expressed as the transfer function in terms of the variable ω (circular frequency). In Eq. (A5), T_0 denotes the sampling period.

$$q = e^{i\omega T_0} \tag{A5}$$
$$G(\omega) = \frac{b_1 e^{-i\omega T_0} + \cdots + b_n e^{-in\omega T_0}}{1 + a_1 e^{-i\omega T_0} + \cdots + a_n e^{-in\omega T_0}} \tag{A6}$$

Define the parameter vector $\boldsymbol{\theta}$ and the data vector $\boldsymbol{\varphi}(k)$ by Eqs (A7) and (A8), respectively. The prediction of the output at time k from the input–output data until the time number $(k-1)$ can be expressed by Eq. (A9).

$$\boldsymbol{\theta} = \{a_1 \quad \cdots \quad a_{n_a} \quad b_1 \quad \cdots \quad b_{n_b}\}^T \tag{A7}$$
$$\boldsymbol{\varphi}(k) = \{-y(k-1) \quad \cdots \quad -y(k-n_a)^T$$
$$u(k-1) \quad \cdots \quad u(k-n_b)\} \tag{A8}$$
$$\hat{y}(k; \boldsymbol{\theta}) = \boldsymbol{\theta}^T \boldsymbol{\varphi}(k) \tag{A9}$$

Using Eq. (A9), the prediction error in the ARX model can be described by:

$$\varepsilon(k, \boldsymbol{\theta}) = y(k) - \hat{y}(k; \boldsymbol{\theta}) = y(k) - \boldsymbol{\theta}^T \boldsymbol{\varphi}(k) \tag{A10}$$

Introduce the objective function by Eq. (A11) for predicting the parameter vector $\boldsymbol{\theta}$. Then the least-squares method can be applied to the parameter prediction problem.

$$J_N(\boldsymbol{\theta}) = \frac{1}{N_d} \sum_{k=1}^{N_d} \varepsilon^2(k; \boldsymbol{\theta}) \tag{A11}$$

In this case, the prediction problem of $\boldsymbol{\theta}$ can be reduced to the following simultaneous equations.

$$\mathbf{R}\boldsymbol{\theta} = \mathbf{f} \tag{A12}$$

where

$$\mathbf{R} = \frac{1}{N_d} \boldsymbol{\Phi}\boldsymbol{\Phi}^T \quad \left(\boldsymbol{\Phi} = \begin{bmatrix} \boldsymbol{\varphi}(1) & \boldsymbol{\varphi}(2) & \cdots & \boldsymbol{\varphi}(N_d) \end{bmatrix}^T\right) \tag{A13}$$
$$\mathbf{f} = \frac{1}{N_d} \boldsymbol{\Phi}^T \mathbf{y} \quad \left(\mathbf{y} = \{y(1) \quad y(2) \quad \cdots \quad y(N_d)\}^T\right) \tag{A14}$$

The least-square estimation of the unknown parameters based on the N_d-pair input–output measured data may then be expressed by:

$$\hat{\boldsymbol{\theta}} = \mathbf{R}^{-1}\mathbf{f} \tag{A15}$$

2

Critical earthquake response of elastic–plastic structures under near-fault ground motions (Part 1: Fling-step input)

Kotaro Kojima and Izuru Takewaki *

Department of Architecture and Architectural Engineering, Graduate School of Engineering, Kyoto University, Kyoto, Japan

The double impulse input is introduced as a substitute of the fling-step near-fault ground motion and a closed-form solution of the elastic–plastic response of a structure by the "critical double impulse input" is derived. Since only the free-vibration appears under such double impulse input, the energy approach plays an important role in the derivation of the closed-form solution of a complicated elastic–plastic response. It is shown that the maximum inelastic deformation can occur either after the first impulse or after the second impulse depending on the input level. The validity and accuracy of the proposed theory are investigated through the comparison with the response analysis to the corresponding one-cycle sinusoidal input as a representative of the fling-step near-fault ground motion. Since the critical input means the resonant case, the present theory dealing with the resonant response should be applied to buildings except very flexible ones.

Keywords: earthquake response, critical response, elastic–plastic response, ductility factor, near-fault ground motion, fling-step input, double impulse

Edited by:
Nikos D. Lagaros,
National Technical University of
Athens, Greece

Reviewed by:
Sameh Samir F. Mehanny,
Cairo University, Egypt
Michalis F. Vassiliou,
ETH Zürich, Switzerland

*Correspondence:
Izuru Takewaki,
Department of Architecture and
Architectural Engineering, Graduate
School of Engineering, Kyoto
University, Kyotodaigaku-Katsura,
Nishikyo, Kyoto 615-8540, Japan
takewaki@archi.kyoto-u.ac.jp

Introduction

The effects of near-fault ground motions on structural response have been investigated extensively (Bertero et al., 1978; Hall et al., 1995; Sasani and Bertero, 2000; Alavi and Krawinkler, 2004; Mavroeidis et al., 2004; Kalkan and Kunnath, 2006, 2007; Xu et al., 2007; Rupakhety and Sigbjörnsson, 2011; Yamamoto et al., 2011; Khaloo et al., 2015; Vafaei and Eskandari, 2015). The fling-step and forward-directivity are widely recognized as special keywords to characterize such near-fault ground motions (Mavroeidis and Papageorgiou, 2003; Bray and Rodriguez-Marek, 2004; Kalkan and Kunnath, 2006; Mukhopadhyay and Gupta, 2013a,b; Zhai et al., 2013; Hayden et al., 2014; Yang and Zhou, 2014). Especially, Northridge earthquake in 1994, Hyogoken-Nanbu (Kobe) earthquake in 1995, and Chi-Chi (Taiwan) earthquake in1999 raised special attention to many earthquake structural engineers.

The fling-step and forward-directivity inputs have been characterized by two or three wavelets. For this class of ground motions, many useful research works have been conducted. Mavroeidis and Papageorgiou (2003) investigated the characteristics of this class of ground motions in detail and proposed some simple models (for example, Gabor wavelet and Berlage wavelet). Xu et al. (2007) employed a kind of Berlage wavelet and applied it to the performance evaluation of passive energy dissipation systems. Takewaki and Tsujimoto (2011) used the Xu's approach and proposed a method for scaling ground motions from the viewpoints of drift and input energy demand. Takewaki et al. (2012) employed a sinusoidal wave for pulse-type waves. In this paper, a new approach based on the double impulse (Kojima et al., 2015a) is proposed and the intrinsic response characteristics by the near-fault ground motion are captured.

Most of the previous works on the near-fault ground motions deal with the elastic response because the number of parameters (e.g., duration and amplitude of pulse, ratio of pulse frequency to structure natural frequency, change of equivalent natural frequency for the increased input level) to be considered on this topic is many and the computation itself of elastic–plastic response is quite complicated.

In order to tackle such important but complicated problem, the double impulse input is introduced as a substitute of the fling-step near-fault ground motion and a closed-form solution of the elastic–plastic response of a structure by the "critical double impulse input" is derived. It is shown that, since only the free-vibration appears under such double impulse input, the energy approach plays an important role in the derivation of the closed-form solution of a complicated elastic–plastic response. It is also shown that the maximum inelastic deformation can occur either after the first impulse or after the second impulse depending on the input level. The validity and accuracy of the proposed theory are investigated through the comparison with the response analysis result to the corresponding one-cycle sinusoidal input as a representative of the fling-step near-fault ground motion. The amplitude of the double impulse is modulated so that its maximum Fourier amplitude coincides with that of the corresponding one-cycle sinusoidal input.

The closed-form or nearly closed-form solutions of the elastic–plastic earthquake response have been obtained so far only for the steady-state response to sinusoidal input or the transient response to an extremely simple sinusoidal input (Caughey, 1960a,b; Roberts and Spanos, 1990; Liu, 2000). In this paper, the following motivation is raised. If a near-fault ground motion can be represented by a double impulse, the elastic–plastic response (continuation of free-vibrations) can be derived by an energy approach without solving directly the differential equation (equation of motion). The input of impulse is expressed by the instantaneous change of velocity of the structural mass.

In the earthquake-resistant design, the resonance is a key word and it has been investigated extensively. While the resonant equivalent frequency has to be computed for a specified input level by changing the excitation frequency in a parametric manner in dealing with the sinusoidal input (Caughey, 1960a,b; Roberts and Spanos, 1990; Liu, 2000), no iteration is required in the proposed method for the double impulse. This is because the resonant equivalent frequency can be obtained directly without the repetitive procedure. In the double impulse, the analysis can be done without the input frequency (timing of impulses) before the second impulse is input. The resonance can be proved by using energy investigation and the timing of the second impulse can be characterized as the time with zero restoring force. The maximum elastic–plastic response after impulse can be obtained by equating the initial kinetic energy computed by the initial velocity to the sum of hysteretic and elastic strain energies. It should be pointed out that only critical response (upper bound) is captured by the proposed method and the critical resonant

frequency can be obtained automatically for the increasing input level of the double impulse.

In the history of seismic-resistant design of building structures, the earthquake input energy has played an important role together with deformation and acceleration [for example, Housner (1959, 1975), Berg and Thomaides (1960), Housner and Jennings (1975), Zahrah and Hall (1984), Akiyama (1985), and Leger and Dussault (1992)]. While deformation and acceleration can predict and evaluate the performance of a building structure mainly for serviceability, the energy can evaluate the performance of a building structure mainly for safety. Especially energy is appropriate for describing the performance of building structures of different sizes in a unified manner because energy is a global index different from deformation and acceleration as local indices. In fact, in Japan, there are three criteria in parallel (force, deformation, and energy). In 1981, force was introduced as a criterion for safety and in 2000 deformation was introduced as a criterion for safety. More recently in 2005, input energy evaluated from the design velocity response spectrum was used as a criterion. These three criteria are used now in parallel (Building Standard Law in Japan, 1981, 2000, 2005).

A theory of earthquake input energy to building structures under single impulse was shown to be useful for disclosing the property of the energy transfer function (Takewaki, 2004, 2007). This property means that the area of the energy transfer function is constant. The property of the energy transfer function similar to the case of a simple single-degree-of-freedom (SDOF) model has also been clarified for a swaying-rocking model. By using this property, the mechanism of earthquake input energy to the swaying-rocking model including the soil amplification has been made clear under the input of single impulse (Kojima et al., 2015b). However, single impulse may be unrealistic because the frequency characteristic of input cannot be expressed by this input. In order to resolve such issue, the double impulse is introduced in this paper. Furthermore, because the elastic–plastic response is treated, the time-domain formulation is introduced in this paper.

Double Impulse Input

It is well accepted that the fling-step input (fault-parallel) of the near-fault ground motion can be represented by a one-cycle sinusoidal wave and the forward-directivity input (fault-normal) of the near-fault ground motion can be expressed by a series of three sinusoidal wavelets (see **Figure 1**). In this paper and a subsequent paper, it is intended to simplify these typical near-fault ground motions by a double impulse (Kojima et al., 2015a) and a triple impulse. This is because the double impulse and triple impulse have a simple characteristic and a straightforward expression of the response can be expected even for elastic–plastic responses based on an energy approach to free vibrations. Furthermore, the double impulse and triple impulse enable us to describe directly the critical timing of impulses (resonant frequency) which is not easy for the sinusoidal and other inputs without a repetitive procedure.

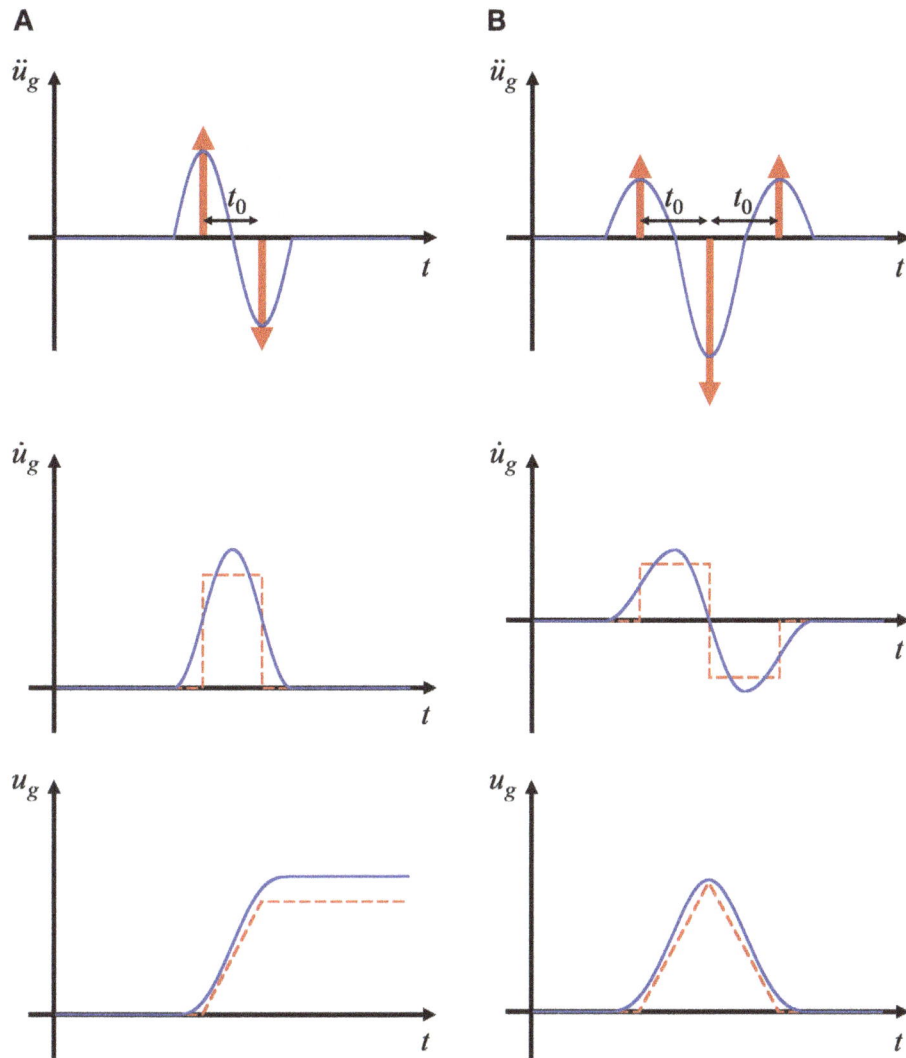

FIGURE 1 | (A) Fling-step input and double impulse, (B) Forward-directivity input and triple impulse.

Consider a ground acceleration $\ddot{u}_g(t)$ as double impulse, as shown in **Figure 1A**, expressed by

$$\ddot{u}_g(t) = V\delta(t) - V\delta(t - t_0) \qquad (1)$$

where V is the given initial velocity and t_0 is the time interval between two impulses. The comparison with the corresponding one-cycle sinusoidal wave as a representative of the fling-step input of the near-fault ground motion (Mavroeidis and Papageorgiou, 2003; Kalkan and Kunnath, 2006) is plotted in **Figure 1A**. The corresponding velocity and displacement of such double impulse and sinusoidal wave are also plotted in **Figure 1A**. It can be understood that the double impulse is a good approximation of the corresponding sinusoidal wave even in the form of velocity and displacement. However, the correspondence in the response should be discussed carefully. This will be conducted in Section "Accuracy Check by Time-history Response Analysis Subjected to the Corresponding One-cycle Sinusoidal Input." The

Fourier transform of $\ddot{u}_g(t)$ of the double impulse input can be derived as

$$\ddot{U}_g(\omega) = \int_{-\infty}^{\infty} \left\{ V\delta(t) - V\delta(t - t_0) \right\} e^{-i\omega t} dt$$

$$= \int_{-\infty}^{\infty} \left\{ V\delta(t)e^{-i\omega t} - V\delta(t - t_0)e^{-i\omega t_0} e^{-i\omega(t - t_0)} \right\} dt$$

$$= V(1 - e^{-i\omega t_0}) \qquad (2)$$

SDOF System

Consider an undamped elastic-perfectly plastic SDOF system of mass m and stiffness k. The yield deformation and yield force are denoted by d_y and f_y (see **Figure 2**). Let $\omega_1 = \sqrt{k/m}$, u, and f denote the undamped natural circular frequency, the displacement of the mass relative to the ground and the restoring force of the model, respectively. The time derivative is denoted by

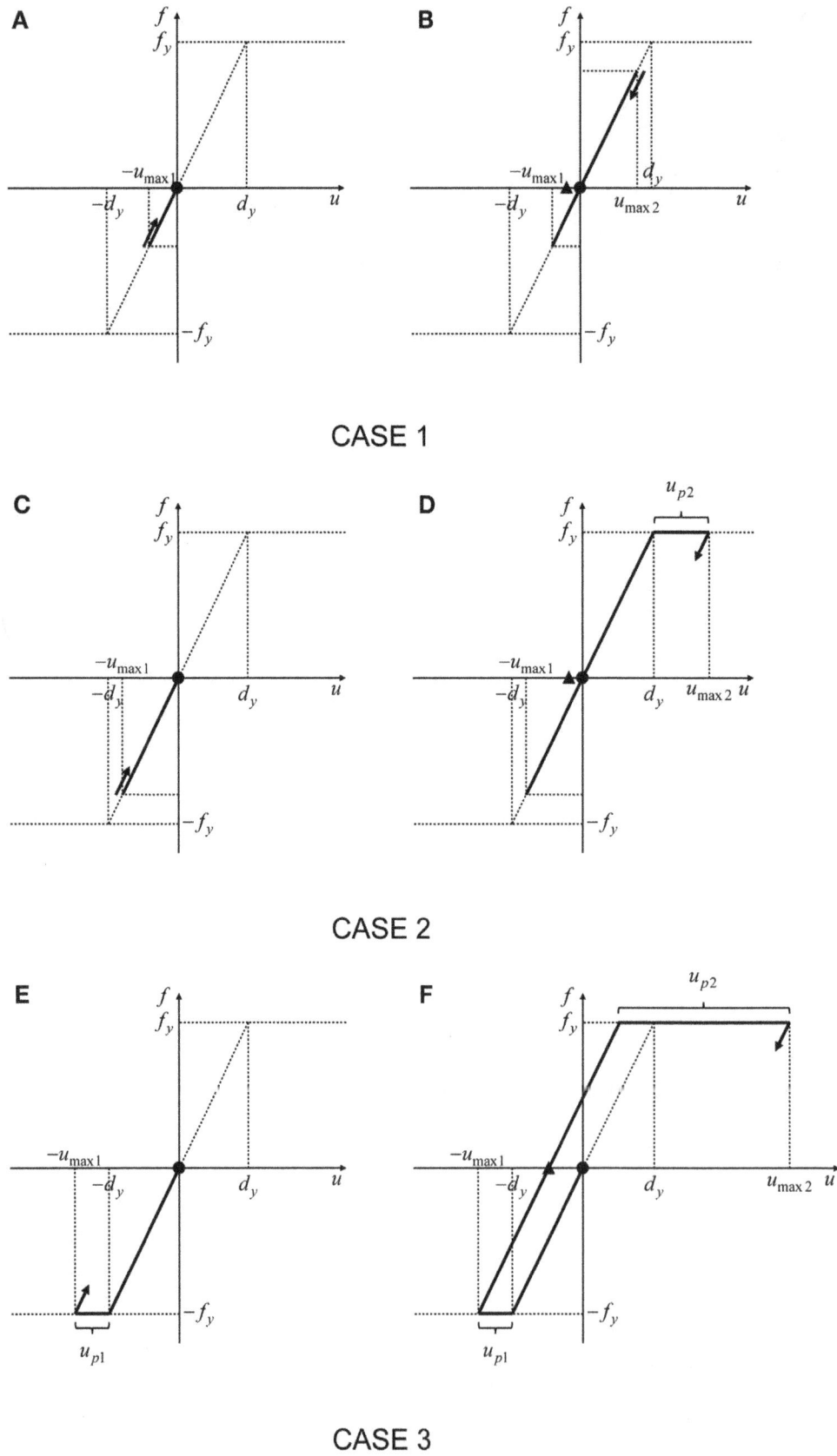

FIGURE 2 | Prediction of maximum elastic–plastic deformation under double impulse based on energy approach: (A,B) Case 1: Elastic response; (C,D) Case 2: Plastic response after the second impulse; (E,F) Case 3: Plastic response after the first impulse (●: first impulse, ▲: second impulse).

an over-dot. In Section "Maximum Elastic–plastic Deformation of SDOF System Subjected to Double Impulse," these parameters will be dealt with in a non-dimensional or normalized form to derive the relation of permanent interest between the input and the elastic–plastic response. However numerical parameters will be introduced partially in Section "Accuracy Check by Time-history Response Analysis Subjected to the Corresponding One-cycle Sinusoidal Input" to demonstrate an example of actual parameters.

Maximum Elastic–Plastic Deformation of SDOF System Subjected to Double Impulse

The elastic–plastic response to the double impulse can be described by the continuation of free-vibrations. The maximum deformation after the first impulse is denoted by $u_{\max 1}$ and that after the second impulse is expressed by $u_{\max 2}$ as shown in **Figure 2**. The input of each impulse is expressed by the instantaneous change of velocity of the structural mass. Such response can be derived by an energy approach without solving directly the differential equation (equation of motion). The kinetic energy given at the initial stage (the time of the first impulse) and at the time of the second impulse is transformed into the sum of the hysteretic energy and the strain energy corresponding to the yield deformation. By using this rule, the maximum deformation can be obtained in a simple manner.

It should be emphasized that, while the resonant equivalent frequency has to be computed for a specified input level by changing the excitation frequency in a parametric manner in dealing with the sinusoidal input (Caughey, 1960a,b; Roberts and Spanos, 1990; Liu, 2000; Moustafa et al., 2010), no iteration is required in the proposed method for the double impulse. This is because the resonant equivalent frequency (resonance can be proved by using energy investigation: see Appendix) can be obtained directly without the repetitive procedure. As a result, the timing of the second impulse can be characterized as the time with zero restoring force.

Only critical response (upper bound) is captured by the proposed method and the critical resonant frequency can be obtained automatically for the increasing input level of the double impulse. One of the original points in this paper is the introduction of the concept of "critical excitation" in the elastic–plastic response (Drenick, 1970; Abbas and Manohar, 2002; Takewaki, 2007; Moustafa et al., 2010). Once the frequency and amplitude of the critical double impulse are computed, the corresponding one-cycle sinusoidal motion as a representative of the fling-step motion can be identified.

Let us explain the evaluation method of $u_{\max 1}$ and $u_{\max 1}$. The plastic deformation after the first impulse is expressed by u_{p1} and that after the second impulse is denoted by u_{p2}. There are three cases to be considered depending on the yielding stage. Let V_y ($=\omega_1 d_y$) denote the input level of velocity of the double impulse at which the SDOF system just attains the yield deformation after the first impulse.

Figures 2A,B show the maximum deformation after the first impulse and that after the second impulse, respectively, for the elastic case (Case 1) during the whole stage. $u_{\max 1}$ can be obtained from the following energy conservation law.

$$mV^2/2 = ku_{\max 1}^2/2 \tag{3}$$

On the other hand, $u_{\max 2}$ can be computed from another energy conservation law.

$$m(2V)^2/2 = ku_{\max 2}^2/2 \tag{4}$$

As explained in the previous part of this section, the critical timing of the second impulse is the time of zero restoring force and the velocity $-V$ by the second impulse is added to the velocity $-V$ induced by the first impulse (full recovery at the zero restoring force due to zero damping).

Consider next the case (Case 2) where the model goes into the yielding stage after the second impulse. **Figures 2C,D** show the schematic diagram of the response in this case. As in Case 1, $u_{\max 1}$ can be obtained from the energy conservation law.

$$mV^2/2 = ku_{\max 1}^2/2 \tag{5}$$

On the other hand, $u_{\max 2}$ can be computed from another energy conservation law by regarding the system as a non-linear elastic system tentatively.

$$m(2V)^2/2 = f_y d_y/2 + f_y u_{p2} = f_y d_y/2 + f_y(u_{\max 2} - d_y) \tag{6}$$

As in the above case, the velocity $-V$ by the second impulse is added to the velocity $-V$ induced by the first impulse.

Consider finally the case (Case 3) where the model goes into the yielding stage even after the first impulse. **Figures 2E,F** show the schematic diagram of the response in this case. $u_{\max 1}$ can be obtained from the following energy conservation law.

$$mV^2/2 = f_y d_y/2 + f_y u_{p1} = f_y d_y/2 + f_y(u_{\max 1} - d_y) \tag{7}$$

On the other hand, $u_{\max 2}$ can be computed from another energy conservation law.

$$m(v_c + V)^2/2 = f_y d_y/2 + f_y u_{p2} \tag{8}$$

where v_c is characterized by $mv_c^2/2 = f_y d_y/2$ and u_{p2} is characterized by $u_{\max 2} + (u_{\max 1} - d_y) = d_y + u_{p2}$. In other words, $u_{\max 2}$ can be obtained from

$$m(v_c + V)^2/2 = f_y d_y/2 + f_y(u_{\max 1} + u_{\max 2} - 2d_y). \tag{9}$$

As in the above case, the velocity $-V$ by the second impulse is added to the velocity $-v_c$ induced by the first impulse (the maximum velocity during the unloading stage).

Figure 3 shows the plot of $u_{\max}/d_y = \max(u_{\max 1}/d_y, u_{\max 2}/d_y)$ with respect to the input level. There are three regions corresponding to Cases 1–3. In Cases 1 and 2, $u_{\max 2}/d_y$ is larger than

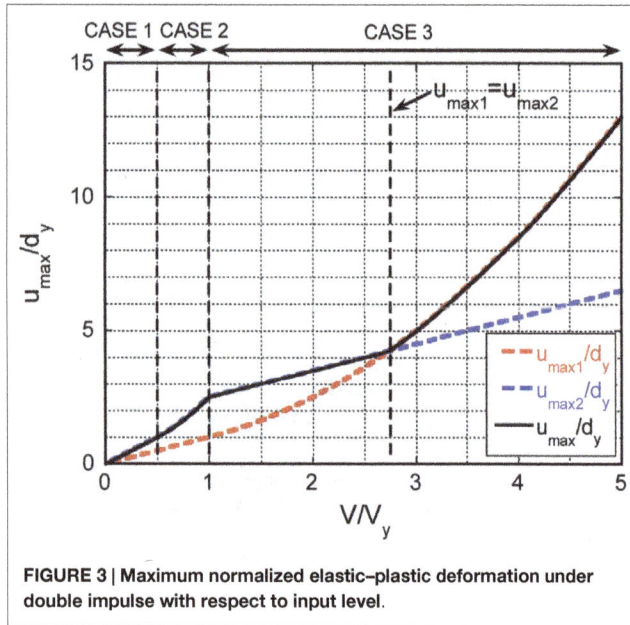

FIGURE 3 | Maximum normalized elastic–plastic deformation under double impulse with respect to input level.

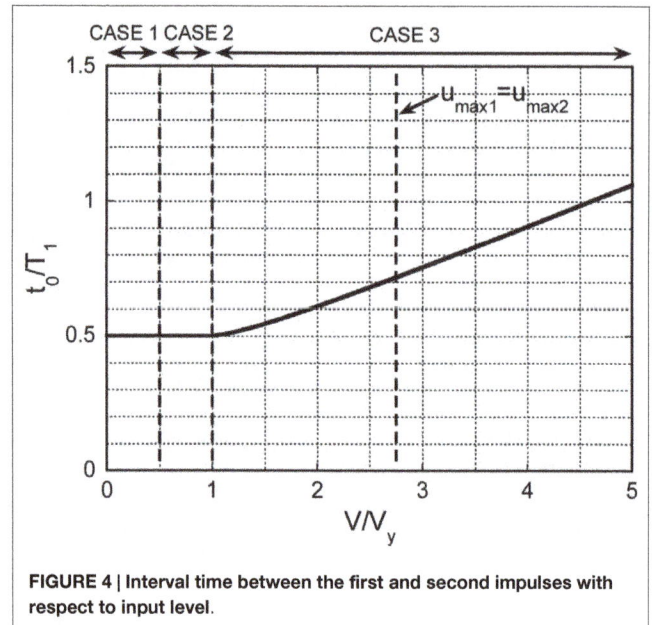

FIGURE 4 | Interval time between the first and second impulses with respect to input level.

$u_{max\,1}/d_y$. On the other hand, in Case 3, two regions exist for the boundary case of $u_{max\,1} = u_{max\,2}$. While $u_{max\,2}/d_y$ is larger than $u_{max\,1}/d_y$ in the smaller input level, $u_{max\,1}/d_y$ is larger than $u_{max\,2}/d_y$ in the larger input level.

Figure 4 presents the normalized timing t_0/T_1 ($T_1 = 2\pi/\omega$) of the second impulse with respect to the input level. As stated before, this timing coincides with the time of zero restoring force after the first unloading (see **Figure 2**). It can be observed that the timing is delayed due to plastic deformation as the input level increases. It seems noteworthy to state again that only critical response giving the maximum value of $u_{max\,2}/d_y$ is sought by the proposed method and the critical resonant frequency is obtained automatically for the increasing input level of the double impulse. One of the original points in this paper is the tracking of the critical elastic–plastic response.

Accuracy Check by Time-History Response Analysis Subjected to the Corresponding One-Cycle Sinusoidal Input

In order to investigate the accuracy of using the double impulse as a substitute of the corresponding one-cycle sinusoidal wave (representative of the fling-step input), the time-history response analysis of the elastic–plastic SDOF model under the one-cycle sinusoidal wave has been conducted.

In the evaluation procedure, it is important to adjust the input level of the double impulse and the corresponding one-cycle sinusoidal wave based on the equivalence of the Fourier amplitude. **Figure 5** shows one example for the input level $V/V_y = 3$. **Figures 6A,B** illustrate the comparison of the ground displacement and velocity between the double impulse and the corresponding one-cycle sinusoidal wave for the input level $V/V_y = 3$. In **Figure 5** and **Figures 6A,B**, $\omega_1 = 2\pi$(rad/s) ($T_1 = 1.0$ s) and $d_y = 0.16$(m) are used.

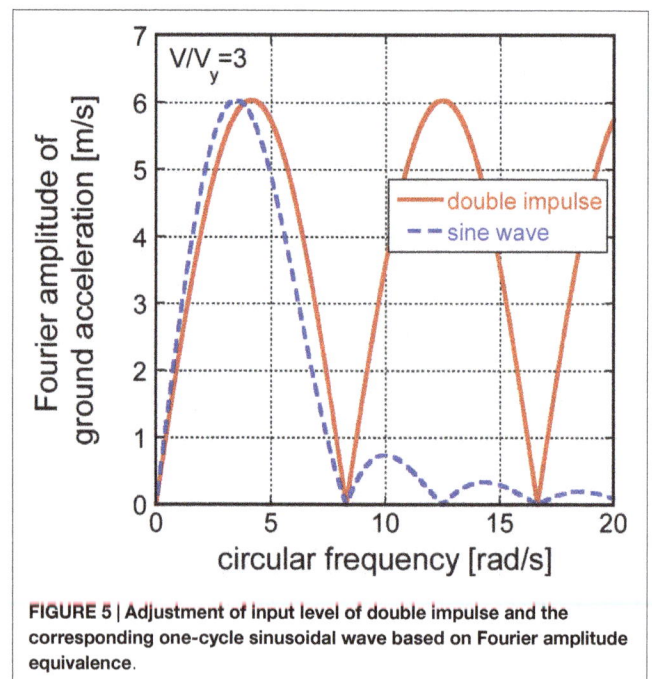

FIGURE 5 | Adjustment of input level of double impulse and the corresponding one-cycle sinusoidal wave based on Fourier amplitude equivalence.

Figure 7 presents the comparison of the ductility (maximum normalized deformation) of the elastic–plastic structure under the double impulse and the corresponding one-cycle sinusoidal wave with respect to the input level. It can be seen that the double impulse provides a fairly good substitute of the one-cycle sinusoidal wave in the evaluation of the maximum deformation if the maximum Fourier amplitude is adjusted appropriately. Although some discrepancy is observed in the large deformation range, that response range is out of interest in the earthquake structural engineering. If desired, other adjustment criterion on input level can be introduced. This is a future issue.

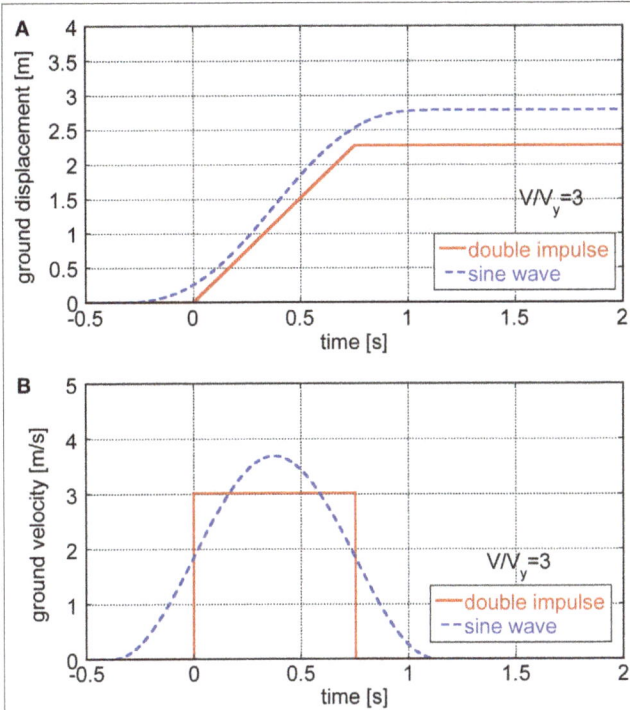

FIGURE 6 | Comparison of ground displacement and velocity between double impulse and the corresponding one-cycle sinusoidal wave: (A) displacement, (B) velocity.

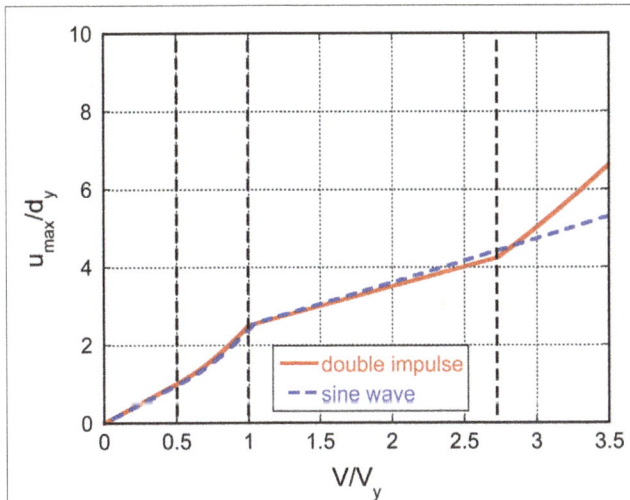

FIGURE 8 | Comparison of earthquake input energies by double impulse and the corresponding one-cycle sinusoidal wave.

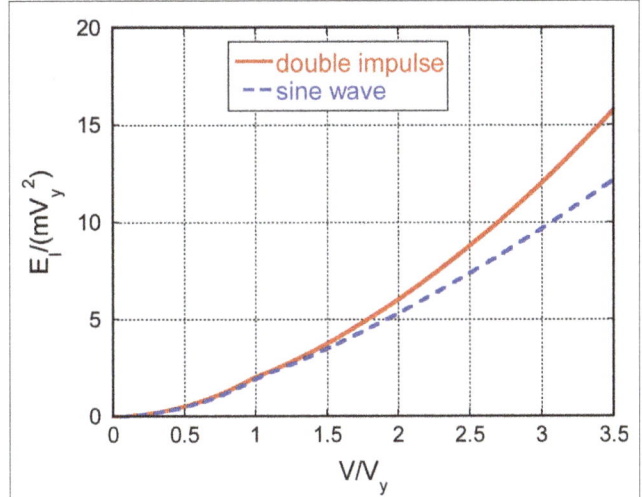

FIGURE 7 | Comparison of ductility of elastic–plastic structure under double impulse and the corresponding one-cycle sinusoidal wave.

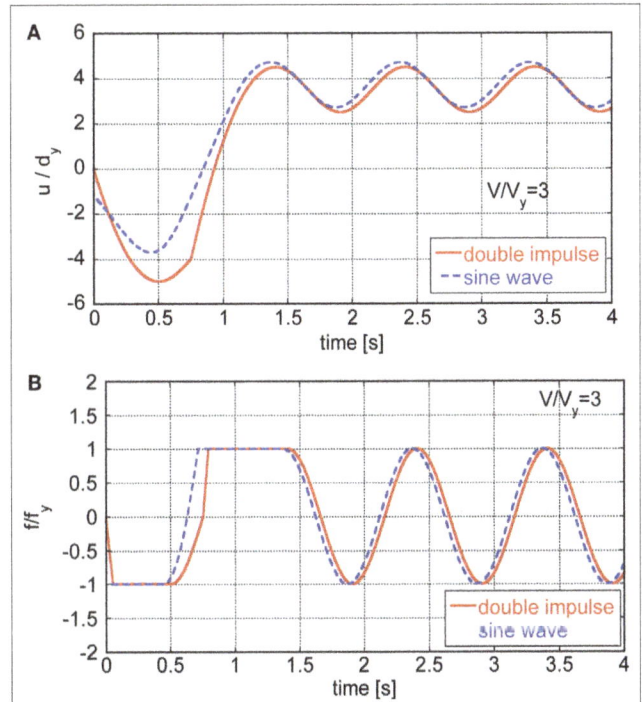

FIGURE 9 | Comparison of response time-history under double impulse and that under the corresponding one-cycle sinusoidal wave: (A) Normalized deformation, (B) Restoring-force.

Figure 8 shows the comparison of the earthquake input energies by the double impulse and the corresponding one-cycle sinusoidal wave. Although a good correspondence can be observed in a lower input level, the double impulse tends to provide a slightly larger upper bound in the larger input level. This property can be understood from the time-history responses shown in **Figures 9** and **10**, i.e., a rather clear difference in deformation after the first impulse.

Figure 9 illustrates the comparison of response time histories (normalized deformation and restoring-force) under the double impulse and those under the corresponding one-cycle sinusoidal wave. The parameters $\omega_1 = 2\pi$ (rad/s) ($T_1 = 1.0$ s), $d_y = 0.16$(m) were also used here. While a rather good correspondence can be seen in the restoring-force, the maximum deformation after the first impulse exhibits a rather larger value in the double impulse. The difference in the initial condition may affect these response discrepancies. However, it is noteworthy that the maximum deformation after the second impulse demonstrates a rather good correspondence. This may be related to the fact that the effect of the initial condition becomes smaller in this stage. **Figure 10** presents the comparison of the restoring-force characteristic under the double impulse and that under the corresponding one-cycle sinusoidal wave. The parameters $\omega_1 = 2\pi$ (rad/s) ($T_1 = 1.0$ s), $d_y = 0.16$(m) are also used here. As seen in **Figure 9**, while the maximum deformation after the first impulse exhibits a rather larger value in the double impulse compared to that of the corresponding one-cycle sinusoidal wave, the maximum deformation after the second impulse demonstrates a rather good correspondence.

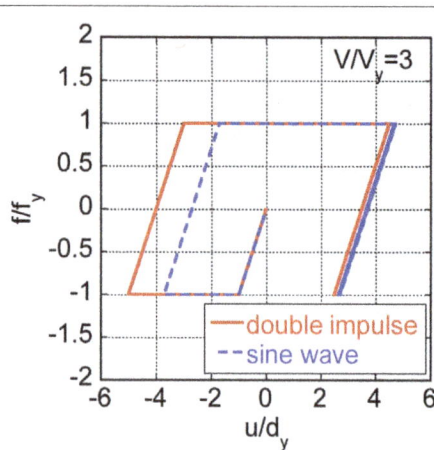

FIGURE 10 | Comparison of restoring-force characteristic under double impulse and that under the corresponding one-cycle sinusoidal wave.

Design of Stiffness and Strength for Specified Velocity and Period of Near-Fault Ground Motion Input and Response Ductility

It is useful to present a flowchart for design of stiffness and strength for the specified velocity and period of the near-fault ground motion input and response ductility. This design concept is based on the philosophy that, if we focus on the worst case of resonance, the safety for other non-resonant cases is guaranteed [see Takewaki (2002)].

Since **Figures 3** and **4** are non-dimensional ones, they can be used for such design. **Figure 11** shows the flowchart for design of stiffness and strength. One example can be drawn as follows:

[*Specified conditions*]: $V = 2$(m/s) (velocity of double impulse), $t_0 = 0.5$(s) (interval of the double impulse and half the period of the corresponding sine wave), $u_{max}/d_y = 4.0$ (ductility), $m = 4.0 \times 10^6$(kg).

[*Design results*]: $V/V_y = 2.5$, $V_y = 0.80$(m/s), $T_1 = 0.74$(s), $d_y = 0.094$(m), $k = 2.9 \times 10^8$(N/m), $f_y = 2.7 \times 10^7$(N).

From **Figure 3**, $V/V_y = 2.5$ can be obtained for the specified ductility $u_{max}/d_y = 4.0$. Then $V_y = 0.80$(m/s) is derived from the specified condition $V = 2$(m/s) and $V/V_y = 2.5$. In the next step, $T_1 = 0.74$(s) is found from **Figure 4** for $V/V_y = 2.5$ and $t_0 = 0.5$(s). In this model, $d_y = 0.094$(m) is determined from $V_y = \omega_1 d_y$ and $T_1(= 2\pi/\omega_1) = 0.74$(s). Finally $k = 2.9 \times 10^8$(N/m) is obtained from $k = \omega_1^2 m$ and $f_y = 2.7 \times 10^7$(N) is derived by $f_y = kd_y$.

It should be reminded that, while most of the previous researches on near-fault ground motions are aimed at disclosing the response characteristics of elastic or elastic–plastic structures with arbitrary stiffness and strength parameters and require tremendous amount of numerical task, the present paper focused

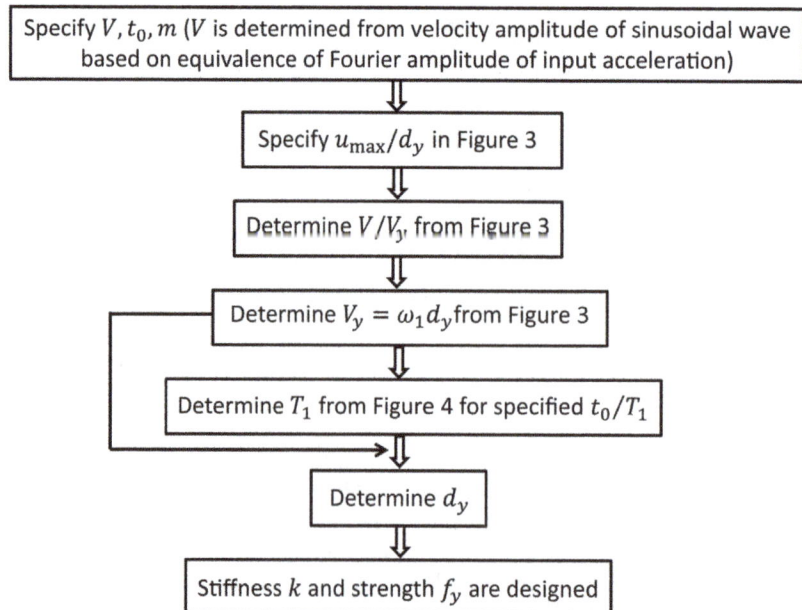

FIGURE 11 | Flowchart for design of stiffness and strength.

on the critical response (resonant response) and enabled the drastic reduction of computational works.

Conclusion

The conclusions may be summarized as follows:

(1) The double impulse input has been introduced as a substitute of the fling-step near-fault ground motion and a closed-form solution of the elastic–plastic response of a structure by the critical double impulse input has been derived.

(2) It has been shown that, since only the free-vibration appears in such double impulse input, the energy approach plays an important role in the derivation of the closed-form solution of a complicated elastic–plastic response. In other words the energy approach enables the derivation of the maximum elastic–plastic seismic response without solving the differential equation (equation of motion). In this process, the input of impulse is expressed by the instantaneous change of velocity of the structural mass. The maximum elastic–plastic response after impulse can be obtained by equating the initial kinetic energy computed by the initial velocity to the sum of hysteretic and elastic strain energies. It has been shown that the maximum inelastic deformation can occur either after the first impulse or after the second impulse depending on the input level.

(3) The validity and accuracy of the proposed theory have been investigated through the comparison with the response analysis result to the corresponding one-cycle sinusoidal input as a representative of the fling-step near-fault ground motion. It has been made clear that, if the level of the double impulse is adjusted so as for its maximum Fourier amplitude to coincide with that of the corresponding one-cycle sinusoidal wave, the maximum elastic–plastic deformation to the double impulse exhibits a good correspondence with that to the one-cycle sinusoidal wave.

(4) While the resonant equivalent frequency has to be computed for a specified input level by changing the excitation frequency in a parametric manner in dealing with the sinusoidal input, no iteration is required in the proposed method for the double impulse. This is because the resonant equivalent frequency can be obtained directly without the repetitive procedure. The resonance has been proved by using energy investigation and it has been made clear that the timing of the second impulse can be characterized as the time with zero restoring force.

(5) Only critical response (upper bound) has been captured by the proposed method and it has been shown that the critical resonant frequency can be obtained automatically for the increasing input level of the double impulse. Once the frequency and amplitude of the critical double impulse are computed, the corresponding one-cycle sinusoidal motion as a representative of the fling-step motion can be identified.

(6) A flowchart for design of stiffness and strength for the specified velocity and period of the near-fault ground motion input and response ductility has been proposed using the newly derived non-dimensional relations among response ductility, input velocity, and input period. It has been demonstrated that this flowchart can provide a useful result for such design. Since the critical input of double impulse means the resonant case, the proposed method may be conservative when applied to flexible structures that do not fall in the resonant region.

Acknowledgments

Part of the present work is supported by the Grant-in-Aid for Scientific Research of Japan Society for the Promotion of Science (No. 24246095, No. 15H04079) and the 2013-MEXT- Supported Program for the Strategic Research Foundation at Private Universities in Japan (No. S1312006). This support is greatly appreciated.

References

Abbas, A. M., and Manohar, C. S. (2002). Investigations into critical earthquake load models within deterministic and probabilistic frameworks. *Earthq. Eng. Struct. Dyn.* 31, 013–032. doi:10.1002/eqe.124.abs

Akiyama, H. (1985). *Earthquake Resistant Limit-State Design for Buildings*. Tokyo: University of Tokyo Press.

Alavi, B., and Krawinkler, H. (2004). Behaviour of moment resisting frame structures subjected to near-fault ground motions. *Earthq. Eng. Struct. Dyn.* 33, 687–706. doi:10.1002/eqe.370

Berg, G. V., and Thomaides, T. T. (1960). "Energy consumption by structures in strong-motion earthquakes," in *Proceedings of 2nd World Conference on Earthquake Engineering*, (Kyoto), 681–696.

Bertero, V. V., Mahin, S. A., and Herrera, R. A. (1978). Aseismic design implications of near-fault San Fernando earthquake records. *Earthq. Eng. Struct. Dyn.* 6, 31–42. doi:10.1002/eqe.4290060105

Bray, J. D., and Rodriguez-Marek, A. (2004). Characterization of forward-directivity ground motions in the near-fault region. *Soil Dyn. Earthq. Eng.* 24, 815–828. doi:10.1016/j.soildyn.2004.05.001

Building Standard Law in Japan. (1981, 2000, 2005). *Earthquake Resistant Design Code*.

Caughey, T. K. (1960a). Sinusoidal excitation of a system with bilinear hysteresis. *J. Appl. Mech.* 27, 640–643. doi:10.1115/1.3644077

Caughey, T. K. (1960b). Random excitation of a system with bilinear hysteresis. *J. Appl. Mech.* 27, 649–652. doi:10.1115/1.3644077

Drenick, R. F. (1970). Model free design of aseismic structures. *J. Eng. Mech. Div.* 96, 483–493.

Hall, J. F., Heaton, T. H., Halling, M. W., and Wald, D. J. (1995). Near-source ground motion and its effects on flexible buildings. *Earthq Spectra* 11, 569–605. doi:10.1193/1.1585828

Hayden, C. P., Bray, J. D., and Abrahamson, N. A. (2014). Selection of near-fault pulse motions. *J. Geotechnical. Geoenviron. Eng.* 140, 04014030. doi:10.1061/(ASCE)GT.1943-5606.0001129

Housner, G. W. (1959). Behavior of structures during earthquakes. *J. Eng. Mech. Div.* 85, 109–129.

Housner, G. W. (1975). "Measures of severity of earthquake ground shaking," in *Proceedings of the US National Conference on Earthquake Engineering*, (Ann Arbor, MI), 25–33.

Housner, G. W., and Jennings, P. C. (1975). "The capacity of extreme earthquake motions to damage structures," in *Structural and Geotechnical Mechanics*: A Volume Honoring N.M. Newmark, ed. W. J. Hall (Englewood Cliff, NJ: Prentice-Hall), 102–116.

Kalkan, E., and Kunnath, S. K. (2006). Effects of fling step and forward directivity on seismic response of buildings. *Earthq. Spectra* 22, 367–390. doi:10.1193/1.2192560

Kalkan, E., and Kunnath, S. K. (2007). Effective cyclic energy as a measure of seismic demand. *J. Earthq. Eng.* 11, 725–751. doi:10.1080/13632460601033827

Khaloo, A. R., Khosravi, H., and Hamidi Jamnani, H. (2015). Nonlinear interstory drift contours for idealized forward directivity pulses using "modified fish-bone" models. *Adv. Struct. Eng.* 18, 603–627. doi:10.1260/1369-4332.18.5.603

Kojima, K., Fujita, K., and Takewaki, I. (2015a). Critical double impulse input and bound of earthquake input energy to building structure. *Front. Built Environ.* 1:5. doi:10.3389/fbuil.2015.00005

Kojima, K., Sakaguchi, K., and Takewaki, I. (2015b). Mechanism and bounding of earthquake energy input to building structure on surface ground subjected to engineering bedrock motion. *Soil Dyn. Earthq. Eng.* 70, 93–103. doi:10.1016/j.soildyn.2014.12.010

Leger, P., and Dussault, S. (1992). Seismic-energy dissipation in MDOF structures. *J. Struct. Eng.* 118, 1251–1269. doi:10.1061/(ASCE)0733-9445(1992)118:5(1251)

Liu, C.-S. (2000). The steady loops of SDOF perfectly elastoplastic structures under sinusoidal loadings. *J. Mar. Sci. Technol.* 8, 50–60.

Mavroeidis, G. P., Dong, G., and Papageorgiou, A. S. (2004). Near-fault ground motions, and the response of elastic and inelastic single-degree-freedom (SDOF) systems. *Earthq. Eng. Struct. Dyn.* 33, 1023–1049. doi:10.1002/eqe.391

Mavroeidis, G. P., and Papageorgiou, A. S. (2003). A mathematical representation of near-fault ground motions. *Bull. Seism. Soc. Am.* 93, 1099–1131. doi:10.1785/0120020100

Moustafa, A., Ueno, K., and Takewaki, I. (2010). Critical earthquake loads for SDOF inelastic structures considering evolution of seismic waves. *Earthq. Struct.* 1, 147–162. doi:10.12989/eas.2010.1.2.147

Mukhopadhyay, S., and Gupta, V. K. (2013a). Directivity pulses in near-fault ground motions – I: identification, extraction and modeling. *Soil Dyn. Earthq. Eng.* 50, 1–15. doi:10.1016/j.soildyn.2013.02.017

Mukhopadhyay, S., and Gupta, V. K. (2013b). Directivity pulses in near-fault ground motions – II: estimation of pulse parameters. *Soil Dyn. Earthq. Eng.* 50, 38–52. doi:10.1016/j.soildyn.2013.02.019

Roberts, J. B., and Spanos, P. D. (1990). *Random Vibration and Statistical Linearization*. New York, NY: Wiley.

Rupakhety, R., and Sigbjörnsson, R. (2011). Can simple pulses adequately represent near-fault ground motions? *J. Earthq. Eng.* 15, 1260–1272. doi:10.1080/13632469.2011.565863

Sasani, M., and Bertero, V. V. (2000). "Importance of severe pulse-type ground motions in performance-based engineering: historical and critical review," in *Proceedings of the Twelfth World Conference on Earthquake Engineering*, (Auckland).

Takewaki, I. (2002). Robust building stiffness design for variable critical excitations. *J. Struct. Eng.* 128, 1565–1574. doi:10.1061/(ASCE)0733-9445(2002)128:12(1565)

Takewaki, I. (2004). Bound of earthquake input energy. *J. Struct. Eng.* 130, 1289–1297. doi:10.1061/(ASCE)0733-9445(2004)130:9(1289)

Takewaki, I. (2007). *Critical Excitation Methods in Earthquake Engineering*, Second Edn. Oxford: Elsevier, 2013.

Takewaki, I., Moustafa, A., and Fujita, K. (2012). *Improving the Earthquake Resilience of Buildings: The Worst Case Approach*. London: Springer.

Takewaki, I., and Tsujimoto, H. (2011). Scaling of design earthquake ground motions for tall buildings based on drift and input energy demands. *Earthq. Struct.* 2, 171–187. doi:10.12989/eas.2011.2.2.171

Vafaei, D., and Eskandari, R. (2015). Seismic response of mega buckling-restrained braces subjected to fling-step and forward-directivity near-fault ground motions. *Struct. Design Tall Spec. Build.* 24, 672–686. doi:10.1002/tal.1205

Xu, Z., Agrawal, A. K., He, W.-L., and Tan, P. (2007). Performance of passive energy dissipation systems during near-field ground motion type pulses. *Eng. Struct.* 29, 224–236. doi:10.1016/j.engstruct.2006.04.020

Yamamoto, K., Fujita, K., and Takewaki, I. (2011). Instantaneous earthquake input energy and sensitivity in base-isolated building. *Struct. Design Tall Spec. Build.* 20, 631–648. doi:10.1002/tal.539

Yang, D., and Zhou, J. (2014). A stochastic model and synthesis for near-fault impulsive ground motions. *Earthq. Eng. Struct. Dyn.* 44, 243–264. doi:10.1002/eqe.2468

Zahrah, T. F., and Hall, W. J. (1984). Earthquake energy absorption in SDOF structures. *J. Struct. Eng.* 110, 1757–1772. doi:10.1061/(ASCE)0733-9445(1984)110:8(1757)

Zhai, C., Chang, Z., Li, S., Chen, Z.-Q., and Xie, L. (2013). Quantitative identification of near-fault pulse-like ground motions based on energy. *Bull. Seism. Soc. Am.* 103, 2591–2603. doi:10.1785/0120120320

Conflict of Interest Statement: The authors declare that the research was conducted in the absence of any commercial or financial relationships that could be construed as a potential conflict of interest.

Appendix

Proof of Critical Timing of the Second Impulse and Numerical Demonstration

Consider the critical timing of the second impulse. Let v_c denote the velocity of the mass passing the zero restoring-force (zero elastic strain energy) after the first unloading and v^*, u^* denote the velocity and the elastic deformation component at an arbitrary point between the first unloading and the second yielding. Since the first unloading starts from the state with zero velocity and the elastic strain energy $f_y d_y/2$, the relation $m v_c^2/2 = f_y d_y/2$ holds. From the energy conservation law between the first unloading and the second yielding, the relation $m v^{*2}/2 + k u^{*2}/2 = f_y d_y/2$ holds. Consider the second impulse at the same time of the state of v^*, u^*. The total mechanical energy can be expressed by $m(v^* + V)^2/2 + k u^{*2}/2$. Since the relation $m(v^* + V)^2/2 + k u^{*2}/2 = m v^{*2}/2 + k u^{*2}/2 + m v^* V + m V^2/2 = f_y d_y/2 + m v^* V + m V^2/2$ holds and the maximum deformation after the second yielding is caused by the maximum total mechanical energy, the maximum velocity v^* causes the maximum deformation after the second yielding. This timing is the zero restoring-force after the first unloading. This completes the proof.

In order to confirm the validity of the critical timing shown above, numerical computation has been conducted. Let t_{0c} denote the critical interval between two impulses. **Figure A1** shows the

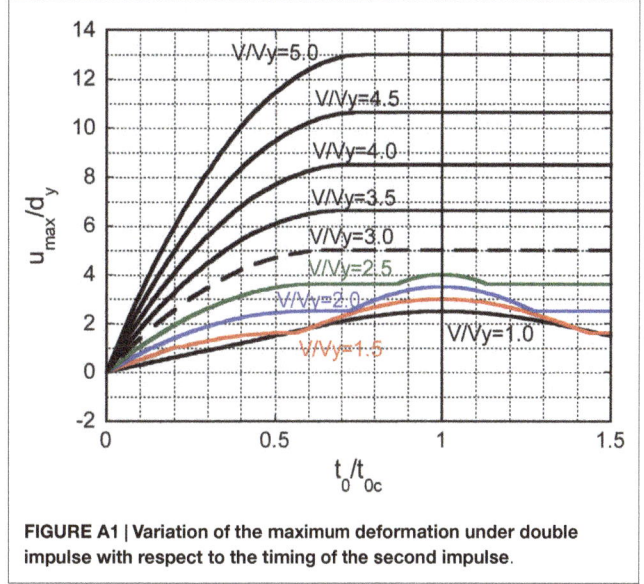

FIGURE A1 | Variation of the maximum deformation under double impulse with respect to the timing of the second impulse.

plot of u_{max}/d_y with respect to t_0/t_{0c}. $t_0/t_{0c} = 1$ indicates the critical timing at zero restoring force. It can be understood that, for larger V/V_y, $t_0/t_{0c} = 1$ is one of the value yielding the maximum value of u_{max}/d_y and, for smaller V/V_y, $t_0/t_{0c} = 1$ certainly gives the maximum value of u_{max}/d_y. This supports the numerical validation of the proof given above.

3

Critical earthquake response of elastic–plastic structures under near-fault ground motions (Part 2: Forward-directivity input)

Kotaro Kojima and Izuru Takewaki *

Department of Architecture and Architectural Engineering, Graduate School of Engineering, Kyoto University, Kyoto, Japan

Edited by:

Nikos D. Lagaros,
National Technical University of
Athens, Greece

Reviewed by:

Johnny Ho,
The University of Queensland,
Australia
Sameh Samir F. Mehanny,
Cairo University, Egypt

*Correspondence:

Izuru Takewaki,
Department of Architecture and
Architectural Engineering, Graduate
School of Engineering, Kyoto
University, Kyotodaigaku-Katsura,
Nishikyo, Kyoto 615-8540, Japan
takewaki@archi.kyoto-u.ac.jp

The triple impulse input is used as a simplified version of the forward-directivity near-fault ground motion and a closed-form solution of the elastic–plastic response of a structure by this triple input is obtained. It is noteworthy that only the free-vibration appears under such triple impulse input. An almost critical excitation is defined and its response is derived. The energy approach plays an important role in the derivation of the closed-form solution of a complicated elastic–plastic response. It is shown that the maximum inelastic deformation can occur after the second impulse or the third impulse depending on the input level. The validity and accuracy of the proposed theory are discussed through the comparison with the response analysis result to the corresponding three wavelets of sinusoidal waves as a representative of the forward-directivity near-fault ground motion.

Keywords: earthquake response, critical response, elastic–plastic response, ductility factor, near-fault ground motion, forward-directivity input, triple impulse

Introduction

The near-fault ground motions have been investigated from various viewpoints (Bertero et al., 1978; Hall et al., 1995; Sasani and Bertero, 2000; Alavi and Krawinkler, 2004; Mavroeidis et al., 2004; Kalkan and Kunnath, 2006, 2007; Xu et al., 2007; Rupakhety and Sigbjörnsson, 2011; Yamamoto et al., 2011; Khaloo et al., 2015; Kojima and Takewaki, 2015b; Vafaei and Eskandari, 2015). Those ground motions are characterized by the well-known phenomena called fling-step and forward-directivity (Mavroeidis and Papageorgiou, 2003; Bray and Rodriguez-Marek, 2004; Kalkan and Kunnath, 2006; Mukhopadhyay and Gupta, 2013a,b; Zhai et al., 2013; Hayden et al., 2014; Yang and Zhou, 2014; Kojima and Takewaki, 2015b).

As pointed out in the previous paper (Kojima and Takewaki, 2015b), the fling-step input is a fault-parallel input and the forward-directivity input are a fault-normal input. Those ground motions have been characterized by two or three wavelets and it is recognized that the forward-directivity input has larger effects on structures in general. Recently, some important research works have been conducted. Mavroeidis and Papageorgiou (2003) summarized the characteristics of this class of ground motions in detail and proposed some simple wavelet models (for example, Gabor wavelet and Berlage wavelet). Xu et al. (2007) employed the Berlage wavelet and applied it to the performance evaluation of passive energy dissipation systems. Takewaki and Tsujimoto (2011) used the Xu's model and proposed a method for scaling ground motions from the viewpoints of drift and input energy demand. Takewaki et al. (2012) employed a sinusoidal wave for pulse-type waves. Kojima and Takewaki (2015b) introduced a simplified input model called "double impulse" (Kojima et al., 2015a) and derived a closed-form solution of the critical elastic–plastic deformation of a single-degree-of-freedom (SDOF) model to the double impulse input. They clarified that (i) a closed-form solution of the critical elastic–plastic deformation can be derived based on a simple energy approach and (ii) the double impulse can be a good substitute of the fling-step input (one-cycle sinusoidal input) under the equivalence assumption of the maximum Fourier amplitude of accelerations. In this paper, the approach by Kojima and Takewaki (2015b) is extended to the forward-directivity input and the intrinsic response characteristics by the forward-directivity are captured.

Most of the previous works on the near-fault ground motions deal with the elastic response because the number of parameters (e.g., duration and amplitude of pulse, ratio of pulse frequency to structure natural frequency, change of equivalent natural frequency for the increased input level) to be considered is tremendous and the computation of elastic–plastic response itself is quite complicated.

In order to tackle such important but complicated problem, a simple input as the double impulse has been employed as a substitute of the fling-step near-fault ground motion in the previous paper (Kojima and Takewaki, 2015b) and a closed-form solution of the critical elastic–plastic response of a structure by this double impulse has been derived. Following the previous paper, the approach is extended to the forward-directivity input. It is shown that, since only the free-vibration appears under such triple impulse input, the energy approach plays an important role in the derivation of the closed-form solution of a complicated elastic–plastic response. An almost critical excitation is defined and its response is derived. It is also shown that the maximum inelastic deformation can occur either after the second impulse or after the third impulse depending on the input level. The validity and accuracy of the proposed theory are investigated through the comparison with the response analysis result to the corresponding three wavelets of sinusoidal input as a representative of the forward-directivity near-fault ground motion. The amplitude of the triple impulse is modulated so that its maximum Fourier amplitude coincides with that of the corresponding three wavelets of sinusoidal input.

The closed-form solutions of the elastic–plastic response have been obtained so far only for the steady-state response to an extremely simple sinusoidal input (Caughey, 1960; Liu, 2000). In the previous paper (Kojima and Takewaki, 2015b) and this paper, the following motivation is posed. If a near-fault ground motion can be represented by double impulse or triple impulse, the critical elastic–plastic response (continuation of free-vibrations) can be derived by an energy approach. The input of impulse is expressed by the instantaneous change of velocity of the structural mass. The restriction of the response to an almost critical one, which may be interesting in the design stage for safety, enables a unique solution of such complicated elastic–plastic responses.

While the resonant equivalent frequency has to be computed for a specified input level by changing the excitation frequency in a parametric manner in dealing with a sinusoidal input (Caughey, 1960; Liu, 2000), no iteration is required in the proposed method for the triple impulse. This is because the resonant equivalent frequency (resonance can be proved by using energy investigation)

can be obtained directly without the repetitive procedure (the timing of the second impulse can be characterized as the time with zero restoring force). In the triple impulse, the analysis can be conducted without the input frequency (timing of impulses) before the second impulse. The criticality is defined only for the response before the third impulse and it is shown that this restriction is a reasonable condition for safety evaluation of structures. The maximum elastic–plastic response after impulse can be obtained by equating the initial kinetic energy computed by the initial velocity to the sum of hysteretic and elastic strain energies. It should be pointed out that only critical response (upper bound) is captured by the proposed method and the critical resonant frequency can be obtained automatically for the increasing input level of the triple impulse.

The significance of using a one-cycle sinusoidal wave and three wavelets of sinusoidal wave as substitutes of fling-step and forward-directivity ground motion inputs have been explained by many researchers (Mavroeidis and Papageorgiou, 2003; Kalkan and Kunnath, 2006) and comparison with recorded ground motions has been conducted. On the other hand, the merit of the present paper is to derive a closed-form solution for even elastic–plastic responses under the critical input, which will reduce the computational load drastically and enhance the safety level of structures under such near-fault ground motions.

Triple Impulse Input

As pointed out in the previous paper (Kojima and Takewaki, 2015b), it is well accepted that the fling-step input (fault-parallel) of the near-fault ground motion can be represented by a one-cycle sinusoidal wave and the forward-directivity input (fault-normal) of the near-fault ground motion can be expressed by three wavelets of sinusoidal input (see **Figure 1**). In the previous paper and this paper, it is intended to simplify these typical near-fault ground motions by double impulse (Kojima et al., 2015a) or triple impulse. This is because the double impulse and triple impulse have a simple characteristic and a straightforward expression of response can be expected even for elastic–plastic responses based on a simple energy approach to free vibrations. Furthermore, the double impulse and triple impulse enable us to describe directly the critical timing of impulses (resonant frequency), which is not possible for the sinusoidal and other inputs without a repetitive procedure.

Consider a ground motion acceleration $\ddot{u}_g(t)$ as triple impulse, as shown in **Figure 1B**, expressed by

$$\ddot{u}_g(t) = 0.5V\delta(t) - V\delta(t - t_0) + 0.5V\delta(t - 2t_0) \qquad (1)$$

where $0.5V$ is the given initial velocity and t_0 is the time interval among three impulses. The comparison with the corresponding three wavelets of sinusoidal waves as a representative of the forward-directivity input of the near-fault ground motion (Mavroeidis and Papageorgiou, 2003; Kalkan and Kunnath, 2006) is also plotted in **Figure 1B**. The corresponding velocity and displacement of such triple impulse and three wavelets of sinusoidal waves are plotted in **Figure 1B**. The Fourier transform of $\ddot{u}_g(t)$ of the triple impulse input can be derived as

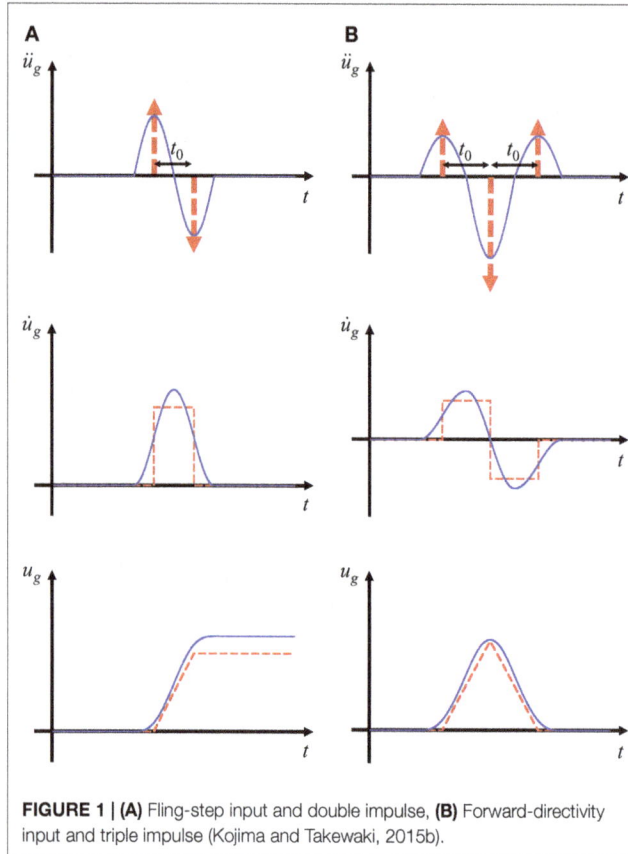

FIGURE 1 | (A) Fling-step input and double impulse, (B) Forward-directivity input and triple impulse (Kojima and Takewaki, 2015b).

$$\ddot{U}_g(\omega) = \int_{-\infty}^{\infty} \left\{ 0.5V\delta(t) - V\delta(t-t_0) + 0.5V\delta(t-2t_0) \right\} e^{-i\omega t} dt$$

$$= \int_{-\infty}^{\infty} \left\{ \begin{array}{c} 0.5V\delta(t)e^{-i\omega t} - V\delta(t-t_0)e^{-i\omega t_0}e^{-i\omega(t-t_0)} \\ + 0.5V\delta(t-2t_0)e^{-i\omega 2t_0}e^{-i\omega(t-2t_0)} \end{array} \right\} dt \qquad (2)$$

$$= V(0.5 - e^{-i\omega t_0} + 0.5e^{-i\omega 2t_0})$$

SDOF System

Consider an undamped elastic-perfectly plastic SDOF system of mass m and stiffness k. The yield deformation and the yield force are denoted by d_y and f_y. Let $\omega_1 = \sqrt{k/m}$, u and f denote the undamped natural circular frequency, the displacement (deformation) of the mass relative to the ground and the restoring force of the model, respectively. The time derivative is denoted by an over-dot.

Maximum Elastic–Plastic Deformation of SDOF System Subjected to Triple Impulse

The elastic–plastic response to the triple impulse can be described by the continuation of free-vibrations. The maximum deformation after the first impulse is denoted by u_{max1}, that after the second impulse is expressed by u_{max2} and that after the third impulse is described by u_{max3} as shown in **Figure 2**. The input of each impulse is expressed by the instantaneous change of velocity of the structural mass. Such response can be derived by the combination of a

simple energy approach and the solution of differential equations (equations of motion). The kinetic energy given at the initial stage (the time of the first impulse), that at the time of the second impulse, and the kinetic energy plus the elastic strain energy at the time of the third impulse are transformed into the sum of the hysteretic energy and the elastic strain energy corresponding to the yield deformation. Using this rule and incorporating the information from the equations of motion, the maximum deformation can be obtained in a simple manner. It should be noted that, while a simple and clear concept of critical input was defined in the case of double impulse (Kojima and Takewaki, 2015b), the criticality can be used only before the third impulse in the present triple impulse. This is because the timing of the third impulse, determined already for the first and second impulses, decreases the maximum deformation u_{max2} after the second impulse and may increase the maximum deformation u_{max3} after the third impulse. However, it is shown that this treatment of setting of timing provides the true criticality in an input level of practical interest.

It should also be emphasized that, while the resonant equivalent frequency has to be computed for a specified input level by changing the excitation frequency in a parametric or mathematical programing manner in dealing with the sinusoidal input (Caughey, 1960; Liu, 2000; Moustafa et al., 2010), no iteration is required in the proposed method for the triple impulse. This is because the resonant equivalent frequency (resonance can be proved by using energy investigation: see Proof of Critical Timing in Appendix) can be obtained directly without the repetitive procedure (the timing of the second impulse can be characterized as the time with zero restoring force). It should be noted again that the resonance is defined before the third impulse.

Only critical response (upper bound) is captured by the proposed method and the critical resonant frequency can be obtained automatically for the increasing input level of the triple impulse. One of the original points in this paper is the introduction of the concept of "critical excitation" in the elastic–plastic response (Drenick, 1970; Abbas and Manohar, 2002; Takewaki, 2004, 2007; Moustafa et al., 2010; Kojima and Takewaki, 2015b). Once the frequency and amplitude of the critical triple impulse are computed, the corresponding three wavelets of sinusoidal waves as a representative of the forward-directivity motion can be identified.

Let us explain the evaluation method of u_{max1}, u_{max2}, and u_{max3}. The plastic deformation after the first impulse is expressed by u_{p1}, that after the second impulse is described by u_{p2}, and that after the third impulse is denoted by u_{p3}. There are four cases to be considered depending on the yielding stage.

Case 1: elastic response during all response stages (u_{max3} is the largest).

Case 2: yielding after the third impulse (u_{max3} is the largest).

Case 3: yielding after the second impulse (u_{max2} or u_{max3} is the largest).

    1: the timing of the third impulse is in the unloading stage.

    2: the timing of the third impulse is in the yielding (loading) stage.

Case 4: yielding after the first impulse (u_{max2} is the largest).

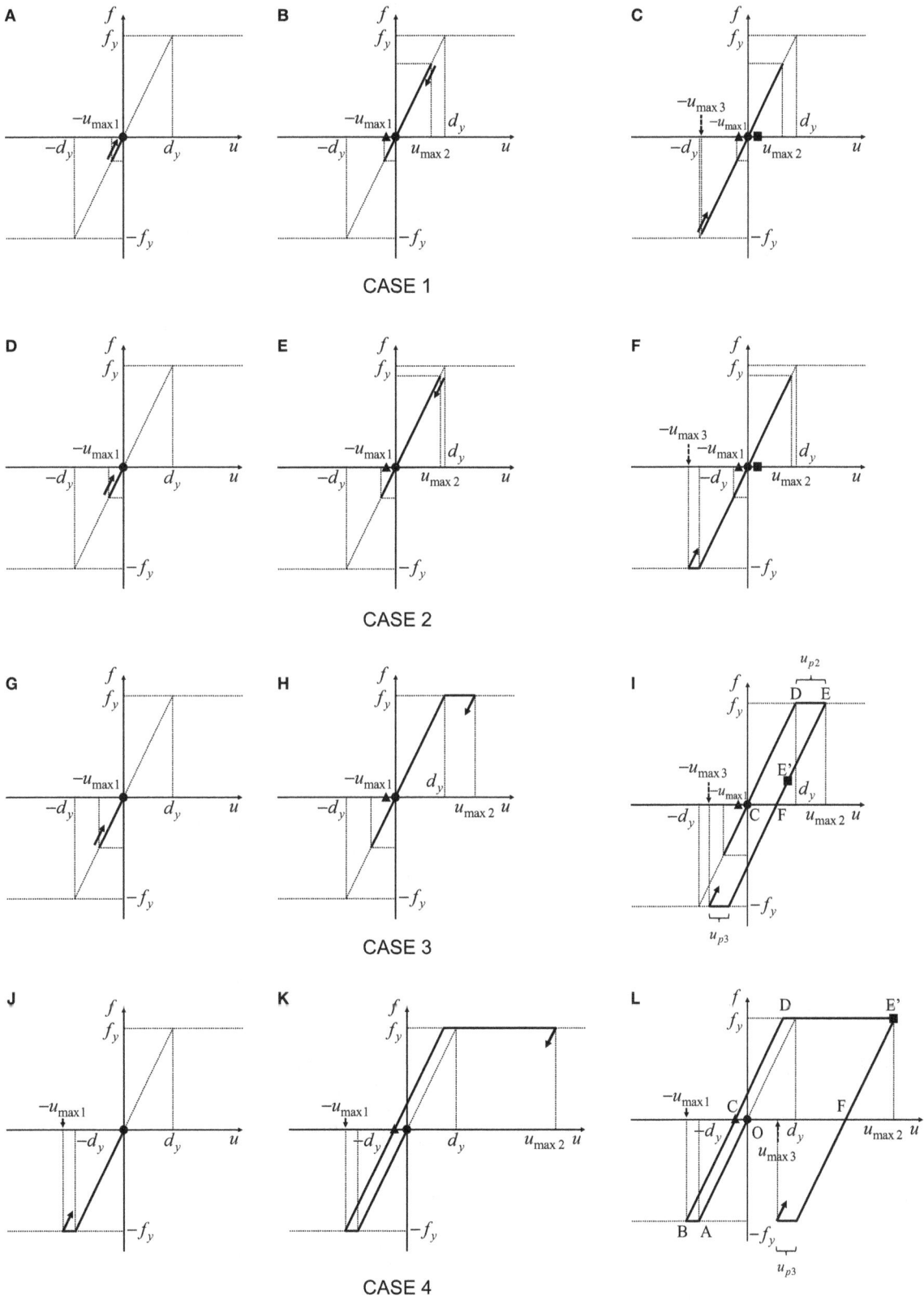

FIGURE 2 | Prediction of maximum elastic–plastic deformation under triple impulse based on energy approach: (A–C) Case 1: elastic response, (D–F) Case 2: plastic response after the third impulse, (G–I) Case 3: plastic response after the second impulse, (J–L) Case 4: plastic response after the first impulse (•: first impulse, ▲: second impulse, ■: third impulse).

In comparison with the double impulse, the triple impulse is quite difficult to derive the critical timing in a general case. This is because the timing of three impulses is fixed and there exist many complicated situations. In this paper, a case is treated where the critical timing is defined only before the third impulse. This means that, if the third impulse does not exist, timing gives the maximum value of u_{max2}.

Case 1

Figures 2A–C show the maximum deformation after the first impulse, that after the second impulse, and that after the third impulse, respectively, for the elastic case (Case 1) during the whole stage. u_{max1} can be obtained from the energy conservation law.

$$m(0.5V)^2 / 2 = ku_{max1}^2 / 2 \tag{3}$$

On the other hand, since it can be proved that the critical timing of the second impulse to produce the maximum deformation u_{max2} is the time of zero restoring force (the proof similar to the Section "Proof of Critical Timing" in Appendix), u_{max2} can be computed from another energy conservation law.

$$m(0.5V + V)^2 / 2 = ku_{max2}^2 / 2 \tag{4}$$

In the elastic case, the critical timing of the second impulse is the time of zero restoring force and the velocity – V by the second impulse is added to the velocity $-0.5V$ induced by the first impulse (full recovery at the zero restoring force due to zero damping).

Furthermore, since the timing of the third impulse is the time of zero restoring force, u_{max3} can be computed from another energy conservation law.

$$m(0.5V + V + 0.5V)^2 / 2 = ku_{max3}^2 / 2 \tag{5}$$

As explained above, the critical timing of the third impulse is the time of zero restoring force and the velocity $0.5V$ by the third impulse is added to the velocity $1.5V$ induced by the first and second impulses (full recovery at the zero restoring force due to zero damping).

It should be noted again that the critical timing t_0 corresponds to the time of zero restoring force in Case 1 (see Proof of Critical Timing in Appendix). As a result, u_{max3} becomes the largest deformation among u_{max1}, u_{max2}, and u_{max3}.

Case 2

Consider next the case (Case 2) where the model goes into the yielding stage after the third impulse. **Figures 2D–F** show the schematic response in this case. As in Case 1, u_{max1} can be obtained from the energy conservation law.

$$m(0.5V)^2 / 2 = ku_{max1}^2 / 2 \tag{6}$$

On the other hand, u_{max2} can be computed from another energy conservation law.

$$m(0.5V + V)^2 / 2 = ku_{max2}^2 / 2 \tag{7}$$

As stated above, the velocity – V by the second impulse is added to the velocity – $0.5V$ induced by the first impulse. Furthermore, u_{max3} can be computed from another energy conservation law.

$$m(0.5V + V + 0.5V)^2 / 2 = f_y d_y / 2 + f_y (u_{max3} - d_y) \tag{8}$$

As explained above, the velocity $0.5V$ by the third impulse is added to the velocity $1.5V$ induced by the first and second impulses. It should be noted that the critical timing t_0 corresponds to the time of zero restoring force also in Case 2.

Case 3

Consider next the case (Case 3) where the model goes into the yielding stage after the second impulse. **Figures 2G–I** show the schematic response in this case. u_{max1} can be obtained from the energy conservation law.

$$m(0.5V)^2 / 2 = ku_{max1}^2 / 2 \tag{9}$$

As in Case 2, the critical timing of the second impulse is the time of zero restoring force. Although a more complicated discussion is needed to show this critical timing depending on the timing of the third impulse, it is omitted here. Case 3-1 and Case 3-2 (**Figure 3**) should be considered in Case 3. In Case 3-1, the timing of the third impulse is in the second unloading stage (**Figures 2I and 3A**).

Case 3-1

In this case (Case 3-1), u_{max2} can be computed from another energy conservation law.

$$m(0.5V + V)^2 / 2 = f_y d_y / 2 + f_y (u_{max2} - d_y) \tag{10a}$$

Then, u_{max2} can be expressed as

$$u_{max2} = d_y + m\{(1.5V)^2 - (\omega_1 d_y)^2\} / (2f_y) \tag{10b}$$

On the other hand, u_{max3} can be computed from another energy conservation law.

$$m(v_{E'} - 0.5V)^2 / 2 + k\Delta u_{E'F} / 2 = f_y d_y / 2 + f_y u_{p3} \tag{11}$$

where $v_{E'}(<0)$ is the velocity at the time of third impulse (point E') and $\Delta u_{E'F} = u_{E'} - u_{p2}$ ($u_{E'}(>0)$: deformation at E'). u_{p2} is characterized by $u_{p2} = u_{max2} - d_y$ and u_{p3} satisfies $u_{max2} + u_{max3} = 2d_y + u_{p3}$. $v_{E'}$ and $u_{E'}$ are characterized by Eqs 12 and 13 by solving the equation of motion.

$$v_{E'} = -\omega_1 d_y \sin \omega_1 t_{EE'} \tag{12}$$

$$u_{E'} = d_y \cos \omega_1 t_{EE'} + u_{p2} = d_y \cos \omega_1 t_{EE'} + m\{(1.5V)^2 - (\omega_1 d_y)^1\} / (2f_y) \tag{13}$$

In these equations, $t_{EE'} = (T_1/2) - (t_{CD} + t_{DE})$ is the time interval between the first yielding termination point (E) and the third impulse (point E'), $t_{CD} = \{Sin^{-1}(2\omega_1 d_y / 3V)\} / \omega_1$ is the time interval between the time of the second impulse and the time of the first yielding initiation (after the second impulse), and $t_{DE} = m\sqrt{(9V^2/4) - (\omega_1 d_y)^2} / f_y$ is the time interval between the time of the first yielding initiation (after the second impulse) and the time of the second unloading initiation. t_{CD} and t_{DE} are computed by solving the equations of motion and substituting the transition conditions (yielding and unloading conditions). In other words, u_{max3} can be obtained from

$$m(v_{E'} - 0.5V)^2 / 2 + k\Delta u_{E'F} / 2 = f_y d_y / 2$$
$$+ f_y (u_{max2} + u_{max3} - 2d_y) \tag{14a}$$

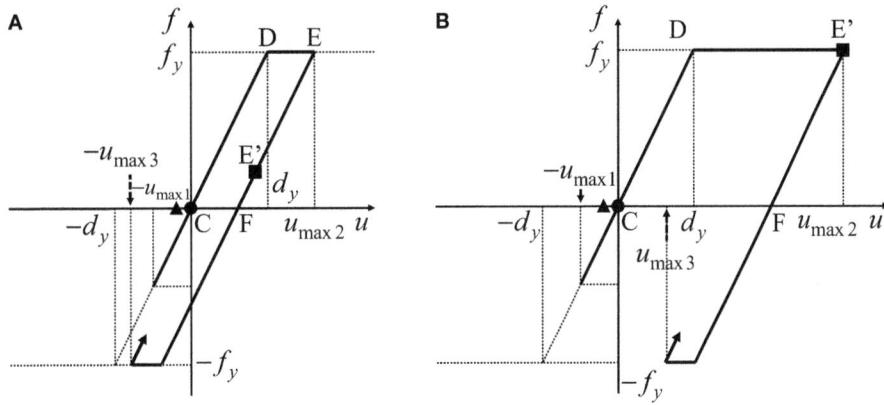

FIGURE 3 | Two different cases in Case 3: (A) Case 3-1, (B) Case 3-2 (•: first impulse, ▲: second impulse, ■: third impulse).

Then, u_{max3} can be expressed as

$$u_{max3} = d_y + m\{(v_{E'}^2 - v_{E'}V - 2V^2) + (\omega_1 \Delta u_{E'F})^2\}/(2f_y) \quad (14b)$$

If the maximum deformation after the third impulse (corresponding to u_{max3}) is positive, u_{p3} is characterized by $u_{max2} - u_{max3} = 2d_y + u_{p3}$. In this case, u_{max3} can be computed from another energy conservation law.

$$m(v_{E'} - 0.5V)^2/2 + k\Delta u_{E'F}/2 = f_y d_y/2 \\ + f_y(u_{max2} - u_{max3} - 2d_y) \quad (14c)$$

Then, u_{max3} can be expressed by

$$u_{max3} = -d_y - m\{(v_{E'}^2 - v_{E'}V - 2V^2) + (\omega_1 \Delta u_{E'F})^2\}/(2f_y) \quad (14d)$$

Case 3-2

In Case 3-2, the timing of the third impulse is in the yielding stage (**Figure 3B**). In this case (Case 3-2), u_{max2} is the deformation at the time of the third impulse ($u_{E'}$) $u_{E'}$ can be computed by solving the equation of motion and can be expressed by Eq. 15.

$$u_{E'} = -(f_y/2m)t_{DE'}^2 + \sqrt{(9V^2/4) - (\omega_1 d_y)^2}\,t_{DE'} + d_y \quad (15)$$

In this case, u_{max2} is given by

$$u_{max2} = u_{E'} \quad (16)$$

On the other hand, u_{max3} can be computed from another energy conservation law.

$$m(v_{E'} - 0.5V)^2/2 + f_y d_y/2 = f_y d_y/2 + f_y u_{p3} \quad (17)$$

where $v_{E'}$ is the velocity at the third impulse and u_{p3} is characterized by $u_{max2} - u_{max3} = 2d_y + u_{p3}$. $v_{E'}$ is characterized by Eq. 18 by solving the equation of motion.

$$v_{E'} = -(f_y/m)t_{DE'} + \sqrt{(9V^2/4) - (\omega_1 d_y)^2} \quad (18)$$

In these equations, $t_{DE'} = (T_1/2) - t_{CD}$ is the time interval between the first yielding initiation and the third impulse and $t_{CD} = \{Sin^{-1}(2\omega_1 d_y/(3V))\}/\omega_1$ is the time interval between the time of the second impulse (zero restoring force) and the time of the

first yielding initiation (after the second impulse). t_{CD} is computed by solving the equation of motion and substituting the transition conditions (yielding and unloading conditions). In other words, u_{max3} can be obtained from

$$m(v_{E'} - 0.5V)^2/2 + f_y d_y/2 \\ = f_y d_y/2 + f_y(u_{max2} - u_{max3} - 2d_y) \quad (19a)$$

Then, u_{max3} can be expressed by

$$u_{max3} = -2d_y + u_{E'} - m(v_{E'} - 0.5V)^2/(2f_y) \quad (19b)$$

Case 4

Consider finally the case (Case 4) where the model goes into the yielding stage even after the first impulse. **Figures 2J–L** show the schematic response in this case. u_{max1} can be obtained from the energy conservation law.

$$m(0.5V)^2/2 = f_y d_y/2 + f_y(u_{max1} - d_y) \quad (20)$$

In this case (Case 4), u_{max2} is the deformation at the third impulse ($u_{E'}$). $u_{E'}$ can be computed by solving the equation of motion and $u_{E'}$ can be obtained from Eq. 21.

$$u_{E'} = -(f_y/2m)t_{DE'}^2 + \sqrt{V^2 + 2\omega_1 d_y V}\,t_{DE'} + d_y \\ - m\{(V^2/4) - (\omega_1 d_y)^2\}/(2f_y) \quad (21)$$

In this case, u_{max2} is given by

$$u_{max2} = u_{E'} \quad (22)$$

On the other hand, u_{max3} can be computed from another energy conservation law.

$$m(v_{E'} - 0.5V)^2/2 + f_y d_y/2 = f_y d_y/2 + f_y u_{p3} \quad (23)$$

where $v_{E'}$ is the velocity at the third impulse and u_{p3} is characterized by $u_{max2} - u_{max3} = 2d_y + u_{p3}$. $v_{E'}$ is characterized by Eq. 24 by solving the equation of motion.

$$v_{E'} = -(f_y/m)t_{DE'} + \sqrt{V^2 + 2\omega_1 d_y V} \quad (24)$$

In these equations, $t_{DE'} = t_0 - t_{CD} = (t_{OA} + t_{AB} + t_{BC}) - t_{CD}$ is the time interval between the second yielding initiation and the third impulse and t_0 is the time interval between the time of the first impulse and the time of the second impulse. $t_{OA} = \{\text{Sin}^{-1}(2\omega_1 d_y/V)\}/\omega_1$ is the time interval between the time of the first impulse and the time of the first yielding initiation, $t_{AB} = m\sqrt{(V^2/4) - (\omega_1 d_y)^2}/f_y$ is the time interval between the time of the first yielding initiation and the time of the first unloading initiation, $t_{BC} = T_1/4$ is the time interval between the time of the first unloading initiation and the time of the second impulse and $t_{CD} = \{\text{Sin}^{-1}(\omega_1 d_y/(\omega_1 d_y + V))\}/\omega_1$ is the time interval between the time of the second impulse and the time of the second yielding initiation. t_{OA}, t_{AB}, and t_{CD}, are computed by solving the equation of motion and substituting the transition conditions (yielding and unloading conditions). In other words, u_{max3} can be obtained from

$$m(v_{E'} - 0.5V)^2 / 2 + f_y d_y / 2 = f_y d_y / 2$$
$$+ f_y(u_{max2} - u_{max3} - 2d_y) \quad (25a)$$

Then, u_{max3} can be expressed by

$$u_{max3} = -2d_y + u_{E'} - m(v_{E'} - 0.5V)^2 / (2f_y) \quad (25b)$$

Figure 4 shows the plot of $u_{max}/d_y = \max(u_{max1}/d_y, u_{max2}/d_y, u_{max3}/d_y)$ with respect to the input level. $2V_y$ is the input level at which the maximum deformation after the first impulse just attain the yield deformation d_y. Here, V_y is expressed by $V_y = \omega_1 d_y$. As stated before, there are four cases (Case 1–4).

Case 1: elastic response during all response stages (u_{max3} is the largest).
Case 2: yielding after the third impulse (u_{max3} is the largest).
Case 3: yielding after the second impulse (u_{max2} or u_{max3} is the largest).
Case 4: yielding after the first impulse (u_{max2} is the largest).

In Case 1 and 2, u_{max3} is the largest. On the other hand, in Case 3, u_{max2} or u_{max3} is the largest and in Case 4, u_{max2} is the largest.

As observed in **Figure 3**, the timing of the third impulse sometimes decreases u_{max2}. It may be useful to assume the timing of the third impulse at the zero restoring force in the unloading process as shown in **Figure 5**. In Case 1 and 2, this assumption is valid (see **Figures 2C,F**). **Figure 6** presents the corresponding figure in which the timing of the third impulse is the time of zero restoring force after the attainment of u_{max2} (in the process of the second unloading). In **Figure 6**, four cases (Case 1*, Case 2*, Case 3*, Case 4*) are introduced corresponding to the previously defined four cases. Case 1* and Case 2* are Case 1 and Case 2 themselves. It can be understood, because the timing of the third impulse defined as the same interval between the first and second impulses decreases u_{max2}, u_{max}/d_y in Case 4 in **Figure 4** becomes smaller than that in **Figure 6**. However, it is noteworthy that u_{max}/d_y in **Figure 6** can be a good upper bound of that in **Figure 4** and u_{max}/d_y (u_{max2}/d_y) to any other timing t_0 (except zero restoring force) does not become larger than that in **Figure 6** (see Proof of Critical Timing and Upper Bound of Response Ductility via Relaxation of Timing of Third Impulse in Appendix).

Figure 7 presents the normalized timing t_0/T_1 ($T_1 = 2\pi/\omega_1$) of the second impulse with respect to the input level. As stated before,

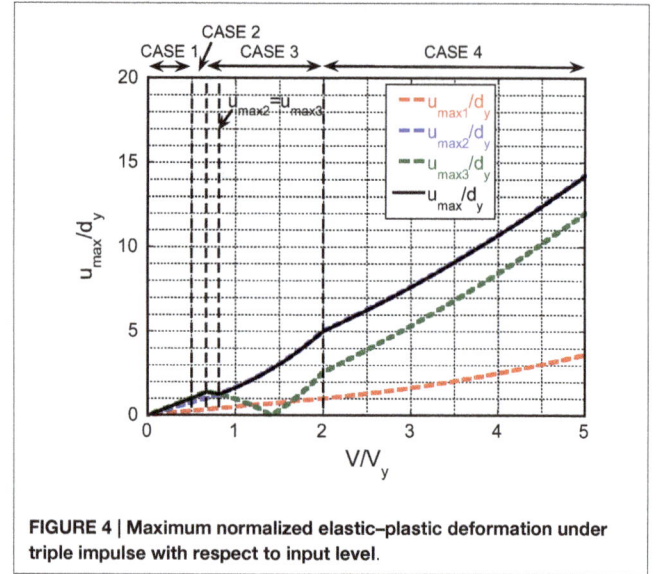

FIGURE 4 | Maximum normalized elastic–plastic deformation under triple impulse with respect to input level.

this timing coincides with the time of zero restoring force after the first unloading (see **Figure 2**). It can be observed that the timing is delayed as the input level increases. It seems noteworthy to state again that only critical response giving the maximum value of u_{max}/d_y (in case of the timing of the third impulse after the second unloading) is sought by the proposed method and the critical resonant frequency is obtained automatically for the increasing input level of the triple impulse. One of the original points in this paper is the tracking of the critical elastic–plastic response.

Accuracy Check by Time–History Response Analysis Subjected to the Corresponding Three Wavelets of Sinusoidal Waves

In order to investigate the accuracy of using the triple impulse as a substitute of the corresponding three wavelets of sinusoidal waves (representative of the forward-directivity input), the time–history response analysis of the elastic–plastic SDOF model under the three wavelets of sinusoidal waves has been conducted.

In the evaluation procedure, it is important to adjust the input level of the triple impulse and the corresponding three wavelets of sinusoidal waves based on the equivalence of the Fourier amplitude. **Figure 8** shows one example for the input level $V/V_y = 3$. **Figures 9A,B** illustrate the comparison of the ground displacement and velocity between the triple impulse and the corresponding three wavelets of sinusoidal waves for the input level $V/V_y = 3$. Only in **Figures 8** and **9A,B**, $\omega_1 = 2\pi$ (rad/s)($T_1 = 1.0$ s) and $d_y = 0.16$(m) are used.

Figure 10 presents the comparison of the ductility (maximum normalized deformation) of the elastic–plastic structure under the triple impulse and the corresponding three wavelets of sinusoidal waves with respect to the input level. It can be seen that the triple impulse provides a good substitute of the three wavelets of sinusoidal waves in the evaluation of the maximum deformation if the maximum Fourier amplitude is adjusted appropriately.

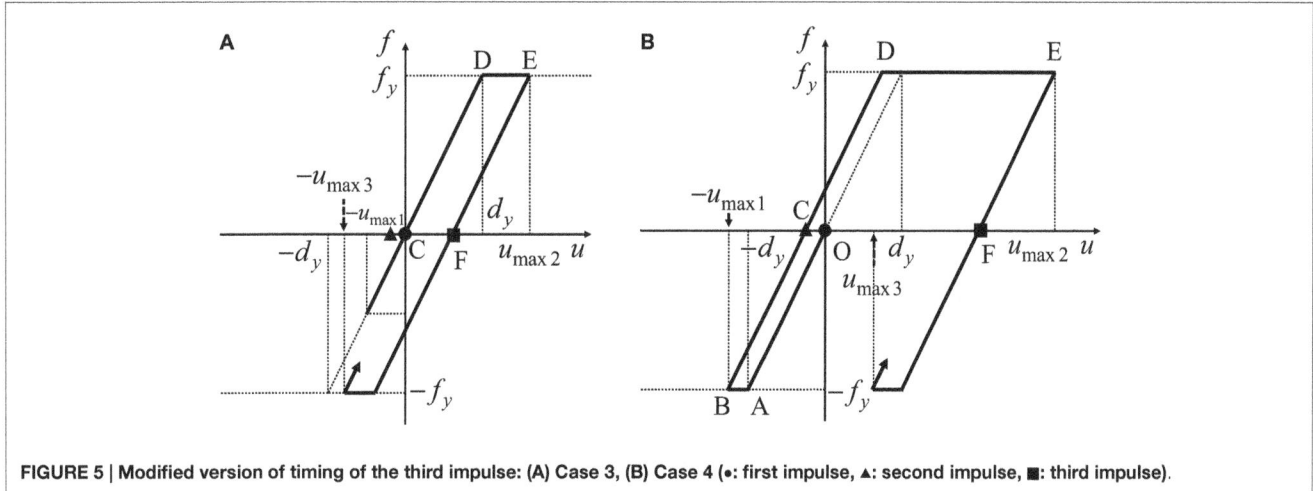

FIGURE 5 | Modified version of timing of the third impulse: (A) Case 3, (B) Case 4 (•: first impulse, ▲: second impulse, ■: third impulse).

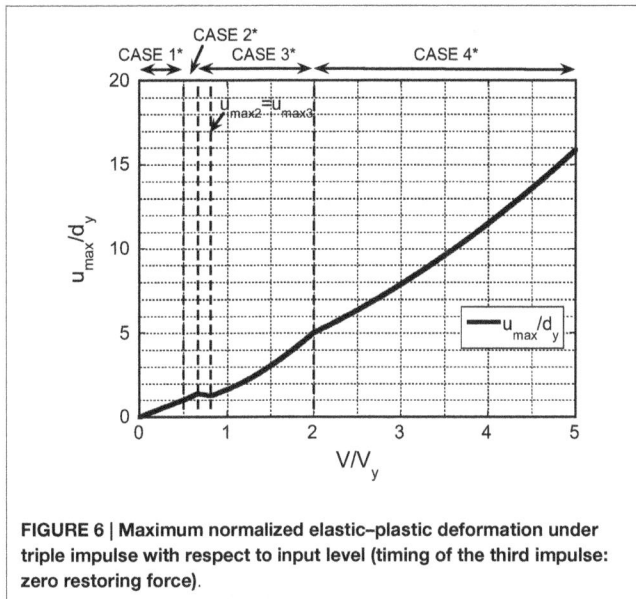

FIGURE 6 | Maximum normalized elastic–plastic deformation under triple impulse with respect to input level (timing of the third impulse: zero restoring force).

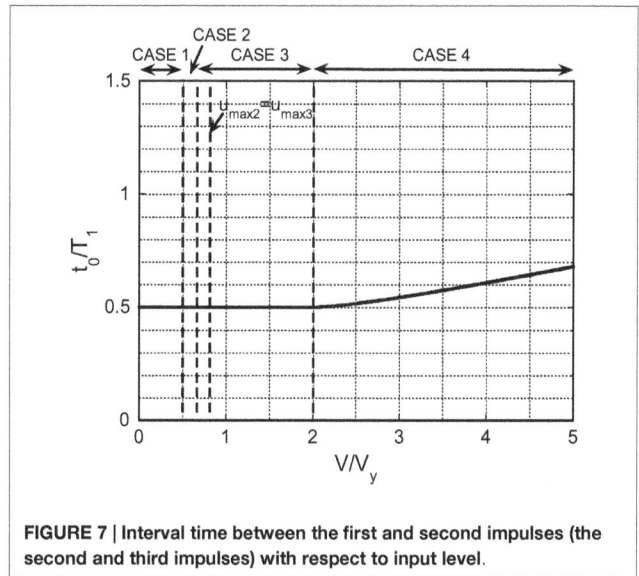

FIGURE 7 | Interval time between the first and second impulses (the second and third impulses) with respect to input level.

Figure 11 shows the comparison of the earthquake input energies by the triple impulse and the corresponding three wavelets of sinusoidal waves. An extremely accurate correspondence can be observed. This supports the validity of the triple impulse as a substitute of the forward-directivity near-fault ground motion.

Figure 12 illustrates the comparison of response time histories (normalized deformation and restoring force) under triple impulse and those under the corresponding three wavelets of sinusoidal waves. The parameters $\omega_1 = 2\pi$ (rad/s) ($T_1 = 1.0$ s), $d_y = 0.16$(m) are used here. While a rather good correspondence can be seen in general, the amplitude of deformation after the third impulse exhibits a slight difference resulting from the difference in timing of the third impulse. At the same time, a difference in phase can be observed both in the deformation and restoring force. This may also result from the difference in timing of the third impulse. The difference in the amount of energy input at the third impulse seems influence the later response.

Figure 13 presents the comparison of the restoring force characteristic under the triple impulse and that under the

corresponding three wavelets of sinusoidal wave. The parameters $\omega_1 = 2\pi$ (rad/s)($T_1 = 1.0$ s), $d_y = 0.16$(m) are also used here. As seen in **Figure 12**, while the maximum deformations after the first and second impulses exhibit a rather good correspondence, the deformation response after the third impulse exhibits an unnegligible difference. However, since the deformation response after the third impulse does not affect the maximum deformation in an overall time range, this difference may not be significant.

Design of Stiffness and Strength for Specified Velocity and Period of Forward-Directivity Near-Fault Ground Motion Input and Response Ductility

As in the case of the double impulse as a substitute of the near-fault fling-step input, it may be meaningful to present a flowchart for design of stiffness and strength for the specified velocity and period of the near-fault forward-directivity input and response ductility. This design concept is based on the philosophy that,

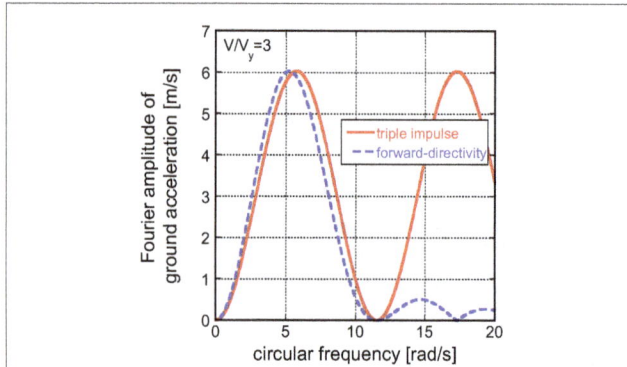

FIGURE 8 | Adjustment of input level of triple impulse and the corresponding three wavelets of sinusoidal waves based on Fourier amplitude equivalence.

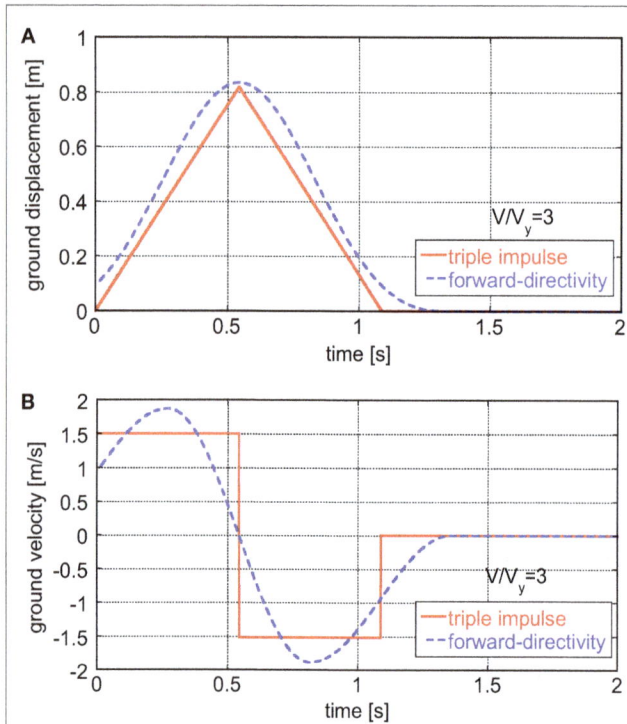

FIGURE 9 | Comparison of ground displacement and velocity between triple impulse and the corresponding three wavelets of sinusoidal waves: **(A)** displacement, **(B)** velocity.

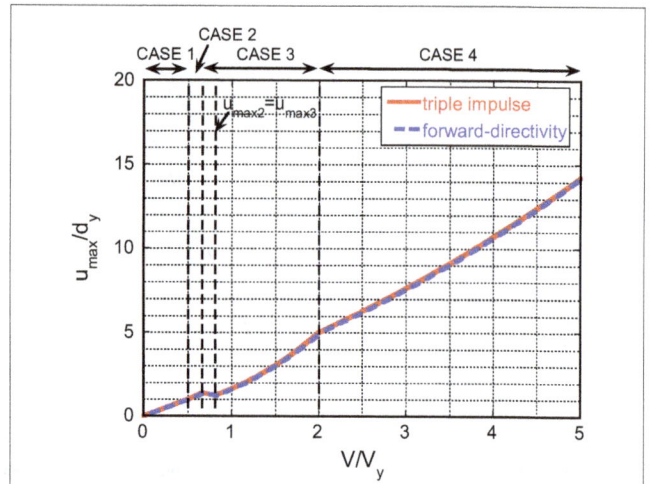

FIGURE 10 | Comparison of ductility of elastic–plastic structure under triple impulse and the corresponding three wavelets of sinusoidal waves.

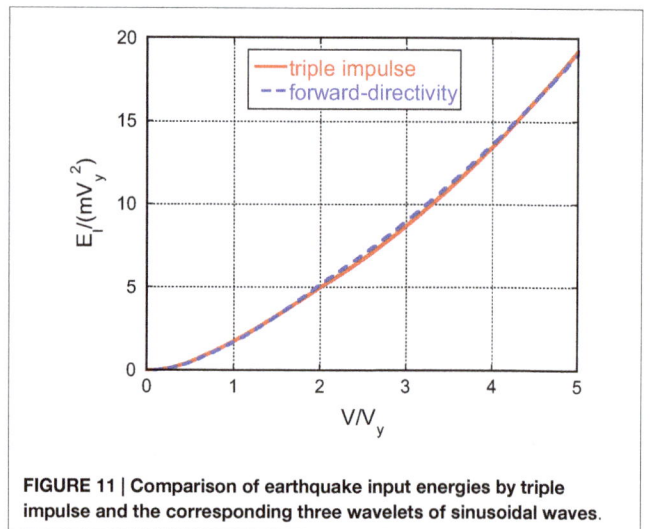

FIGURE 11 | Comparison of earthquake input energies by triple impulse and the corresponding three wavelets of sinusoidal waves.

if we focus on the worst case of resonance, the safety for other non-resonant cases is guaranteed (Takewaki, 2002). This fact will be explained in the following section.

Since **Figures 4** and **7** are non-dimensional ones, they can be used for such design. **Figure 14** shows the flowchart for design of stiffness and strength. One example can be drawn as follows:

[*Specified conditions*] $V = 2.00$(m/s), $t_0 = 0.500$(s), $u_{max}/d_y = 4.00$, $m = 4.00 \times 10^6$(kg)

[*Design results*] $V/V_y = 1.70$, $V_y = 1.18$(m/s), $T_1 = 1.00$(s), $d_y = 0.188$(m), $k = 1.58 \times 10^8$(N/m), $f_y = 2.97 \times 10^7$(N)

From **Figure 4**, $V/V_y = 1.70$ can be obtained for the specified ductility $u_{max}/d_y = 4.0$. Then, $V_y = 1.18$(m/s) is derived from the specified condition $V = 2.00$(m/s) and $V/V_y = 1.70$. In the next step, $T_1 = 1.00$(s) is found from **Figure 7** for $V/V_y = 1.70$ and $t_0 = 0.5$(s). In this model, $d_y = 0.188$(m) is determined from $V_y = \omega_1 d_y$ and $T_1 = (2\pi/\omega_1) = 1.00$(s). Finally, $k = 1.58 \times 10^8$(N/m) is obtained from $k = \omega_1^2 m$ and $f_y = 2.97 \times 10^7$(N) is derived by $f_y = k d_y$.

Approximate Prediction of Response Ductility for Specified Design of Stiffness and Strength and Specified Velocity and Period of Near-Fault Ground Motion Input

Until Section "Accuracy Check by Time–History Response Analysis Subjected to the Corresponding Three Wavelets of Sinusoidal Waves," only the critical set of velocity and period of

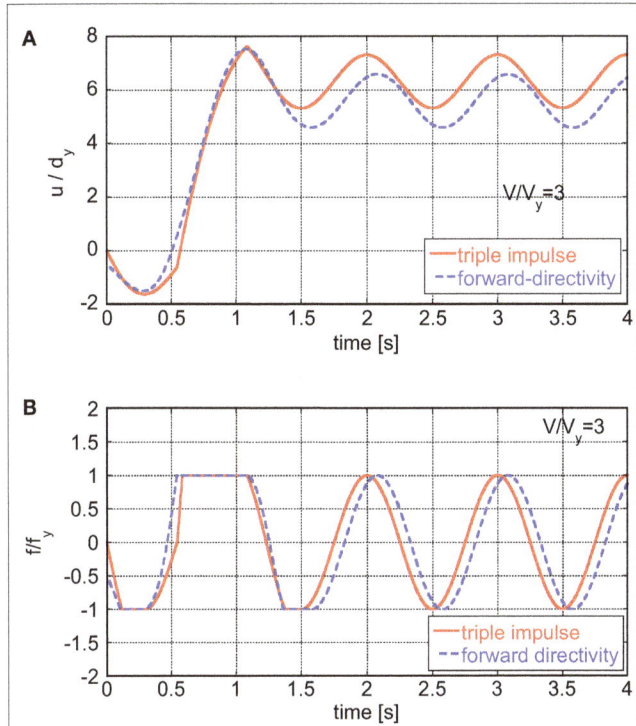

FIGURE 12 | Comparison of response time history under triple impulse and that under the corresponding three wavelets of sinusoidal waves: (A) normalized deformation, (B) normalized restoring force.

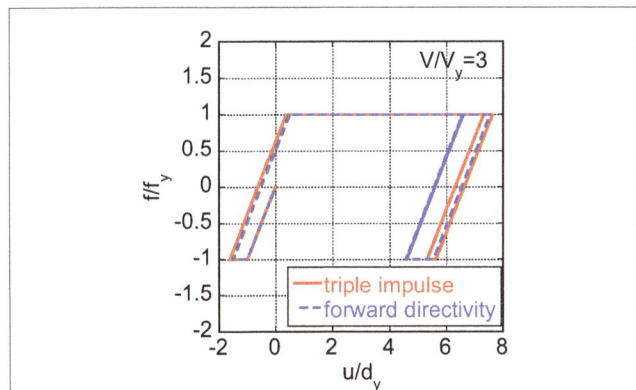

FIGURE 13 | Comparison of restoring force characteristic under triple impulse and that under the corresponding three wavelets of sinusoidal waves.

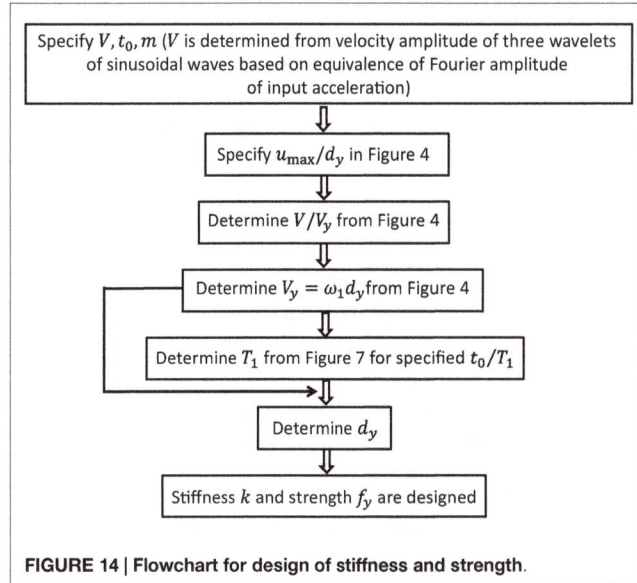

FIGURE 14 | Flowchart for design of stiffness and strength.

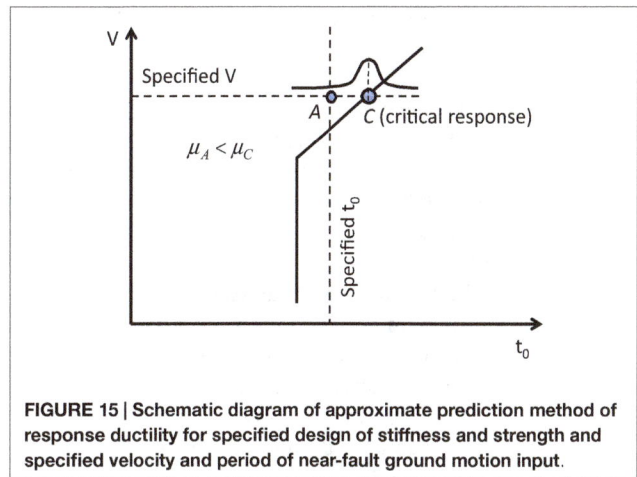

FIGURE 15 | Schematic diagram of approximate prediction method of response ductility for specified design of stiffness and strength and specified velocity and period of near-fault ground motion input.

and period of near-fault ground motion input is explained. If a more exact response is desired then the response analysis for an arbitrary timing of impulses and an arbitrary input level can be done.

Figure 15 shows a schematic diagram of the approximate prediction method (only prediction of upper bound) of response ductility $\mu = u_{max}/d_y$ for a specified design of stiffness and strength and a specified velocity and period of near-fault ground motion input using the corresponding critical response. Generally, the specified set of velocity and period of near-fault ground motion input is not the critical set for a given structure. In such case, consider the critical set (point C) of velocity and period of input corresponding to the specified set of velocity. Let μ_A and μ_C denote the response ductilities corresponding to point A and C. From the Section "Proof of Critical Timing" in Appendix, $\mu_A < \mu_C$ can be drawn directly. This enables an approximate prediction of response ductility (only upper bound) for a specified design of stiffness and strength and a specified velocity and period of near-fault ground motion input.

near-fault ground motion input and the corresponding critical response have been treated for a specified design of stiffness and strength. On the other hand, in Section "Design of Stiffness and Strength for Specified Velocity and Period of Forward-Directivity Near-Fault Ground Motion Input and Response Ductility," the design flowchart of stiffness and strength for the specified velocity and period of the near-fault ground motion input and specified response ductility has been presented. In this section, an approximate prediction method of response ductility for specified design of stiffness and strength and specified velocity

Conclusion

The conclusions may be summarized as follows:

(1) The triple impulse input has been introduced as a simplified version of the forward-directivity near-fault ground motion and a closed-form solution of the critical elastic–plastic response of a structure by this triple impulse input has been derived.

(2) It has been shown that, since only the free-vibration appears under such triple impulse input, the energy approach plays an important role in the derivation of the closed-form solution of a complicated elastic–plastic response. In other words, the energy approach enables the derivation of the maximum elastic–plastic seismic response. In this process, the input of impulse is expressed by the instantaneous change of velocity of the structural mass. The maximum elastic–plastic response after impulse can be obtained by equating the initial kinetic energy computed by the initial velocity to the sum of hysteretic and elastic strain energies. It has been shown that the maximum inelastic deformation can occur after either the second or third impulse depending on the input level.

(3) The validity and accuracy of the proposed theory have been investigated through the comparison with the response analysis result to the corresponding three wavelets of sinusoidal input as a representative of the forward-directivity near-fault ground motion. It has been made clear that, if the level of the triple impulse is adjusted so that its maximum Fourier amplitude coincides with that of the corresponding three wavelets of sinusoidal input, the maximum elastic–plastic deformation to the triple impulse exhibits a good correspondence with that to the three wavelets of sinusoidal wave.

(4) While the resonant equivalent frequency has to be computed for a specified input level by changing the excitation frequency in a parametric manner in dealing with the sinusoidal input, no iteration is required in the proposed method for the triple impulse. This is because the resonant equivalent frequency (resonance can be proved by using energy investigation) can be obtained directly without the repetitive procedure

(the timing of the second impulse can be characterized as the time with zero restoring force). In the triple impulse, the analysis can be conducted without the input frequency (timing of impulses) before the second impulse. It should be noted that, while a simple and clear concept of critical input was defined in the case of double impulse (Kojima and Takewaki, 2015b), the criticality can be used only before the third impulse in the present triple impulse. This is because the timing of the third impulse, determined already for the first and second impulses, decreases the maximum deformation u_{max2} after the second impulse and may increase the maximum deformation u_{max3} after the third impulse.

(5) Only critical response (upper bound) is captured by the proposed method and the critical resonant frequency can be obtained automatically for the increasing input level of the triple impulse. Once the frequency and amplitude of the critical triple impulse are computed, the corresponding three wavelets of sinusoidal motion as a representative of the forward-directivity motion can be identified.

(6) A flowchart for design of stiffness and strength for the specified velocity and period of the near-fault ground motion input and response ductility has been proposed using the newly derived non-dimensional relations among response ductility, input velocity, and input period. It has been demonstrated that this flowchart can provide a useful result for such design.

(7) An approximate prediction method of response ductility (only prediction of upper bound) using the corresponding critical response can be developed for a specified design of stiffness and strength and a specified velocity and period of near-fault ground motion input.

Acknowledgments

Part of the present work is supported by the Grant-in-Aid for Scientific Research of Japan Society for the Promotion of Science (No. 24246095, No. 15H04079) and the 2013-MEXT-Supported Program for the Strategic Research Foundation at Private Universities in Japan (No. S1312006). This support is greatly appreciated.

References

Abbas, A. M., and Manohar, C. S. (2002). Investigations into critical earthquake load models within deterministic and probabilistic frameworks. *Earthq. Eng. Struct. Dyn.* 31, 813–832. doi:10.1002/eqe.124.abs

Alavi, B., and Krawinkler, H. (2004). Behaviour of moment resisting frame structures subjected to near-fault ground motions. *Earthq. Eng. Struct. Dyn.* 33, 687–706. doi:10.1002/eqe.370

Bertero, V. V., Mahin, S. A., and Herrera, R. A. (1978). Aseismic design implications of near-fault San Fernando earthquake records. *Earthq. Eng. Struct. Dyn.* 6, 31–42. doi:10.1002/eqe.4290060105

Bray, J. D., and Rodriguez-Marek, A. (2004). Characterization of forward-directivity ground motions in the near-fault region. *Soil Dyn. Earthq. Eng.* 24, 815–828. doi:10.1016/j.soildyn.2004.05.001

Caughey, T. K. (1960). Sinusoidal excitation of a system with bilinear hysteresis. *J. Appl. Mech.* 27, 640–643. doi:10.1115/1.3644077

Drenick, R. F. (1970). Model-free design of aseismic structures. *J. Eng. Mech. Div.* 96, 483–493.

Hall, J. F., Heaton, T. H., Halling, M. W., and Wald, D. J. (1995). Near-source ground motion and its effects on flexible buildings. *Earthq. Spectra* 11, 569–605. doi:10.1193/1.1585828

Hayden, C. P., Bray, J. D., and Abrahamson, N. A. (2014). Selection of near-fault pulse motions. *J. Geotech. Geoenviron. Eng.* 140:04014030. doi:10.1061/(ASCE) GT.1943-5606.0001129

Kalkan, E., and Kunnath, S. K. (2006). Effects of fling step and forward directivity on seismic response of buildings. *Earthq. Spectra* 22, 367–390. doi:10.1193/1.2192560

Kalkan, E., and Kunnath, S. K. (2007). Effective cyclic energy as a measure of seismic demand. *J. Earthq. Eng.* 11, 725–751. doi:10.1080/13632460601033827

Khaloo, A. R., Khosravi1, H., and Hamidi Jamnani, H. (2015). Nonlinear inter-story drift contours for idealized forward directivity pulses using "Modified Fish-Bone" models. *Adv. Struct. Eng.* 18, 603–627. doi:10.1260/1369-4332. 18.5.603

Kojima, K., Fujita, K., and Takewaki, I. (2015a). Critical double impulse input and bound of earthquake input energy to building structure. *Front. Built Environ.* 1:5. doi:10.3389/fbuil.2015.00005

Kojima, K., and Takewaki, I. (2015b). Critical earthquake response of elastic-plastic structures under near-fault ground motions (Part 1: Fling-step input). *Front. Built Environ.* 1:12. doi:10.3389/fbuil.2015.00012

Liu, C.-S. (2000). The steady loops of SDOF perfectly elastoplastic structures under sinusoidal loadings. *J. Mar. Sci. Technol.* 8, 50–60.

Mavroeidis, G. P., Dong, G., and Papageorgiou, A. S. (2004). Near-fault ground motions, and the response of elastic and inelastic single-degree-freedom (SDOF) systems. *Earthq. Eng. Struct. Dyn.* 33, 1023–1049. doi:10.1002/eqe.391

Mavroeidis, G. P., and Papageorgiou, A. S. (2003). A mathematical representation of near-fault ground motions. *Bull. Seismol. Soc. Am.* 93, 1099–1131. doi:10.1785/0120020100

Moustafa, A., Ueno, K., and Takewaki, I. (2010). Critical earthquake loads for SDOF inelastic structures considering evolution of seismic waves. *Earthq. Struct.* 1, 147–162. doi:10.12989/eas.2010.1.2.147

Mukhopadhyay, S., and Gupta, V. K. (2013a). Directivity pulses in near-fault ground motions – I: identification, extraction and modeling. *Soil Dyn. Earthq. Eng.* 50, 1–15. doi:10.1016/j.soildyn.2013.02.017

Mukhopadhyay, S., and Gupta, V. K. (2013b). Directivity pulses in near-fault ground motions – II: estimation of pulse parameters. *Soil Dyn. Earthq. Eng.* 50, 38–52. doi:10.1016/j.soildyn.2013.02.019

Rupakhety, R., and Sigbjörnsson, R. (2011). Can simple pulses adequately represent near-fault ground motions? *J. Earthq. Eng.* 15, 1260–1272. doi:10.1080/136324 69.2011.565863

Sasani, M., and Bertero, V. V. (2000). "Importance of severe pulse-type ground motions in performance-based engineering: historical and critical review," in Proceedings of the Twelfth World Conference on Earthquake Engineering (Auckland).

Takewaki, I. (2002). Robust building stiffness design for variable critical excitations. *J. Struct. Eng. ASCE* 128, 1565–1574. doi:10.1061/(ASCE)0733-9445 (2002)128:12(1565)

Takewaki, I. (2004). Bound of earthquake input energy. *J. Struct. Eng. ASCE* 130, 1289–1297. doi:10.1061/(ASCE)0733-9445(2004)130:9(1289)

Takewaki, I. (2007). *Critical Excitation Methods in Earthquake Engineering*, 2nd Edn. Oxford: Elsevier, 2013.

Takewaki, I., Moustafa, A., and Fujita, K. (2012). *Improving the Earthquake Resilience of Buildings: The Worst Case Approach.* London: Springer.

Takewaki, I., and Tsujimoto, H. (2011). Scaling of design earthquake ground motions for tall buildings based on drift and input energy demands. *Earthq. Struct.* 2, 171–187. doi:10.12989/eas.2011.2.2.171

Vafaei, D., and Eskandari, R. (2015). Seismic response of mega buckling-restrained braces subjected to fling-step and forward-directivity near-fault ground motions. *Struct. Des. Tall Spec. Build.* 24, 672–686. doi:10.1002/tal.1205

Xu, Z., Agrawal, A. K., He, W.-L., and Tan, P. (2007). Performance of passive energy dissipation systems during near-field ground motion type pulses. *Eng. Struct.* 29, 224–236. doi:10.1016/j.engstruct.2006.04.020

Yamamoto, K., Fujita, K., and Takewaki, I. (2011). Instantaneous earthquake input energy and sensitivity in base-isolated building. *Struct. Des. Tall Spec. Build.* 20, 631–648. doi:10.1002/tal.539

Yang, D., and Zhou, J. (2014). A stochastic model and synthesis for near-fault impulsive ground motions. *Earthq. Eng. Struct. Dyn.* 44, 243–264. doi:10.1002/eqe.2468

Zhai, C., Chang, Z., Li, S., Chen, Z.-Q., and Xie, L. (2013). Quantitative identification of near-fault pulse-like ground motions based on energy. *Bull. Seismol. Soc. Am.* 103, 2591–2603. doi:10.1785/0120120320

Conflict of Interest Statement: The authors declare that the research was conducted in the absence of any commercial or financial relationships that could be construed as a potential conflict of interest.

Appendix

Proof of Critical Timing

In comparison with the double impulse, the triple impulse is quite difficult to derive the critical timing in a general case. This is because the timing of three impulses is fixed and there exist many complicated situations. In this paper, a case is treated where the critical timing is defined only before the third impulse. This means that, if the third impulse does not exist, timing gives the maximum value of u_{max2}. The following explanation is the same as in the previous paper for the double impulse.

Consider the critical timing of the second impulse. Let v_c denote the velocity of the mass passing the zero restoring force (zero elastic strain energy) after the first unloading and v^*, u^* denote the velocity and the elastic deformation component at an arbitrary point between the first unloading and the second yielding. Since the first unloading starts from the state with zero velocity and the elastic strain energy $f_y d_y / 2$, the relation $m v_c^2 / 2 = f_y d_y / 2$ holds. From the energy conservation law between the first unloading and the second yielding, the relation $m v^{*2}/2 + k u^{*2} = f_y d_y/2$ holds. Consider the second impulse at the same time of the state of v^*, u^*. The total mechanical energy can be expressed by $m(v^* + V)^2/2 + k u^{*2}/2$. Since the relation $m(v^* + V)^2/2 + k u^{*2}/2 = m v^{*2}/2 + k u^{*2}/2 + m v^* V + m V^2/2 = f_y d_y/2 + m v^* V + m V^2/2$ holds and the maximum deformation after the second yielding is caused by the maximum total mechanical energy, the maximum velocity v^* causes the maximum deformation after the second yielding. This timing is the zero restoring force after the first unloading. This completes the proof.

Upper Bound of Response Ductility via Relaxation of Timing of Third Impulse

The case is treated here where the third impulse acts at the timing of zero restoring force in the second unloading process. It can be shown that this case provides the larger maximum deformation u_{max2} than the case treated before (the same timing between the first and second impulses).

Consider the case where the model goes into the yielding stage even after the first impulse. The case where the model goes into the yielding stage after the second impulse (elastic after the first impulse) can be explained in almost the same manner. **Figure 5B** shows the schematic response in this case. u_{max1} can be obtained from the energy conservation law.

$$m(0.5V)^2 / 2 = f_y d_y / 2 + f_y u_{p1} = f_y d_y / 2 + f_y (u_{max1} - d_y) \quad (A1)$$

On the other hand, u_{max2} can be computed from another energy conservation law.

$$m(v_c + V)^2 / 2 = f_y d_y / 2 + f_y u_{p2} \quad (A2)$$

where v_c is characterized by $m v_c^2 / 2 = f_y d_y / 2$ and u_{p2} is characterized by $u_{max2} + (u_{max1} - d_y) = d_y + u_{p2}$. In other words, u_{max2} can be obtained from

$$m(v_c + V)^2 / 2 = f_y d_y / 2 + f_y (u_{max1} + u_{max2} - 2d_y) \quad (A3)$$

Furthermore, u_{max3} can be computed from another energy conservation law.

$$m(v_c + 0.5V)^2 / 2 = f_y d_y / 2 + f_y u_{p3} \quad (A4)$$

where u_{p3} is characterized by $-u_{max3} + (u_{max2} - d_y) = d_y + u_{p3}$. In other words, u_{max3} can be obtained from

$$m(v_c + 0.5V)^2 / 2 = f_y d_y / 2 + f_y (u_{max2} - u_{max3} - 2d_y) \quad (A5)$$

4

Empirical assessment of non-linear seismic demand of mainshock–aftershock ground-motion sequences for Japanese earthquakes

Katsuichiro Goda[1]*, Friedemann Wenzel[2] and Raffaele De Risi[1]

[1] *Department of Civil Engineering, University of Bristol, Bristol, UK,* [2] *Geophysical Institute, Karlsruhe Institute of Technology, Karlsruhe, Germany*

This study investigates the effects of earthquake types, magnitudes, and hysteretic behavior on the peak and residual ductility demands of inelastic single-degree-of-freedom systems and evaluates the effects of major aftershocks on the non-linear structural responses. An extensive dataset of real mainshock–aftershock sequences for Japanese earthquakes is developed. The constructed dataset is large, compared with previous datasets of similar kinds, and includes numerous sequences from the 2011 Tohoku earthquake, facilitating an investigation of spatial aspects of the aftershock effects. The empirical assessment of peak and residual ductility demands of numerous inelastic systems having different vibration periods, yield strengths, and hysteretic characteristics indicates that the increase in seismic demand measures due to aftershocks occurs rarely but can be significant. For a large mega-thrust subduction earthquake, a critical factor for major aftershock damage is the spatial occurrence process of aftershocks.

Keywords: peak ductility, residual ductility, Japanese earthquakes, mainshock and aftershocks, 2011 Tohoku earthquake

Edited by:
Nikos D. Lagaros,
National Technical University of
Athens, Greece

Reviewed by:
Carmine Galasso,
University College London, UK
Bing Qu,
California Polytechnic State
University, USA

*Correspondence:
Katsuichiro Goda,
Queen's Building, University Walk,
Bristol BS8 1TR, UK
katsu.goda@bristol.ac.uk

Introduction

Ground-motion records are the main source of uncertainty in predicting non-linear responses of structures subjected to earthquake loading. Key record features can be represented by amplitude, duration, frequency content, and their temporal evolution. They are influenced by physical environments and characteristics, such as earthquake type (crustal/interface/inslab), moment magnitude (M_w), faulting mechanism, stress drop, seismic wave propagation, and local site condition (Stein and Wysession, 2003). In the last decade, observation networks of strong motion around the world have been expanded significantly, and numerous recordings have been made available publicly, e.g., K-NET/KiK-net in Japan, TSMIP in Taiwan, GeoNet in New Zealand, and ITACA in Italy. These databases facilitate the development of new generations of empirical ground-motion prediction equations that are essential for probabilistic seismic hazard analysis (e.g., Morikawa and Fujiwara, 2013). Moreover, they are useful for developing inelastic seismic demand prediction models (e.g., Ruiz-García and Miranda, 2003; Vamvatsikos and Cornell, 2004; Federal Emergency Management Agency, 2005; Iervolino and Cornell, 2005). The integration of seismic hazard and ground-motion models with seismic vulnerability models results in a comprehensive performance-based earthquake engineering (PBEE) framework that accounts for main sources of uncertainty related to seismic damage assessment and loss estimation (Cornell et al., 2002; Goulet et al., 2007).

Recent earthquake disasters highlight that a cluster of major aftershocks causes incremental damage to structures whose seismic capacities may have been reduced by a mainshock and poses significant risk to evacuees and residents in a post-disaster situation. For instance, the 2011 Christchurch aftershock sequence (notably the 22 February 2011 $M_w 6.2$ event), initiated by the 2010 $M_w 7.0$ Darfield event, caused extensive damage to buildings and infrastructure in downtown Christchurch (Smyrou et al., 2011). After the 2011 $M_w 9.0$ Tohoku earthquake in Japan, numerous aftershocks as large as $M_w 7.9$ were observed, and additional structural damage and disruption to utility services were caused by major aftershocks (Goda et al., 2013). In Indonesia, regional seismic activities have been heightened since the 2004 $M_w 9.3$ Sumatra earthquake (Shcherbakov et al., 2013). Numerous moderate-to-large earthquakes occurred and caused major seismic damage to structures in Sumatra (e.g., 2005 $M_w 8.6$ and 2007 $M_w 8.5$ events). To evaluate seismic responses of different structures (i.e., steel, concrete, and wood-frame buildings) due to mainshock–aftershock (MS–AS) sequences, various models, such as single-degree-of-freedom (SDOF) and multi-degree-of-freedom systems with different hysteretic models, have been used (e.g., Li and Ellingwood, 2007; Moustafa and Takewaki, 2010; Goda, 2012; Ruiz-García, 2012; Zhai et al., 2013). The developed seismic demand models for MS–AS sequences can be incorporated into the PBEE framework to account for seismic damage and loss caused by aftershocks (Salami and Goda, 2014).

In Japan, national and regional strong-motion networks, K-NET/KiK-net[1] and SK-net[2], have been established aftermath the 1995 Kobe earthquake. The availability of strong-motion records in Japan has increased drastically and numerous invaluable data have been recorded. One of the events that are extremely well-recorded is the 2011 $M_w 9.0$ Tohoku earthquake; more than 1000 high-quality recordings are available from these networks for ground motion and seismic vulnerability studies. Because numerous aftershocks were triggered by the 11 March 2011 mainshock, an extensive set of MS–AS sequence data can be developed. The new dataset for MS–AS sequences in Japan offers a new opportunity to compare the non-linear seismic demand potential due to different earthquake types (e.g., crustal versus interface events, which are often distinguished in seismic design codes). Moreover, for the 2011 Tohoku mainshock, the aftershock effects can be evaluated from not only temporal/sequential but also spatial viewpoints of the major aftershock occurrence, providing with valuable insights into the aftershock hazard processes.

The main objectives of this study are to investigate the non-linear seismic demand potential of inelastic SDOF systems due to real MS–AS sequences in Japan, and to establish an empirical benchmark for the non-linear seismic demand assessment for Japanese earthquakes. To draw generic conclusions, 112 inelastic SDOF systems having four intact vibration periods ($T = 0.2$, 0.5, 1.0, and 2.0 s), seven yield strengths, and four hysteretic characteristics (which are approximated by the Bouc–Wen model; Wen, 1976; Foliente, 1993; Goda and Atkinson, 2009), are considered. The yield strengths of the inelastic systems are expressed in terms of spectral acceleration, and their values are selected such that the considered yield capacities broadly represent those of typical building stock in Japan (Nagato and Kawase, 2004). As the non-linear response metrics, peak and residual ductility demands are focused upon. The latter parameter is relevant for PBEE-based seismic performance assessment where excessive residual displacements prohibit residents from reoccupation and result in demolishing non-collapse buildings (Ruiz-García and Miranda, 2006; Ramirez and Miranda, 2012). It is noted that the investigations carried out in this study (constant strength approach) differ from the constant R approach (where R is the strength reduction factor; Ruiz-García and Miranda, 2003), as carried out in the previous investigations (Goda and Atkinson, 2009; Goda, 2012). In the constant R approach, seismic excitation levels of ground-motion records are kept constant with respect to the yield strength of a structural system, whereas in the constant strength approach, the yield strength of a structural system is varied relative to a set of selected ground-motion records (Galasso et al., 2012). A novelty of this study is that an extensive dataset of as-recorded MS–AS sequences for Japanese earthquakes is compiled and employed for the non-linear seismic demand potential evaluation. The new dataset contains 531 MS–AS sequences from 20 mainshock events (note: each sequence consists of two horizontal components). The statistical analysis is performed to relate the non-linear seismic demand potential and aftershock effects to key seismological parameters. Among the 531 sequences, 304 sequences are from the 2011 Tohoku event. This facilitates a rigorous assessment of the aftershock effects with regard to the spatial distribution of major aftershocks. This paper is organized as follows. First, the construction of the real MS–AS sequence database based on the K-NET, KiK-net, and SK-net is explained, which is the main innovative feature of this study. Second, non-linear structural models with Bouc–Wen hysteresis are introduced, and non-linear structural responses due to the constructed real MS–AS sequence records for Japanese earthquakes are compared in terms of earthquake type, magnitude, and hysteretic behavior. Subsequently, the aftershock effects on the non-linear seismic demand are discussed by focusing upon the key seismological parameters for the increased ductility demands. Moreover, spatial aspect of the aftershock effects is evaluated for the 2011 Tohoku sequences.

Mainshock–Aftershock Sequence Records for Japanese Earthquakes

A new ground-motion database *2012 KKiKSK* is developed for the purpose of ground-motion prediction studies. It combines recordings from the K-NET, KiK-net, and SK-net up to the end of 2012. Records from different networks are first integrated by matching event information (occurrence time, location, earthquake size, etc.). Subsequently, duplicates and erroneous data (typically, SK-net recordings that contain spurious spikes, discontinuities, and base-line shift) are identified and removed from the database. A set of broad record selection criteria is then applied to determine records that are included in the database: (i) minimum Japan

[1]http://www.kyoshin.bosai.go.jp/

[2]http://www.sknet.eri.u-tokyo.ac.jp/

Meteorological Agency (JMA) magnitude M_{JMA} is 3.0; (ii) maximum focal depth is 500 km; (iii) maximum hypocentral distance is 1500 km; (iv) minimum horizontal peak ground acceleration (PGA) is 1.0 cm/s^2; and (v) at least 10 records are available for each seismic event (satisfying the preceding four conditions). This has led to a set of 555,750 records from 6261 earthquakes. Further checks are conducted to improve the quality of the database.

Subsequently, metadata, such as M_w, fault mechanism (normal/reverse/strike-slip), and earthquake type (crustal/inslab/interface), are assigned to seismic events with M_{JMA} greater than or equal to 5.5 individually by referring to the Harvard Centroid Moment Tensor (CMT) solutions[3] and the F-net CMT solutions[4]. In calculating representative source-to-site distances for moderate-to-large earthquakes, finite fault plane information for 57 events are gathered from the Geospatial Institute Authority of Japan webpages[5] and the EIC/NGY seismological notes by Kikuchi and Yamanaka[6,7]. Using the finite fault plane models, rupture distance (i.e., shortest distance from a site to a fault plane) is calculated. Note that the majority of significant earthquakes are associated with the finite fault plane models (exceptions include moderate-to-large events that occur off-shore regions). Site information for the K-NET and KiK-net is obtained from the NIED webpages (see text footnote 1); for the K-NET, relocation information is taken into account. For assigning site information to the SK-net sites, an approach adopted by Goda and Atkinson (2010) is implemented, which combines various kinds of site information, such as geomorphological classification, micro-tremor measurements, and borehole-logging. By reflecting the availability of site information, usability of record components is determined for the SK-net. In total, the usable record set contains 528,022 records from 6259 earthquakes. Individual components in the record set are processed uniformly (i.e., tapering, zero-padding, and band-pass filtering; Boore, 2005). Various elastic ground-motion parameters, such as PGA, peak ground velocity, and 5%-damped elastic response spectra at vibration periods ranging from 0.05 to 10.0 s, are computed using the processed record components.

The development of MS–AS record sequences based on the *2012 KKiKSK* database is carried out in two stages. In the first stage, the record database is downsized by eliminating weak ground motions. The record selection criteria that are applied are: (i) $M_w \geq 5.0$, (ii) focal depth is less than 150 km, (iii) average shear-wave velocity in the uppermost 30 m V_{S30} is between 100 and 1500 m/s, (iv) source-to-site distance is less than 300 km, and (v) average PGA of the two horizontal components (geometric mean) is greater than 75 cm/s^2 (such a criterion is typically applied in inelastic demand estimation studies; Ruiz-García and Miranda, 2003; Goda and Atkinson, 2009). The application of the above five criteria has resulted in 5000 records, consisting of 367 events.

In the second stage, a list of MS–AS sequences is developed using the reduced dataset of 5000 records. Initially, a candidate mainshock, or reference event, is identified as event having $M_w > 5.9$. For a given reference event, a time-space window is applied to identify possible candidate aftershock events; the length of the time window is set to 100 days before and after the date of occurrence of the reference event (note: for the 2011 Tohoku mainshock, the post-event time window is extended to 600 days), while the spatial window is circular in shape around the epicenter of the reference event and the radius is calculated by d (km) $= 0.02 \times 10^{0.5 \times \min(Mw,ref,8.5)}$ (Kagan, 2002), where $M_{w,ref}$ is the moment magnitude of the reference event (i.e., initially $M_{w,ref}$ equals the magnitude of the candidate mainshock and is changed to magnitudes of reference events). In addition, the difference of focal depths of the reference event and a candidate aftershock is used to determine inclusion/exclusion of the candidate aftershock by considering a threshold of 30 km. The above search process is repeated for all events included in the identified MS–AS sequence; after the completion of the search process for the candidate mainshock, the reference event is changed to one of the identified aftershocks, and this process is continued until all candidate aftershocks are examined exhaustively. For instance, the process starts with a mainshock, and then when additional aftershock events are identified, they are included in the MS–AS sequence. The same screening process (i.e., space-time window) is applied to all events in the sequence (note: the size of the sequence usually grows and the radius of the spatial window varies). This process has led to the identification of 20 MS–AS sequences. Subsequently, for each sequence, eligible records are reorganized on a station basis, and time-history data for individual sequences are constructed by inserting 30 s of zeros between records. This has resulted in 531 MS–AS record sequences. In each sequence, an event with the largest magnitude is designated as *mainshock*, whereas an event with the second largest M_w is determined as *major aftershock*, consistent with the definitions adopted by Goda (2012). A summary of the mainshock characteristics of the identified 20 sequences is given in **Table 1**. The MS–AS sequences for the 2011 Tohoku earthquake comprise of about 57% of the database. This database is considered for record selection to be used in empirical assessment of inelastic seismic demand potential due to real MS–AS sequences.

Figure 1 shows the locations of mainshocks, magnitude–distance plots of mainshocks and major aftershocks, and histogram of V_{S30} for the sites included in the database. In the map (**Figure 1A**), the sequences are divided into four subsets: 2011 Tohoku event (304 sequences), 2003 Tokachi event (36 sequences), crustal events (122 sequences), and interface/inslab events (69 sequences, excluding those for the 2011 Tohoku and 2003 Tokachi events). The magnitudes for the 2011 Tohoku and 2003 Tokachi events are significantly greater than other events (**Table 1**), whereas the magnitudes for crustal events and interface/inslab events are broadly similar (in the range between M_w6 and M_w7) but their locations are different (i.e., on-shore versus off-shore, indicating different propagation paths). This classification is used for comparing the elastic and inelastic seismic demands of different earthquake types in the following. The magnitude–distance plots indicate that the magnitudes of

---

[3]http://www.globalcmt.org/

[4]http://www.fnet.bosai.go.jp/

[5]http://www.gsi.go.jp/bousai.html

[6]http://www.eri.u-tokyo.ac.jp/sanchu/Seismo_Note/

[7]http://www.seis.nagoya-u.ac.jp/sanchu/Seismo_Note/

TABLE 1 | Summary of the mainshock characteristics of the 20 mainshock–aftershock sequences.

Sequence ID	Date	Event type	Latitude	Longitude	Depth (km)	M_w	Number of sequences
1	1996/08/11	Crustal	38.920	140.630	10.0	5.92	1
2	1996/10/19	Inslab	31.803	131.998	22.0	6.70	4
3	1997/03/26	Crustal	31.986	130.365	10.0	6.10	11
4	2000/07/30	Crustal	33.965	139.397	10.0	6.45	1
5	2000/10/06	Crustal	35.278	133.345	10.0	6.65	3
6	2001/03/24	Inslab	34.123	132.705	50.0	6.80	3
7	2002/11/03	Inslab	38.896	142.138	39.0	6.40	1
8	2003/07/26	Crustal	38.405	141.170	6.0	6.04	10
9	2003/09/26	Interface	41.781	144.074	27.0	8.26	36
10	2004/04/04	Inslab	36.390	141.154	31.0	5.93	1
11	2004/09/05	Inslab	33.146	137.139	10.0	7.37	17
12	2004/10/23	Crustal	37.291	138.867	16.0	6.56	53
13	2004/11/29	Inslab	42.946	145.274	39.0	6.98	35
14	2005/03/20	Crustal	33.738	130.175	10.0	6.58	13
15	2007/03/25	Crustal	37.220	136.685	8.0	6.67	5
16	2007/07/16	Crustal	37.557	138.608	12.0	6.62	7
17	2008/06/14	Crustal	39.028	140.880	7.8	6.87	7
18	2010/03/14	Inslab	37.723	141.817	32.0	6.53	8
19	2011/03/11	Interface	38.103	142.860	24.4	9.08	304
20	2011/03/12	Crustal	36.985	138.597	9.3	6.30	11

A supplementary spreadsheet, which contains detailed record information of the mainshock–aftershock sequences, is provided as part of this paper.

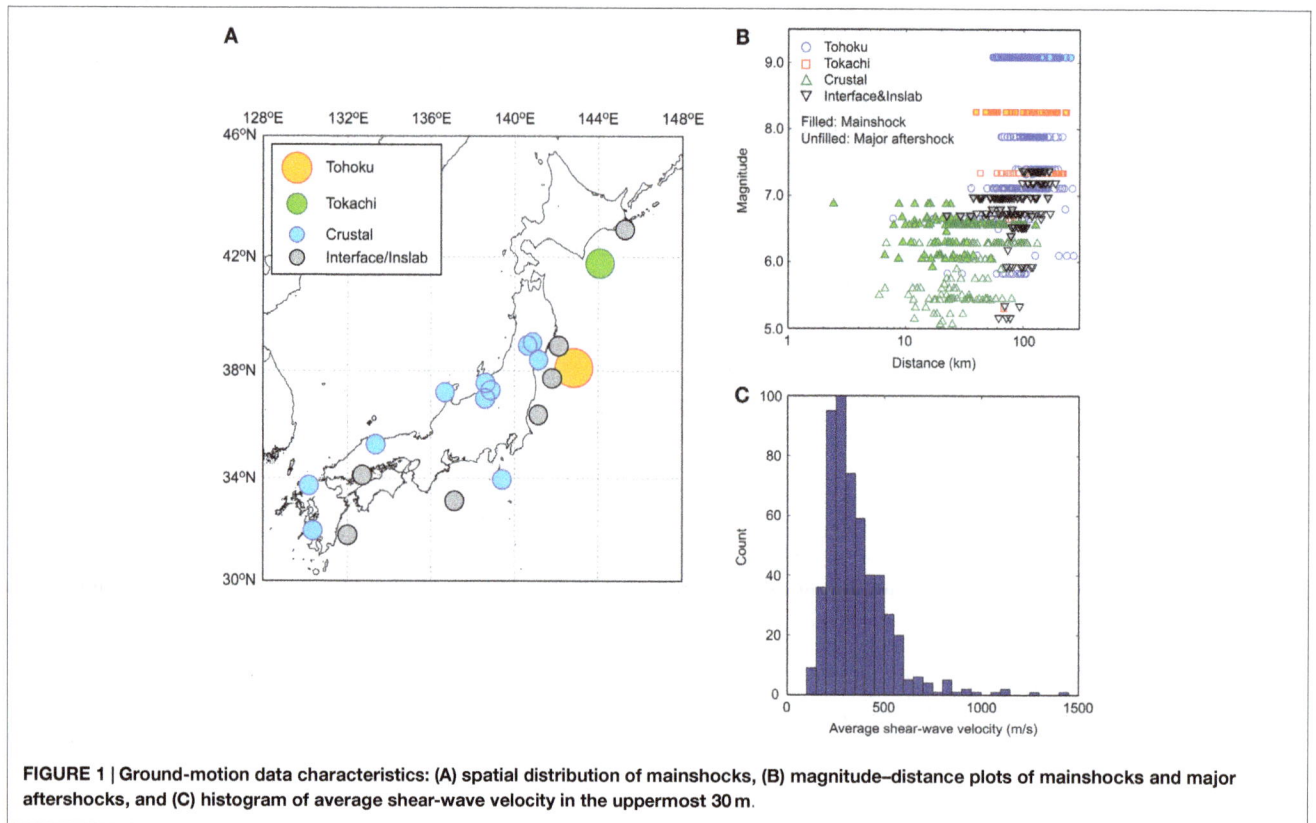

FIGURE 1 | Ground-motion data characteristics: (A) spatial distribution of mainshocks, (B) magnitude–distance plots of mainshocks and major aftershocks, and (C) histogram of average shear-wave velocity in the uppermost 30 m.

mainshocks are greater (by approximately one magnitude unit) than those of major aftershocks, which is expected and is broadly consistent with the empirical Bath's law (Shcherbakov et al., 2005). An implication of these differences is that frequency/spectral content of mainshock records and aftershock records differ significantly (on average). This is important when record scaling is implemented in seismic vulnerability assessment (e.g.,

incremental dynamic analysis; Goda, 2015). The histogram of V_{S30} indicates that the majority of sites included in the developed database are NEHRP site class C or D, and recordings at NEHRP site class A/B or E are rare.

Figure 2A compares the statistics (median, 16th percentile, and 84th percentile) of the 5%-damped response spectra that are calculated using the four datasets (i.e., Tohoku, Tokachi,

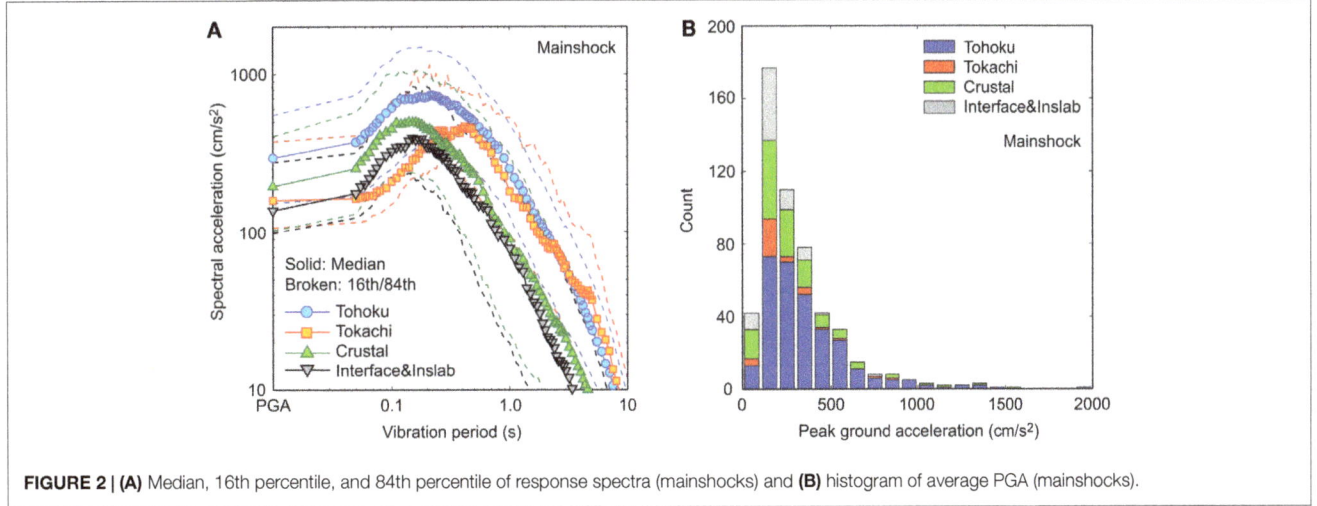

FIGURE 2 | (A) Median, 16th percentile, and 84th percentile of response spectra (mainshocks) and (B) histogram of average PGA (mainshocks).

Crustal, and Interface and Inslab). **Figure 2B** shows the histogram of PGA for the four datasets. Two key observations from **Figure 2A** are: (i) at short vibration periods ($T < 0.5$ s), response spectra for the Tohoku dataset are greater than the other three, whereas (ii) at moderate-to-long vibration periods ($T > 0.5$ s), response spectra for the Tohoku and Tokachi datasets are similar and are significantly greater than those for the Crustal and Interface and Inslab datasets. The seismic intensity parameters (for example, PGA as shown in **Figure 2B**) vary within the dataset significantly. Although the direct comparisons of the response spectra are not readily applicable due to different record features of these datasets (**Figure 1**), the former observation can be attributed to the complex source process of the 2011 Tohoku mainshock with high stress drop and low attenuation path (Goda et al., 2013). The latter can be explained by the differences of the earthquake magnitude (i.e., $M_w 8$–9 versus $M_w 6$–7; the source spectra tend to contain richer low-frequency content with increasing magnitude; Stein and Wysession, 2003). The important point is that the damage potential of ground-motion records can be associated with physical features of the source and path effects, and the developed database for MS–AS sequences is useful for investigating the effects of such features on the inelastic seismic demand statistically. This is the main focus of the subsequent sections.

Bouc–Wen Hysteretic Model

Hysteretic features of structures significantly affect the assessment of non-linear damage potential in a complex way and are important for inelastic seismic demand estimation. The Bouc–Wen model facilitates the flexible hysteresis representation, including degradation and pinching. In normalized displacement space, the equations of motion can be expressed as:

$$\ddot{\mu} + 2\xi\omega\dot{\mu} + \alpha\omega^2\mu + (1-\alpha)\omega^2\mu_z = -\ddot{u}_g(t)/u_y$$

$$\dot{\mu}_z = \frac{h(\mu_z,\varepsilon_n)}{1+\delta_\eta\varepsilon_n}[\dot{\mu} - (1+\delta_v\varepsilon_n)(\beta\,|\dot{\mu}|\,|\mu_z|^{n-1}\mu_z + \gamma\dot{\mu}|\mu_z|^n)]$$

$$h(\mu_z,\varepsilon_n) = 1 - \zeta_s(1 - e^{-p\varepsilon_n})\exp$$
$$\times\left(-\left(\frac{\mu_z sgn(\dot{\mu}) - q/[(1+\delta_v\varepsilon_n)(\beta+\gamma)]^{1/n}}{(\lambda+\zeta_s[1 - e^{-p\varepsilon_n}])(\psi+\delta_\psi\varepsilon_n)}\right)^2\right) \quad (1)$$
$$\dot{\varepsilon}_n = (1-\alpha)\dot{\mu}\mu_z$$

where μ and μ_z are the displacement and hysteretic displacement, respectively, normalized by the yield displacement capacity of an inelastic SDOF system u_y (i.e., $\mu = u/u_y$ and $\mu_z = z/u_y$, in which u and z are the displacement and hysteretic displacement, respectively); a dot represents the differential operation with respect to time; ξ is the damping ratio; ω is the natural vibration frequency (rad/s); $\ddot{u}_g(t)$ is the ground acceleration time-history; $h(\mu_z,\varepsilon_n)$ is the pinching function; ε_n is the normalized hysteresis energy; α, β, γ, and n are the shape parameters; δ_v and δ_η are the degradation parameters; ζ_s, p, q, ψ, δ_ψ, and λ are the pinching parameters; and $sgn(\bullet)$ is the signum function. The main characteristics of the Bouc–Wen hysteretic systems are defined by the second relationship in Eq. 1, where non-linear restoring force is a function of the imaginary hysteretic displacement. More detailed explanations of the Bouc–Wen parameters can be found in Foliente (1993).

Inelastic seismic demand potential can be quantified using various damage measures. For the case of inelastic SDOF systems, choice of damage measures can be reduced to a few popular ones, such as peak ductility demand and residual ductility demand (Ruiz-García and Miranda, 2003, 2006). The peak ductility demand μ_{max} is defined as $\mu_{max} = \max(|\mu(t)|)$ for all t, while the residual ductility demand μ_{res} is defined as $\mu_{res} = \mu(t=\infty)$. For a given ground-motion record, μ_{max} can be evaluated for a combination of the natural vibration period T ($=2\pi/\omega$) and the yield displacement capacity u_y. For convenience, the yield displacement capacity of a system is specified in terms of spectral acceleration at yielding S_{ay}, rather than spectral displacement at yielding S_{dy} [i.e., $S_{ay} = S_{dy}(2\pi/T)^2$].

In total, 112 inelastic SDOF systems (combinations of four vibration periods, seven yield strengths, and four hysteresis models) are considered for assessing the non-linear seismic demand parameters (i.e., μ_{max} and μ_{res}) subjected to the 531 MS–AS sequences. The intact vibration periods are: $T = 0.2$, 0.5, 1.0,

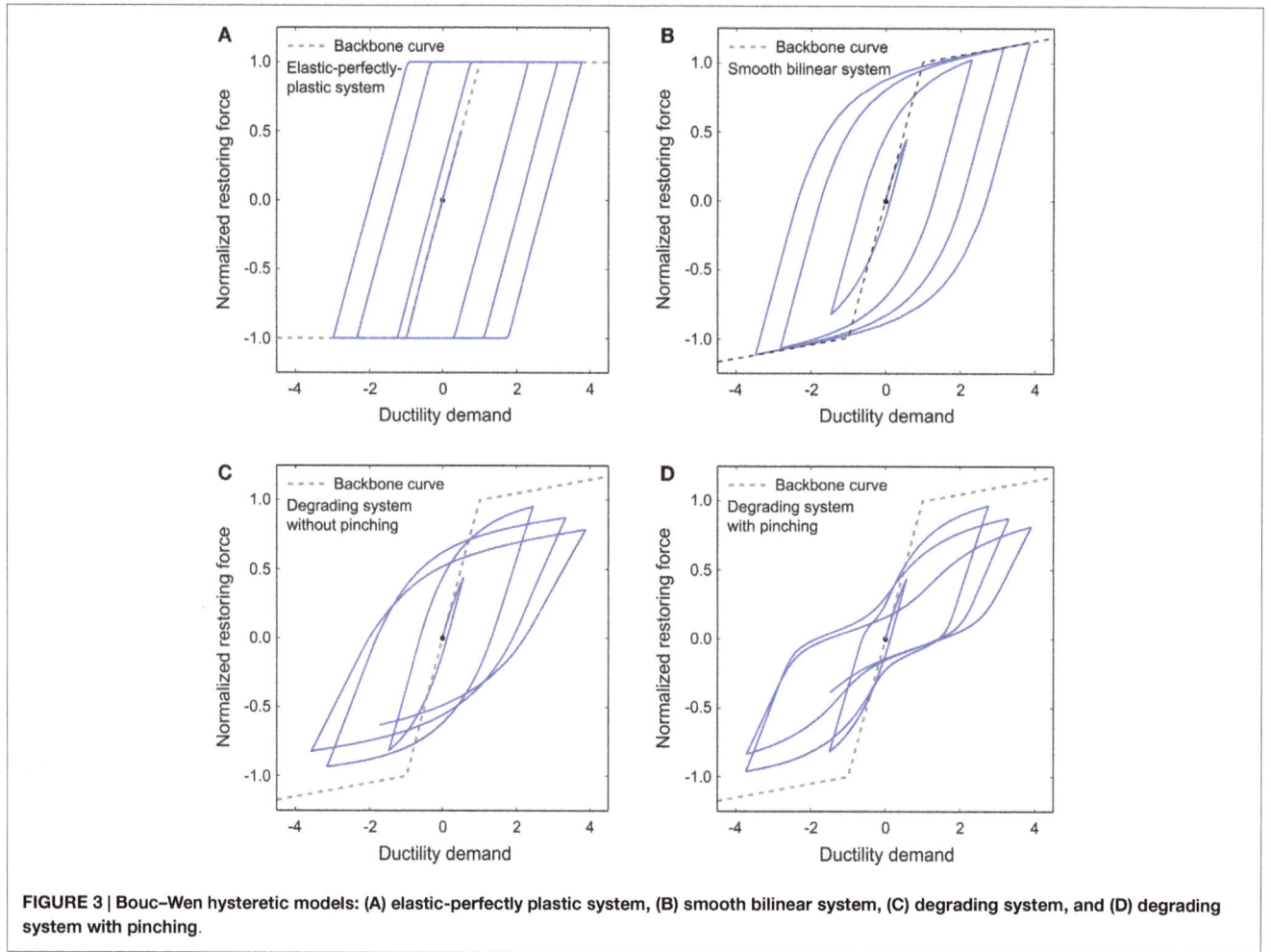

FIGURE 3 | Bouc–Wen hysteretic models: (A) elastic-perfectly plastic system, (B) smooth bilinear system, (C) degrading system, and (D) degrading system with pinching.

and 2.0 s (which cover a typical range for the first vibration mode dominated structures). The yield spectral acceleration levels are varied from 0.05 to 1.0 g: $S_{ay} = 0.05$, 0.1, 0.15, 0.2, 0.3, 0.5, and 1.0 g, which cover a range of existing structures broadly. For a given set of ground-motion records, systems with larger S_{ay} values are expected to behave linearly, while systems with smaller S_{ay} values tend to behave non-linearly. It is also instructive to compare the considered values of S_{ay} with the response spectra of the record data (**Figure 2A**). Four hysteretic models are considered: elastic-perfectly plastic (EPP) model ($\alpha = 0.0$, $\beta = \gamma = 0.5$, $n = 25$, $\delta_v = \delta_\eta = \zeta_s = 0.0$), smooth bilinear model ($\alpha = 0.05$, $\beta = \gamma = 0.5$, $n = 1$, $\delta_v = \delta_\eta = \zeta_s = 0.0$), degrading model without pinching ($\alpha = 0.05$, $\beta = \gamma = 0.5$, $n = 1$, $\delta_v = 0.1$, $\delta_\eta = 0.05$, $\zeta_s = 0.0$), and degrading model with pinching ($\alpha = 0.05$, $\beta = \gamma = 0.5$, $n = 1$, $\delta_v = 0.1$, $\delta_\eta = 0.05$, $\zeta_s = 0.9$, $p = 2.5$, $q = 0.15$, $\psi = 0.1$, $\delta_\psi = 0.005$, and $\lambda = 0.5$). **Figure 3** illustrates normalized displacement μ versus normalized restoring force $\alpha\mu + (1 - \alpha)\mu_z$, for the four Bouc–Wen hysteretic models.

Regarding the selected values of S_{ay} in this study, Nagato and Kawase (2004) estimated seismic capacities of reinforced concrete (RC), steel, and wooden structures using damage statistics from the 1995 Kobe earthquake. The methodology was to calibrate a yield base shear coefficient of an inelastic structural system (i.e., total shear force at base divided by total weight) such that

the predicted damage statistics from the set of structural models approximately match actual damage statistics from the 1995 Kobe earthquake. Their results indicate: (i) for RC buildings (3-story to 12-story), natural vibration periods are around 0.3–0.8 s and average yield base shear coefficients are around 0.3–0.7 (depending on the number of stories; generally, low-rise structures have shorter vibration periods and greater base shear coefficients), (ii) for steel buildings (3-story to 5-story), natural vibration periods are around 0.5–0.9 s and average yield base shear coefficients are around 0.4–0.7, and (iii) for wooden buildings (2-story), natural vibration periods are about 0.3 s and average yield base shear coefficients are about 0.4–0.7. The drift ratios corresponding to the yield base shear coefficients are about 0.007–0.01, 0.005–0.008, and 0.01–0.015 for RC, steel, and wooden structures, respectively. It is noteworthy that the definition of the yield capacity point depends on the specifics of the adopted structural models; for instance, Nagato and Kawase (2004) used a trilinear force-deformation curve to characterize the hysteretic behavior. If a bilinear representation is considered, instead of the trilinear one, the yield point typically is located somewhere between the first and second yield points of the trilinear curve. Moreover, the calibrated structural models should be only regarded as representative, whereas actual structures have significant variability/uncertainty with regard to their yield (and ultimate) capacities;

according to Nagato and Kawase (2004), factors of 0.5 and 2.0 are possible. Based on the above information, it is thus possible to associate the inelastic SDOF systems that are considered in this study with typical buildings in Japan.

Non-Linear Seismic Demand Assessment

The main objectives of this section are: (i) to investigate the effects of earthquake types, magnitudes, and hysteretic behavior on the peak and residual ductility demands and (ii) to evaluate the effects of major aftershocks on the non-linear structural responses. In addition, spatial aspect of the aftershock effects is evaluated for the 2011 Tohoku sequences. In the following, MS–AS sequences having V_{S30} between 150 and 600 m/s (most prevalent site conditions in Japan) are focused upon (see **Figure 1C**); the total number of MS–AS sequences is 492. Initially, EPP models are used as base case and later other hysteretic models are considered (**Figure 3**). In the following, the discussion is focused upon structural systems having vibration periods of 0.2 and 1.0 s due to limitations of space. The results obtained for these two periods can be interpolated/extrapolated to structural systems with vibration periods of 0.5 and 2.0 s by taking into account input ground motion and structural characteristics. The detailed results for systems that are not presented in detail are available upon request.

Effects of Earthquake Types

First, subsets of the entire MS–AS database are focused upon to examine the similarity or dissimilarity of the non-linear seismic demand potential for different earthquake types. They are obtained by limiting sequences having the average PGA between 100 and 200 cm/s² (see **Figure 2B**). This criterion is selected such that homogenous datasets (to the extent possible) can be obtained for the Tohoku, Tokachi, Crustal, and Interface and Inslab events. The number of sequences is 69, 21, 38, and 36 for the Tohoku, Tokachi, Crustal, and Interface and Inslab subsets, respectively. These are considered as sufficient to obtain the statistics of the structural responses, noting that each sequence consists of two horizontal components. **Figure 4A** compares the median, 16th percentile, and 84th percentile of the response spectra for the

four datasets. The result indicates that the response spectra for the Tohoku and Tokachi subsets are similar in terms of median and 16th/84th percentiles (i.e., red versus blue); the same can be observed for the Crustal and Interface and Inslab subsets (i.e., green versus black). On the other hand, the response spectra for the Tohoku and Tokachi subsets are significantly different from those for the Crustal and Interface and Inslab subsets (i.e., red/blue versus green/black). The main reason for the different elastic response spectra is the earthquake magnitude. It is noted that the differences of the response spectra in the short-period range for the Tohoku and Tokachi datasets that are observed in **Figure 2A** (by considering the entire database) disappear when more homogeneous datasets are considered.

Figure 5 shows the cumulative probability distributions of the peak and residual ductility demands of two EPP models with $T = 0.2$ s and $S_{ay} = 0.2\,g$ and with $T = 1.0$ s and $S_{ay} = 0.1\,g$ due to mainshock records only by considering the four subsets having the average PGA between 100 and 200 cm/s². The two systems are selected to illustrate the interesting results clearly and concisely (among many cases), and they correspond to structures with low seismic capacities among the existing building stock in Japan. The results shown in **Figure 5** indicate that both peak and residual ductility demands for the Tohoku and Tokachi subsets are greater than those for the Crustal and Interface and Inslab datasets. The differences of the non-linear structural responses are greater for $T = 1.0$ s and for residual ductility demands. The differences can be attributed to the response spectral characteristics of these subsets, shown in **Figure 4A**. Another attribute that has influence on residual ductility demand is the duration. The seismological source parameter that affects the spectral content and duration of ground motions is the earthquake magnitude.

To cover the parameter space of the calculated cases more widely, peak as well as residual ductility demand curves for EPP models ($T = 0.2$ and 1.0 s) with different yield spectral accelerations are compared in **Figure 6** by considering the four subsets. The peak ductility demand curves gradually decrease with increasing yield spectral acceleration (i.e., stronger systems), whereas the slopes of the residual ductility demand curves are steeper than those of the peak ductility demand curves. These

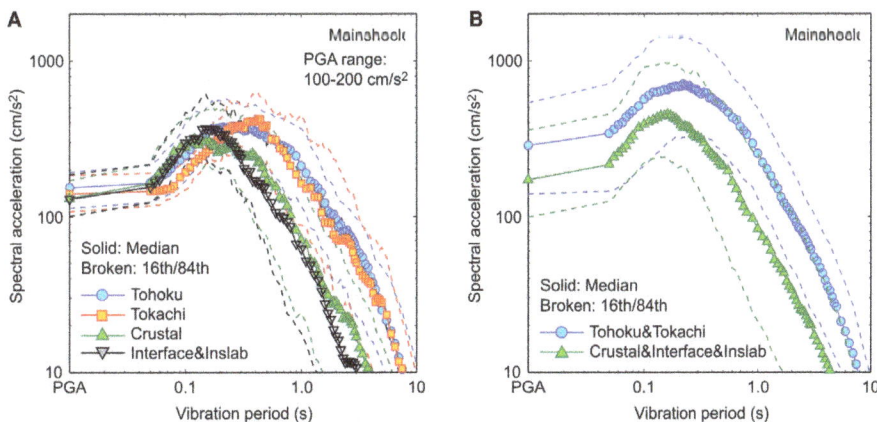

FIGURE 4 | Comparison of median, 16th percentile, and 84th percentile of response spectra (mainshocks): (A) Tohoku, Tokachi, Crustal, and Interface and Inslab datasets having average PGA between 100 and 200 cm/s² and (B) Tohoku and Tokachi and Crustal, Interface, and Inslab datasets.

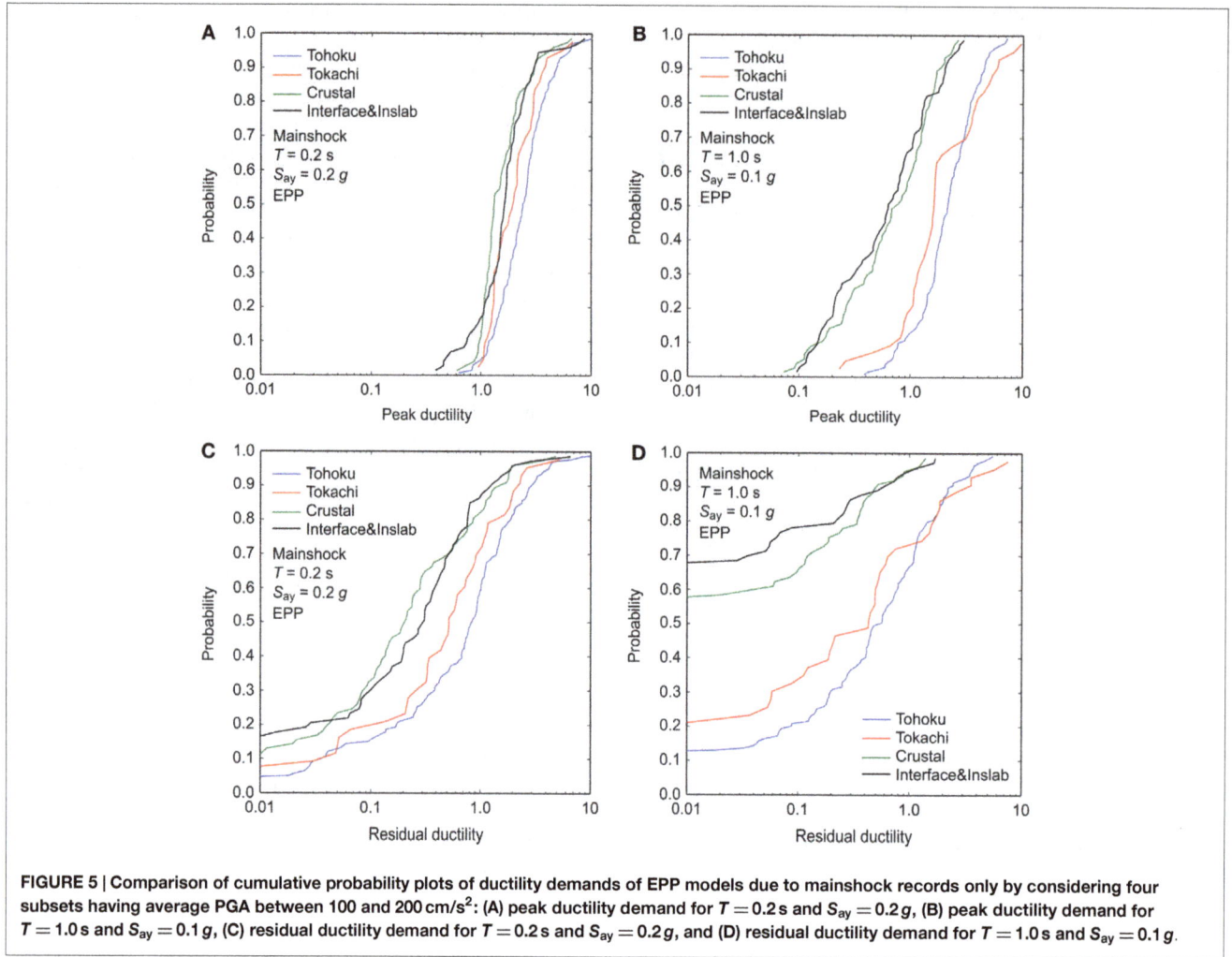

FIGURE 5 | Comparison of cumulative probability plots of ductility demands of EPP models due to mainshock records only by considering four subsets having average PGA between 100 and 200 cm/s^2: **(A)** peak ductility demand for $T = 0.2$ s and $S_{ay} = 0.2\,g$, **(B)** peak ductility demand for $T = 1.0$ s and $S_{ay} = 0.1\,g$, **(C)** residual ductility demand for $T = 0.2$ s and $S_{ay} = 0.2\,g$, and **(D)** residual ductility demand for $T = 1.0$ s and $S_{ay} = 0.1\,g$.

suggest that for the considered EPP models, seismic damage due to transient peak demands can occur for relatively moderate ground motions, whereas seismic damage due to permanent residual demands occurs when severe ground motions affect the structures. Importantly, the results confirm the similarity of peak and residual ductility demands for the Tohoku and Tokachi datasets and for the Crustal and Interface and Inslab datasets, and that the former is greater than the latter. The conclusions are applicable to different hysteretic models as well as subsets with different selection criteria.

Effects of Magnitudes and Hysteretic Behavior

Based on the above results, one of the controlling features of the ductility demands is the earthquake magnitude. To further investigate the key features that affect the non-linear seismic demand potential (i.e., hysteretic characteristics and major aftershocks), the entire MS–AS dataset is divided into two subsets according to the magnitude ranges: the Tohoku and Tokachi (T&T), or large-magnitude, dataset (319 sequences) and the Crustal, Interface, and Inslab (C&I&I), or moderate-magnitude, dataset (173 sequences). **Figure 4B** compares the median, 16th percentile, and 84th percentile of the response spectra for the two datasets. The response spectra for the large-magnitude dataset are greater than those

for the moderate-magnitude dataset. The response spectral shape for the former dataset has richer long-period spectral content, in comparison with that for the latter dataset.

To inspect the results for specific systems, data points of peak/residual ductility demands and corresponding spectral acceleration values at the intact vibration periods are plotted in **Figure 7**. The considered systems are two EPP models with $T = 0.2$ s and $S_{ay} = 0.2\,g$ and with $T = 1.0$ s and $S_{ay} = 0.1\,g$ subjected to MS–AS sequences. In the figure, individual data points are displayed with small markers, whereas larger markers with a line show the median trend of the individual data. In the context of the PBEE methodology, ductility demands are the engineering demand parameters (EDP) and spectral accelerations are the intensity measures (IM). The results shown in **Figure 7** are the assessments of inelastic seismic demand (i.e., empirical IM–EDP relationships) based on cloud analysis (Jalayer and Cornell, 2009). Note that the main objective of this study is not the development of the (generic) inelastic seismic demand prediction models (e.g., constant R approach). Rather, it is focused upon identifying the key factors that result in different inelastic seismic demand predictions, and thus these parameters should be incorporated in developing such prediction models for specific structures.

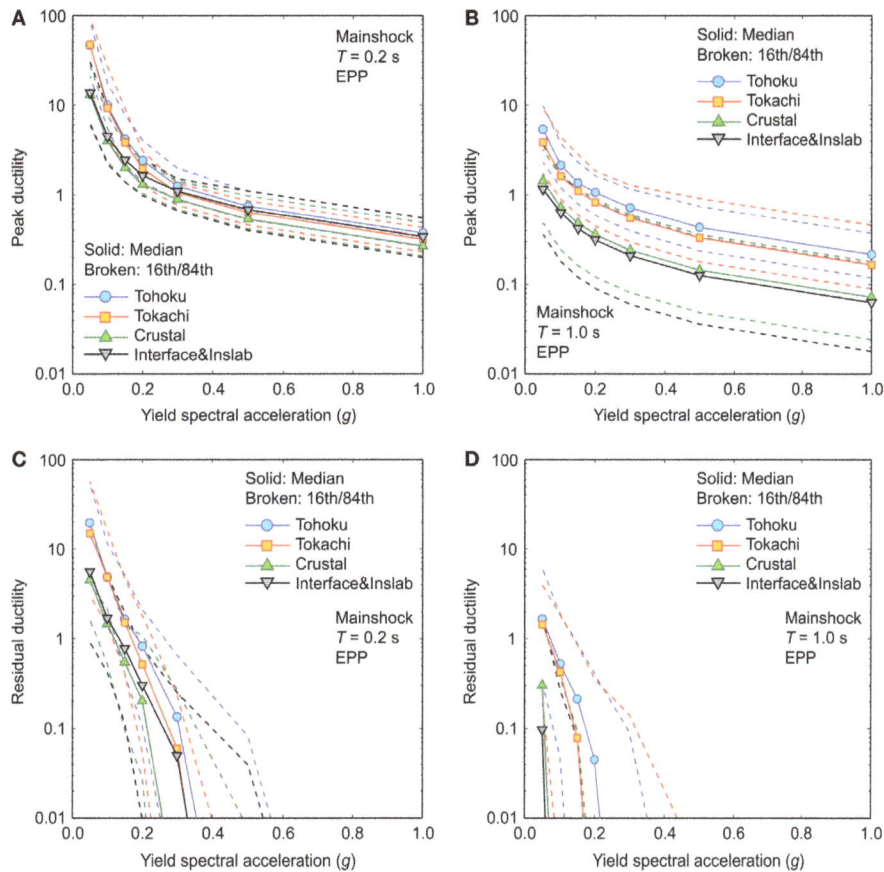

FIGURE 6 | Comparison of ductility demand curves for EPP models with different yield spectral accelerations due to mainshock records only by considering four subsets having average PGA between 100 and 200 cm/s²: (A) peak ductility demand for $T = 0.2\,\text{s}$, (B) peak ductility demand for $T = 1.0\,\text{s}$, (C) residual ductility demand for $T = 0.2\,\text{s}$, and (D) residual ductility demand for $T = 1.0\,\text{s}$.

A notable trend of the results shown in **Figure 7** related to the magnitude ranges of the ground-motion data is that for $T = 0.2\,\text{s}$ (**Figures 7A,C**), the median ductility demand curves (both peak and residual) for the large-magnitude dataset are greater than those for the moderate-magnitude dataset. On the other hand, such differences are not observed for $T = 1.0\,\text{s}$ (**Figures 7B,D**). This may appear to be inconsistent with the results shown in **Figures 5** and **6**. The different trends are caused because in **Figure 7**, the base parameters for describing the seismic hazard intensity (i.e., IM) are the spectral accelerations at the intact vibration periods, while in **Figures 5** and **6**, the base IM parameter is the PGA (note: PGA is a popular parameter for record selection purposes). For the considered systems, spectral accelerations at the intact vibration period are more efficient than PGA (i.e., an IM–EDP relationship is characterized by smaller variability of the relationship; Luco and Cornell, 2007), and it is customary to adopt more efficient IMs in evaluating the values of EDP (however, full exploration of efficient IMs is beyond the scope of this study). More specifically, when the response spectra of the large-magnitude dataset and of the moderate-magnitude dataset are matched at $T = 0.2\,\text{s}$ (see **Figure 4B**), the former has the richer spectral content than the latter in the vibration period range greater than $T = 0.2\,\text{s}$ and when the structural systems go into

the inelastic response domain, inelastic responses of the systems are strongly affected by ground motions in the vibration period range longer than the intact vibration period (Luco and Bazzurro, 2007). When the matching of response spectra is carried out at $T = 1.0\,\text{s}$, the matched response spectra in the vibration period range longer than 1.0 s become similar (note: in this case, major differences appear in the vibration period range shorter than 1.0 s; however, the inelastic SDOF systems considered in this study are not sensitive to ground motions in this period range). Further to note, although no results are presented and discussed in this study, results for inelastic seismic demand estimation based on the constant R approach using the same MS–AS sequence datasets indicate that the magnitude effects on the ductility demands are significant for short-period structures.

Returning to the original focus of this study (i.e., empirical assessment of ductility demands), **Figure 8** compares peak ductility demand curves for EPP models subjected to MS–AS sequences with those for smooth bilinear models, degrading models, and degrading models with pinching (**Figure 3**). Both large- and moderate-magnitude datasets are considered. The intension of this figure is to present the effects of hysteretic characteristics of the inelastic SDOF systems on the peak ductility demands; it is not to compare the peak ductility demands for the two datasets (which

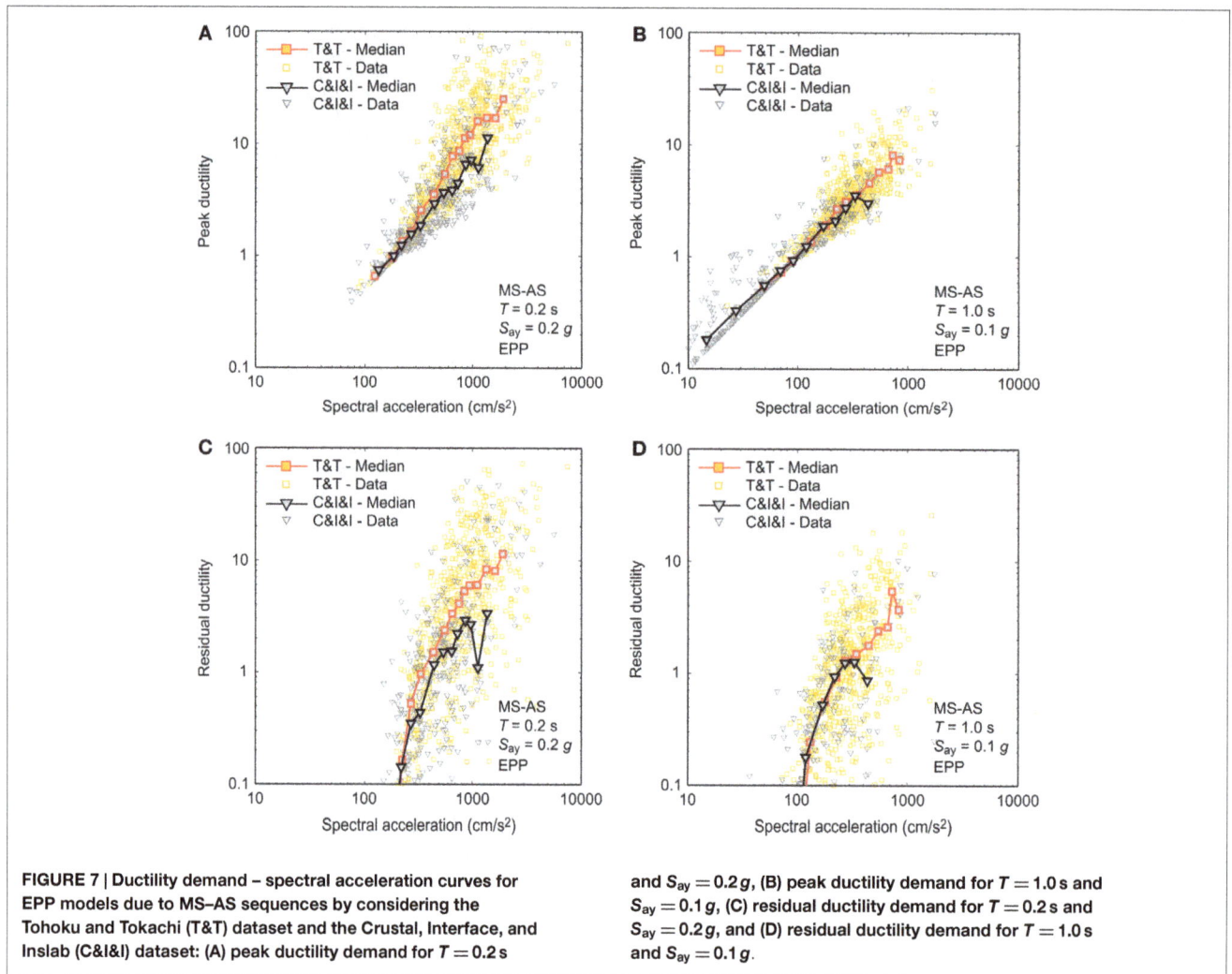

FIGURE 7 | Ductility demand – spectral acceleration curves for EPP models due to MS–AS sequences by considering the Tohoku and Tokachi (T&T) dataset and the Crustal, Interface, and Inslab (C&I&I) dataset: (A) peak ductility demand for $T = 0.2$ s and $S_{ay} = 0.2\,g$, (B) peak ductility demand for $T = 1.0$ s and $S_{ay} = 0.1\,g$, (C) residual ductility demand for $T = 0.2$ s and $S_{ay} = 0.2\,g$, and (D) residual ductility demand for $T = 1.0$ s and $S_{ay} = 0.1\,g$.

is not of interest because the seismic excitation levels are different). In these comparisons, EPP systems are used as reference and thus their results are shown in all figure panels.

Figures 8A,B suggest that the consideration of smooth bilinear systems (α and n are changed from EPP systems) leads to the decreased peak ductility demand, and that the extent of reduction of the peak ductility demand is greater for $T = 0.2$ s than for $T = 1.0$ s. The key factor for the decreased peak ductility demand is α (Ma et al., 2004). The consideration of degrading effects (**Figures 8C,D**) results in the increased peak ductility demand. The influence of degradation is more significant for $T = 0.2$ s than $T = 1.0$ s. For $T = 0.2$ s, the peak ductility demand curves for the degrading systems become greater than those for the EPP systems (i.e., overcoming the reduction due to the positive post-yield stiffness ratio), whereas for $T = 1.0$ s, the increase is minimal. The pinching behavior affects the structural systems having short vibration periods, whereas its effect on systems with long vibration periods is not significant (**Figures 8E,F**). For $T = 0.2$ s, the effect due to pinching behavior is particularly large, increasing the peak ductility demands in the low-to-moderate ranges significantly. It is noted that the effects of hysteretic behavior, as demonstrated above, depend on the vibration period as well as seismic excitation

level. The above-mentioned observations are in agreement with Goda and Atkinson (2009).

The similar comparisons for the residual ductility demands for different hysteretic models are omitted for brevity. It is observed that when the hysteretic behavior is changed from EPP systems to other systems having positive post-yield stiffness ratios (i.e., $\alpha = 0.0$ versus $\alpha = 0.05$), the absolute values of the residual ductility demand decrease dramatically. For instance, the overall trends of the residual ductility demand curves for the EPP systems ($T = 0.2$ and 1.0 s) by considering the large-magnitude and moderate-magnitude datasets are similar to those shown in **Figure 6**. When the bilinear and degrading systems without/with pinching are considered, the absolute values of the residual ductility demand curves become significantly less (median as well as 84th percentile curves rarely exceed the ductility demand of 0.1, which is of no engineering significance). These results are in agreement with Ruiz-García and Miranda (2006).

Effects of Major Aftershocks

The effects of major aftershocks on the peak and residual ductility demands are evaluated by considering the large- and moderate-magnitude datasets. To inspect the impact of major aftershocks

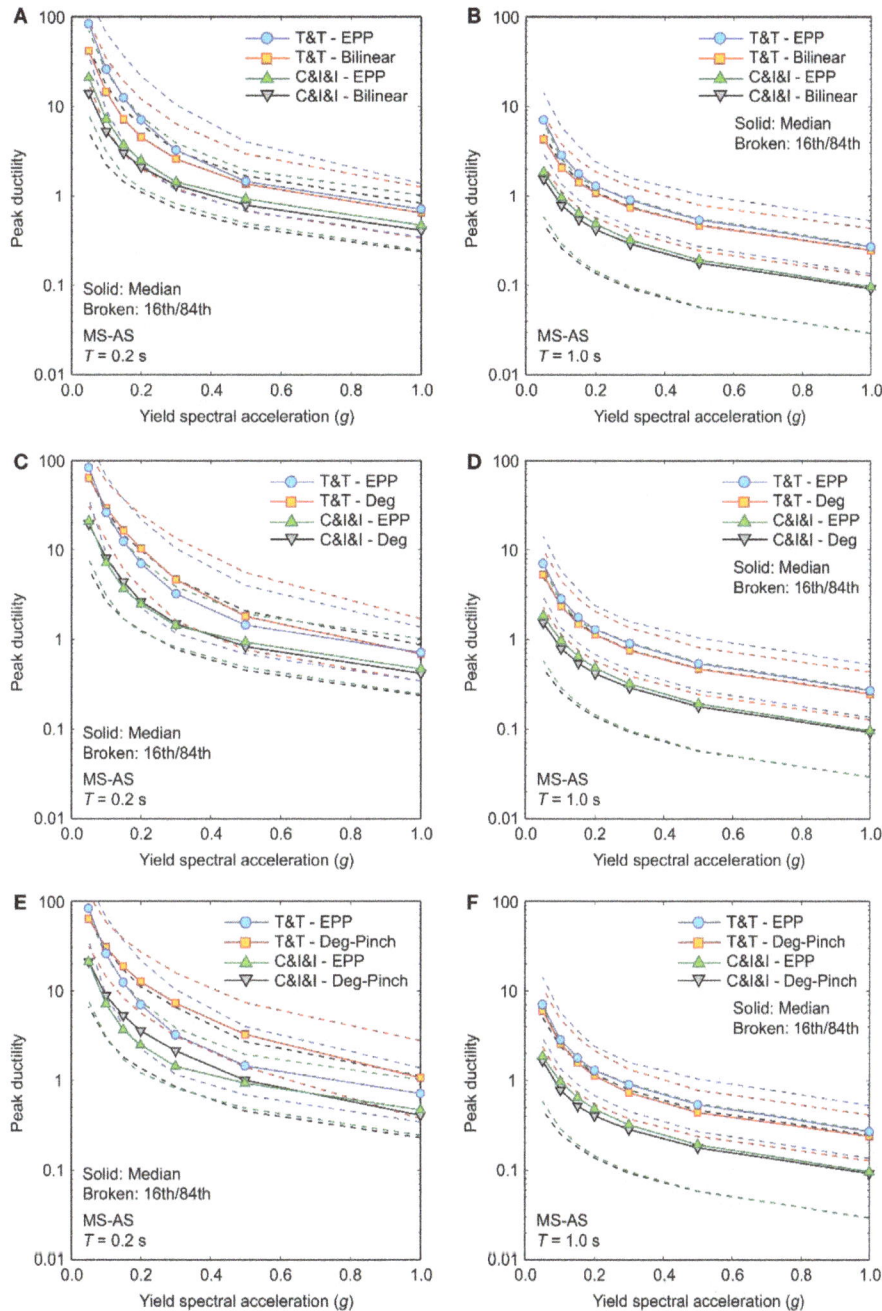

FIGURE 8 | Comparison of peak ductility demand curves for different hysteretic models with different yield spectral accelerations due to MS–AS sequences by considering the Tohoku and Tokachi (T&T) dataset and the Crustal, Interface, and Inslab (C&I&I) dataset: **(A,B)** bilinear models for $T = 0.2$ and 1.0 s, **(C,D)** degrading models for $T = 0.2$ and 1.0 s, and **(E,F)** degrading models with pinching for $T = 0.2$ and 1.0 s.

visually, median, 84th percentile, and 98th percentile of the MS–AS to mainshock ductility demand ratios (i.e., MS–AS to MS ratios) for EPP models ($T = 0.2$ and 1.0 s) are presented in **Figure 9**. Because the MS–AS to MS ratios can be extremely large when ductility demands for mainshock records only are small (this is particularly applicable to residual ductility demands) and such cases are of little engineering interests, the MS–AS to MS ratios are computed using peak/residual ductility demands due to

mainshocks greater than 0.1. **Figure 9** shows that for the majority of the cases, the median ratios are 1 (both peak and residual), indicating that more than 50% of the cases, the major aftershocks do not increase the seismic demand levels caused by the mainshocks. However, in rare cases, the major aftershocks can increase the seismic damage extent significantly. The extent of the aftershock effects is greater for the moderate-magnitude dataset than for the large-magnitude dataset. For instance, the 98th percentile

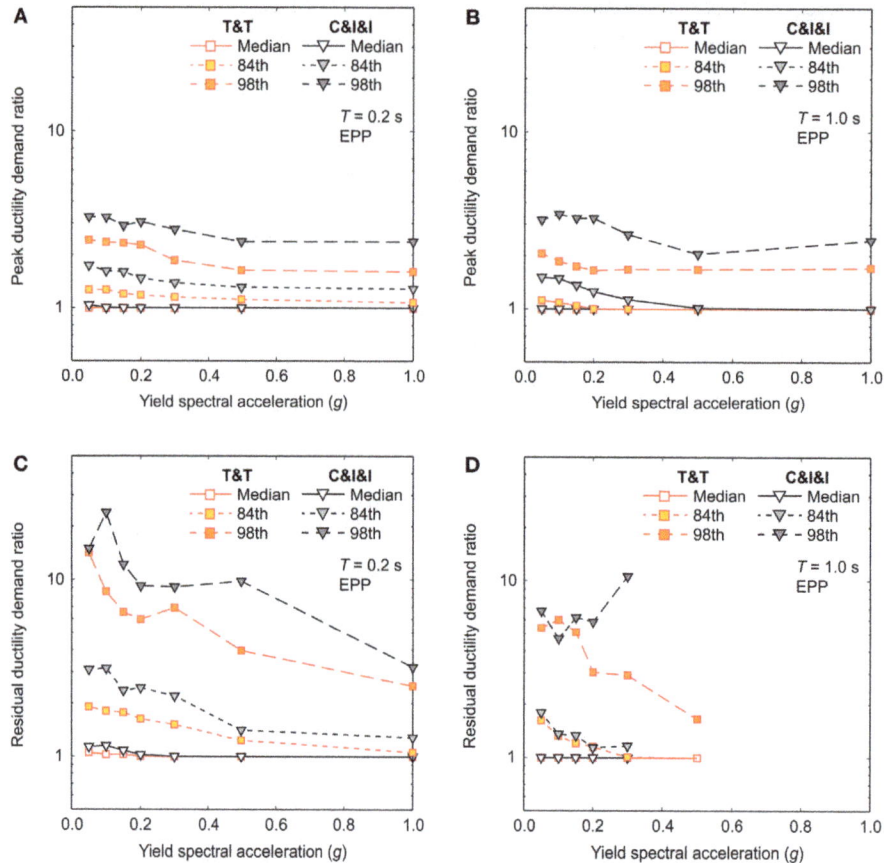

FIGURE 9 | Statistics of the MS–AS to mainshock ductility demand ratios for EPP models by considering the Tohoku and Tokachi (T&T) dataset and the Crustal, Interface, and Inslab (C&I&I) dataset: (A) peak ductility demand for $T = 0.2$ s, (B) peak ductility demand for $T = 1.0$ s, (C) residual ductility demand for $T = 0.2$ s, and (D) residual ductility demand for $T = 1.0$ s.

curves of the MS–AS to MS peak ductility demand ratios for the moderate- and large-magnitude datasets range around 2–3 and 1.5–2, respectively. The comparison of the results for the peak and residual ductility demands indicates that the MS–AS to MS ratios for the residual ductility demands are more sensitive than those for the peak ductility demands; these are partly attributed to the fact that for EPP systems the absolute values of the residual ductility demands are smaller than those of the peak ductility demands and the residual ductility demands tend to increase more rapidly with the yield spectral acceleration (**Figure 6**).

To further investigate the aftershock effects in terms of hysteretic behavior, **Figure 10** compares the 84th percentile and 98th percentile curves of the MS–AS to MS peak ductility demand ratios for different hysteretic models (note: 50th percentile curves are not shown as they are equal to 1 for most of the cases). Both large- and moderate-magnitude datasets are considered. Similarly to **Figure 8**, the results for EPP systems are used as reference. The consideration of bilinear systems with positive post-yield stiffness ratios results in slightly smaller MS–AS to MS peak ductility demand ratios (e.g., 84th percentile curves for $T = 0.2$ s), however, the overall impact is not significant (**Figures 10A,B**). The results for the degrading systems without/with pinching indicate that the MS–AS to MS peak ductility demand ratios for $T = 0.2$ s are

slightly more influenced by hysteretic behavior than the ratios for $T = 1.0$ s (**Figures 10C–F**). Noticeable increases of the MS–AS to MS ratios are observed due to pinching behavior for $T = 0.2$ s (**Figure 10E**). Overall, it can be concluded that the effects of hysteretic characteristics on the MS–AS to MS peak ductility demand ratios are not particularly large. The similar results for the MS–AS to MS residual ductility demand ratios are omitted because the residual ductility demands for bilinear and degrading systems without/with pinching are small (the majority of the data are below the threshold of 0.1).

Finally, dependency of the MS–AS to MS ratios (both peak and residual) of EPP systems on various seismological parameters is investigated using the large- and moderate-magnitude datasets. The considered explanatory parameters are: mainshock peak/residual ductility demand, average shear-wave velocity (V_{S30}), mainshock magnitude, mainshock distance, aftershock magnitude, aftershock distance, mainshock PGA, mainshock spectral acceleration at the intact vibration period, aftershock PGA, and aftershock spectral acceleration at the intact vibration period. By visually inspecting the scatter plots of the MS–AS to MS ratios with respect to the examined parameters and by carrying out linear regression analysis (note: regression analyses are performed in log–log space, except for

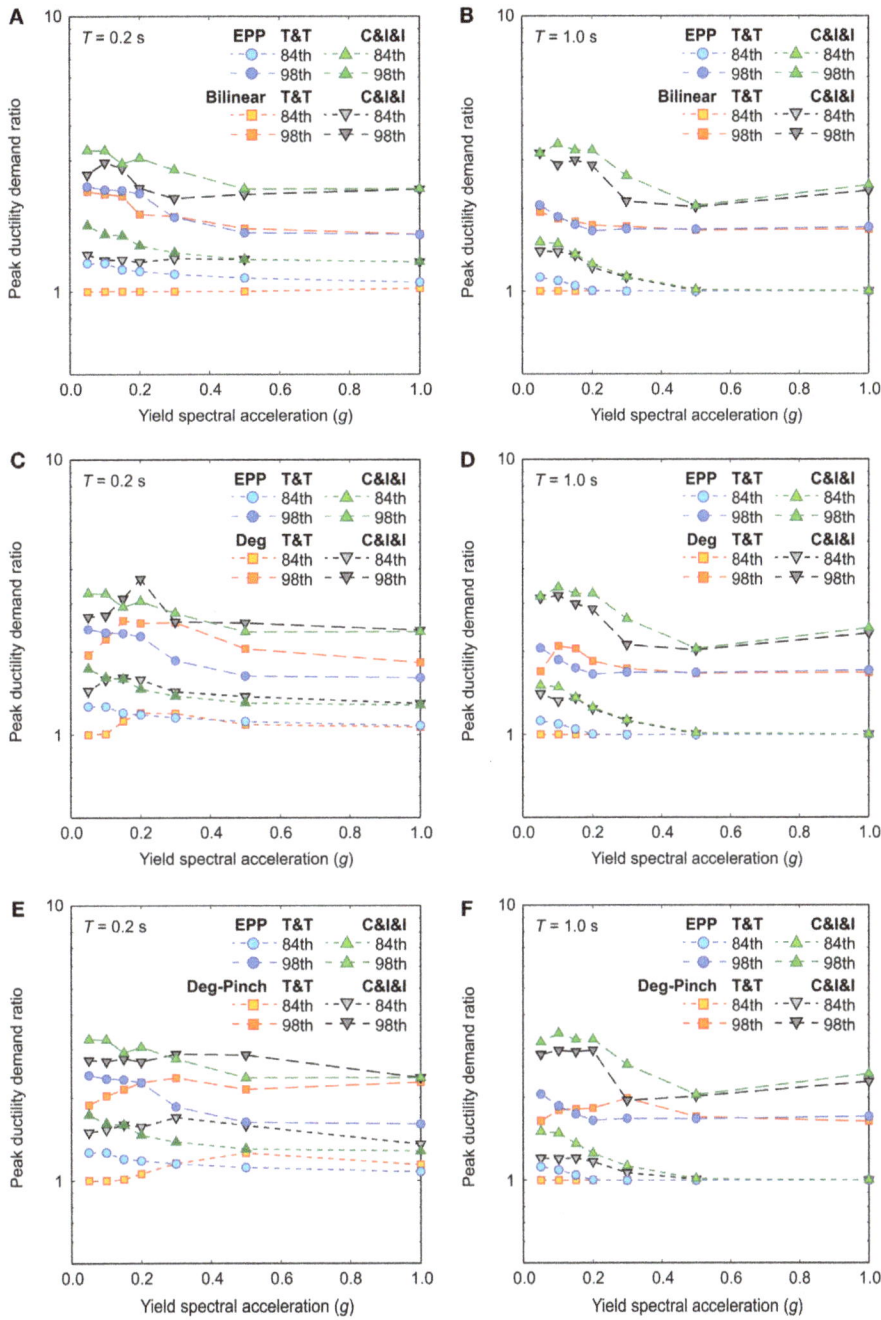

FIGURE 10 | Statistics of MS–AS to mainshock peak ductility demand ratios for different hysteretic models by considering the Tohoku and Tokachi (T&T) dataset and the Crustal, Interface, and Inslab (C&I&I) dataset: (A,B) bilinear models for $T = 0.2$ and 1.0 s, (C,D) degrading models for $T = 0.2$ and 1.0 s, and (E,F) degrading models with pinching for $T = 0.2$ and 1.0 s.

the mainshock/aftershock magnitude), their dependency is evaluated. The dependency between the MS–AS to MS ratios and the parameters is judged to be significant when the 95% confidence intervals of the slope coefficient do not include zero (i.e., confidence intervals are either both positive or both negative). The regression analysis results suggest that the MS–AS to MS peak ductility demand ratios clearly depend on aftershock PGA and spectral acceleration at the intact vibration period, while they are

weakly dependent on the mainshock peak ductility demand. The former is simply interpreted that stronger aftershocks have greater potential to cause additional seismic damage, whereas the latter can be understood that relative effects due to major aftershocks become less critical when the mainshock causes large seismic damage to structures. Note that minor trends can be recognized for aftershock magnitude and distance; however, the trends are not consistent for the majority of cases and these parameters are

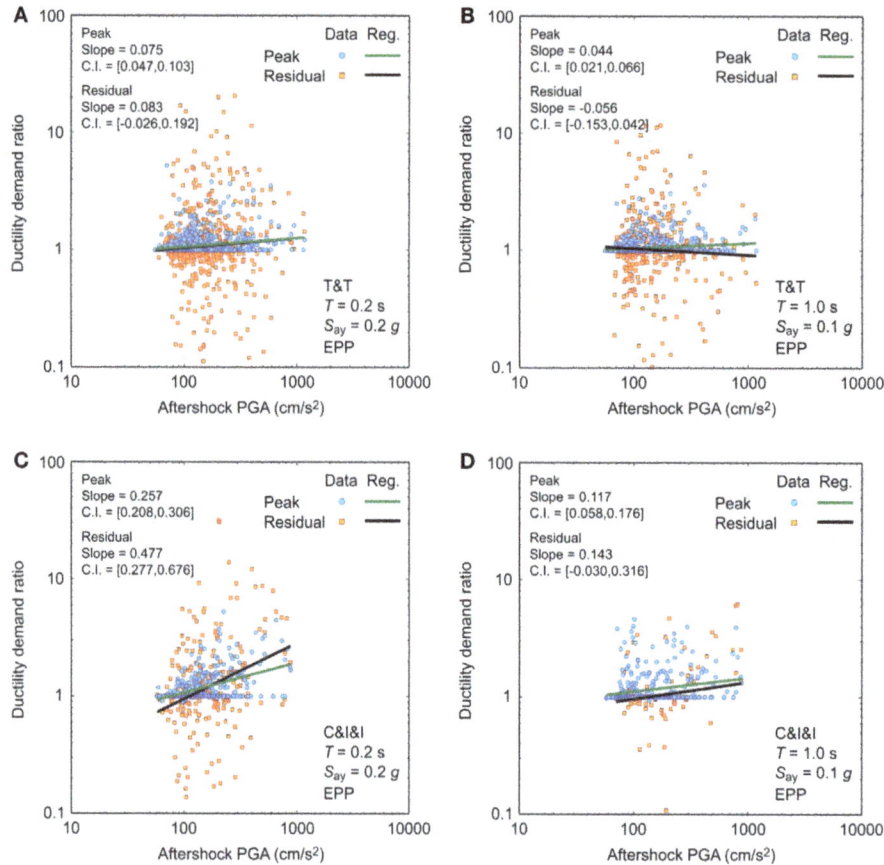

FIGURE 11 | Dependency of MS–AS to mainshock peak and residual ductility demand ratios for EPP systems by considering: (A) Tohoku and Tokachi (T&T) dataset for $T = 0.2$ s and $S_{ay} = 0.2\,g$, (B) Tohoku and Tokachi (T&T) dataset for $T = 1.0$ s and $S_{ay} = 0.1\,g$, (C) Crustal, Interface, and Inslab (C&I&I) dataset for $T = 0.2$ s and $S_{ay} = 0.2\,g$, and (D) Crustal, Interface, and Inslab (C&I&I) dataset for $T = 1.0$ s and $S_{ay} = 0.1\,g$.

regarded as secondary factors that affect aftershock PGA and spectral accelerations. On the other hand, the results for the MS–AS to MS residual ductility demand ratios are less clear because of large scatter of the data points. Therefore, it is concluded that the dependency between the MS–AS to MS residual ductility demand ratios and the aftershock elastic response parameters is too weak. To illustrate the above-mentioned observations, **Figure 11** presents the scatter plots of the MS–AS to MS ratio and the aftershock PGA for EPP systems with $T = 0.2$ s and $S_{ay} = 0.2\,g$ and with $T = 1.0$ s and $S_{ay} = 0.1\,g$ by considering the large- and moderate-magnitude datasets. In the figure panels, the regression lines as well as the slope value and its confidence intervals are included. For the peak ductility demands, clear positive trends are observed for $T = 0.2$ s, whereas such trends become weak for $T = 1.0$ s. Generally, aftershock spectral accelerations at the intact vibration period are more correlated with the MS–AS to MS ratios. **Figure 11** also shows that the scatter of the data points for the residual ductility demands is significantly greater than that for the peak ductility demands.

Spatial Distribution of Major Aftershocks for the 2011 Tohoku Sequence

Figure 12 shows the spatial distribution of the peak ductility demands, peak ductility demand ratios, residual ductility

demands, and residual ductility demand ratios for two EPP systems with $T = 0.2$ s and $S_{ay} = 0.2\,g$ and with $T = 1.0$ s and $S_{ay} = 0.1\,g$. The intensity of the ductility demands and the ductility demand ratios are color-coded (see the captions in **Figure 12**); the ranges of the demand values and ratios are chosen to represent different seismic damage severities (e.g., peak ductility demand of 10 is considered to be major damage). In the figure panels for the peak/residual ductility demands (**Figures 12A,C,E,G**), strong-motion generation areas, which are characterized as areas with large slip velocities within a total rupture plane, are indicated. These areas are estimated by Kurahashi and Iikura (2013) via strong-motion source inversion of the 2011 Tohoku mainshock data. Whereas in the figure panels for the peak/residual ductility demand ratios (**Figures 12B,D,F,H**), locations of the major aftershocks for the 2011 Tohoku sequences are shown.

Inspection of the results for the peak ductility demands of EPP systems having $T = 0.2$ s and $S_{ay} = 0.2\,g$ (**Figure 12A**) indicates that the seismic damage potential due to the mainshock is high at sites around 38–39°N (Miyagi Prefecture) and at sites around 36–37°N (Fukushima and Ibaraki Prefectures). The sites in Miyagi Prefecture are affected by the two overlapping strong-motion generation areas (which resulted in noticeable multiple-shock features of the recorded ground motions), whereas the sites in Fukushima

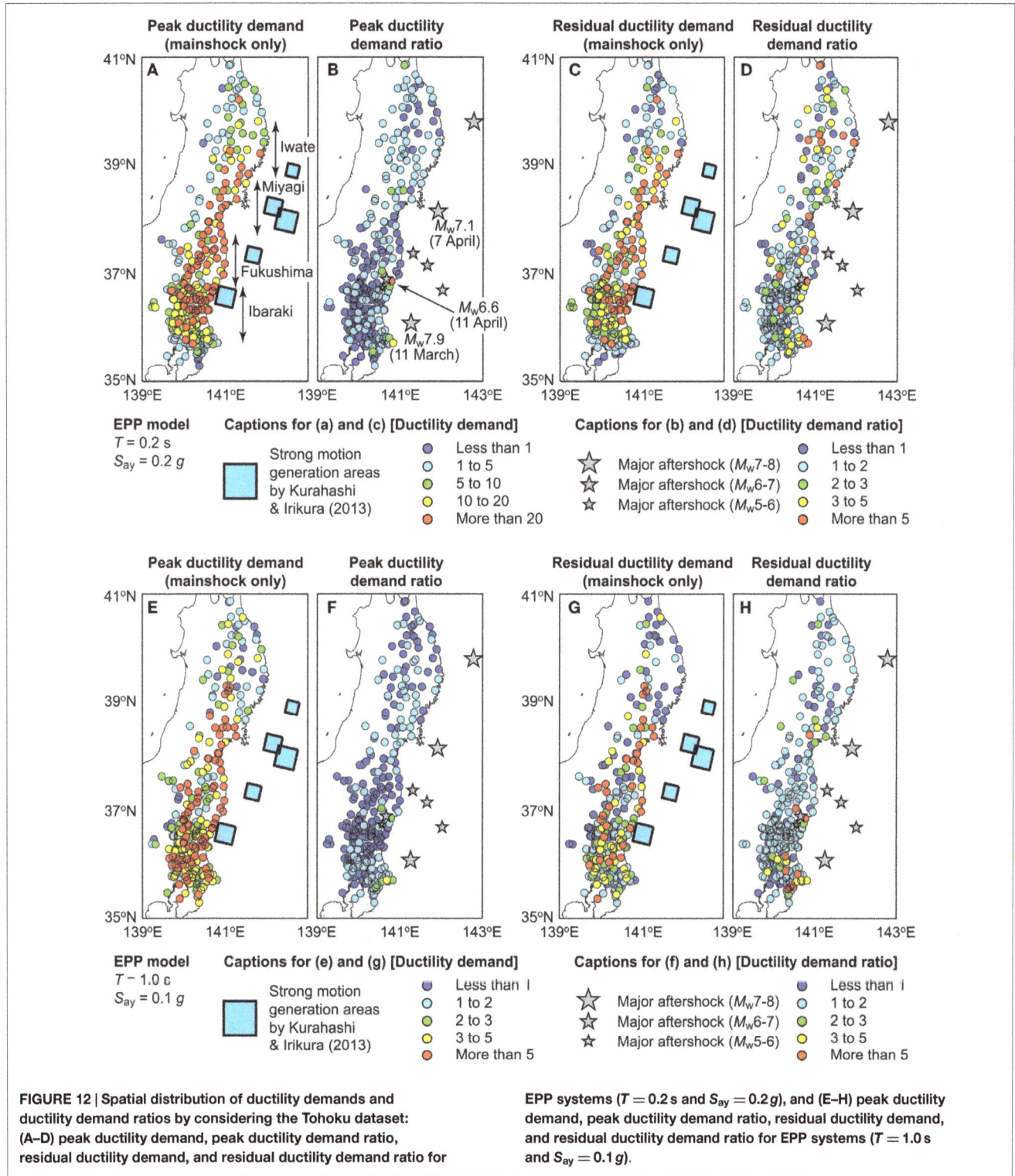

FIGURE 12 | Spatial distribution of ductility demands and ductility demand ratios by considering the Tohoku dataset: (A–D) peak ductility demand, peak ductility demand ratio, residual ductility demand, and residual ductility demand ratio for EPP systems ($T = 0.2$ s and $S_{ay} = 0.2 g$), and (E–H) peak ductility demand, peak ductility demand ratio, residual ductility demand, and residual ductility demand ratio for EPP systems ($T = 1.0$ s and $S_{ay} = 0.1 g$).

and Ibaraki Prefectures are influenced by the southernmost strong-motion generation area, which is located near the coastline. Furthermore, many of these structures are located along the coast, and therefore are likely to be subjected to the tsunami actions between mainshock and aftershock. Such actions may further degrade the structural behavior and make the buildings weaker,

being unable to resist the next seismic excitation. Moreover, it can be observed from **Figure 12B** that the additional seismic damage occurs in the vicinity of major aftershocks. In particular, the $M_w 7.9$ aftershock off Ibaraki Prefecture that occurred 30 min after the mainshock increases the peak ductility demands at sites in the southern part of Ibaraki Prefecture (green-to-yellow circles in **Figure 12B**). The $M_w7.1$ aftershock that occurred on 7 April off Miyagi Prefecture causes small-to-moderate increase of the peak seismic demands in Miyagi Prefecture (light-blue-to-green circles in **Figure 12B**). Notably, the $M_w6.6$ aftershock that occurred on 11 April in the upper crust causes a major increase of the peak seismic demands at a nearby location (red circle in **Figure 12B**). The causal relationship between major aftershocks and increased seismic demands can be understood physically and intuitively; simply, when a major aftershock strikes near a site of interest, the seismic demand potential due to the aftershock becomes greater. The above explanations are applicable to the residual ductility demands (**Figures 12C,D**) as well as the results for the other EPP system (**Figures 12E–H**). The results shown in **Figure 12** are consistent with the results shown in **Figure 11**.

From seismic risk-management perspectives, critical situations arise when moderate-to-severe damage is caused by a mainshock and major aftershocks occur nearby, aggravating the conditions of the mainshock-damaged structures. The results shown in **Figure 12** highlight that the spatial occurrence process of aftershocks is important. This is a major source of uncertainty in ensuring the safe evacuation and deciding upon the reoccupation of buildings in a post-disaster environment. Moreover, by reflecting upon the observations made regarding **Figure 9** (i.e., aftershock effects for the moderate-magnitude dataset is greater than those for the large-magnitude dataset), the reason for less frequent occurrence of damaging aftershocks for mega-thrust subduction earthquakes may be attributed to the fact that mainshock seismic damage is caused at many locations over a larger geographical region but aftershock-triggered seismic damage is concentrated at more local areas. To study such aspects, spatial modeling of aftershock occurrence needs to be incorporated in generating artificial MS–AS sequences (Goda, 2012).

Conclusion

This study aimed at evaluating the peak and residual ductility demands of inelastic SDOF systems due to real MS–AS sequences from an empirical perspective. For this purpose, an extensive dataset of as-recorded MS–AS sequences for Japanese earthquakes was developed (containing 531 sequences; each with two horizontal components). The constructed dataset is large, compared with previous datasets of similar kinds, and thus more rigorous investigations regarding the non-linear seismic demand potential for MS–AS sequences can be carried out. To draw generic conclusions, numerous inelastic SDOF systems having different vibration periods, yield strengths, and hysteretic characteristics that are represented by the Bouc–Wen model, were considered. Such assessment is useful in two aspects. Firstly, it serves as a benchmark, when non-linear structural responses due to large mainshocks having different record characteristics and due to major aftershocks are evaluated using artificial MS–AS data. Back-to-back applications of (scaled) mainshock records as aftershocks often lead to overestimation of the aftershock seismic demand potential (Goda, 2015), and thus careful construction of artificial MS–AS sequences is important. Secondly, investigations of the relationships between seismic demands of inelastic SDOF systems and key seismological parameters of MS–AS sequences

provide useful guidance as to which parameters should be taken into account in developing seismic demand prediction models for more realistic structural models. Moreover, the developed MS–AS sequence dataset facilitates the assessment of the aftershock effects in relation to the spatial distribution of major aftershocks. Numerical analysis was set up to investigate the above-mentioned problems.

Based on the analysis results, the following conclusions can be drawn:

1. One of the controlling factors for determining the severity of peak and residual ductility demands (for the same level of seismic excitation) is earthquake magnitude. For inelastic seismic demand estimation, earthquake magnitude or a surrogate measure, such as earthquake event type, may need to be included (depending on regional seismic hazard characteristics).
2. Hysteretic behavior of structural systems can have major influence on the estimation of the inelastic seismic demand. The consideration of the positive post-yield stiffness ratio, in comparison with zero post-yield stiffness ratio as in EPP systems, reduces peak and residual ductility demands (particularly significant impact on the residual ductility demand). Moreover, both degradation and pinching behavior have moderate effects on the peak ductility demand.
3. The aggravation of the inelastic seismic demand due to major aftershocks is not common, because the mainshock often causes severer damage to structures. However, in rare cases, major aftershocks can increase the seismic damage severity significantly. Moreover, hysteretic behavior does not affect the MS–AS to MS peak ductility demand ratios significantly. The key factors for damaging aftershocks are: the ground-motion intensity of major aftershocks and the severity of damage caused by the mainshock.
4. The causal relationship between major aftershocks and increased seismic demands can be understood physically; greater seismic demand potential results in greater seismic demand. For mega-thrust subduction events, such as the 2011 Tohoku earthquake, spatial occurrence process of aftershocks is critical, because the size of major aftershocks is significantly smaller than the mainshock and thus their impact are much more localized. Improved spatial modeling of major aftershocks needs to be incorporated in generating artificial MS–AS sequences for seismic vulnerability assessment.

Acknowledgments

Strong-motion data used in this study were obtained from the K-NET and KiK-net (http://www.kyoshin.bosai.go.jp/) and the SK-net (http://www.sknet.eri.u-tokyo.ac.jp/). KG is supported by the Alexander von Humboldt Fellowship for Experienced Researchers. RR is funded by the Engineering and Physical Sciences Research Council (EP/M001067/1).

References

Boore, D. M. (2005). On pads and filters: processing strong-motion data. *Bull. Seismol. Soc. Am.* 95, 745–750. doi:10.1785/0120040160

Cornell, C. A., Jalayer, F., Hamburger, R. O., and Foutch, D. A. (2002). Probabilistic basis for 2000 SAC Federal Emergency Management Agency steel moment frame guidelines. *J. Struct. Eng.* 128, 526–533. doi:10.1061/(ASCE)0733-9445(2002)128:4(526)

Federal Emergency Management Agency. (2005). *Improvement of Nonlinear Static Seismic Analysis Procedures.* Washington, DC: Federal Emergency Management Agency.

Foliente, G. C. (1993). *Stochastic Dynamic Response of Wood Structural Systems.* Ph.D. dissertation, Virginia Polytechnic Institute and State University, Blacksburg, VA.

Galasso, C., Zarain, F., Iervolino, I., and Graves, R. W. (2012). Validation of ground motion simulations for historical events using SDOF systems. *Bull. Seismol. Soc. Am.* 102, 2727–2740. doi:10.1785/0120120018

Goda, K. (2012). Nonlinear response potential of mainshock-aftershock sequences from Japanese earthquakes. *Bull. Seismol. Soc. Am.* 102, 2139–2156. doi:10.1785/0120110329

Goda, K. (2015). Record selection for aftershock incremental dynamic analysis. *Earthquake Eng. Struct. Dyn.* 44, 1157–1162. doi:10.1002/eqe.2513

Goda, K., and Atkinson, G. M. (2009). Seismic demand estimation of inelastic SDOF systems for earthquakes in Japan. *Bull. Seismol. Soc. Am.* 99, 3284–3299. doi:10.1785/0120090107

Goda, K., and Atkinson, G. M. (2010). Intraevent spatial correlation of ground-motion parameters using SK-net data. *Bull. Seismol. Soc. Am.* 100, 3055–3067. doi:10.1785/0120100031

Goda, K., Pomonis, A., Chian, S. C., Offord, M., Saito, K., Sammonds, P., et al. (2013). Ground motion characteristics and shaking damage of the 11th March 2011 M_w9.0 Great East Japan earthquake. *Bull. Earthquake Eng.* 11, 141–170. doi:10.1007/s10518-012-9371-x

Goulet, C. A., Haselton, C. B., Mitrani-Reiser, J., Beck, J. L., Deierlein, G. G., Porter, K. A., et al. (2007). Evaluation of the seismic performance of a code-conforming reinforced-concrete frame building – from seismic hazard to collapse safety and economic losses. *Earthquake Eng. Struct. Dyn.* 36, 1973–1997. doi:10.1002/eqe.694

Iervolino, I., and Cornell, C. A. (2005). Record selection for nonlinear seismic analysis of structures. *Earthquake Spectra* 21, 685–713. doi:10.1193/1.1990199

Jalayer, F., and Cornell, C. A. (2009). Alternative nonlinear demand estimation methods for probability-based seismic assessments. *Earthquake Eng. Struct. Dyn.* 38, 951–972. doi:10.1002/eqe.876

Kagan, Y. Y. (2002). Aftershock zone scaling. *Bull. Seismol. Soc. Am.* 92, 641–655. doi:10.1785/0120010172

Kurahashi, S., and Iikura, K. (2013). Short-period source model of the 2011 M_w 9.0 off the Pacific coast of Tohoku earthquake. *Bull. Seismol. Soc. Am.* 103, 1373–1393. doi:10.1785/0120120157

Li, Q., and Ellingwood, B. R. (2007). Performance evaluation and damage assessment of steel frame buildings under main shock-aftershock earthquake sequences. *Earthquake Eng. Struct. Dyn.* 36, 405–427. doi:10.1002/eqe.667

Luco, N., and Bazzurro, P. (2007). Does amplitude scaling of ground motion records result in biased nonlinear structural drift responses? *Earthquake Eng. Struct. Dyn.* 36, 1813–1835. doi:10.1002/eqe.695

Luco, N., and Cornell, C. A. (2007). Structure-specific scalar intensity measures for near-source and ordinary earthquake ground motions. *Earthquake Spectra* 23, 357–392. doi:10.1193/1.2723158

Ma, F., Zhang, H., Bockstedte, A., Foliente, G. C., and Paevere, P. (2004). Parameter analysis of the differential model of hysteresis. *ASME J. Appl. Mech.* 71, 342–349. doi:10.1115/1.1668082

Morikawa, N., and Fujiwara, H. (2013). A new ground motion prediction equation for Japan applicable up to M9 mega-earthquake. *J. Disaster Res.* 8, 878–888.

Moustafa, A., and Takewaki, I. (2010). Modeling critical ground-motion sequences for inelastic structures. *J. Adv. Struct. Eng.* 13, 665–680. doi:10.1260/1369-4332.13.4.665

Nagato, K., and Kawase, H. (2004). Damage evaluation models of reinforced concrete buildings based on the damage statistics and simulated strong motions during the 1995 Hyogo-ken Nanbu earthquake. *Earthquake Eng. Struct. Dyn.* 33, 755–774. doi:10.1002/eqe.376

Ramirez, C. M., and Miranda, E. (2012). Significance of residual drifts in building earthquake loss estimation. *Earthquake Eng. Struct. Dyn.* 41, 1477–1493. doi:10.1002/eqe.2217

Ruiz-García, J. (2012). Mainshock-aftershock ground motion features and their influence in building's seismic response. *J. Earthquake Eng.* 16, 719–737. doi:10.1080/13632469.2012.663154

Ruiz-García, J., and Miranda, E. (2003). Inelastic displacement ratios for evaluation of existing structures. *Earthquake Eng. Struct. Dyn.* 32, 1237–1258. doi:10.1002/eqe.271

Ruiz-García, J., and Miranda, E. (2006). Evaluation of residual drift demands in regular multi-storey frames for performance-based seismic assessment. *Earthquake Eng. Struct. Dyn.* 35, 1609–1629. doi:10.1002/eqe.593

Salami, M. R., and Goda, K. (2014). Seismic loss estimation of residential wood-frame buildings in southwestern British Columbia considering mainshock-aftershock sequences. *J. Perform. Constr. Facil.* 28, A4014002. doi:10.1061/(ASCE)CF.1943-5509.0000514

Shcherbakov, R., Goda, K., Ivanian, A., and Atkinson, G. M. (2013). Aftershock statistics of major subduction earthquakes. *Bull. Seismol. Soc. Am.* 103, 3222–3234. doi:10.1785/0120120337

Shcherbakov, R., Turcotte, D. L., and Rundle, J. B. (2005). Aftershock statistics. *Pure Appl. Geophys.* 162, 1051–1076. doi:10.1007/s00024-004-2661-8

Smyrou, E., Tasioupolou, P., Bal, IE., Gazetas, G., and Vintzileou, E. (2011). Ground motions versus geotechnical and structural damage in the February 2011 Christchurch earthquake. *Seismol. Res. Lett.* 82, 882–892. doi:10.1785/gssrl.82.6.882

Stein, S., and Wysession, M. (2003). *An Introduction to Seismology, Earthquakes, and Earth Structure.* Hoboken, NJ: Wiley-Blackwell.

Vamvatsikos, D., and Cornell, C. A. (2004). Applied incremental dynamic analysis. *Earthquake Spectra* 20, 523–553. doi:10.1193/1.1737737

Wen, Y. K. (1976). Method for random vibration of hysteretic systems. *J. Eng. Mech.* 102, 249–263.

Zhai, C. H., Wen, W. P., Chen, Z. Q., Li, S., and Xie, L. L. (2013). Damage spectra for the mainshock-aftershock sequence-type ground motions. *Soil Dyn. Earthquake Eng.* 45, 1–12. doi:10.1016/j.soildyn.2012.10.001

Conflict of Interest Statement: The authors declare that the research was conducted in the absence of any commercial or financial relationships that could be construed as a potential conflict of interest.

5

Reliability of system identification technique in super high-rise building

Ayumi Ikeda, Kohei Fujita and Izuru Takewaki*

Department of Architecture and Architectural Engineering, Kyoto University, Kyoto, Japan

A smart physical-parameter-based system identification (SI) method has been proposed in the previous paper (Ikeda et al., 2014a). This method deals with time-variant non-parametric identification of natural frequencies and modal damping ratios using Auto-Regressive eXogenous (ARX) models and has been applied to high-rise buildings during the 2011 off the Pacific coast of Tohoku earthquake. In this perspective article, the current state of knowledge in this class of SI methods is explained briefly, and the reliability of this smart method is discussed through the comparison with the result by a more confident technique.

Keywords: system identification, super high-rise building, time-varying identification, reliability of system identification, ARX model, noise reduction

Edited by:
Solomon Tesfamariam,
The University of British Columbia,
Canada

Reviewed by:
Xinzheng Lu,
Tsinghua University, China
Marie-José Nollet,
École de Technologie Supérieure,
Canada

*Correspondence:
Izuru Takewaki,
Department of Architecture and
Architectural Engineering, Kyoto
University, Nishikyo,
Kyoto 615-8540, Japan
takewaki@archi.kyoto-u.ac.jp

System identification (SI) techniques are very popular and widely used worldwide. SI techniques may be classified into a modal-parameter SI and a physical-parameter SI (Housner et al., 1997; Takewaki et al., 2011a). SI techniques are usually used as tools for damage detection after extreme natural disasters (e.g., earthquakes) or methods for the investigation of accuracy of models utilized in the design stage.

In damage detection, the monitoring of healthy states is inevitable. This procedure requires a large amount of preparing work before damage occurs, and it is necessary to gather data for the healthy state. Unless we know the healthy state, it is not possible to know whether the state after an event is healthy or damaged. Although it may be possible to monitor the structural behavior in real time, it will require a large amount of cost. As far as a shear building model and a shear-bending model are concerned, the recording at limited floors (not real-time monitoring) may be realistic and possible (Kuwabara et al., 2013; Minami et al., 2013). These methods are based on a pioneering non-parametric approach (Udwadia et al., 1978) and the subsequent realistic extensions (Takewaki and Nakamura, 2000, 2005, 2009; Takewaki et al., 2011a). Kuwabara et al. (2013) developed a physical-parameter damage detection technique of a shear-bending model for the first time and Minami et al. (2013) opened the door for the physical-parameter SI techniques of a shear-bending model. On the other hand, some approaches using full data and a great deal of repeated calculations have been proposed for shear buildings (Zhang and Johnson, 2012, 2013a,b; Johnson and Wojtkiewicz, 2014).

For the reduction of influence of noise, some approaches using ARX models have been developed (Takewaki and Nakamura, 2009; Takewaki et al., 2011a; Minami et al., 2013; Ikeda et al., 2014a). Especially the method by Takewaki and Nakamura (2009), Takewaki et al. (2011a) deals with time-variant non-parametric identification of natural frequencies and modal damping ratios using ARX models. The method has been applied successfully to an actual base-isolated building (Takewaki and Nakamura, 2009; Takewaki et al., 2011a) and has also been applied to high-rise buildings during 2011 off the Pacific coast of Tohoku earthquake (Minami et al., 2013; Ikeda et al., 2014a).

Figure 1 shows a super high-rise steel building of 55 stories at Osaka bay area which was the highest in the western Japan at that time and was constructed in 1995 just after the Hyogoken-Nanbu earthquake (Kobe earthquake) (Takewaki et al., 2011b; Celebi et al., 2014). An important recording

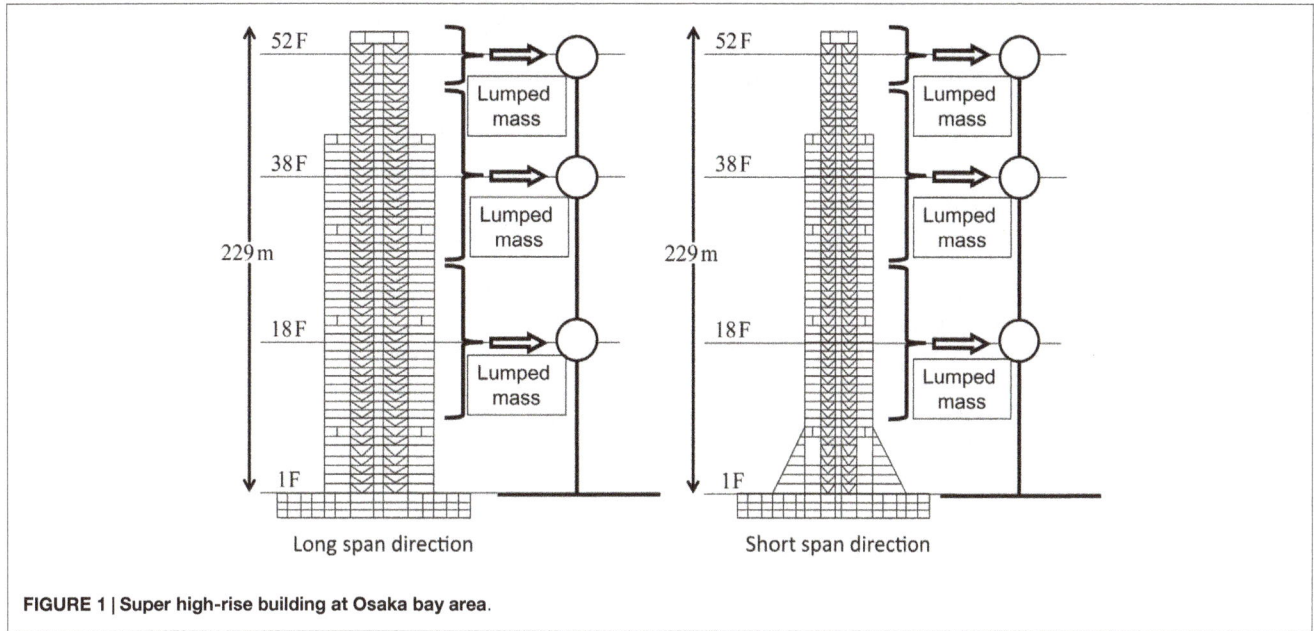

FIGURE 1 | Super high-rise building at Osaka bay area.

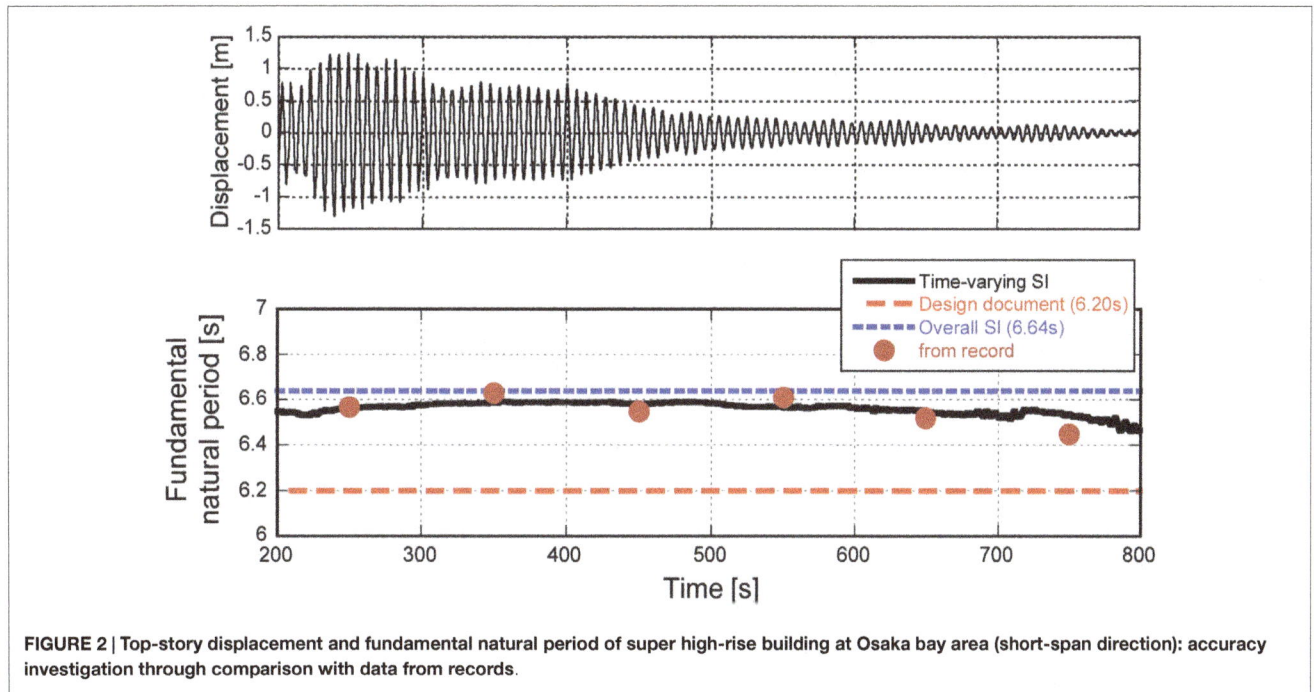

FIGURE 2 | Top-story displacement and fundamental natural period of super high-rise building at Osaka bay area (short-span direction): accuracy investigation through comparison with data from records.

was made in this building during 2011 off the Pacific coast of Tohoku earthquake. The building height is 256 m. The fundamental natural period is 5.8 s in the long-span direction and 5.3 s in the short-span direction. This building was shaken severely even though Osaka is located far from the epicenter (about 800 km) and the Japan Meteorological Agency (JMA) instrumental intensity was 3 in Osaka. It is remarkable that the level of velocity response spectra of ground motions observed here (first floor) is almost the same as that at the Shinjuku station (K-NET) in Tokyo and the top-story displacement are about 1.4 m (short-span direction) and 0.9 m (long-span direction). It was understood that this building

was just resonant with the surface ground fundamental natural period of about 6.4 s [4 × (depth of deep ground)/(shear wave velocity) = 4 × 1.6 (km)/1.0 (km/s)] and the fundamental-mode vibration component was predominant.

A smart physical-parameter based SI has been conducted in this building (Minami et al., 2013; Ikeda et al., 2014a). The outline of the time-varying identification, called the moving-window batch least-squares method, has been explained in the reference (Ikeda et al., 2014a). The duration of 30 s was used as the evaluation time for the batch least-squares method and this process was repeated by moving the window progressively by 1 s. The top-story displacement and the time-varying fundamental natural period of this building (short-span direction) have been identified (Ikeda et al., 2014a), which was evaluated by the moving-window batch least-squares method. The time-varying SI indicates the result by the moving-window batch least-squares method and the overall SI means the result, which uses the batch least-squares method in the whole duration. The time dependency of the fundamental natural period may result from its amplitude dependence and reported slight damage to some non-structural elements (ceiling walls, etc.). The accuracy of the method can be confirmed by the comparison with data from record shown in **Figure 2**. The six points in **Figure 2** have been plotted by investigating the period between zero-crossing points in the top-story displacement. As pointed out before, the fundamental vibration mode was predominant in this building during the 2011 Tohoku earthquake, and the top-story displacement was governed mostly by the fundamental vibration mode. The good

correspondence of the fundamental natural period between the time-varying SI, and the data analysis using the top-story displacement demonstrates the reliability of the time-varying SI (Ikeda et al., 2014a).

The reduction of influence of noise may be a central issue in the field of SI. The applicability to ambient vibration data and the development of more reliable methods should be investigated in the future (Ikeda et al., 2014b; Fujita et al., 2015).

Acknowledgments

The use of ground motion records from Building Research Institute of Japan and Osaka Prefectural Office is appreciated. Part of the present work is supported by the Grant-in-Aid for Scientific Research of Japan Society for the Promotion of Science (Nos. 24246095 and 15H04079). This support is greatly appreciated.

References

Celebi, M., Okawa, I., Kashima, T., Koyama, S., and Iiba, M. (2014). Response of a tall building far from the epicenter of the 11 March 2011 M9.0 great east Japan earthquake and aftershocks. *Struct. Des. Tall Spec. Build.* 23, 427–441. doi:10.1002/tal.1047

Fujita, K., Ikeda, A., and Takewaki, I. (2015). Application of story-wise shear building identification method to actual ambient vibration. *Front. Built Environ.* 1:2. doi:10.3389/fbuil.2015.00002

Housner, G. W., Bergman, L. A., Caughey, T. K., Chassiakos, A. G., Claus, R. O., Masri, S. F., et al. (1997). Special issue, structural control: past, present, and future. *J. Eng. Mech.* 123, 897–971. doi:10.1061/(ASCE)0733-9399(1997)123:9(897)

Ikeda, A., Minami, Y., Fujita, K., and Takewaki, I. (2014a). Smart system identification of super high-rise buildings using limited vibration data during the 2011 Tohoku earthquake. *Int. J. High Rise Build.* 3, 255–271.

Ikeda, A., Fujita, K., and Takewaki, I. (2014b). Story-wise system identification of shear building using ambient vibration data and ARX model. *Earthq. Struct.* 7, 1093–1118. doi:10.12989/eas.2014.7.6.1093

Johnson, E., and Wojtkiewicz, S. (2014). Efficient sensitivity analysis of structures with local modifications. II: transfer functions and spectral densities. *J. Eng. Mech.* 140, 04014068. doi:10.1061/(ASCE)EM.1943-7889.0000769

Kuwabara, M., Yoshitomi, S., and Takewaki, I. (2013). A new approach to system identification and damage detection of high-rise buildings. *Struct. Control Health Monit.* 20, 703–727. doi:10.1002/stc.1486

Minami, Y., Yoshitomi, S., and Takewaki, I. (2013). System identification of super high-rise buildings using limited vibration data during the 2011 Tohoku (Japan) earthquake. *Struct. Control Health Monit.* 20, 1317–1338.

Takewaki, I., and Nakamura, M. (2000). Stiffness-damping simultaneous identification using limited earthquake records. *Earthq. Eng. Struct. Dyn.* 29, 1219–1238. doi:10.1002/1096-9845(200008)29:8<1219::AID-EQE968>3.0.CO;2-X

Takewaki, I., and Nakamura, M. (2005). Stiffness-damping simultaneous identification under limited observation. *J. Eng. Mech.* 131, 1027–1035. doi:10.1061/(ASCE)0733-9399(2005)131:10(1027)

Takewaki, I., and Nakamura, M. (2009). Temporal variation of modal properties of a base-isolated building during an earthquake. *J. Zhejiang Univ. Sci. A* 11, 1–8. doi:10.1631/jzus.A0900462

Takewaki, I., Nakamura, M., and Yoshitomi, S. (2011a). *System Identification for Structural Health Monitoring*. Southampton: WIT Press.

Takewaki, I., Murakami, S., Fujita, K., Yoshitomi, S., and Tsuji, M. (2011b). The 2011 off the Pacific coast of Tohoku earthquake and response of high-rise buildings under long-period ground motions. *Soil Dyn. Earthq. Eng.* 31, 1511–1528. doi:10.1016/j.soildyn.2011.06.001

Udwadia, F., Sharma, D., and Shah, C. (1978). Uniqueness of damping and stiffness distributions in the identification of soil and structural systems. *J. Appl Mech.* 45, 181–187. doi:10.1115/1.3424224

Zhang, D., and Johnson, E. (2012). Substructure identification for shear structures: cross power spectral density method. *Smart Mater. Struct.* 21, 055006. doi:10.1088/0964-1726/21/5/055006

Zhang, D., and Johnson, E. (2013a). Substructure identification for shear structures I: substructure identification method. *Struct. Control Health Monit.* 20, 804–820. doi:10.1002/stc.1497

Zhang, D., and Johnson, E. (2013b). Substructure identification for shear structures with nonstationary structural responses. *J. Eng. Mech.* 139, 1769–1779. doi:10.1061/(ASCE)EM.1943-7889.0000626

6

Application of a direct procedure for the seismic retrofit of a R/C school building equipped with viscous dampers

Tomaso Trombetti, Michele Palermo*, Antoine Dib, Giada Gasparini, Stefano Silvestri and Luca Landi

Department of Civil, Chemical, Environmental, and Materials Engineering, University of Bologna, Bologna, Italy

Edited by:
Oren Lavan,
Technion – Israel Institute of
Technology, Israel

Reviewed by:
Izuru Takewaki,
Kyoto University, Japan
Shinta Yoshitomi,
Ritsumeikan University, Japan

*Correspondence:
Michele Palermo,
Department of Civil, Chemical,
Environmental, and Materials
Engineering, University of Bologna,
Viale del Risorgimento 2, 40136
Bologna, Italy
michele.palermo7@unibo.it

Several design methods aimed at sizing the viscous dampers to be inserted in building structures have been proposed in the last decades. Among others, the authors proposed a five-step procedure that guides the practical design from the choice of a target reduction in the seismic response of the structural system (with respect to the response of a structure without any additional damping devices) to the identification of the corresponding damping ratio and the mechanical characteristics of the commercially available viscous dampers. The original procedure requires, also at the preliminary design stage, the development of linear seismic time-history analyses for the dampers working velocities, necessary for the evaluation of the non-linear damping coefficient. In the present paper, the original five-step procedure is further simplified leading to a direct (i.e., fully analytical) procedure, which can be very useful in a preliminary design phase. The proposed direct procedure is then applied to design the added viscous dampers to be inserted in a real school building in order to improve its seismic capacity, and compared with the well-known MCEER procedure.

Keywords: peak inter-storey velocities, added viscous dampers, design procedure

Introduction

Manufactured viscous dampers are hydraulic devices, which can be inserted in building structures in order to mitigate the seismic effects through the dissipation of part of the kinetic energy by the earthquake to the structure (Chopra, 1995; Soong and Dargush, 1997; Constantinou et al., 1998; Hart and Wong, 2000; Christopoulos and Filiatrault, 2006). The effectiveness of such devices in reducing the seismic demand on the structural elements has been demonstrated by a number of research works since the 1980s (Constantinou and Tadjbakhsh, 1983; Constantinou and Symans, 1992, 1993; Trombetti and Silvestri, 2004, 2006, 2007; Silvestri and Trombetti, 2007; Occhiuzzi, 2009; Takewaki, 2009; Silvestri et al., 2011; Diotallevi et al., 2012; Hwang et al., 2013; Landi et al., 2013, 2014a; Palermo et al., 2013b). Most of the research works on viscous dampers (Takewaki, 1997, 2000, 2009, 1997; Shukla and Datta, 1999; Lopez Garcia, 2001; Singh and Moreschi, 2002; Levy and Lavan, 2006) basically propose sophisticated numerical algorithms for dampers optimization, i.e., damper size and location, sometimes leading to complex design procedures. Nevertheless, the application of such algorithms often requires computational expertise and time (beyond the typical availabilities of the designers) and relies mainly upon numerical results, which do not provide physical insight into the matter.

In 1992, report NCEER-92-0032 (Constantinou and Symans, 1992) first investigated the problem of selecting the damping coefficients of linear viscous dampers in an elastic system to provide a specific damping ratio. In 2000, report MCEER-00-0010 (Ramirez et al., 2000) proposed an analytical relationship between the viscous damping ratio in a given mode of vibration and the damping coefficients on the basis of an energetic approach, assuming a given undamped mode shape. Later on, other methods have been proposed. Among others, the most affordable for practitioners are likely to be the following ones: (i) Lopez Garcia (2001) developed a simple algorithm for optimal damper configuration in MDOF structures, assuming a constant inter-storey height and a straight-line first modal shape and (ii) ASCE 7 (American Society of Civil Engineers, 1999) absorbed the MCEER-00-0010 approach.

Nonetheless, other alternative approaches leading to practical design procedures for the sizing of the viscous dampers have been proposed in the last years: (i) Christopoulos and Filiatrault (2006) suggested a design approach for estimating the damping coefficients of added viscous dampers consisting in a trial and error procedure and (ii) Silvestri et al. (2010) proposed a direct design approach, referred to as the "five-step procedure."

This latter five-step procedure aims at guiding the professional engineer from the choice of the target objective performance (reduction of significant response quantities with respect to a 5% damped system) to the identification of the mechanical characteristics (i.e., damping coefficient, oil stiffness, maximum damper forces) of commercially available viscous dampers. The original procedure (Silvestri et al., 2010, 2011; Palermo et al., 2013a) although mostly based on analytical expressions, still requires the development of numerical time-history analyses of FE models in order to estimate the maximum inter-storey velocity, necessary to obtain the maximum forces in the added dampers. This step inhibits the completion of the damper sizing relying on analytical results only, useful for preliminary design and for subsequent check of numerical results. The identification of an analytical expression of the inter-storey velocity profiles would allow the designer to directly obtain the maximum dampers forces (often a key parameter for the evaluation of the dampers cost), without performing numerical simulations. In this regard, a recent work by Adachi et al. (2013) acknowledged that the distribution of the maximum inter-storey velocities is a key index in order to evaluate the along-height demand on viscous dampers and exhibits specific characteristics depending on the number of the storeys of the building. Alternatively, the authors recently proposed simple formulas for the estimation of the peak inter-storey velocities developed in shear-type frame structures equipped with inter-storey viscous dampers under earthquake excitation (Silvestri et al., 2014).

In the present paper, the simple formulas for the estimation of the peak inter-storey velocities are used to further simplify the five-step procedure leading to a direct (i.e., fully analytical) procedure useful for the preliminary design of viscous dampers to be inserted in frame structures.

An Estimation of the Peak First Mode Inter-Storey Velocities

It is of common belief that the effectiveness of dampers allocation is closely related to the inter-storey drift demand. However, a recent work by Adachi et al. (2013) mentioned that "*while this understanding is almost true in rather low or medium-rise buildings, the distribution of the maximum interstory velocities plays a critical role in super high-rise buildings*." In fact, the distribution of the maximum inter-storey velocities may substantially differ (in terms of shape) with respect to that of the maximum inter-storey drifts due to a more significant higher modes contribution. In the same work, the authors also introduced approximate predictions for the maximum horizontal force of linear oil dampers. In detail, correction factors are introduced to account for the contribution of the higher modes.

The same problem has been recently faced by the authors (Silvestri et al., 2014) through a semi-analytical approach. In more details, by assuming the following analytical first mode shape:

$$\{\phi^1\} = \beta \left\{ \begin{array}{c} \frac{N(N+1)}{2} \\ \cdots \\ i\frac{N(N+1)}{2} - \sum_{j=0}^{i-1} \frac{j(j+1)}{2} \\ \cdots \\ \frac{N^2(N+1)}{2} - \sum_{j=0}^{N-1} \frac{j(j+1)}{2} \end{array} \right\} \tag{1}$$

where i represents the i-th storey, N is the total number of storeys, and β is an arbitrary constant. The deformed shape of Eq. 1 corresponds to the first Rayleigh–Ritz eigenvector (first iteration starting from a linear distribution of static forces along the building height) of a uniform shear-type building (i.e., constant floor mass m and storey stiffness k). By assuming that the base shear is given entirely by the first mode (conservative assumption reasonable for regular frame structures):

$$m_{tot}S_a = \{m\}^T \cdot \{\ddot{\phi}^1\} \tag{2}$$

where $\{\ddot{\phi}^1\} = \omega_1^2 \{\phi^1\}$ is the pseudo-acceleration vector, m_{tot} is the building seismic mass, ω_1 is the fundamental frequency of the structure, and S_a is the ordinate of the pseudo-acceleration spectrum at the fundamental period. After obtaining β from Eq. 2 and evaluating the pseudo-velocity vector as $\{\dot{\phi}^1\} = \omega_1 \{\phi^1\}$, it is possible to derive the peak inter-storey velocity profile under seismic excitation:

$$\{\dot{\delta}\} = \frac{S_a}{\omega_1} \frac{24}{(N+1)(2+5N+5N^2)} \left\{ \begin{array}{c} \frac{N(N+1)}{2} \\ \cdots \\ \frac{N(N+1)}{2} - \frac{i(i-1)}{2} \\ \cdots \\ \frac{N(N+1)}{2} - \frac{N(N-1)}{2} \end{array} \right\} \tag{3}$$

It has been shown that Eq. 3 leads to good predictions for short period structures (say $T < 0.5\,s$), since the higher modes have a limited contributions to the inter-storey velocities. On the contrary, for long-period structures, it becomes un-conservative due to a significant higher modes contribution. The interested reader may find additional details in the work by Silvestri et al. (2014), where a correction factor has been also introduced to include the higher modes influence for structures with $N > 5$.

If a linear deformed shape is assumed (i.e., constant inter-storey drifts), the application of the above described procedure leads to the following estimation of the peak inter-storey velocity (equal at all floors):

$$\dot{\delta} = \frac{S_a}{\omega_1} \frac{2}{(N+1)} \tag{4}$$

The Direct Five-Step Procedure for the Dimensioning of Added Viscous Dampers

In 2010, some of the authors proposed the so-called five-step procedure for the design of frame structures equipped with added viscous dampers (Silvestri et al., 2010). A summary of the five-step procedure is here provided.

Step 1: identification of the target damping ratio $\overline{\xi}$ leading to a certain target performance $\overline{\eta}$ (e.g., base shear, maximum inter-storey drift, ...);

Step 2: identification of the linear damping coefficients c_L for preliminary design purposes, by using the following formula:

$$c_L = \overline{\xi} \cdot \omega_1 \cdot m_{tot} \cdot \left(\frac{N+1}{n} \right) \cdot \frac{1}{\cos^2\theta} \tag{5}$$

where n is the total number of equally sized viscous dampers placed at each storey in a given direction and θ is the inclination of the dampers with respect to the horizontal direction.

Step 3: development of linear numerical time-history analyses of the building structure equipped with the linear viscous dampers identified in Step 2. The aim is to identify the range of "working" velocities, v_{max}, for the linear dampers.

Step 4: identification of the target characteristics of the actual non-linear viscous dampers characterized by a constitutive behavior of the type $F_d = c_{NL} sgn(\dot{u})|\dot{u}|^\alpha$ (damping coefficient c_{NL}, exponent α), which is equivalent to the linear damper. Several approaches have been proposed in literature (Peckan et al., 1999; Diotallevi et al., 2012; Landi et al., 2014b; Tubaldi et al., 2014). Here, the following formula, based on the energy equivalence over a full cycle of harmonic motion, is used:

$$c_{NL} = c_L \cdot (0.8 \cdot v_{max})^{1-\alpha} \tag{6}$$

Step 5: verification of the performances of the structure equipped with the non-linear viscous dampers sized in Step 4 through non-linear seismic time-history analyses.

For more details and for all the notations which will be used hereafter, the interested reader is referred to the work by Silvestri et al. (2010). It is worth noticing that, in the light of the results presented in the previous section, the original procedure can be now updated by using estimation of the dampers working velocities given by Eqs 3 and 4. In this way, the preliminary design

(Steps 1–4) of viscous dampers can be fully developed by means of analytical formulas, which can be also summarized in a single direct formula for the maximum damper force (i.e., the force at the ground floor) estimation:

For the deformed shape of Eq. 1:

$$F_d = 0.8^{1-\alpha} \cdot \overline{\xi} \cdot m_{tot} \cdot \frac{1}{n \cdot \cos^2\theta} \cdot S_a \left(T_1, \overline{\xi} \right) \cdot \frac{12N(N+1)}{(2+5N+5N^2)} \tag{7}$$

For the linear deformed shape:

$$F_d = 2 \cdot 0.8^{1-\alpha} \cdot \overline{\xi} \cdot m_{tot} \cdot \frac{1}{n \cdot \cos^2\theta} \cdot S_a \left(T_1, \overline{\xi} \right) \tag{8}$$

It is clear that the final design verification (Step 5) still should be carried out by means of non-linear seismic time-history analyses. For regular frame structures, the direct five-step procedure generally leads to a conservative design of the added dampers. In the next section, the procedure is applied to a real case study and compared with the MCEER approach (Ramirez et al., 2000). In particular, Eqs 7–29 and 7–32 in Ramirez et al. (2000) for the first mode supplemental damping provided by linear and non-linear viscous dampers have been considered. It can be shown that for a linear first mode profile and a uniform damper distribution the MCEER procedure and the direct five-step procedure can be related such as

$$\frac{c_{L,MCEER}}{c_{L,5\text{-STEP}}} = \frac{2N+1}{3N} \tag{9}$$

$$\frac{\dot{\delta}_{MCEER}}{\dot{\delta}_{5\text{-STEP}}} = \frac{2}{N+1} \cdot \frac{2N+1}{3} \tag{10}$$

For $N = 1$, the two design methods lead to the same result, while for N approaching to infinite the direct five-step procedure leads to a more conservative result ($c_{L,5\text{-STEP}} = 1.5 \cdot c_{L,MCEER}$ and $\dot{\delta}_{5\text{-STEP}} = 1.33 \cdot \dot{\delta}_{MCEER}$).

Applicative Example

The Case Study and the Retrofitting Strategy

A three storey R/C school building located in Bisignano (Cosenza, Southern Italy) and constructed in 1983 is analyzed in this section. The building was selected as a benchmark structure for a Research Project financed by the Italian Department of Civil Protection, with the aim of studying its seismic behavior, as well as of proposing and comparing alternative retrofitting solutions based on different seismic protection strategies. The building plan is rectangular with dimensions equal to $21 \, m \times 15 \, m$ (**Figure 1**). The columns have rectangular cross section ($40 \, cm \times 50 \, cm$) with the long side along the longitudinal direction (X direction). The interior beams have $40 \, cm \times 60 \, cm$ cross section, while the perimeter beams have $50 \, cm \times 40 \, cm$ cross section. The structure was designed according to the building code in force in the 1970s in Italy for a medium risk seismic region, but prior modern seismic design methodologies. Additional details regarding the building structure can be found in the work by Mazza and Vulcano (2014) and Sorace and Terenzi (2014).

The numerical simulations have been developed using the software SAP 2000 NL v16. A group of seven artificial records have been generated by SIMQKE (Vanmarcke et al., 1990) in order to

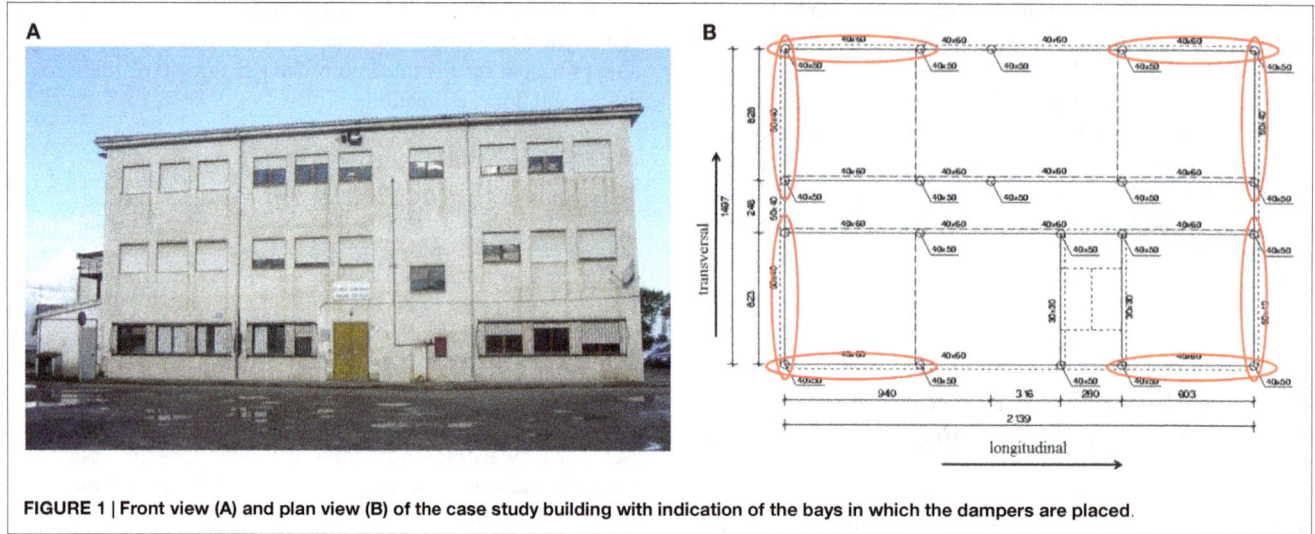

FIGURE 1 | Front view (A) and plan view (B) of the case study building with indication of the bays in which the dampers are placed.

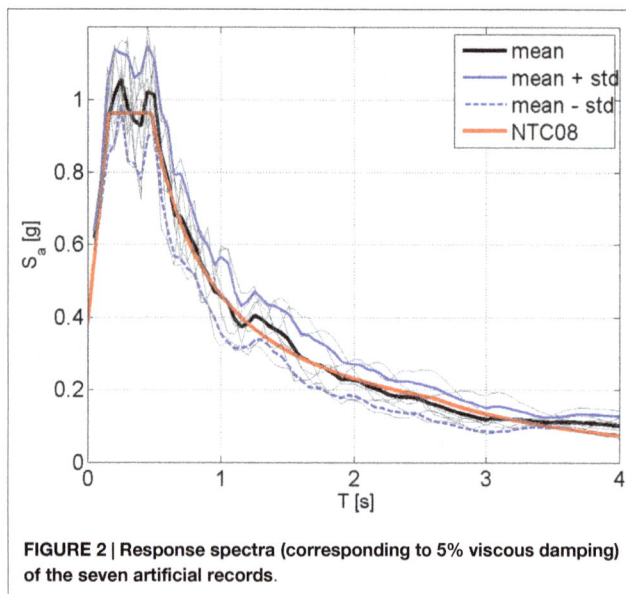

FIGURE 2 | Response spectra (corresponding to 5% viscous damping) of the seven artificial records.

TABLE 1 | Non-linear dampers characteristics along the X direction.

	c_L [kN·s/m]	c_{NL} [kN·s$^{0.15}$/m$^{0.15}$]	v_{max} [m/s]	F_d [kN]
5 Step	5330	940	0.16	717
MCEER	4090	636	0.14	473

TABLE 2 | Non-linear dampers characteristics along the Y direction.

	c_L [kN·s/m]	c_{NL} [kN·s$^{0.15}$/m$^{0.15}$]	v_{max} [m/s]	F_d [kN]
5 Step	3000	630	0.20	497
MCEER	2300	427	0.18	328

- the naked structure, i.e., the building structure without added dampers (UND model);
- the building structure equipped with inter-storey linear viscous dampers (D-L model);
- the building structure equipped with inter-storey non-linear viscous dampers (D-NL model).

It has to be noted that in all models, beams and columns are modeled through linear-elastic frame elements. In the D-L and D-NL model, the viscous dampers are modeled as damper-exponential link element with value of the damper exponent equal to 1.0 (i.e., linear damper) and 0.15, respectively. The linear and non-linear damping coefficients are evaluated according to both the direct five-step procedure and the MCEER procedure (see **Tables 1** and **2**). A 3D view of the D-NL SAP model is shown in **Figure 3**.

The Design of the Added Viscous Dampers

A target damping ratio $\overline{\xi} = 0.30$ has been chosen in order to design the system of added viscous dampers. The dampers are sized assuming a fundamental period $T_1 = 0.45$ s and a spectral acceleration $S_a\left(T_1, \overline{\xi}\right) = 0.52$ g along the X direction and a fundamental period $T_1 = 0.8$ s and a spectral acceleration $S_a\left(T_1, \overline{\xi}\right) = 0.36$ g along the Y directions. A damping exponent $\alpha = 0.15$ is assumed. The design is performed according to both

match the design spectrum according to the Italian building code (NTC, 2008) (**Figure 2**).

The seismic weight of the building W_{tot} is approximately equal to 17,240 kN, while the first three periods (two mainly translational and one mainly rotational) of the naked frame structure (without the non-structural masonry infills) are equal to 0.8, 0.45, and 0.52 s, respectively.

The retrofitting strategy adopted here is based on the insertion of inter-storey viscous dampers. In detail, four inter-storey viscous dampers are supposed to be located at each floor along both the longitudinal and the transversal directions, as shown in the plan view of the building (**Figure 1**). The damper system is therefore composed of 24 viscous dampers and is designed according to both the direct five-step procedure and the MCEER procedure (see next section). For the final verification of the effectiveness of the designed damper system (i.e., Step 5 of the procedure), three numerical models have been developed:

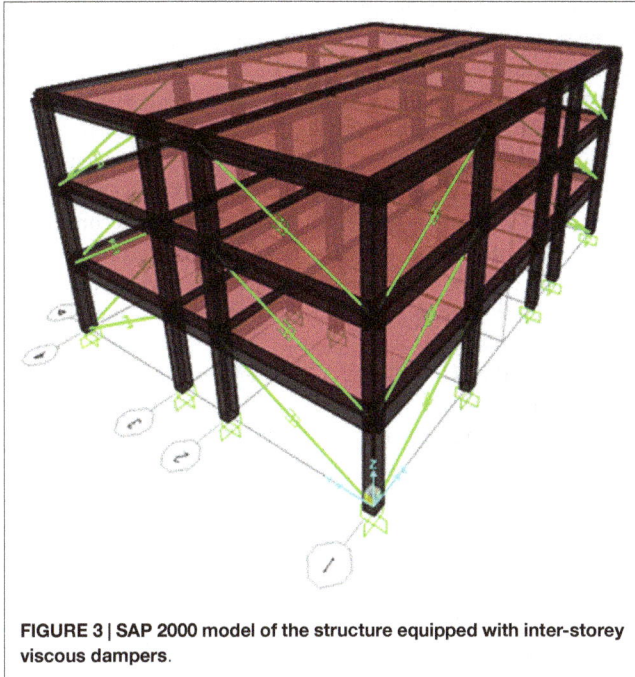

FIGURE 3 | SAP 2000 model of the structure equipped with inter-storey viscous dampers.

TABLE 3 | Reduction in the base shear (direct five-step procedure).

		UND	D-L	D-NL
Along X direction	V_{base} [kN]	9777	3661	2618
	η		0.37	0.27
Along Y direction	V_{base} [kN]	5258	1870	2104
	η		0.36	0.40

TABLE 4 | Reduction in the base shear (MCEER procedure).

		UND	D-L	D-NL
Along X direction	V_{base} [kN]	9777	4064	3400
	η		0.42	0.35
Along Y direction	V_{base} [kN]	5258	2189	2269
	η		0.42	0.43

the direct five-step procedure and the MCEER procedure assuming a linear deformed shape and equal dampers at all storeys. The results of the two design procedures are summarized in **Tables 1** and **2**, which give the values of the linear and non-linear damping coefficients (c_L and c_{NL}), the expected working velocities (v_{max}) and maximum damper force (F_d).

Verification Through Non-Linear Time-History Analyses

The averages (over the seven accelerograms) of the maximum values of the base shear, as obtained for the UND, the D-L, and the D-NL models are reported in **Tables 3** and **4** for the damping systems designed according to the five-step procedure and the MCEER procedure, respectively. Also, the damping reduction factor of the base shear ($\eta = \frac{V_{base,D}}{V_{base,UND}}$) is reported. Note that, for a target damping ratio of $\overline{\xi} = 0.30$, the widely used formulation by Bommer et al. (2000) leads to $\overline{\eta} = 0.53$.

TABLE 5 | Maximum damper forces in the non-linear dampers (direct five-step procedure).

	From numerical simulations [kN]	Prediction [kN]	Relative error (%)
Along X direction	580	717	+19
Along Y direction	427	497	+14

TABLE 6 | Maximum damper forces in the non-linear dampers (MCEER procedure).

	From numerical simulations [kN]	Prediction [kN]	Relative error (%)
Along X direction	430	473	+9
Along Y direction	312	328	+5

TABLE 7 | Maximum damper forces in the linear dampers (direct five-step procedure).

	From numerical simulations [kN]	Prediction [kN]	Relative error (%)
Along X direction	425	860	+50
Along Y direction	306	600	+49

TABLE 8 | Maximum damper forces in the linear dampers (according to the MCEER procedure).

	From numerical simulations [kN]	Prediction [kN]	Relative error (%)
Along X direction	383	570	+33
Along Y direction	270	395	+31

First, it can be noted that both the design procedures are conservative provided that the obtained reductions of the base shear are larger than the expected ones. As expected, the five-step procedure is more conservative than the MCEER procedure, especially in the non-linear case. This is again an expected result provided that in the non-linear case both the discrepancies in the evaluation of the linear damping coefficients and the inter-storey velocities are present (see Eqs 9 and 10).

Tables 5 and **6** compare the averages (over the seven accelerograms) of the maximum values of the damper forces with the corresponding predictions according to the direct five-step procedure and the MCEER procedure, respectively. Again, as expected, the forces derived with the direct five-step procedure-based FE model are larger than the ones obtained with the MCEER-based FE model. The average relative errors in the estimation of the damper forces according to the five-step procedure are of the order of 15–20%, while the average relative errors given by the MCEER procedure are around 5–10%. However, it has to be noted that, due to the high non-linear constitutive behavior of the dampers (damping exponent of 0.15), a relative small error in the estimation of the maximum damper forces would result in larger errors in the estimation of the maximum strokes and displacements, as shown by the results of the linear time-history simulations summarized in **Tables 7** and **8** in terms of maximum (average values over the seven accelerograms) damper forces in the linear

TABLE 9 | Reduction in the base shear (original vs. direct five-step procedure).

		UND	Direct	Original
Along X direction	V_{base} [kN]	9777	2618	3575
	η		0.27	0.37
Along Y direction	V_{base} [kN]	5258	2104	2490
	η		0.40	0.47

dampers (relative errors around 50 and 30% for the direct five-step procedure and the MCEER procedure, respectively). The level of accuracy in the estimation of the structural response can be improved in the verification phase, i.e., in the time-history analyses of the Step 5. Otherwise, in the design phase, it can be improved by employing the original formulation of the five-step procedure, at the cost of additional numerical simulations required to evaluate the working velocities of the linear dampers (Silvestri et al., 2010). For instance, **Table 9** provides a comparison between the original five-step procedure and the direct five-step procedure in terms of the reduction of the base shear. It can be noted that the original five-step procedure leads to similar results in terms of base shear reduction (slightly less conservative) with respect to the MCEER procedure.

Conclusion

In this paper, the original five-step procedure for the seismic design of viscous dampers to be added in multi-storey framed structures, as proposed by Silvestri et al. (2010), is further simplified leading to a direct (i.e., fully analytical) procedure. The new developments allow to carry out the preliminary design with analytical formulas only, therefore not requiring the development of the linear seismic time-history analyses required by the original procedure to evaluate the working velocities of the dampers. First, it is noted that the direct five-step procedure leads to a more conservative design of the added dampers with respect to the MCEER procedure. In detail for a one-storey building, the two design procedures lead to the same design, while as the storey number approach to infinite the linear damping coefficient calculated according to the direct five-step procedure approach to 1.5 the corresponding one calculated according to the MCEER procedure, and the damper inter-storey velocities calculated according to the direct five-step procedure approach to 1.33 the corresponding one calculated according to the MCEER procedure.

Finally, the direct five-step procedure has been applied to design added viscous dampers to be inserted in a real building, i.e., a RC school building located in southern Italy. It is confirmed that the direct procedure, besides allowing an easy and quick identification of the mechanical characteristics of the viscous dampers, leads to a conservative achievement of the target performances (i.e., reduction in seismic response with respect to the reference structure with no additional damping system) and to a quite good estimation (from an engineering point of view) of the damper forces, even though less accurate than the MCEER procedure. With regard to the strokes and deformations, an improved estimate can be obtained in the verification time-history analyses of Step 5 or, in the design phase, using the original five-step procedure (or alternatively the MCEER procedure).

Acknowledgments

Financial support from the Department of Civil Protection (DPC-Reluis 2014–2018 Grant – Research line 6: "Seismic isolation and dissipation") is gratefully acknowledged.

References

Adachi, F., Fujita, K., Tsuji, M., and Takewaki, I. (2013). Importance of inter-storey velocity on optimal along-height allocation of viscous oil dampers in super high-rise buildings. *Eng. Struct.* 56, 489–500. doi:10.1016/j.engstruct.2013.05.036

American Society of Civil Engineers (ASCE). (1999). *Minimum Design Loads for Buildings and Other Structures*, ASCE 7-98. Reston, VA: ASCE.

Bommer, J. J., Elnashai, A. S., and Weir, A. G. (2000). "Compatible acceleration and displacement spectra for seismic design codes," in *Proceedings of the 12th World Conference on Earthquake Engineering* (Auckland).

Chopra, A. K. (1995). *Dynamics of Structures. Theory and Applications to Earthquake Engineering*. Upper Saddle River, NJ: Prentice-Hall.

Christopoulos, C., and Filiatrault, A. (2006). *Principles of Passive Supplemental Damping and Seismic Isolation*. Pavia: IUSS Press.

Constantinou, M. C., Soong, T. T., and Dargush, G. F. (1998). *Passive Energy Dissipation Systems for Structural Design and Retrofit, Monograph No. 1*. Buffalo, NY: Multidisciplinary Center for Earthquake Engineering Research.

Constantinou, M. C., and Symans, M. D. (1992). *Experimental and Analytical Investigation of Seismic Response of Structures with Supplemental Fluid Viscous Dampers*. NCEER-92-0032. Technical Report. Buffalo, NY: National Center for Earthquake Engineering Research.

Constantinou, M. C., and Symans, M. D. (1993). Seismic response of structures with supplemental damping. *Struct. Des. Tall Build.* 2, 77–92. doi:10.1002/tal.4320020202

Constantinou, M. C., and Tadjbakhsh, I. G. (1983). Optimum design of a first story damping system. *Comput. Struct.* 17, 305–310. doi:10.1016/0045-7949(83)90019-6

Diotallevi, P. P., Landi, L., and Dellavalle, A. (2012). A methodology for the direct assessment of the damping ratio of structures equipped with nonlinear viscous dampers. *J. Earthq. Eng.* 16, 350–373. doi:10.1080/13632469.2011.618521

Hart, G. C., and Wong, K. (2000). *Structural Dynamics for Structural Engineers*. New York, NY: Wiley.

Hwang, J. S., Lin, W. C., and Wu, N. J. (2013). Comparison of distribution methods for viscous damping coefficients to buildings. *Struct. Infrastruct. Eng.* 9, 28–41. doi:10.1080/15732479.2010.513713

Landi, L., Diotallevi, P. P., and Castellari, G. (2013). On the design of viscous dampers for the rehabilitation of plan-asymmetric buildings. *J. Earthq. Eng.* 17, 1141–1161. doi:10.1080/13632469.2013.804893

Landi, L., Lucchi, S., and Diotallevi, P. P. (2014a). A procedure for the direct determination of the required supplemental damping for the seismic retrofit with viscous dampers. *Eng. Struct.* 71, 137–149. doi:10.1016/j.engstruct.2014.04.025

Landi, L., Fabbri, O., and Diotallevi, P. P. (2014b). A two-step direct method for estimating the seismic response of nonlinear structures equipped with nonlinear viscous dampers. *Earthq. Eng. Struct. Dyn.* 43, 1641–1659. doi:10.1002/eqe.2415

Levy, R., and Lavan, O. (2006). Fully stressed design of passive controllers in framed structures for seismic loadings. *Struct. Multidiscip. Optim.* 32, 485–498. doi:10.1007/s00158-005-0558-5

Lopez Garcia, D. (2001). A simple method for the design of optimal damper configurations in MDOF structures. *Earthq. Spectra* 17, 387–398. doi:10.1193/1.1586180

Mazza, F., and Vulcano, A. (2014). Equivalent viscous damping for displacement-based seismic design of hysteretic damped braces for retrofitting framed buildings. *Bull. Earthquake Eng.* 12, 2797–2819. doi:10.1007/s10518-014-9601-5

Norme Tecniche per le Costruzioni (NTC). (2008). Italian building code, adopted with D.M. 14/01/2008, Published on S.O. n. 30 G.U. n. 29.

Occhiuzzi, A. (2009). Additional viscous dampers for civil structures: analysis of design methods based on effective evaluation of modal damping ratios. *Eng. Struct.* 31, 1093–1101. doi:10.1016/j.engstruct.2009.01.006

Palermo, M., Muscio, S., Silvestri, S., Landi, L., and Trombetti, T. (2013a). On the dimensioning of viscous dampers for the mitigation of the earthquake-induced effects in moment-resisting frame structures. *Bull. Earthq. Eng.* 11, 2429–2446. doi:10.1007/s10518-013-9474-z

Palermo, M., Silvestri, S., Trombetti, T., and Landi, L. (2013b). Force reduction factor for building structures equipped with added viscous dampers. *Bull. Earthq. Eng.* 11, 1661–1681. doi:10.1007/s10518-013-9458-z

Peckan, G., Mander, J. B., and Chen, S. S. (1999). Fundamental considerations for the design of non-linear viscous dampers. *Earthq. Eng. Struct. Dyn.* 28, 1405–1425. doi:10.1002/(SICI)1096-9845(199911)28:11<1405::AID-EQE875>3.3.CO;2-1

Ramirez, O. M., Constantinou, M. C., Kircher, C. A., Whittaker, A. S., Johnson, M. W., and Gomez, J. D. (2000). *Development and Evaluation of Simplified Procedures for Analysis and Design of Buildings with Passive Energy Dissipation Systems*. MCEER Report 00-0010. Buffalo, NY: Multidisciplinary Center for Earthquake Engineering Research, University at Buffalo, State University of New York.

Shukla, A. K., and Datta, T. K. (1999). Optimal use of viscoelastic dampers in building frames for seismic force. *J. Struct. Eng.* 125, 401–409. doi:10.1061/(ASCE)0733-9445(1999)125:4(401)

Silvestri, S., Gasparini, G., and Trombetti, T. (2010). A five-step procedure for the dimensioning of viscous dampers to be inserted in building structures. *J. Earthq. Eng.* 14, 417–447. doi:10.1080/13632460903093891

Silvestri, S., Gasparini, G., and Trombetti, T. (2011). Seismic design of a precast r. c. structure equipped with viscous dampers. *Earthq. Struct.* 2, 297–321. doi:10.12989/eas.2011.2.3.297

Silvestri, S., Palermo, M., Landi, L., Gasparini, G., and Trombetti, T. (2014). "Estimation of maximum damper forces in shear-type buildings subjected to seismic input," in *Proceedings of the 2nd European Conference on Earthquake Engineering and Seismology* (Istanbul), 24–29.

Silvestri, S., and Trombetti, T. (2007). Physical and numerical approaches for the optimal insertion of seismic viscous dampers in shear-type structures. *J. Earthq. Eng.* 11, 787–828. doi:10.1080/13632460601034155

Singh, M. P., and Moreschi, L. M. (2002). Optimal placement of dampers for passive response control. *Earthq. Eng. Struct. Dyn.* 31, 955–976. doi:10.1002/eqe.132.abs

Soong, T. T., and Dargush, G. F. (1997). *Passive Energy Dissipation Systems in Structural Engineering*. Chichester: Wiley.

Sorace, S., and Terenzi, G. (2014). Motion control-based seismic retrofit solutions for a R/C school building designed with earlier Technical Standards. *Bull. Earthquake Eng.* 12, 2723–2744. doi:10.1007/s10518-014-9616-y

Takewaki, I. (1997). Optimal damper placement for minimum transfer functions. *Earthq. Eng. Struct. Dyn.* 26, 1113–1124. doi:10.1002/(SICI)1096-9845(199711)26:11<1113::AID-EQE696>3.0.CO;2-X

Takewaki, I. (2000). Optimal damper placement for critical excitation. *Prob. Eng. Mech.* 15, 317–325. doi:10.1016/S0266-8920(99)00033-8

Takewaki, I. (2009). *Building Control with Passive Dampers: Optimal Performance-Based Design for Earthquakes*. Singapore: Wiley.

Trombetti, T., and Silvestri, S. (2004). Added viscous dampers in shear-type structures: the effectiveness of mass proportional damping. *J. Earthq. Eng.* 8, 275–313. doi:10.1080/13632460409350490

Trombetti, T., and Silvestri, S. (2006). On the modal damping ratios of shear-type structures equipped with Rayleigh damping systems. *J. Sound Vib.* 292, 21–58. doi:10.1016/j.jsv.2005.07.023

Trombetti, T., and Silvestri, S. (2007). Novel schemes for inserting seismic dampers in shear-type systems based upon the mass proportional component of the Rayleigh damping matrix. *J. Sound Vib.* 302, 486–526. doi:10.1016/j.jsv.2006.11.030

Tubaldi, E., Ragni, L., and Dall'Asta, A. (2014). Probabilistic seismic response assessment of linear systems equipped with nonlinear viscous dampers. *Earthq. Eng. Struct. Dyn.* 44, 101–120. doi:10.1002/eqe.2461

Vanmarcke, E. H., Cornell, C. A., Gasparini, D. A., and Hou, S. (1990). "SIMQKE-I: simulation of earthquake ground motions," ed. T. F. Blake (Cambridge, MA: Department of Civil Engineering, Massachusetts Institute of Technology).

Conflict of Interest Statement: The authors declare that the research was conducted in the absence of any commercial or financial relationships that could be construed as a potential conflict of interest.

7

Critical input and response of elastic–plastic structures under long-duration earthquake ground motions

Kotaro Kojima and Izuru Takewaki*

Department of Architecture and Architectural Engineering, Graduate School of Engineering, Kyoto University, Kyoto, Japan

The multiple impulse input is introduced as a substitute of the long-duration earthquake ground motion, mostly expressed in terms of harmonic waves, and a closed-form solution is derived of the elastic–plastic response of a single-degree-of-freedom structure under the "critical multiple impulse input." Since only the free vibration appears under such multiple impulse input, the energy approach plays an important role in the derivation of the closed-form solution of a complicated elastic–plastic response. It is shown that the critical inelastic deformation and the corresponding critical input frequency can be captured depending on the input level by the substituted multiple impulse input in the form of original and modified input sequence. The validity and accuracy of the proposed theory are investigated through the comparison with the response analysis to the corresponding sinusoidal input as a representative of the long-duration earthquake ground motion.

Edited by:
*Nikos D. Lagaros,
National Technical University of
Athens, Greece*

Reviewed by:
*Johnny Ho,
The University of Queensland,
Australia
Peng Pan,
Tsinghua University, China*

*Correspondence:
*Izuru Takewaki,
Department of Architecture and
Architectural Engineering, Graduate
School of Engineering, Kyoto
University, Kyotodaigaku-Katsura,
Nishikyo, Kyoto 615-8540, Japan
takewaki@archi.kyoto-u.ac.jp*

Keywords: earthquake response, critical input, critical response, elastic–plastic response, ductility factor, long-duration ground motion, resonance, multiple impulse

Introduction

There are several types of earthquake ground motions. One is a near-fault ground motion, which is getting much interest recently, another is a random ground motion, which is represented by El Centro NS etc., and the other is a long-duration, long-period ground motion, which was observed rather recently [see Takewaki et al. (2011, 2012)]. The effects of near-fault ground motions on structural response have been investigated extensively (Bertero et al., 1978; Hall et al., 1995; Sasani and Bertero, 2000; Alavi and Krawinkler, 2004; Mavroeidis et al., 2004; Kalkan and Kunnath, 2006, 2007; Xu et al., 2007; Rupakhety and Sigbjörnsson, 2011; Yamamoto et al., 2011; Khaloo et al., 2015; Vafaei and Eskandari, 2015). The fling-step and forward-directivity are widely recognized as special keywords to characterize such near-fault ground motions (Mavroeidis and Papageorgiou, 2003; Bray and Rodriguez-Marek, 2004; Kalkan and Kunnath, 2006; Mukhopadhyay and Gupta, 2013a,b; Zhai et al., 2013; Hayden et al., 2014; Yang and Zhou, 2014). Especially, Northridge earthquake in 1994, Hyogoken-Nanbu (Kobe) earthquake in 1995, and Chi-Chi (Taiwan) earthquake in1999 raised special attention to many earthquake structural engineers.

The fling-step and forward-directivity inputs have been characterized by two or three wavelets. For this class of ground motions, many useful research works have been conducted. Mavroeidis and Papageorgiou (2003) investigated the characteristics of this class of ground motions in detail and proposed some simple models (e.g., Gabor wavelet and Berlage wavelet). Xu et al. (2007) employed a kind of Berlage wavelet and applied it to the performance evaluation of passive energy dissipation systems. Takewaki and Tsujimoto (2011) used the Xu's approach and proposed a method for scaling ground motions from the viewpoints of drift and input energy demand. Takewaki et al. (2012) employed a sinusoidal wave for pulse-type waves.

Most of the previous works on the near-fault ground motions deal with the elastic response because the number of parameters (e.g., duration and amplitude of pulse, ratio of pulse frequency to structure natural frequency, and change of equivalent natural frequency for the increased input level) to be considered on this topic is many and the computation itself of elastic–plastic response is quite complicated.

In order to tackle such important but complicated problem, the double impulse input was introduced by Kojima and Takewaki (2015a) as a substitute of the fling-step near-fault ground motion and a closed-form solution of the elastic–plastic response of a structure by the "critical double impulse input" is derived. It was shown that, since only the free vibration appears under such double impulse input, the energy approach plays an important role in the derivation of the closed-form solution of a complicated elastic–plastic response. It was also shown that the maximum inelastic deformation can occur either after the first impulse or after the second impulse depending on the input level. The validity and accuracy of the proposed theory are investigated through the comparison with the response analysis result to the corresponding one-cycle sinusoidal input as a representative of the fling-step near-fault ground motion. The amplitude of the double impulse was modulated so that its maximum Fourier amplitude coincides with that of the corresponding one-cycle sinusoidal input. The extension of the theory for the fling-step near-fault ground motion to the forward-directivity near-fault ground motion was made by Kojima and Takewaki (2015b).

It was pointed by Takewaki (1996, 1997) that, when considering the upper bound of response to the random earthquake ground motions, it is appropriate to introduce the response spectrum method and the bounding theories [see Takewaki (1996, 1997)].

The closed-form or nearly closed-form solutions of the elastic–plastic earthquake response have been obtained so far only for the steady-state response to sinusoidal input or the transient response to an extremely simple sinusoidal input (Caughey, 1960a,b; Roberts and Spanos, 1990; Liu, 2000). In this article, the following motivation is raised. If a long-duration ground motion can be represented by a multiple impulse, the elastic–plastic response (continuation of free vibrations) can be derived by an energy approach without solving directly the differential equation (equation of motion). The input of impulse is expressed by the instantaneous change of velocity of the structural mass. A closed-form expression of plastic-deformation amplitude is derived by using an energy approach. An approximate expression of residual displacement is also provided by using the multiple impulse.

In the earthquake-resistant design, the resonance is a key word and it has been investigated extensively. While the resonant equivalent frequency has to be computed for a specified input level by changing the excitation frequency in a parametric manner in dealing with the sinusoidal input (Caughey, 1960a,b; Iwan, 1961, 1965a,b; Roberts and Spanos, 1990; Liu, 2000), no

iteration is required in the proposed method for the multiple impulse. This is because the resonant equivalent frequency can be obtained directly without the repetitive procedure. In the multiple impulse, the analysis can be done without the input frequency (timing of impulses) before the second impulse is input. The resonance can be proved by using energy investigation and the timing of the second and third impulses can be characterized as the time with zero restoring force. The maximum elastic–plastic response after impulse can be obtained by equating the initial kinetic energy computed by the initial velocity to the sum of hysteretic and elastic strain energies. It should be pointed out that only critical response (upper bound) is captured by the proposed method and the critical resonant frequency can be obtained automatically for the increasing input level of the multiple impulse.

Figure 1 shows an actual resonant response of a super high-rise building in Osaka, Japan, during the 2011 off the Pacific coast of Tohoku earthquake. This phenomenon clearly indicates the necessity and requirement of consideration of response under long-duration ground motion.

Multiple Impulse Input

It has been shown that the fling-step input (fault-parallel) of the near-fault ground motion can be represented by a one-cycle sinusoidal wave, and the forward-directivity input (fault-normal) of the near-fault ground motion can be expressed by a series of three sinusoidal wavelets (Kalkan and Kunnath, 2006; Khaloo et al., 2015). In the works by Kojima and Takewaki (2015a,b), it was demonstrated that these typical near-fault ground motions can be simplified by a double impulse (Kojima et al., 2015) and a triple impulse. This is because the double impulse and triple impulse have a simple characteristic and a straightforward expression of the response can be expected even for elastic–plastic responses based on an energy approach to free vibrations. Furthermore, the double impulse and triple impulse enabled us to describe directly the critical timing of impulses (resonant frequency), which is not easy for the sinusoidal and other inputs without a repetitive procedure (Caughey, 1960a,b; Iwan, 1961, 1965a,b).

Consider a ground acceleration $\ddot{u}_g(t)$ as the multiple impulse, as shown in **Figure 2A**, expressed by

$$\ddot{u}_g(t) = V\delta(t) - V\delta(t-t_0) + V\delta(t-2t_0) - V\delta(t-3t_0) + \cdots \quad (1a)$$

where V is the given initial velocity and t_0 is the time interval between two consecutive impulses. Its velocity and displacement are shown in **Figure 2A**. In view of the realistic point of view, the following modified multiple input is introduced and treated principally in this article (see **Figure 2B**).

$$\ddot{u}_g(t) = 0.5V\delta(t) - V\delta(t-t_0) + V\delta(t-2t_0) - V\delta(t-3t_0) + \cdots \quad (1b)$$

The comparison with the corresponding multicycle sinusoidal wave as a representative of the long-duration earthquake ground motion input is plotted in **Figure 2B**. The corresponding velocity and displacement of such multiple impulse and sinusoidal wave are also plotted in **Figure 2B**. It can be understood that the multiple impulse is a good approximation of the corresponding sinusoidal wave even in the form of velocity and displacement. However, the

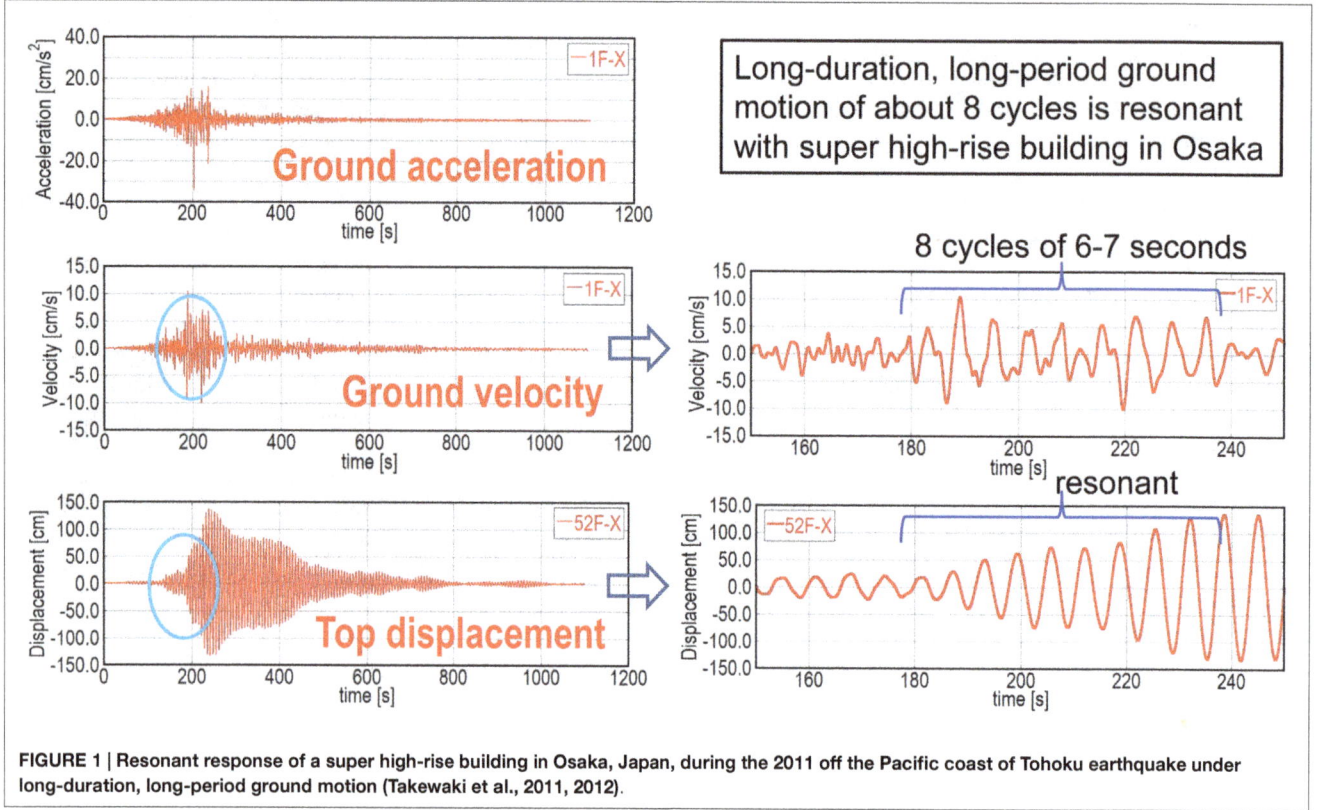

FIGURE 1 | Resonant response of a super high-rise building in Osaka, Japan, during the 2011 off the Pacific coast of Tohoku earthquake under long-duration, long-period ground motion (Takewaki et al., 2011, 2012).

correspondence in the response should be discussed carefully. This will be conducted later (see Section: Maximum Elastic–Plastic Deformation of SDOF System Subjected to Multiple Impulse; **Figure 5**).

Figure 3A shows the Input Sequence 1 (original input with equal interval) corresponding to **Figure 2B**. The points of impulses in the force–deformation relation converge to two points. On the other hand, **Figure 3B** presents the Input Sequence 2 (critical timing with residual deformation). The acting points of impulses in the force–deformation relation indicate the points with zero restoring force. It can be found that the time interval between the first and second impulses is different from those between the later consecutive two impulses. It is interesting to note that, if we consider the case as shown in **Figure 3C**, we can set the residual displacement to zero by changing the magnitude of the first impulse. It is also interesting that, if we employ the critical timing t_0^c in **Figure 3B** [criticality can be shown by the same reason as proved in Appendix in Kojima and Takewaki (2015a)] as the timing of the multiple impulse in **Figure 3A**, the acting points of impulses in the force–deformation relation converge to two points with zero restoring force. This fact supports the significance of introducing the Input Sequence 2 in order to find the critical interval of the multiple impulse for the Input Sequence 1 without repetition. This is the most original aspect in this article.

The Fourier transform of $\ddot{u}_g(t)$ of the multiple impulse input (Input Sequence 1 expressed by Eq. 1b) can be derived as

$$
\begin{aligned}
\ddot{U}_g(\omega) &= \int_{-\infty}^{\infty} \left\{ 0.5V\delta(t) - V\delta(t - t_0) + V\delta(t - 2t_0) \right. \\
&\quad \left. - V\delta(t - 3t_0) + \cdots \right\} e^{-i\omega t} dt \\
&= \int_{-\infty}^{\infty} \left\{ 0.5V\delta(t)e^{-i\omega t} - V\delta(t - t_0)e^{-i\omega t_0}e^{-i\omega(t - t_0)} \right. \\
&\quad + V\delta(t - 2t_0)e^{-i\omega 2t_0}e^{-i\omega(t - 2t_0)} \\
&\quad \left. - V\delta(t - 3t_0)e^{-i\omega 3t_0}e^{-i\omega(t - 3t_0)} + \cdots \right\} dt \\
&= V(0.5 - e^{-i\omega t_0} + e^{-i\omega 2t_0} - e^{-i\omega 3t_0} + \cdots)
\end{aligned}
\tag{2}
$$

The absolute value of Eq. 2 leads to

$$
\begin{aligned}
\left| \ddot{U}_g(\omega) \right| &= V \left| 0.5 - e^{-i\omega t_0} + e^{-i\omega 2t_0} - e^{-i\omega 3t_0} + \cdots \right| \\
&= V \left| 0.5 + \sum_{n=1}^{N-1} (-1)^n e^{-i\omega n t_0} \right|
\end{aligned}
\tag{3}
$$

SDOF System

Consider an undamped elastic–perfectly plastic single-degree-of-freedom (SDOF) system of mass m and stiffness k. The yield deformation and yield force are denoted by d_y and f_y. Let $\omega_1 = \sqrt{k/m}, u$, and f denote the undamped natural circular frequency, the displacement of the mass relative to the ground (deformation of the system), and the restoring force of the model, respectively. The time derivative is denoted by an over-dot. In Section "Maximum Elastic–Plastic Deformation of SDOF System Subjected to Multiple Impulse," these parameters will be dealt with in a non-dimensional or normalized form to derive the relation

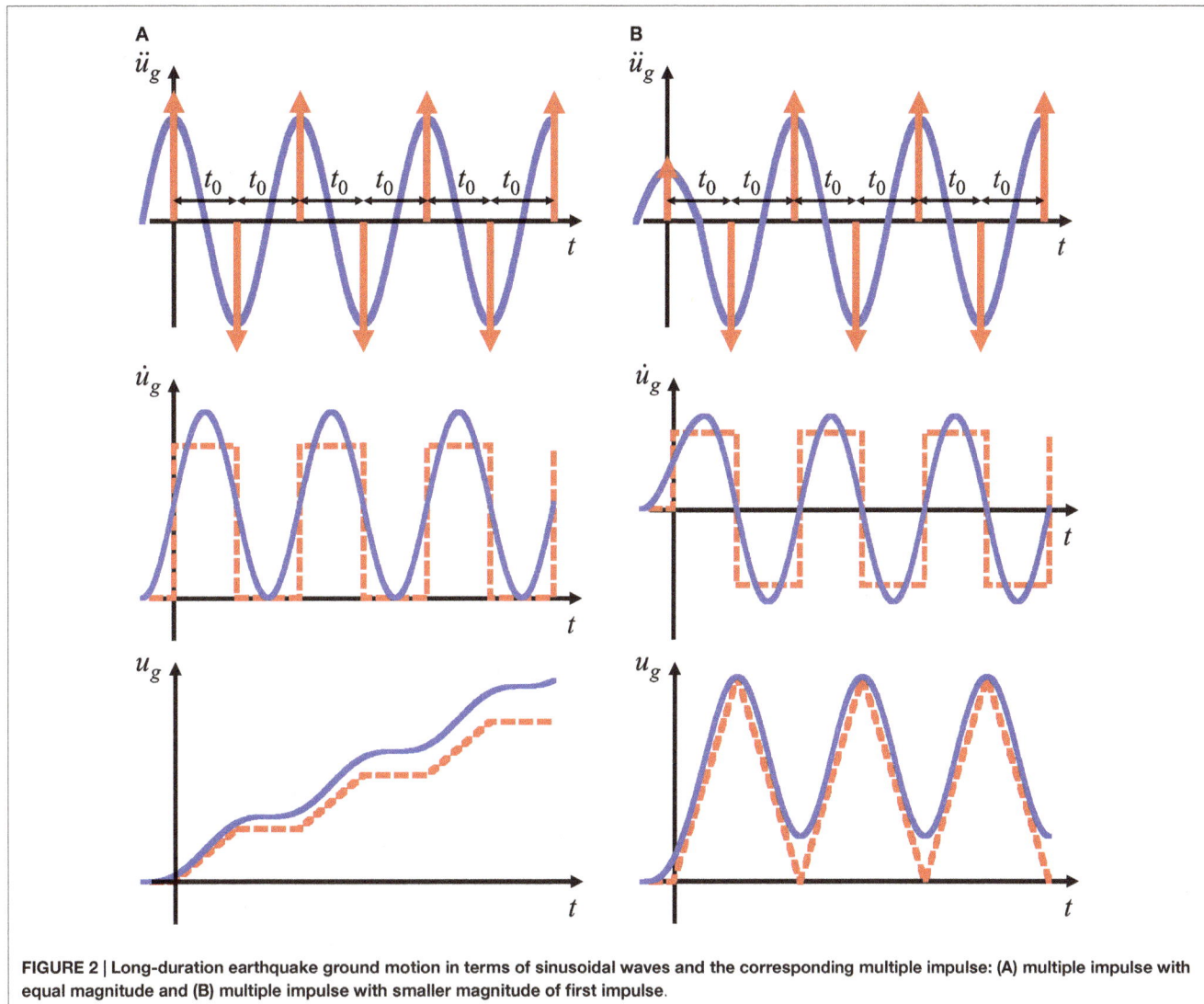

FIGURE 2 | Long-duration earthquake ground motion in terms of sinusoidal waves and the corresponding multiple impulse: (A) multiple impulse with equal magnitude and (B) multiple impulse with smaller magnitude of first impulse.

of permanent interest between the input and the elastic–plastic response. However, numerical parameters will be introduced partially in Sections "Maximum Elastic–Plastic Deformation of SDOF System Subjected to Multiple Impulse" and "Accuracy Check by Time-History Response Analysis Subjected to the Corresponding Multicycle Sinusoidal Input" to demonstrate an example of actual parameters.

Maximum Elastic–Plastic Deformation of SDOF System Subjected to Multiple Impulse

Non-Iterative Determination of Critical Timing and Critical Plastic Deformation by Using Modified Input Sequence

Consider Input Sequence 1 in **Figure 3A** at first. If the SDOF system is elastic, the critical timing t_0 is half the natural period of the SDOF system. However, if the SDOF system goes into a plastic region, the critical set of input amplitude and input frequency

(timing of impulse) has to be computed iteratively. This situation is the same for the multicycle sinusoidal wave (Caughey, 1960a,b; Iwan, 1961, 1965a,b).

In order to overcome this difficulty, consider Input Sequence 2 in **Figure 3B,** which introduces a modified input (only the timing between the first and second impulses is modified so that the second impulse is given at the zero restoring force). Input Sequence 2 is based on the assumption that, if the steady state exists in which the impulse is given at zero restoring-force timing, impulse provides the maximum steady-state plastic deformation. This assumption is verified by giving the critical timing obtained from Input Sequence 2 to Input Sequence 1. In other words, if the critical timing obtained from Input Sequence 2 is given to Input Sequence 1, the timing of impulse converges to zero restoring-force timing. This verification is also supported by the one-to-one correspondence between the input amplitude and its critical timing of impulses (impulses have to be given at zero restoring-force points). It is also possible to derive the Input Sequence 3 and its response with zero residual displacement (see **Figure 3C**).

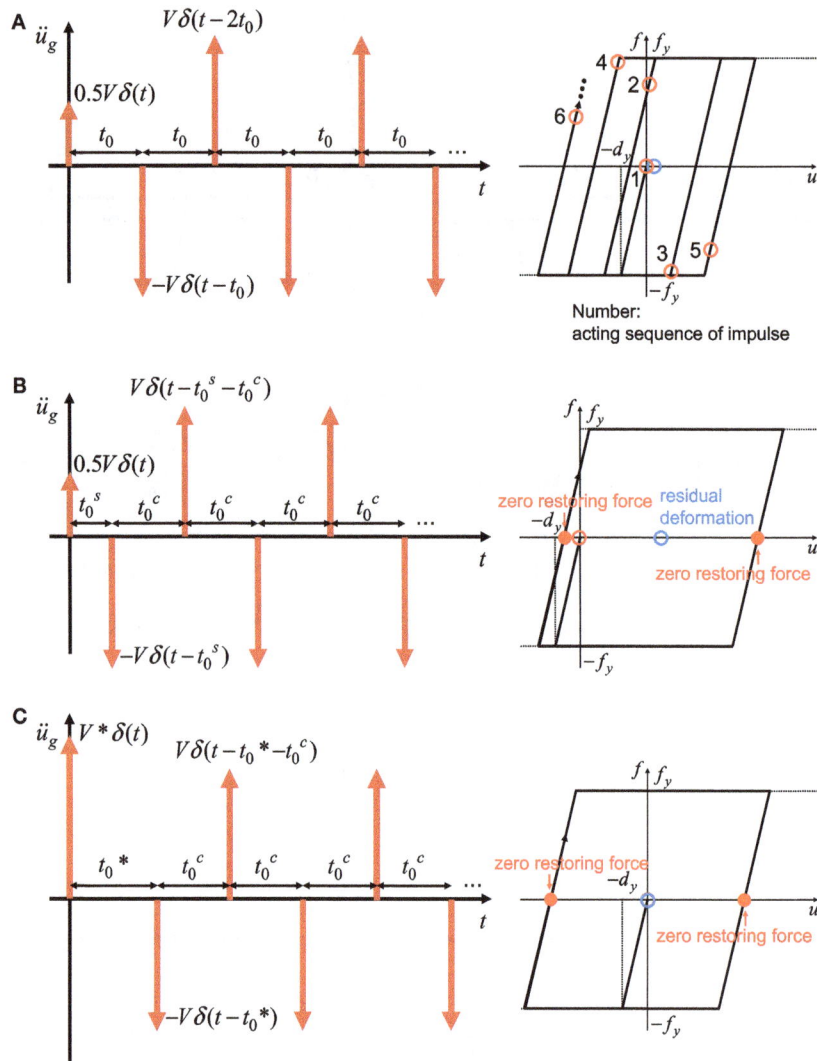

FIGURE 3 | Three input sequences: (A) Input Sequence 1: multiple impulse input with equal interval; (B) Input Sequence 2: multiple impulse input with modification of first impulse timing; (C) Input Sequence 3: multiple impulse input with modification of first impulse timing and first impulse amplitude.

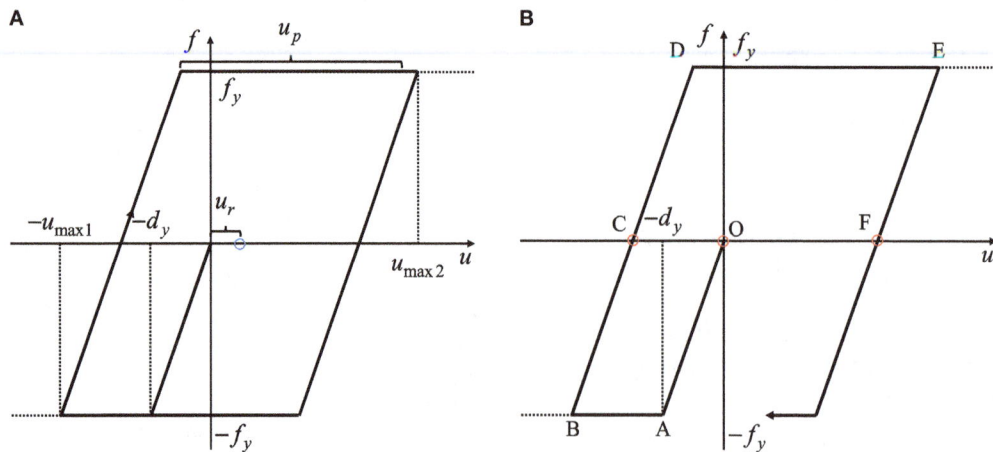

FIGURE 4 | Definition of response quantities and response transition: (A) schematic diagram of deformation quantities in force–deformation relation and (B) transition of response process and impulse timing (Input Sequence 2).

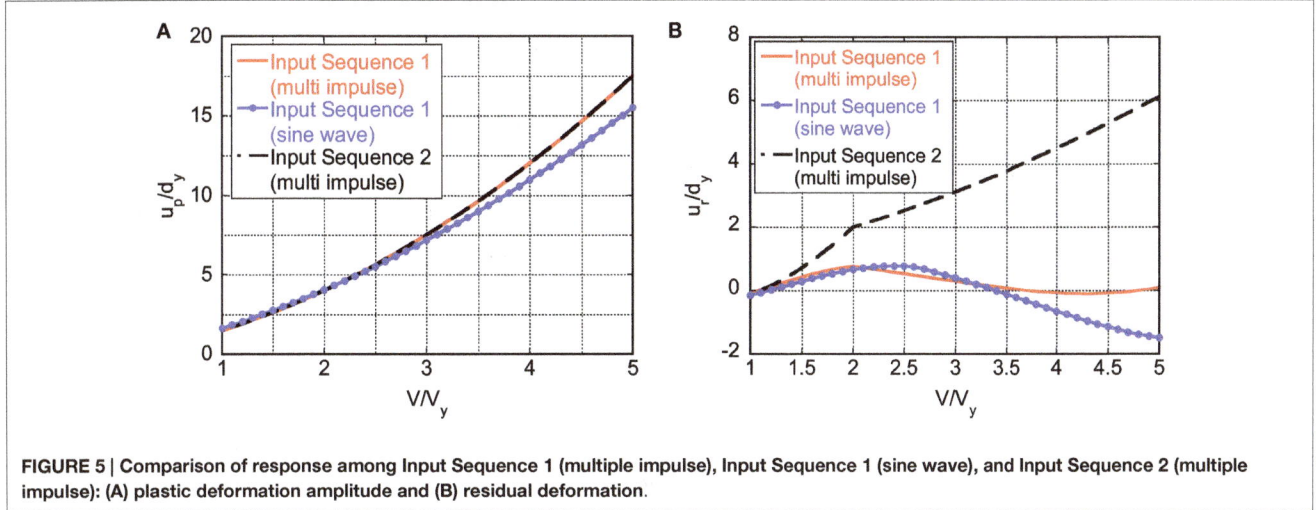

FIGURE 5 | Comparison of response among Input Sequence 1 (multiple impulse), Input Sequence 1 (sine wave), and Input Sequence 2 (multiple impulse): (A) plastic deformation amplitude and (B) residual deformation.

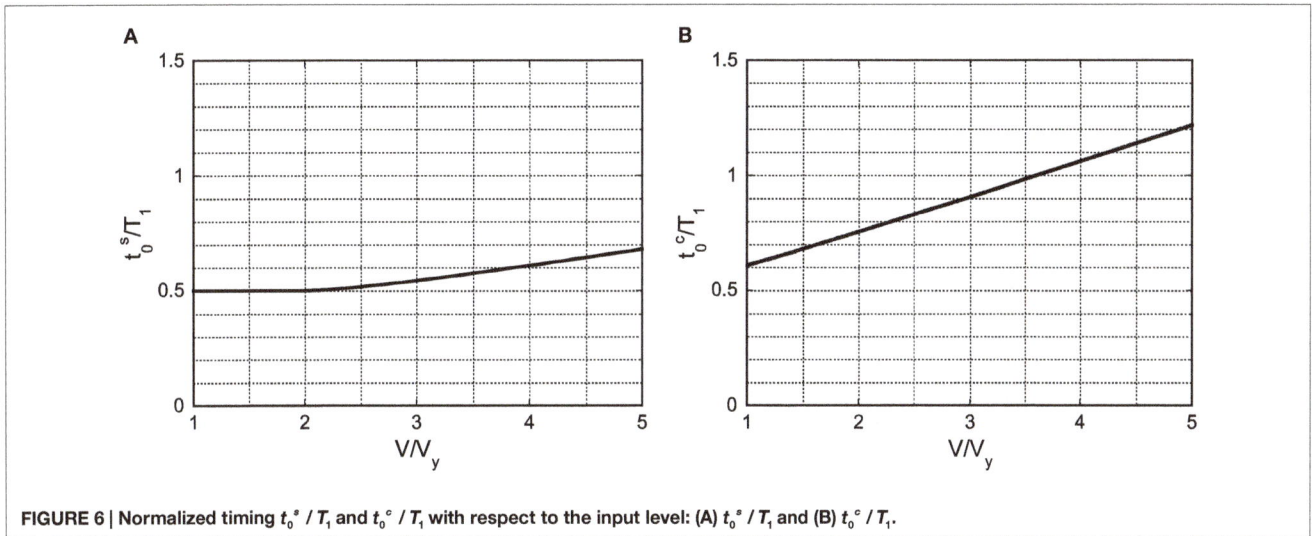

FIGURE 6 | Normalized timing t_0^s / T_1 and t_0^c / T_1 with respect to the input level: (A) t_0^s / T_1 and (B) t_0^c / T_1.

The elastic–plastic response to the multiple impulse can be described by the continuation of free vibrations.

The maximum deformation after the first impulse is denoted by $u_{max 1}$ and that after the second impulse is expressed by $u_{max 2}$ as shown in **Figure 4**. The input of each impulse is expressed by the instantaneous change of velocity of the structural mass. Such response can be derived by an energy approach without solving directly the differential equation (equation of motion). The kinetic energy given at the initial stage (the time of the first impulse) and at the time of the second impulse is transformed into the sum of the hysteretic energy and the strain energy corresponding to the yield deformation. By using this rule, the maximum deformation can be obtained in a simple manner.

It should be emphasized that, while the resonant equivalent frequency has to be computed for a specified input level by changing the excitation frequency in a parametric manner in dealing with the sinusoidal input (Caughey, 1960a,b; Iwan, 1961, 1965a,b; Roberts and Spanos, 1990; Liu, 2000; Moustafa et al., 2010), no iteration is required in the proposed method for the

multiple impulse. This is because the resonant equivalent frequency [resonance can be proved by using energy investigation: see Appendix in Kojima and Takewaki (2015a)] can be obtained directly without the repetitive procedure. As a result, the timing of the second impulse can be characterized as the time with zero restoring force.

Only critical response (upper bound) is captured by the proposed method and the critical resonant frequency can be obtained automatically for the increasing input level of the multiple impulse. One of the original points in this article is the introduction of the concept of "critical excitation" in the elastic–plastic response (Drenick, 1970; Abbas and Manohar, 2002; Takewaki, 2002, 2007; Moustafa et al., 2010). Once the frequency and amplitude of the critical multiple impulse are computed, the corresponding multicycle sinusoidal motion as a representative of the long-duration earthquake ground motion can be identified.

Let us explain the evaluation method of $u_{max 1}$ and $u_{max 2}$. The plastic deformation after the first impulse is expressed by u_{p1} and that after the second impulse is denoted by u_{p2}. There are

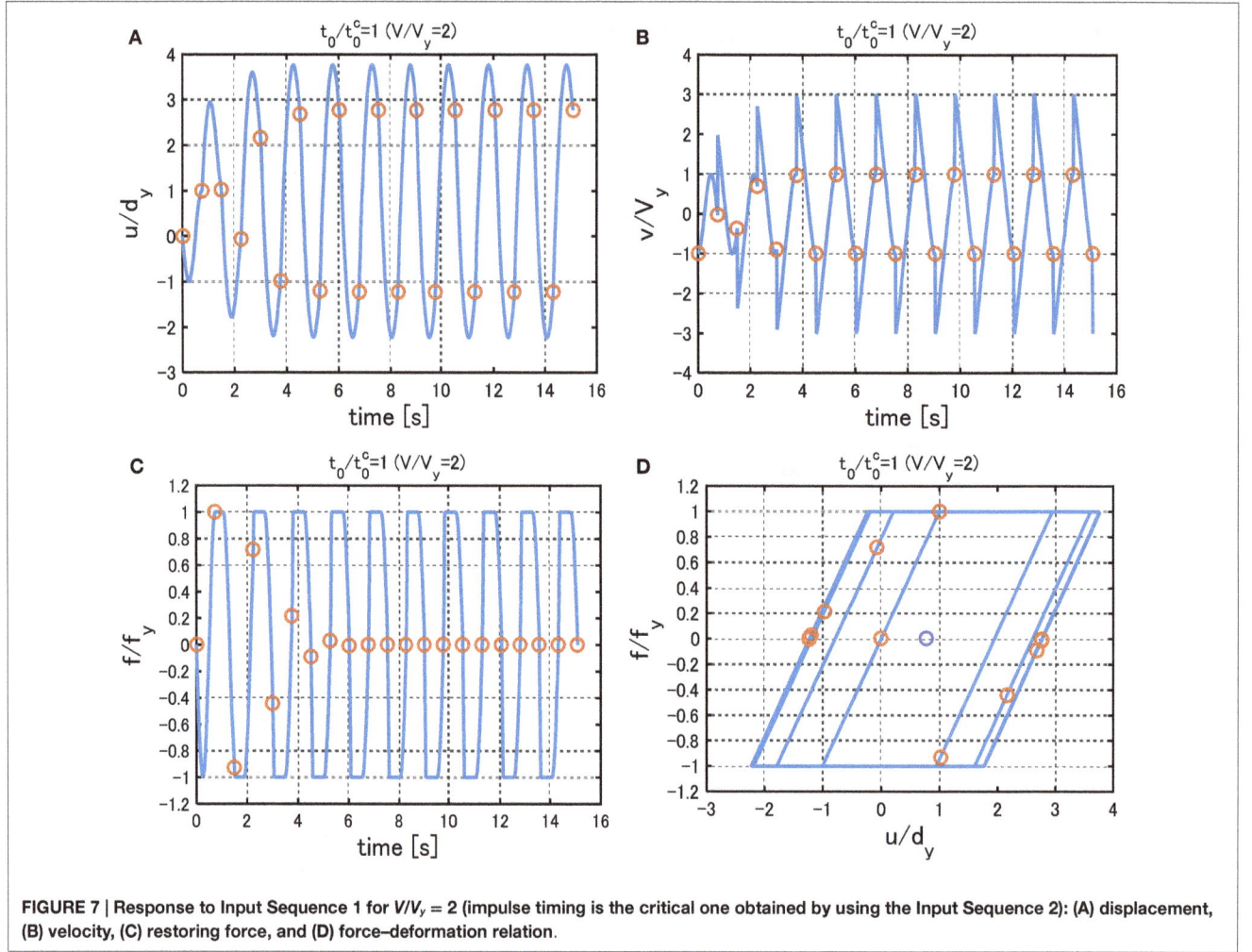

FIGURE 7 | Response to Input Sequence 1 for $V/V_y = 2$ (impulse timing is the critical one obtained by using the Input Sequence 2): (A) displacement, **(B)** velocity, **(C)** restoring force, and **(D)** force–deformation relation.

three cases to be considered depending on the yielding stage (Kojima and Takewaki, 2015a). Let $V_y (= \omega_1 d_y)$ denote the input level of velocity of the impulse at which the SDOF system at rest just attains the yield deformation after the impulse of such velocity.

Consider the case where the model goes into the yielding stage even after the first impulse. This case corresponds to (CASE 3) in the problem of double impulse (Kojima and Takewaki, 2015a). **Figure 4A** shows the schematic diagram of the response in this case. $u_{max\,1}$ can be obtained from the following energy conservation law.

$$m(0.5V)^2 / 2 = f_y d_y / 2 + f_y u_{p1} = f_y d_y / 2 + f_y (u_{max1} - d_y) \quad (4)$$

On the other hand, $u_{max\,2}$ can be computed from another energy conservation law.

$$m(v_c + V)^2 / 2 = f_y d_y / 2 + f_y u_{p2} \quad (5)$$

where v_c is characterized by $m v_c^2 / 2 = f_y d_y / 2$ and u_{p2} is characterized by $u_{max\,2} + (u_{max\,1} - d_y) = d_y + u_{p2}$. In other words, $u_{max\,2}$ can be obtained from

$$m(v_c + V)^2 / 2 = f_y d_y / 2 + f_y (u_{max1} + u_{max2} - 2d_y). \quad (6)$$

As in the above case, the velocity V induced by the second impulse is added to the velocity v_c introduced by the first impulse (the maximum velocity during the unloading stage). Although only CASE 3 in the double impulse (Kojima and Takewaki, 2015a) has been considered here, CASE 2 (yielding only after the second impulse) can be treated by replacing v_c in Eqs 5, 6 by $0.5V$.

Figure 5 shows the plot of the plastic deformation amplitude u_p (u_{p2} in this case) and the residual deformation u_r, shown in **Figure 4A**, with respect to the input level V/V_y for Input Sequence 1 and 2. While the plastic deformation amplitude is the same for Sequence 1 and 2, the residual deformations are different. This results from the difference in the initial disturbances in Input Sequence 1 and 2.

Determination of Critical Timing of Impulses

Consider the Input Sequence 2 in this section. The time between two consecutive impulses can be obtained by solving the differential equations (equations of motion) and substituting the continuation conditions at the transition points. The time t_0^s between the first and second impulses and the time t_0^c between

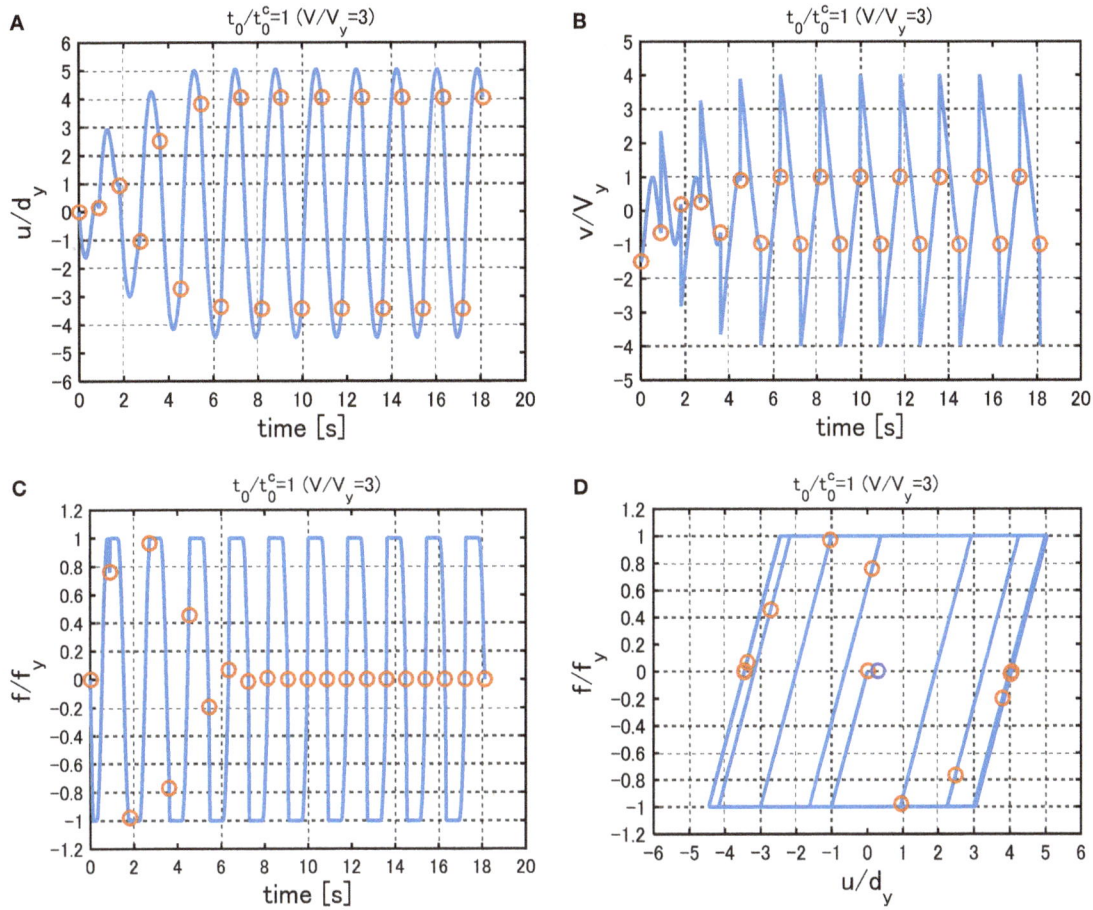

FIGURE 8 | Response to Input Sequence 1 for $V/V_y = 3$ (impulse timing is the critical one obtained by using the Input Sequence 2): (A) displacement, (B) velocity, (C) restoring force, and (D) force–deformation relation.

two consecutive impulses after the second impulse can be expressed as follows:

$$t_0^s / T_1 = (t_{OA} + t_{AB} + t_{BC}) / T_1 \quad (7a)$$

$$t_0^c / T_1 = (t_{CD} + t_{DE} + t_{EF}) / T_1 \quad (7b)$$

where t_{OA}, t_{AB}, t_{BC}, t_{CD}, t_{DE}, and t_{EF} are the times between two consecutive transition points shown in **Figure 4B**. If $V/V_y < 2$, $t_0^s / T_1 = 1/2$.

$$t_{OA} / T_1 = \{\arcsin(2 / \bar{V})\} / (2\pi) \quad (8a)$$

$$t_{AB} / T_1 = \sqrt{(\bar{V} / 2)^2 - 1} / (2\pi) \quad (8b)$$

$$t_{BC} / T_1 = 1 / 4 \quad (8c)$$

$$t_{CD} / T_1 = \{\arcsin(1 / (1 + \bar{V}))\} / (2\pi) \quad (8d)$$

$$t_{DE} / T_1 = \sqrt{(\bar{V})^2 + 2\bar{V}} / (2\pi) \quad (8e)$$

$$t_{EF} / T_1 = 1 / 4 \quad (8f)$$

In Eqs 8a–f, $\bar{V}$ denotes V/V_y.

Figures 6A,B present the normalized timing t_0^s / T_1 and t_0^c / T_1 with respect to the input level. These timings coincide with the time intervals between the points with zero restoring force (see **Figure 4**). It can be observed that the timing is delayed due to plastic deformation as the input level increases. It seems noteworthy to state again that only the critical response giving the maximum value of u_p/d_y is sought by the proposed method and the critical resonant frequency is obtained automatically without repetition for the increasing input level of the multiple impulse. One of the original points in this article is the tracking of the critical elastic–plastic response.

Correspondence of Responses Between Input Sequence 1 (Original One) and Input Sequence 2 (Modified One)

Figures 7 and **8** show the time histories of relative displacement (relative to base motion), relative velocity, restoring force, and the force–deformation relation under Input Sequence 1 with $t_0 / t_0^c = 1.0$ for $V/V_y = 2$ and $V/V_y = 3$, respectively. In **Figures 7** and **8**, $\omega_1 = 2\pi(\text{rad/s})(T_1 = 1.0 \text{ s})$ and $d_y = 0.16$ m are used. Since the steady state is very sensitive to the time increment in the time-history response analysis using an elastic–perfectly plastic

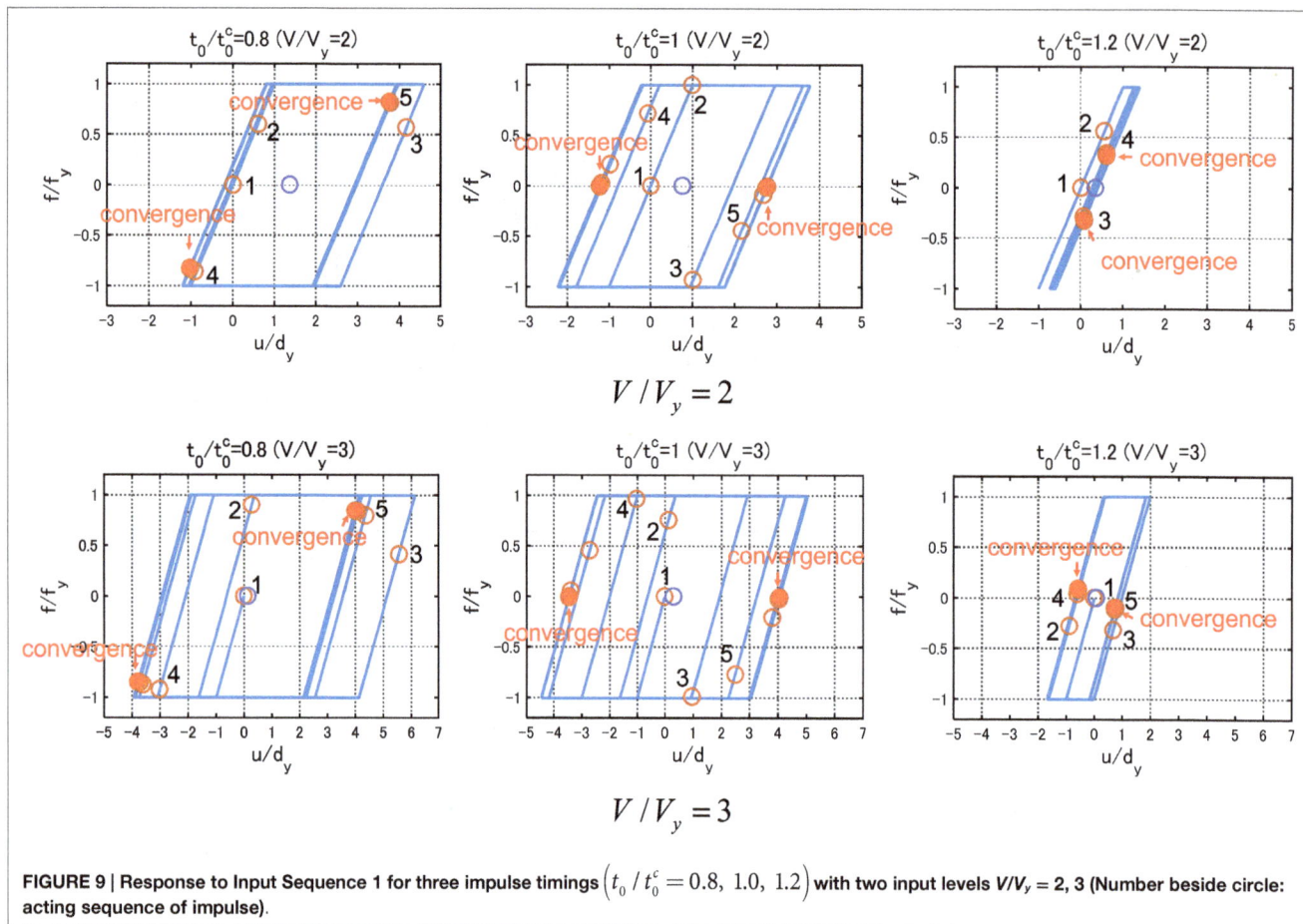

FIGURE 9 | Response to Input Sequence 1 for three impulse timings $\left(t_0 / t_0^c = 0.8,\ 1.0,\ 1.2\right)$ with two input levels $V/V_y = 2, 3$ (Number beside circle: acting sequence of impulse).

model, the time increment has been chosen as 1.0×10^{-6} s. In fact, an elastic–perfectly plastic model was not treated in most works (Caughey, 1960a,b; Iwan, 1961, 1965a,b) for its difficult treatment. It should be noted that the impulse timing is the critical one obtained by using the Input Sequence 2. The circles in **Figures 7** and **8** indicate the acting points of impulses. It can be observed that, although some irregularities appear at first, the response converges to a state with the timing of impulse at zero restoring-force point irrespective of the input level.

Figure 9 summarizes the force–deformation relation under the multiple impulse of Input Sequence 1 with the time interval $t_0 / t_0^c = 0.8, 1.0, 1.2$ for two input levels $V/V_y = 2, 3$. While in **Figures 7** and **8** only the case of $t_0 / t_0^c = 1.0$ is treated, three cases $t_0 / t_0^c = 0.8, 1.0, 1.2$ of time intervals are dealt with in **Figure 9**. It can be confirmed that the response converges to a steady state irrespective of the impulse timing and $t_0 / t_0^c = 1.0$ certainly gives the maximum plastic deformation amplitude u_p. This demonstrates the validity of introducing the Input Sequence 2 for finding the critical timing of multiple impulse even for the Input Sequence 1.

On the other hand, **Figures 10** and **11** show the time histories of relative displacement (relative to base motion), relative velocity, restoring force, and the force–deformation relation under Input Sequence 2 for $V/V_y = 2$ and $V/V_y = 3$, respectively. In **Figures 10** and **11**, $\omega_1 = 2\pi(\text{rad/s})(T_1 = 1.0\ \text{s})$

and $d_y = 0.16$ m are used. It can be observed that the realized response exhibits a steady state from the initial stage and corresponds to a state with the timing of impulse at zero restoring-force point irrespective of the input level. The critical timing t_0^c of multiple impulse computed by Eq. 7b can be obtained without repetition and can be used as the critical timing even for the Input Sequence 1.

Accuracy Check by Time-History Response Analysis Subjected to the Corresponding Multicycle Sinusoidal Input

In order to investigate the accuracy of using the multiple impulse (Input Sequence 1) as a substitute of the corresponding multicycle sinusoidal wave (representative of the long-duration ground motion input), the time-history response analysis of the elastic–plastic SDOF model under the multicycle sinusoidal wave has been conducted.

In the evaluation procedure, it is important to adjust the input level of the multiple impulse and the corresponding multicycle sinusoidal wave. This adjustment is made by using the equivalence of the maximum Fourier amplitude and a modification based on the response equivalence at some points with different input levels.

Figures 12A,B illustrate the comparison of the ground displacement and velocity between the multiple impulse and the

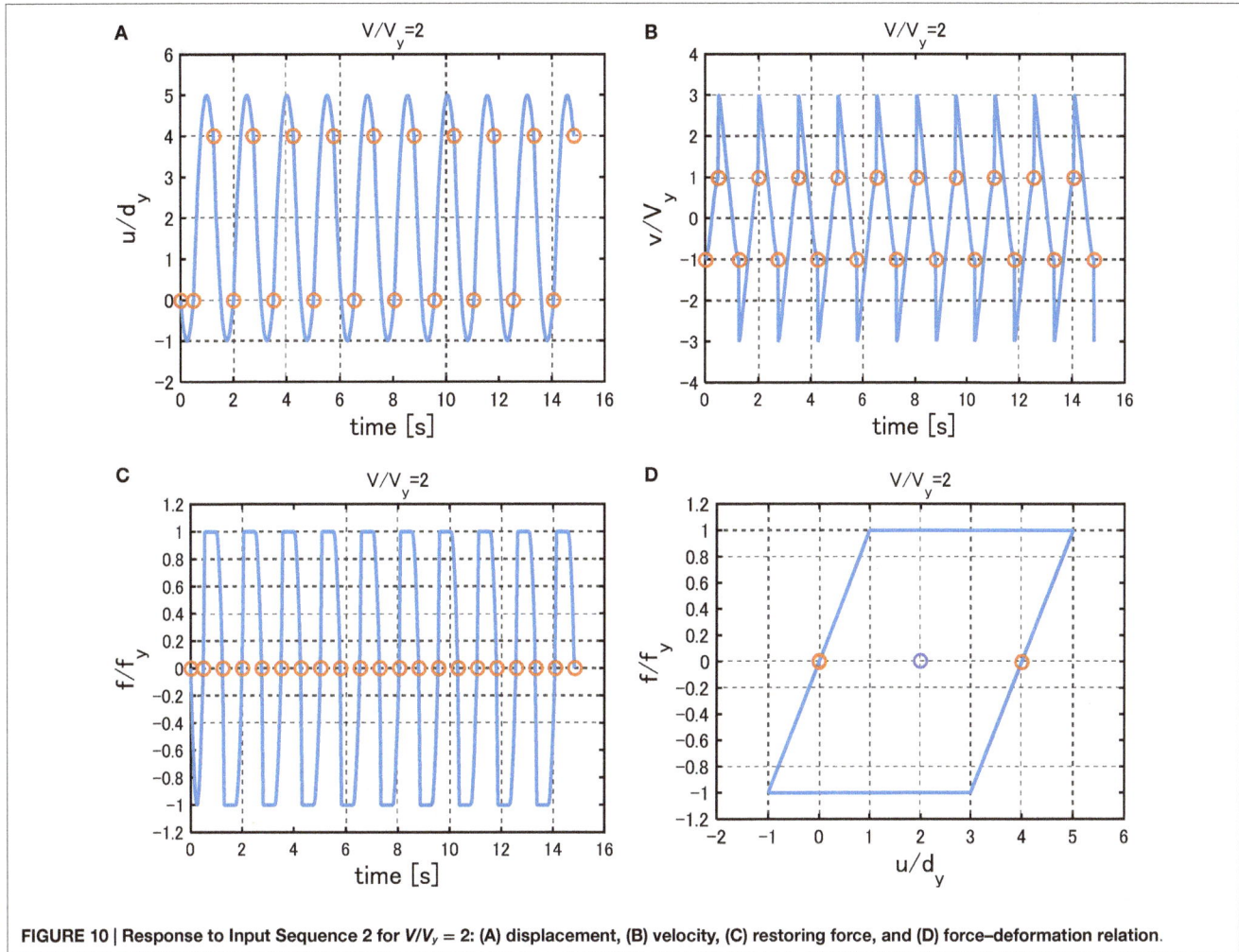

FIGURE 10 | Response to Input Sequence 2 for $V/V_y = 2$: (A) displacement, (B) velocity, (C) restoring force, and (D) force–deformation relation.

corresponding multicycle sinusoidal wave for the input level $V/V_y = 3$ In **Figures 12A,B**, $\omega_1 = 2\pi(\text{rad/s})(T_1 = 1.0 \text{ s})$ and $d_y = 0.16$ m are used. The amplitude of the sinusoidal wave has been amplified by 1.15 after both Fourier amplitudes of the sinusoidal wave and the multiple impulse are adjusted (10 cycles). This amplification factor 1.15 has been set based on the response equivalence at some points with different input levels. It should be remarked that the information on critical timing shown in **Figure 6** is incorporated in **Figure 12**.

Figure 5 presents the comparison of the maximum plastic deformation u_p/d_y and the residual displacement u_r/d_y of the elastic–plastic structure under the multiple impulse and the corresponding multicycle sinusoidal wave with respect to the input level. It can be seen that the multiple impulse provides a fairly good substitute of the multicycle sinusoidal wave in the evaluation of the maximum plastic deformation u_p/d_y if the amplitudes of both inputs are adjusted appropriately. Although the residual displacement exhibits a rather good correspondence between the multiple impulse (Input Sequence 1) and the corresponding multicycle sinusoidal wave, the Input Sequence 2 shows somewhat larger residual displacement. However, since the Input Sequence 2 is used

mainly for getting the critical timing, this discrepancy does not cause any problem.

Figure 13 shows the comparison of displacement responses to the multiple impulse (Input Sequence 1) and the corresponding sinusoidal wave for $V/V_y = 2, 3$ and $t_0 / t_0^c = 1.0$. In **Figure 13**, $\omega_1 = 2\pi(\text{rad/s})(T_1 = 1.0 \text{ s})$ and $d_y = 0.16$ m are used. It should be noted that the phase lag has been adjusted for ease in comparison. The ground displacement and velocity of the corresponding sinusoidal wave for $V/V_y = 3$ are shown in **Figure 12**. It can be observed that, although a slight difference exists in the first cycle, both responses show a fairly good correspondence in the steady state. If desired, the residual displacement can be evaluated from **Figure 5B**. As is well known, the residual displacement is sensitive to the irregularity in the input in the case of the elastic–perfectly plastic system. This issue is beyond the scope of this article.

Proof of Critical Timing

Figure 14 shows the normalized plastic deformation amplitude u_p/d_y with respect to timing of multiple impulse input for various input levels $V/V_y = 1, 2, 3, 4, 5$ (Input Sequence 1). It can be

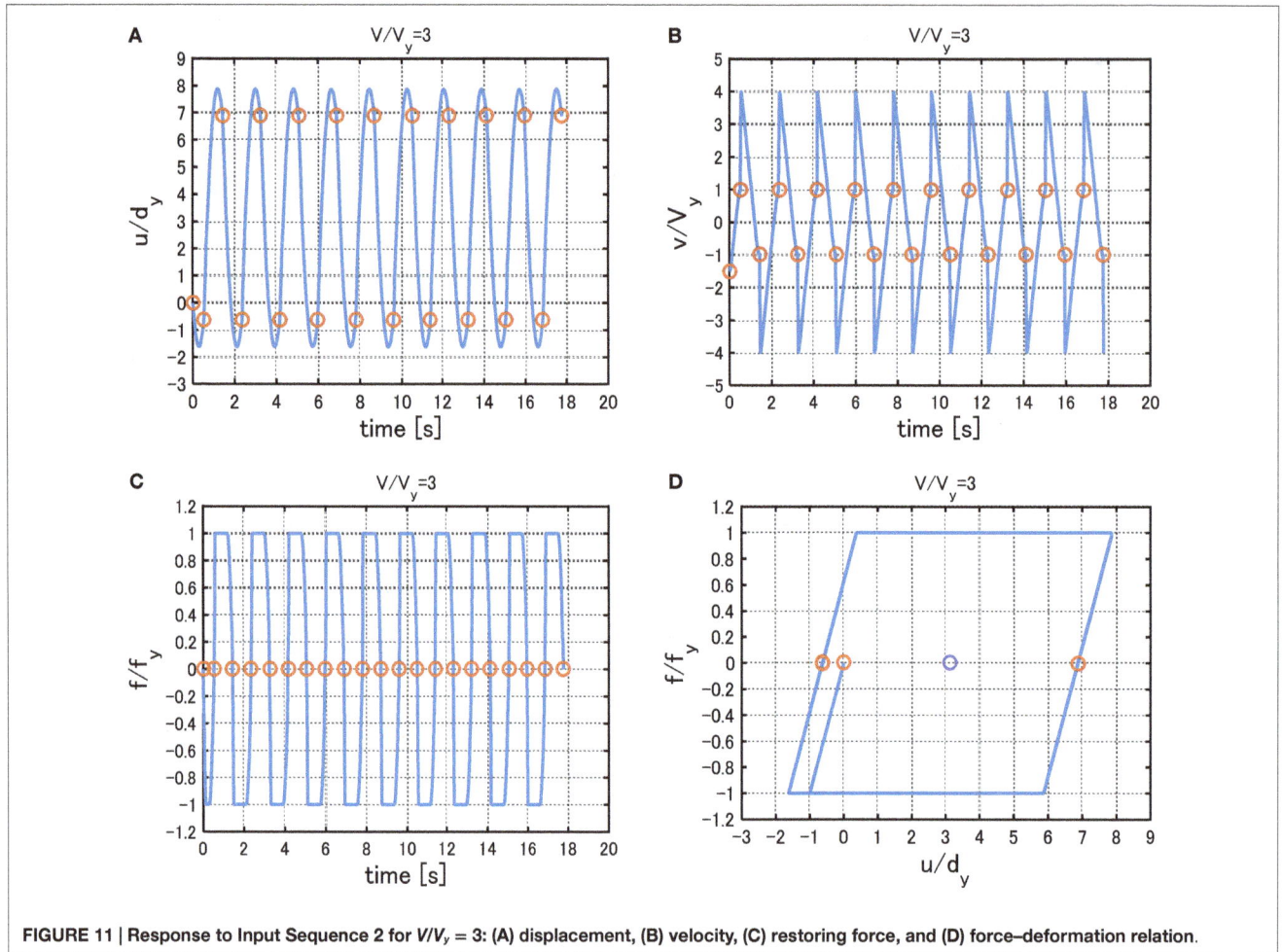

FIGURE 11 | Response to Input Sequence 2 for $V/V_y = 3$: (A) displacement, (B) velocity, (C) restoring force, and (D) force–deformation relation.

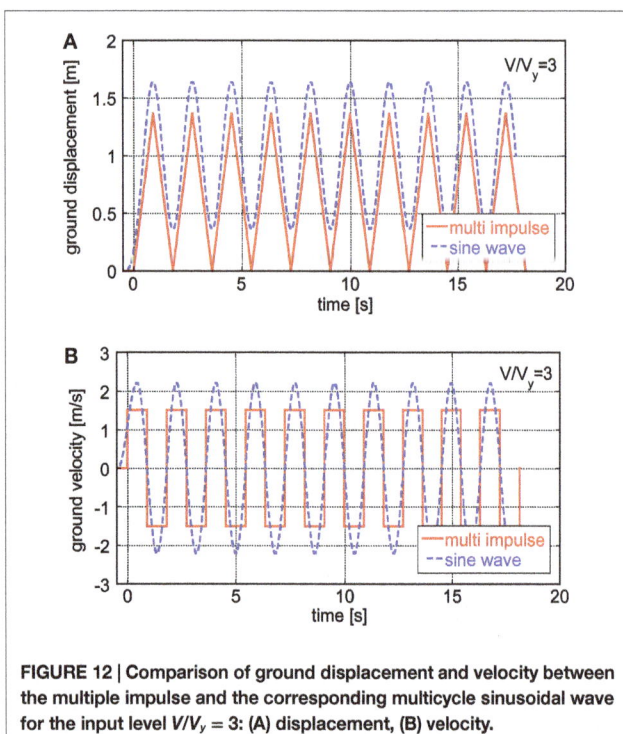

FIGURE 12 | Comparison of ground displacement and velocity between the multiple impulse and the corresponding multicycle sinusoidal wave for the input level $V/V_y = 3$: (A) displacement, (B) velocity.

confirmed that the critical timing $t_0 = t_0^c$ derived from the Input Sequence 2 provides the critical case even under Input Sequence 1. Repetitive appearance of peaks with the same amplitude indicates the existence of multiple solutions. However, the lowest timing $t_0 / t_0^c = 1.0$ is meaningful from the viewpoint of occurrence possibility of such ground motion with long duration. It is noted that the peak at the value of t_0 larger than t_0^c ($t_0 / t_0^c > 1.0$) implies the action of the second impulse after the point with zero restoring force. As the input level becomes smaller, the value t_0 / t_0^c attaining the peak becomes larger.

Conclusion

The conclusions may be summarized as follows:

(1)   The multiple impulse input has been introduced as a substitute of the long-duration earthquake ground motion, mostly expressed in terms of harmonic waves, and a closed-form solution has been derived of the elastic–plastic response of an SDOF structure under the critical multiple impulse input. It should be mentioned that the critical elastic–plastic response is treated mainly in this article.

(2)   While the critical set of input amplitude and input frequency (timing of impulse) have to be computed iteratively for the

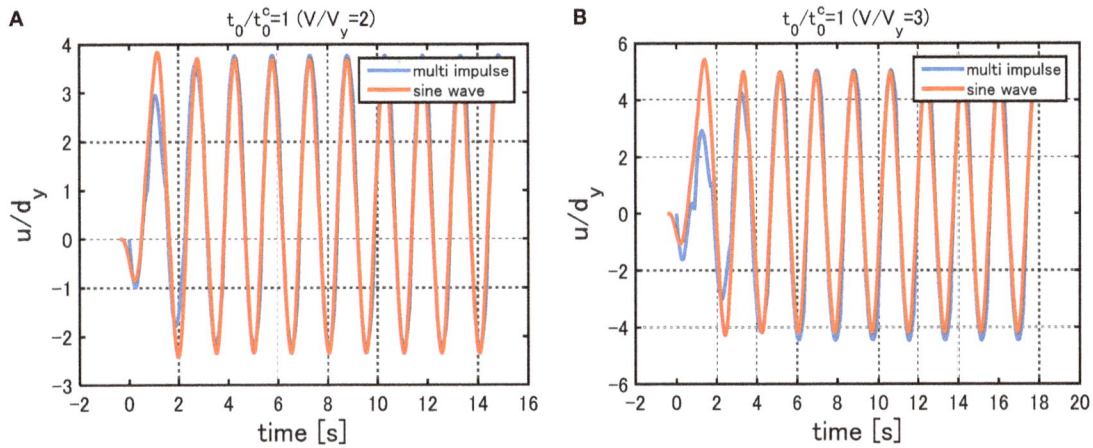

FIGURE 13 | Comparison of responses to multiple impulse and sinusoidal wave (phase lag has been adjusted): (A) $V/V_y = 2$ and (B) $V/V_y = 3$.

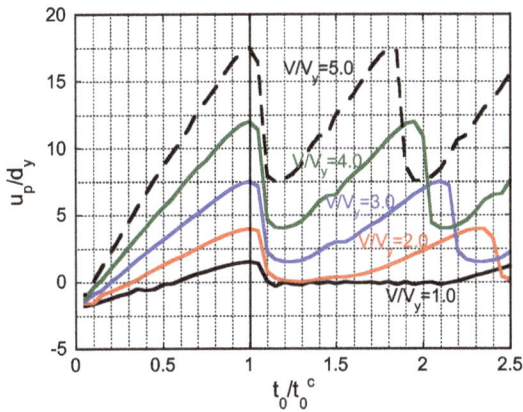

FIGURE 14 | Plastic deformation amplitude with respect to timing of multiple impulse input for various input levels (Input Sequence 1).

multicycle sinusoidal wave that can be obtained directly without iteration for the multiple impulse input by introducing a modified version (only the timing between the first and second impulses is modified so that the second impulse is given at the zero restoring force). The resonance has been proved by using energy investigation, and it has been made clear that the critical timing of the multiple impulses can be characterized as the time with zero restoring force. This decomposition of input amplitude and input frequency has overcome the long-time difficulty in finding the resonant frequency without repetition. This is one of the most original contributions in this article.

(3) It has been shown that, since only the free vibration appears in such multiple impulse input, the energy approach plays an important role in the derivation of the closed-form solution of a complicated elastic–plastic critical response. In other words, the energy approach enables the derivation of the maximum critical elastic–plastic seismic response

without solving the differential equation (equation of motion). In this process, the input of impulse is expressed by the instantaneous change of velocity of the structural mass. The maximum elastic–plastic response after impulse can be obtained by equating the initial kinetic energy computed by the initial velocity to the sum of hysteretic and elastic strain energies. It has been shown that the critical inelastic deformation and the corresponding critical input frequency can be captured by the substituted multiple impulse input depending on the input level. This is the second one of the most original contributions in this article.

(4) The validity and accuracy of the proposed theory have been investigated through the comparison with the response analysis result to the corresponding multicycle sinusoidal input as a representative of the long-duration earthquake ground motion. It has been made clear that, if the adjustment of both inputs is made by using the equivalence of the Fourier amplitude and a modification based on the response equivalence at some points with different input levels, the maximum elastic-plastic deformation to the multiple impulse exhibits a good correspondence with that to the multicycle sinusoidal wave.

(5) While the previous approaches (Caughey, 1960a,b; Iwan, 1961, 1965a,b) are aimed at constructing an equivalent linear structural model to an unchanged input (sinusoidal input) in order to enable the simple approximate computation of complicated elastic–plastic responses, the present article is aimed at finding an equivalent input model for an unchanged exact elastic–plastic model. The most significant difference between two approaches is that, while the previous approaches require the repetition both in the computation of equivalent model parameters for one input frequency and the computation of the resonant frequency giving the maximum response, the present approach does not require any repetition in the

computation of the critical input timing (resonant frequency) and the critical response. The present approach also enables the computation of the steady-state response for an elastic–perfectly plastic model which cannot be treated by the previous approaches.

References

Abbas, A. M., and Manohar, C. S. (2002). Investigations into critical earthquake load models within deterministic and probabilistic frameworks. *Earthq. Eng. Struct. Dyn.* 31, 813–832. doi:10.1002/eqe.124.abs

Alavi, B., and Krawinkler, H. (2004). Behaviour of moment resisting frame structures subjected to near-fault ground motions. *Earthq. Eng. Struct. Dyn.* 33, 687–706. doi:10.1002/eqe.370

Bertero, V. V., Mahin, S. A., and Herrera, R. A. (1978). Aseismic design implications of near-fault San Fernando earthquake records. *Earthq. Eng. Struct. Dyn.* 6, 31–42. doi:10.1002/eqe.4290060105

Bray, J. D., and Rodriguez-Marek, A. (2004). Characterization of forward-directivity ground motions in the near-fault region. *Soil Dyn. Earthq. Eng.* 24, 815–828. doi:10.1016/j.soildyn.2004.05.001

Caughey, T. K. (1960a). Sinusoidal excitation of a system with bilinear hysteresis. *J. Appl. Mech.* 27, 640–643. doi:10.1115/1.3644077

Caughey, T. K. (1960b). Random excitation of a system with bilinear hysteresis. *J. Appl. Mech.* 27, 649–652. doi:10.1115/1.3644077

Drenick, R. F. (1970). Model-free design of aseismic structures. *J. Eng. Mech. Div.* 96, 483–493.

Hall, J. F., Heaton, T. H., Halling, M. W., and Wald, D. J. (1995). Near-source ground motion and its effects on flexible buildings. *Earthq. Spectra.* 11, 569–605. doi:10.1193/1.1585828

Hayden, C. P., Bray, J. D., and Abrahamson, N. A. (2014). Selection of near-fault pulse motions. *J. Geotechnical Geoenvironmental Eng.* 140, doi:10.1061/(ASCE)GT.1943-5606.0001129

Iwan, W. D. (1961). *The Dynamic Response of Bilinear Hysteretic Systems*. Pasadena, CA: California Institute of Technology.

Iwan, W. D. (1965a). The steady-state response of a two-degree-of-freedom bilinear hysteretic system. *J. Appl. Mech.* 32, 151–156. doi:10.1115/1.3625711

Iwan, W. D. (1965b). "The dynamic response of the one-degree-of-freedom bilinear hysteretic system," in *Proceedings of the Third World Conference on Earthquake Engineering*, Vol. II. 783–796.

Kalkan, E., and Kunnath, S. K. (2006). Effects of fling step and forward directivity on seismic response of buildings. *Earthq. Spectra* 22, 367–390. doi:10.1193/1.2192560

Kalkan, E., and Kunnath, S. K. (2007). Effective cyclic energy as a measure of seismic demand. *J. Earthq. Eng.* 11, 725–751. doi:10.1080/13632460601033827

Khaloo, A. R., Khosravi1, H., and Hamidi Jamnani, H. (2015). Nonlinear interstory drift contours for idealized forward directivity pulses using "modified fishbone" models. *Adv. Struct. Eng.* 18, 603–627. doi:10.1260/1369-4332.18.5.603

Kojima, K., Fujita, K., and Takewaki, I. (2015). Critical double impulse input and bound of earthquake input energy to building structure. *Front. Built Environ.* 1:5. doi:10.3389/fbuil.2015.00005

Kojima, K., and Takewaki, I. (2015a). Critical earthquake response of elastic-plastic structures under near-fault ground motions (part 1: fling-step input). *Front. Built Environ.* 1:12. doi:10.3389/fbuil.2015.00012

Kojima, K., and Takewaki, I. (2015b). Critical earthquake response of elastic-plastic structures under near-fault ground motions (part 2: forward-directivity input). *Front. Built Environ.* 1:13. doi:10.3389/fbuil.2015.00013

Liu, C.-S. (2000). The steady loops of SDOF perfectly elastoplastic structures under sinusoidal loadings. *J. Mar. Sci. Technol.* 8, 50–60.

Mavroeidis, G. P., Dong, G., and Papageorgiou, A. S. (2004). Near-fault ground motions, and the response of elastic and inelastic single-degree-freedom (SDOF) systems. *Earthq. Eng. Struct. Dyn.* 33, 1023–1049. doi:10.1002/eqe.391

Mavroeidis, G. P., and Papageorgiou, A. S. (2003). A mathematical representation of near-fault ground motions. *Bull. Seism. Soc. Am.* 93, 1099–1131. doi:10.1785/0120020100

Moustafa, A., Ueno, K., and Takewaki, I. (2010). Critical earthquake loads for SDOF inelastic structures considering evolution of seismic waves. *Earthq. Struct.* 1, 147–162. doi:10.12989/eas.2010.1.2.147

Acknowledgments

Part of the present work is supported by the Grant-in-Aid for Scientific Research of Japan Society for the Promotion of Science (No.15H04079). This support is greatly appreciated.

Mukhopadhyay, S., and Gupta, V. K. (2013a). Directivity pulses in near-fault ground motions – I: Identification, extraction and modeling. *Soil Dyn. Earthq. Eng.* 50, 1–15. doi:10.1016/j.soildyn.2013.02.017

Mukhopadhyay, S., and Gupta, V. K. (2013b). Directivity pulses in near-fault ground motions – II: estimation of pulse parameters. *Soil Dyn. Earthq. Eng.* 50, 38–52. doi:10.1016/j.soildyn.2013.02.017

Roberts, J. B., and Spanos, P. D. (1990). *Random Vibration and Statistical Linearization*. New York, NY: Wiley.

Rupakhety, R., and Sigbjörnsson, R. (2011). Can simple pulses adequately represent near-fault ground motions? *J. Earthq. Eng.* 15, 1260–1272. doi:10.1080/136324 69.2011.565863

Sasani, M., and Bertero, V. V. (2000). "Importance of severe pulse-type ground motions in performance-based engineering: historical and critical review," in *Proceedings of the Twelfth World Conference on Earthquake Engineering* (Auckland, New Zealand).

Takewaki, I. (1996). Design-oriented approximate bound of inelastic responses of a structure under seismic loading. *Comput. Struct.* 61, 431–440. doi:10.1016/0045-7949(96)00086-7

Takewaki, I. (1997). Design-oriented ductility bound of a plane frame under seismic loading. *J. Vib. Control* 3, 411–434. doi:10.1177/107754639700300404

Takewaki, I. (2002). Robust building stiffness design for variable critical excitations. *J. Struct. Eng.* 128, 1565–1574. doi:10.1061/(ASCE)0733-9445(2002)128:12(1565)

Takewaki, I. (2007). *Critical Excitation Methods in Earthquake Engineering*, 2nd Edn in 2013. Oxford: Elsevier.

Takewaki, I., Moustafa, A., and Fujita, K. (2012). *Improving the Earthquake Resilience of Buildings: The Worst Case Approach*. London: Springer.

Takewaki, I., Murakami, S., Fujita, K., Yoshitomi, S., and Tsuji, M. (2011). The 2011 off the Pacific coast of Tohoku earthquake and response of high-rise buildings under long-period ground motions. *Soil Dyn. Earthq. Eng.* 31, 1511–1528. doi:10.1016/j.soildyn.2011.06.001

Takewaki, I., and Tsujimoto, H. (2011). Scaling of design earthquake ground motions for tall buildings based on drift and input energy demands. *Earthq. Struct.* 2, 171–187. doi:10.12989/eas.2011.2.2.171

Vafaei, D., and Eskandari, R. (2015). Seismic response of mega buckling-restrained braces subjected to fling-step and forward-directivity near-fault ground motions. *Struct. Des. Tall Spec. Build.* 24, 672–686. doi:10.1002/tal.1205

Xu, Z., Agrawal, A. K., He, W.-L., and Tan, P. (2007). Performance of passive energy dissipation systems during near-field ground motion type pulses. *Eng. Struct.* 29, 224–236. doi:10.1016/j.engstruct.2006.04.020

Yamamoto, K., Fujita, K., and Takewaki, I. (2011). Instantaneous earthquake input energy and sensitivity in base-isolated building. *Struct. Des. Tall Spec. Build.* 20, 631–648. doi:10.1002/tal.539

Yang, D., and Zhou, J. (2014). A stochastic model and synthesis for near-fault impulsive ground motions. *Earthq. Eng. Struct. Dyn.* 44, 243–264. doi:10.1002/eqe.2468

Zhai, C., Chang, Z., Li, S., Chen, Z.-Q., and Xie, L. (2013). Quantitative identification of near-fault pulse-like ground motions based on energy. *Bull. Seism. Soc. Am.* 103, 2591–2603. doi:10.1785/0120120320

Conflict of Interest Statement: The authors declare that the research was conducted in the absence of any commercial or financial relationships that could be construed as a potential conflict of interest.

8

On the Value of Earthquake Scenario: The Kathmandu Recent Lesson

Philippe Guéguen[1]*, Hugo Yepes[1,2] and Ismael Riedel[1]

[1] *Institute of Earth Science, Université Grenoble Alpes/CNRS/IFSTTAR, Grenoble, France,* [2] *Instituto Geofísico, Escuela Politécnica Nacional, Quito, Ecuador*

Keywords: Nepal earthquake, urban seismology, economic losses, scenarios, seismic risk

Edited by:
Gian Paolo Cimellaro,
University of California Berkeley, USA

Reviewed by:
Sean Wilkinson,
Newcastle University, UK
Fabrizio Mollaioli,
Sapienza University of Rome, Italy

*Correspondence:
Philippe Guéguen
philippe.gueguen@ujf-grenoble.fr

The past two decades have been punctuated by large-scale natural events that produced huge losses whether related to hydrological, atmospheric, or even rare geological hazards. Over the second half of the last century, the total cost of such catastrophes has been multiplied by a factor of 15, clocking up economic losses of around 66 billion dollars per year during the 1990s (Benson and Clay, 2004). Among these phenomena, there are those that cause disasters, i.e., corresponding to infrequent events that have major consequences on the well-being of the region's population, environment, institutions, and financial equilibrium. The predisposition of a region to suffer an infrequent natural disaster is measured by the event's capacity to generate losses that exceed 1% of GNP, thus resulting in a slow, difficult economic recovery (Munich Re, 2002). According to this definition, geological-related disasters stand out from other natural events: they represent approximately 15% of the world's natural disasters but account for one-third of all victims and economic losses (World Conference on the Disaster Reduction, 2004). The 2000s were not spared either, with the Indonesian earthquake in 2004, and the earthquakes in Chile and Haiti in 2010. Together, these two quakes generated losses of around 40 billion dollars, and more than 280,000 victims, i.e., 31 and 80% of economic and human losses caused by natural events respectively, even though earthquakes only represented 6% of disasters in 2010 (Daniell, 2010). After the 1995 Kobe earthquake that caused economic losses of 178 billion dollars (IFRC, 2002), the 2011 Tohoku earthquake in Japan is known as being the event that caused the greatest direct and indirect costs, on a level to match the scale of the earthquake itself (Mw = 9): a direct economic impact of around 187 billion dollars was estimated, while indirect sanitary, ecological and economic costs related to the ensuing nuclear disaster are expected to reach a long-standing record level. Such observations should serve as a reminder that although public policies are paying more attention to phenomena related to global climate change, earthquakes still remain the natural events that are most likely to have disastrous consequences. Unlike floods or storms that, although likely to increase in frequency and intensity in the years to come in parallel with the climate change, leave us time to analyze future scenarios, earthquakes are already causing huge disasters now. Looking from a different perspective, Brauman (2010) has analyzed recent years human emergencies from a medical emergency actor's point of view and concluded that earthquakes pose the first order threat to human life that ranks far above acute climatic events occurring near densely populated areas.

The Mw 7.8 Kathmandu earthquake on April 25, 2015 has reminded us of the inequality of populations facing earthquakes (Coburn and Spence, 2002). Economic and human losses obviously not only depend upon the amplitude and severity of the seismic vibrations but also upon the quality of constructions and the financial investment put into direct efforts to design buildings and to improve knowledge of seismic hazard (Ohta et al., 1986). In Nepal, the preliminary economic estimate predicts losses representing approximately 18% of its GNP (CEDIM Forensic Disaster Analysis Group, CATDAT, and Earthquake-Report. com, 2015), which classes this earthquake as a disaster. The region of Kathmandu is exposed to a high seismic hazard (Sapkota et al., 2013; Bilham et al., 2001). The earthquake on April 25, 2015 was a major event that followed on from a long series of events affecting the region along the Himalayan ranges, each time causing numerous fatalities: e.g., the 1905 earthquake (Mw = 7.8) in the Kangra region of India, north-west of Nepal caused 20,000 deaths, 15,000 deaths were accounted for in 1934 (Mw = 8.1) in the Bihar Nepal region (in the south-eastern part of the country), and the Kashmir earthquake in 2005 (Mw = 7.6) killed 75,000 people. Low probability/high consequences events are still expected. These events can be at the first order compared with the black swan theory proposed by Taleb (2010), which illustrates a cognitive bias that leads us to the erroneous conclusion that rare phenomena will not happen. People and policy makers may pass over their existence because the return periods are not of the same order of magnitude as the time span of the human life. Moreover, if such events are identified, a monetary cost/benefit analysis of earthquake engineering practices may not be in a positive balance. Incorrectly considering these events as black swans can bring important effects in terms of protection, regulation, and resilience. In fact, only observation over periods longer than the characteristic recurrence time of the phenomena, i.e., on the scale of geological time, can confirm the non-existence of such events (Bilham et al., 2001; Bollinger et al., 2014). However, this is not feasible and, by default, our reasoning is built upon incomplete information, resulting in erroneous predictions. As time passes, we realize that we are getting closer and closer to the appearance of a black swan, particularly since larger earthquakes are possible with even longer return periods.

At the same time, the urban population in the Kathmandu valley has been increasing by around 3.6% per year (Fort, 2014) over the past decade, reaching 1.5 million people. Since the return period of major earthquakes is so much longer than the periods of recent rapid urbanization, and Jackson (2006) and Holzer and Savage (2013) proclaim that major catastrophes lie ahead. In the Himalaya region, Bilham et al. (2001) projected the urban population growth to one of the possibly overdue Himalayan earthquakes with the same characteristics as the 1905 historical earthquake that yields 200,000 predictable fatalities. Rapid urbanization also amplifies the risk of disasters due to the complex combination of demographic concentration, social exclusion and poverty, accompanied by ignorance of the risks (Fort, 2014). The physical vulnerability of the constructions that ultimately causes fatalities becomes all the greater as the migratory influx presses for new dwellings and infrastructure, resulting in a strong demand for large numbers of buildings to be erected quickly, often at the expense of quality and safety. This pressure thus results in an inappropriate use of space, a low level of compliance with regulations and a good practice that would minimize damage and operation interruptions, and fewer possibilities for transferring or smoothing the risk as consequence

of the concentration of goods and decision-making bodies in a single city, which most limits country recovery.

Faced with this situation, and in an attempt to reduce risk, the city of Kathmandu, like many others before (such as Quito, for example), has been analyzed in terms of its seismic risk. At the initiative of GeoHazard International, and with backing from the World Bank and UNESCO, a seismic risk management project was launched (Dixit et al., 1998), with the objectives of educating the public, producing a seismic scenario to simulate losses and operational problems, and implementing an action plan to manage seismic risk. The concept of community for risk assessment and resilience improvement is largely discussed in social sciences (Marsh and Buckle, 2001) and these scenarios participate to this scientific concept. Like Corneiro (2006) who comments that funding is mostly devoted to enhancing numerical modeling of structures rather than to actual and effective solutions of reinforcements, we could question the wisdom of concentrating resources on modeling and representing phenomena rather than implementing building reinforcement actions. However, a study by the USGS claims that a 40-million dollar investment in worldwide prevention measures in the 1990s could have reduced economic losses by 280 million dollars (Benson and Twigg, 2004). Other examples concerning natural disasters also provide favorable ratios [cost of prevention actions to loss-reduction benefits] of around 1:3. Scenarios enable the transfer of scientific knowledge in an understandable manner to local decision-makers and let populations to become aware of the risks and to realize that it is preferable to anticipate natural disaster rather than just to respond to a dramatic event. They also support the view of emergency specialists, who are increasingly insistent on the need to invest in preparation, prevention, and disaster attenuation, particularly in view of the disproportional amounts that international organizations are prepared to spend on emergency rescue and recovery operations conducted hurriedly. Moreover, seismic risk analyses are useful to study the best investment framework for the seismic retrofit of buildings. Several cost-benefit explorations for earthquake damage mitigation showed the attractiveness of retrofitting actions on buildings for long return periods events (Smyth et al., 2004). In general, such studies suggest that retrofitting is desirable (cost-effective) in high seismic hazard regions for all but the very shortest time horizons. These reinforcement investments are on average relatively small compared to the repair and replacement cost of the physical damage subsequently avoided. The economic losses and fatalities following Nepal's earthquake could have been largely reduced if retrofitting actions would have been taken in advance, which is certainly the most difficult action to take. The savings in term of economic losses and human casualties might have been enormous by investing a small portion of the global emergency aid to recovery after the disaster.

Based on the 1934 earthquake, the Kathmandu scenario (Dixit et al., 1998) predicted 40,000 victims, 95,000 people injured, 600,000 left homeless, and the destruction of approximately 60% of buildings. In spite of the intensity of the last April earthquake, observations suggest that the actual figures are substantially lower. We know that the uncertainty of the ground motion prediction model controls the uncertainties of the risk model. The only ground motion record in Kathmandu from the Mw 7.8 earthquake yielded around 0.2 g, i.e., a relatively low value in comparison with the predictions used for the earthquake scenario. It is also characterized by maximum amplitude at a vibration period (about 2 s) far from most of the assumed resonance periods of the city's buildings. Very often, seismic ground motions recorded during major earthquakes by seismic networks surprise seismologists and ask about the physical reasons behind them. On the other hand, same reasons are always claimed to explain observations as for Kathmandu (Goda et al., 2015): the most vulnerable buildings suffer more damage, the poorest suburbs without engineering design are the most damaged areas, and design defects are at the origin of most of damaged buildings. However, when the Mw 7.8 earthquake was announced, based on knowledge of the Dixit et al. (1998) scenario, the scientific community expected a disaster at a similar scale as that of the Haiti earthquake of 2010. The Mw 7.3 earthquake located 25 km from Port-au-Prince caused an economic and human disaster, resulting in more than 230,000 deaths, 60% of buildings destroyed, and total losses estimated at 120% of the GNP according to the IMF. Five years later, the country remains dependent upon international aid and the political situation is still equally fragile, due to the postponement of the presidential elections, which are to be funded by foreign countries, threatening the autonomy, independence and democratic sovereignty of the country. The future remains uncertain for Haitians: most schools and universities were destroyed compromising the renewal of the country's active population in the time to come. There are many similarities between these two disasters: in 2010, the human development index in Haiti and Nepal were similar [0.45 for the former and 0.458 for the latter, source United Nations Development Programme (2010)], both had experienced considerable population growth over recent decades and seismic rates are equivalent in both countries. Once again the earthquake was not considered, as in the black swan theory, in the sense that several past earthquakes had already occurred while less densely populated at the time. Unlike Kathmandu, however, no exhaustive scenario had been produced in Haiti, and it is much too easy to postulate that this could be one reason for the smaller scale of the Nepalese disaster. Moreover, it is still too soon to fully understand what happened but this earthquake failed to entirely rupture the locked fault close to Kathmandu and a large earthquake appears to be inevitable in future (Bilham, 2015; Avouac et al., 2015).

A different factor that may influence the resilience of a community faced with a natural disaster is the degree of corruption. Ambraseys and Bilham (2011) calculated that 83% of all deaths from building collapse during earthquakes over the past 30 years occurred in countries that are anomalously corrupt. The Corruption Perceptions Index (Transparency International, 2014) ranks countries and territories based on how corrupt their public sector is perceived to be. It is difficult to make an accurate link of the impact of corruption in Nepal in terms of

building collapse but it is surprising to note that for these two countries with equivalent development index, according to the corruption perception index CPI of Transparency International (2014) Haiti in 2010 was ranked 146th in the World while Nepal in 2014 was ranked 126th. Corruption prevents that the money spent in the construction process goes to earthquake-resistant practices and at the end one wonders if the solution is rather to invest in education and training of populations as the most efficient solution for reducing earthquake disasters as described by Twigg (2009).

Over the last century, urbanization growth has dramatically increased the risk in seismic prone areas and many cities in the world have not sufficiently improved their resilience. In this context, we may indeed wonder about the need of producing catastrophe scenarios. In recent years, probability-based prediction solutions have become more popular than deterministic solutions; they integrate all the uncertainties related to the phenomenon in terms of occurrence, including intensity and location, as well as the expected ground motion. However, practical implementation in a country like Nepal is problematic and so representing a specific event (i.e., deterministic approach) remains an essential vector of information and education. The shortcomings of seismic scenarios in that they may never successfully predict the observed consequences of earthquakes (due to the inherent uncertainties of the hazard) may cause maladaptive behavior. For example, the Dixit et al. (1998) scenario predicted more victims: as fewer were observed, as fewer were observed, it is a concern that the population may reject the scientific evidence and decision makers may lose confidence in such tools, while seismological analysis suggests that significant seismic risk remains (Bilham, 2015); however, these scenarios remain essential tools for risk management, preparedness and for increasing the resilience of communities. Based on these scenarios, local initiatives at the community level (Twigg, 2009) are launched, and their representation provides essential support for the long-term reduction of seismic risk in public policy on time spans that are much longer than political mandates. The full implementation of mitigation strategies must however face the challenges of insufficient resources, which is generally cited as being the major obstacle to any prevention policy. The role of scenario in determining impact level is a critical issue (Alexander, 2000). In addition to our current investment in emergency measures, we need to increase investment in resilience, which in time will allow a reduction in relief expenditures and better means of ensuring the economic, political, democratic, and social stability of a country. We need to ensure that investments in disaster reduction measures are on a scale that matches the risks and in developing countries seismic risk scenarios are still important tools in achieving this.

ACKNOWLEDGMENTS

This work has been supported by a grant from Labex OSUG@2020 (Investissements d'avenir – ANR10 LABX56).

REFERENCES

Alexander, D. (2000). *Confronting Catastrophe – New Perspectives on Natural Disasters*. New York, NY: Oxford University Press.

Ambraseys, N., and Bilham, R. (2011). Corruption kills. *Nature* 469, 153–155. doi:10.1038/469153a

Avouac, J. P., Meng, L., Wei, S., Wang, T., and Ampuero, J. P. (2015). Lower edge of locked main Himalayan thrust unzipped by the 2015 Gorkha earthquake. *Nat. Geosci.* 8, 708–711. doi:10.1038/ngeo2518

Benson, C., and Clay, J. E. (2004). *Understanding the Economic and Financial Impacts of Natural Disasters, Disaster Risk Management Series 4*. The World Bank Edition, Washington, DC, 134.

Benson, C., and Twigg, J. (2004). *Measuring Mitigation: Methodologies for Assessing Natural Hazard Risks and the Net Benefits of Mitigation – A Scoping Study*. Synthesis Report IFRC/ProVention Consortium. Geneva, 154.

Bilham, R. (2015). Seismology: raising Kathmandu. *Nat. Geosci.* 8, 582–584. doi:10.1038/ngeo2498

Bilham, R., Gaur, V. K., and Molnar, P. (2001). Himalayan seismic hazard. *Science* 293, 1442–1444. doi:10.1126/science.1062584

Bollinger, L., Sapkota, S. M., Tapponnier, P., Kilinger, Y., Rizza, M., and Van der Woerd, J. (2014). Estimating the return times of great Himalayan earthquakes in Eastern Nepal: evidence from the Patu and Bardibas strands of the main frontal thrust. *J. Geophys. Res. Solid Earth* 119, 7123–7163. doi:10.1002/2014JB010970

Brauman, R. (2010). *Natural Disasters: Do Something! (Entretiens)*. Available at: http://msf-crash.org/livres/agir-a-tout-prix/catastrophes-naturelles-do-something

CEDIM Forensic Disaster Analysis Group, CATDAT, and Earthquake-Report. com. (2015). *Nepal Earthquake – Report #1 27.04.2015–Situation Report No. 1*. Available at: https://www.cedim.de/download/CEDIM_ImpactSummary_EarthquakeNepal2015_Report1.pdf

Coburn, A., and Spence, R. (2002). *Earthquake Protection*, Second Édition. Bognor Regis: John Wiley and Sons.

Corneiro, M. C. (2006). Can buildings be made earthquake-safe? *Science* 312, 204. doi:10.1126/science.1126302

Daniell, J. (2010). "Damaging Earthquakes Database 2010–The Year in Review," in *CEDIM Earthquake Loss Estimation Series Research Report 2011-01*, 41. Available at: http://earthquake-report.com/wp-content/uploads/2011/03/CATDAT-EQ-Data-1st-Annual-Review-2010-James-Daniell-03-03-2011.pdf

Dixit, A. M., Dwelly-Samant, L., Nakarmi, M., Tucker, B., and Pradhanang, S. B. (1998). *The Kathmandu Valley Earthquake Management Action Plan*. Lalitpur: National Society for Earthquake Technology-Nepal (NSET). Available at: http://geohaz.org/wp/wp-content/uploads/2010/04/KathmanduValleyEQRiskMgtActionPlan.pdf

Fort, M. (2014). *La difficile gestion des risques naturels en Himalaya: une question d'échelle? Le cas du Népal*. BAGF – Géographies, 3, 241–256. Available at: https://www.researchgate.net/publication/268802067_Monique_Fort_BAGF_2014-3_pp_241-256

Goda, K., Kiyota, T., Pokhrel, R., Chiaro, G., Katagiri, T., Sharma, K., et al. (2015). The 2015 Gorkha Nepal earthquake: insights from earthquake damage survey. *Front. Built. Environ* 1:1–15. doi:10.3389/fbuil.2015.00008

Holzer, T. L., and Savage, J. C. (2013). Global earthquake fatalities and population. *Earthquake Spectra* 29, 155–175. doi:10.1193/1.4000106

IFRC. (2002). *World Disasters Report 2001: Focus on Community Resilience*. Geneva: International Federation of Red Cross and Red Crescent Societies (IFRC), 244.

Jackson, J. (2006). Fatal attraction: living with earthquakes, the growth of villages into megacities, and earthquake vulnerability in the modern world. *Philos. Trans. R. Soc.* 364, 1911–1925. doi:10.1098/rsta.2006.1805

Marsh, G., and Buckle, P. (2001). Community: the concept of community in the risk and emergency management context. *Aust. J. Emerg. Manage.* 16, 5–7.

Munich Re. (2002). *Topics Annual Review: Natural Catastrophes 2002*. Munich. Available at: http://ipcc-wg2.gov/njlite_download.php?id=6219

Ohta, Y., Ohashi, H., and Kagami, H. (1986). "A semi-empirical equation for estimating occupant casualty in an earthquake," in *Proceedings of the 8th European Conference on Earthquake Engineering*, Lisbon, 2–3, 81–88.

Sapkota, S., Bollinger, L., Klinger, Y., Tapponnier, P., Gaudemer, Y., and Tiwari, D. (2013). Primary surface ruptures of the great Himalayan earthquake in 1934 and 1255. *Nat. Geosci.* 6, 71–76. doi:10.1038/NGEO1669

Smyth, A. W., Altay, G., Deodatis, G., Erdik, M., Franco, G., Gulkan, P., et al. (2004). Probabilistic benefit-cost analysis for earthquakevdamage mitigation: evaluating measures for apartment houses in Turkey. *Earthquake Spectra* 20, 171–203. doi:10.1193/1.1649937

Taleb, N. N. (2010). *The Black Swan: the Impact of the Highly Improbable*. New York, NY: Random House Ed, A Random House trade paperback Coll.

Transparency International. (2014). *The Corruption Perceptions Index 2014*. Available at: http://www.transparency.org/cpi2014/results

Twigg, J. (2009). *Characteristics of a Disaster-Resilient Community: A Guidance Note (version 2)*. Available at: http://discovery.ucl.ac.uk/1346086/

United Nations Development Programme. (2010). *Human Development Report 2010*. New York, NY: Palgrave Macmillan. Available at: http://hdr.undp.org/sites/default/files/reports/270/hdr_2010_en_complete_reprint.pdf

World Conference on the Disaster Reduction. (2004). "Draft Annotated Outline of the Review of the Yokohama Strategy and Plan of Action," in *Report A/CONF.206/PC(II)/3 World Conference on the Disaster Reduction* (Kobe). Available at: http://www.unisdr.org/2005/wcdr/preparatory-process/prepcom1/pc1-Draft-annotated-outline-review-yokohama-strategy-english.pdf

Conflict of Interest Statement: The authors declare that the research was conducted in the absence of any commercial or financial relationships that could be construed as a potential conflict of interest.

9

Seismic behavior and design of wall–EDD–frame systems

Oren Lavan * *and David Abecassis*

Faculty of Civil and Environmental Engineering, Technion – Israel Institute of Technology, Haifa, Israel

Walls and frames have different deflection lines and, depending on the seismic mass they support, may often possess different natural periods. In many cases, wall–frame structures present an advantageous behavior. In these structures, the walls and the frames are rigidly connected. Nevertheless, if the walls and the frames were not rigidly connected, an opportunity for an efficient passive control strategy would arise: connecting the two systems by energy dissipation devices (EDDs) to result in wall–EDD–frame systems. This, depending on the parameters of the system, is expected to lead to an efficient energy dissipation mechanism. This paper studies the seismic behavior of wall–EDD–frame systems in the context of retrofitting existing frame structures. The controlling non-dimensional parameters of such systems are first identified. This is followed by a rigorous and extensive parametric study that reveals the pros and cons of the new system versus wall–frame systems. The effects of the controlling parameters on the behavior of the new system are analyzed and discussed. Finally, tools are given for initial design of such retrofitting schemes. These enable both choosing the most appropriate retrofitting alternative and selecting initial values for its parameters.

Keywords: seismic retrofitting, energy dissipation devices, passive control, viscous dampers, wall–frame systems

Edited by:
Nikos D. Lagaros,
National Technical University of
Athens, Greece

Reviewed by:
Iolanda-Gabriela Craifaleanu,
Technical University of Civil
Engineering Bucharest, Romania
Vagelis Plevris,
School of Pedagogical and
Technological Education, Greece

*Correspondence:
Oren Lavan,
Faculty of Civil and Environmental
Engineering, Technion – Israel Institute
of Technology, 837 Rabin Building,
Haifa 32000, Israel
lavan@tx.technion.ac.il

Introduction

Many of the relatively new buildings located in seismic regions were designed according to stringent seismic codes. These are expected to perform relatively well in seismic events. Contrariwise, many older existing buildings have known deficiencies. These buildings are still expected to be a part of the landscape for many years to come. Seismic retrofitting of such buildings may reduce their probability of collapse as well as the level of damage expected to them in seismic events. This, in turn, may shorten the time required to bring them to normal functionality (Nakashima et al., 2014).

A very efficient and promising approach for seismic retrofitting and damage control makes use of energy dissipation devices (EDDs) [see, e.g., Soong and Dargush (1997), Christopoulos and Filiatrault (2006), and Takewaki (2009)]. Out of those, fluid viscous dampers (FVDs) have been shown to effectively reduce both displacement and force related seismic responses of structures (Constantinou and Symans, 1992; Lavan and Dargush, 2009; Lavan, 2012). Hence, the use of FVDs seems natural for seismic retrofitting where both displacements and forces due to earthquakes are to be decreased. Indeed, optimal design of such dampers for the purpose of seismic retrofitting of frame structures received much attention (Zhang and Soong, 1992; Gluck et al., 1996; Takewaki, 1997; Singh and Moreschi, 2001; Lopez-Garcia and Soong, 2002; Dargush and Sant, 2005; Lavan and Levy, 2005, 2006, 2009, 2010; Aydin et al., 2007; Lavan et al., 2008; Lavan and Dargush, 2009; Adachi et al., 2013; Aguirre et al., 2013; Kanno, 2013; Martínez et al., 2013; Sonmez et al., 2013; Gidaris and Taflanidis, 2014; Hatzigeorgiou and Pnevmatikos, 2014; Lavan and Amir, 2014; Lin et al., 2014;

Lavan, 2015). To allow a quick examination of whether such devices are a good alternative, simple methodologies for initial design were proposed as well (FEMA 356, 2000; Palermo et al., 2013; Landi et al., 2014; Rama Raju et al., 2014). Uncertainty in structural and/or ground motion parameters was also evaluated (Lavan and Avishur, 2013; Peng et al., 2014; Tubaldi et al., 2015) or taken into account as part of the design process (Gidaris and Taflanidis, 2014).

When seismic retrofitting of frame structures is the sole concern, the abovementioned research presents a wide arsenal of design tools in the hands of the engineer. Nevertheless, in Israel, seismic retrofitting of old buildings is usually done in parallel to adding shelter rooms to protect the tenants from potential threats from various weapons. These shelter rooms are made of reinforced concrete and their in-plan location in the various stories is consistent so as to form reinforced concrete cores. These new cores are rigidly connected to the existing floors, thus, they supply a new lateral load resisting system (see **Figure 1A**). This considerably reduces the displacements of the existing structure, hence the internal forces in some existing structural elements. However, the high stiffness of the cores results in a relatively short natural period. Thus, for most ground motion records, i.e., relatively short predominant period ones, much larger seismic inertia forces are attracted. These forces that act on, and are transferred by the existing slabs, are potentially larger than they can take. Furthermore, the path that the inertia forces take from the mass to the new lateral load resisting system is different from before. As the new cores are located at the edges of the building, the path now is much longer. Thus, the slabs need to transfer loads between points that are at a large distance, i.e., the distance between the center of mass and the lateral load resisting elements. This may require strengthening of the existing slabs, which is not practical. In addition, a major portion of the large inertia forces is to be transferred through the connections between the new cores and the existing slabs. As the existing slabs are usually thin and their concrete quality is relatively low, these connections may be problematic. Furthermore, very large horizontal forces

are to be carried by the new core system. This may result in large base moments. These grow considerably with the number of stories. As the new cores support gravity loads due to their self-weight only, large moments may lead to considerable tension in their foundations. Depending on the soil type, this may lead to additional considerable expenses. Similarly, the combination of large moments and small compression forces in the cores leads to larger reinforcement ratios. Finally, when it comes to important buildings that are expected to function after the earthquake, the large accelerations expected due to the considerable increase in stiffness is also an issue due to the potential damage they could induce.

The new cores, as they are rigidly connected to the existing building, considerably increase the inertia forces. Nevertheless, if they were not rigidly connected, their natural period would have been much different from that of the existing building. Furthermore, their deflection line would have been different in nature from that of the frames: the deflection lines of wall structures possess larger drifts at the top stories while those of frame structures possess larger drifts at the bottom stories. This opens an opportunity for an efficient passive control strategy. If the two systems are not rigidly connected (e.g., by using seismic gaps), the relative displacements and velocities between them are expected to be large. Thus, connecting the two systems by EDDs to result in wall–EDD–frame systems (see **Figure 1B** and the detailed description in Section "Structural Systems") is expected to lead to an efficient energy dissipation mechanism. This, of course, depends on the parameters of the system (wall and frame stiffnesses; height, distribution of mass between the two systems, etc.) as well as on the parameters of the EDDs used (damping coefficients and stiffness). While the motivation for research on this system stems from the particular environment in Israel, the wall–EDD–frame system and the research presented herein are fully applicable in other cases. It should be noted that a related system, where the walls were assumed infinitely rigid, was proposed and studied by Trombetti and Silvestri (2004, 2006) and Silvestri and Trombetti (2007). They noted

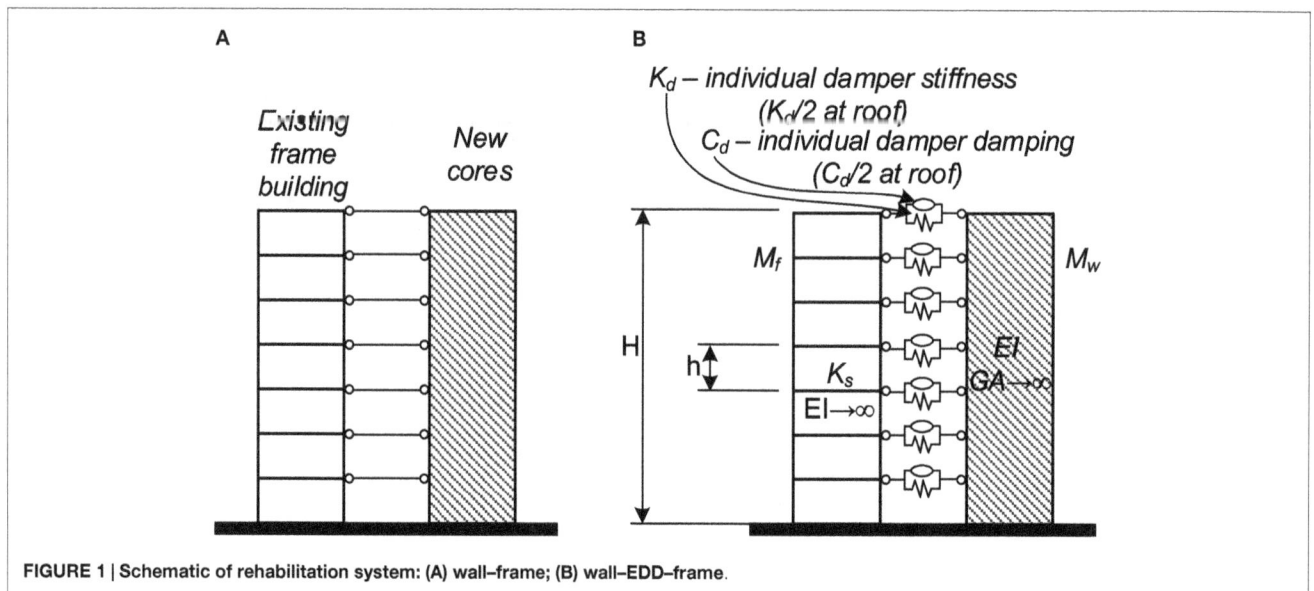

FIGURE 1 | Schematic of rehabilitation system: (A) wall–frame; (B) wall–EDD–frame.

that, in their system, a mass-proportional added damping matrix is attained for the frame. This is in contrast to the stiffness-proportional added damping matrix attained when dampers are assigned within the frame stories. They further emphasized the superiority of mass-proportional damping. Another related system was presented by Murase et al. (2013). In this novel structural system, a base isolated frame structure was connected with oil dampers to a free wall. Through frequency domain and time domain analyses, they demonstrated the effectiveness of this structural system for both pulse-type and long period ground motions.

The idea of connecting adjacent structures by controllers to reduce their responses is not new [see Cimellaro and Lopez-Garcia (2011), and references therein]. In the context of using FVDs or viscoelastic dampers (VEDs), Aida et al. (2001) proposed an approximate analysis model for two buildings connected by a VED. The MDOF model of each building was replaced by its first mode representation, thus, resulting in a 2-DOF model for the analysis. They further indicated that VEDs are only efficient if the buildings connected are of different natural periods. In that context, optimum viscous damping values were proposed for a damper connecting two SDOF systems excited by either harmonic or white-noise ground motions (Bhaskararao and Jangid, 2007). Closed-form equations for the frequency response characteristics of two SDOF systems connected by a spring and a dashpot in parallel where derived as well (Richardson et al., 2013). A 2-DOF reduced model was also used by Huang and Zhu (2013) to investigate the optimal design of VEDs connecting adjacent frame buildings. They further proposed a design method for the optimal locations and sizes of the dampers. Sun et al. (2014) derived expressions for stochastic characteristics and responses of a twin-tower structure linked by a sky-bridge. They assumed a white-noise input and a simplified 3-DOF model (one DOF for the mass of the bridge). Subsequently, they investigated the optimal parameters of the connections.

In other research works, a full MDOF system was considered. Enrique Luco and De Barros (1998) investigated the optimal viscous damping connecting two adjacent shear beams of different heights. For that purpose, they derived the continuous equations (in space) assuming uniformly distributed parameters for each of the beams. The optimal parameters of the dampers were found based on frequency response parameters of the taller structure. Zhang and Xu (1999) adopted the complex modal superposition method using a random seismic input. They further identified the optimal parameters of VEDs connecting adjacent frame structures using a parametric study. In another work, they proposed an analysis procedure for adjacent frame structures connected by VEDs while considering a Maxwell model to represent the dampers (Zhang and Xu, 2000). A parametric study on six story frame buildings connected by FVDs was presented by Trombetti and Silvestri (2007). They highlighted the superiority of this arrangement of dampers that leads to mass-proportional damping over an inter-story placing of viscous dampers. A very unique approach was taken by Takewaki (2007). He developed a frequency domain method to evaluate the input seismic energy to a system of two frames connected by viscous dampers. This enabled him to come to a very strong conclusion that the input energy to the system considered is insensitive to the quantity of the dampers and their locations. Thus, the larger the energy absorbed by the dampers, the smaller the energy input to the frames themselves. Patel and Jangid (2010) investigated several damping distributions between two similar frames. They found that viscous dampers are very efficient in such structures and that there is an optimum damping value to minimize the structural responses. Similar observations had been presented by Kim et al. (2006). Cimellaro and Lopez-Garcia (2011) applied, for the design of damping between adjacent buildings, an algorithm that was originally proposed for the design of dampers in frame structures (Gluck et al., 1996) as well as for the damping and weakening of frame structures (Lavan et al., 2008). Recently, Tubaldi (2015) derived the non-dimensional continuous equations of two shear frames of different heights connected by a VED at the top of the shorter frame. Those were then analytically analyzed to lead to the natural periods and modes. An approximate reduced order model and closed-form solutions for the modal parameters were then proposed and compared to the analytical results with good agreement. Another recent research presented by Bigdeli et al. (2015) proposed a bi-level optimization approach for the optimal design of dampers connecting two frame structures. In the first level of optimization, the combinatorial problem of damper placement was solved, while in the second level the continuous design of each damper was handled.

Most of the abovementioned works investigated the response of either SDOF representations of two buildings connected by a single VED or two MDOF *frames* connected by several VEDs. A main conclusion was that in these cases dampers are expected to efficiently reduce the responses only if the periods of the two SDOF systems, or the two frames, are sufficiently different. It is argued here that if two buildings having different lateral load resisting systems (i.e., frames and walls) are coupled by dampers, a reduction of the responses is expected even if the periods of the buildings are similar. This is due to the differences in their deflection shapes.

Research on frames connected to infinitely rigid walls by dampers has been presented in the literature. Trombetti and Silvestri (2004, 2006) identified the superior effect of mass-proportional damping over stiffness-proportional damping in frame structures. They further proposed various approaches for achieving such damping characteristics by using FVDs. One of the approaches connected horizontal dampers from the floors of the frame to an infinitely rigid wall.

As can be seen, the efficiency of connecting two systems with different periods by means of EDDs has been highlighted many times. This has been investigated using either SDOF representations of two buildings connected by a single VED, two MDOF frames connected by several VEDs or an MDOF frame connected to a rigid wall by several FVDs. The behavior of frame–EDD–wall systems, however, has not been examined while considering the flexibility of the wall. Furthermore, no tools for initial design of such systems are available.

It is the aim of this paper to gain some insight to the seismic behavior of wall–EDD–frame systems as part of retrofitting existing frame structures. This was done by means of a rigorous parametric study. The controlling parameters of such systems were first identified, and their effect on various responses of interest was assessed. Furthermore, the results presented could be used for initial design of such systems in two levels. At the first level, the engineer could assess the feasibility of such a strategy for a given building and its pros and cons compared to rigidly connecting the new walls to the existing building. At the second level, based on the graphs presented herein, the engineer could choose appropriate values for the initial design of the controlling parameters.

Structural Systems Considered and Modeling Assumptions

Structural Systems

The focus of this paper is the behavior and design of wall–EDD–frame systems as part of retrofitting existing frame structures. Such a system is presented in **Figure 1B**. The system consists of a frame and a wall that are connected by viscous dampers and springs. The seismic mass per floor taken by the frame is M_f, equal for all floors, and the story stiffness of the frame is K_s (shear force required to result in a unit inter-story drift). The seismic mass per floor taken by the wall is M_w and the bending stiffness of the wall is EI. The damping coefficient of each viscous damper is C_d while the stiffness of each spring is K_d. The total height of the N story building is H with a typical story height of $h = H/N$. For the sake of comparison, a similar system where the frame and the wall are rigidly connected is also considered. This system is presented in **Figure 1A**. In the models described above a few assumptions are embedded. Those are:

- A plane model is considered, indicating that the parametric study results are applicable to structures where the torsional response is limited.
- A fixed base is assumed for both frame and walls.
- A uniform distribution of structural and damping properties was assumed along the height. While this may not necessarily be the case, in many cases, the use of some average properties may still lead to very good approximations.
- The frame is assumed to behave like a shear beam while the wall is assumed to behave like a flexural beam. That is, in the frame, axial deformations of the columns are neglected while, in the wall, shear deformations are neglected.
- The inherent damping of the frame is considered by a Rayleigh damping matrix constructed for the bare frame with 5% damping in the first and third modes. A similar procedure is used to consider the inherent damping of the wall.
- It is assumed that the behavior of the system could be approximated by use of linear analysis. That is, either the behavior of the system is actually linear, or a linear analysis leads to a good approximation of the displacement related responses. This assumption holds reasonably well for regular structures with either a medium to long period or with low-ductility demands. It should be noted that, when viscous dampers are utilized, the behavior of the system is indeed expected to be linear or close to that. Thus, the ductility demands are expected to be low. Furthermore, in such cases, the fundamental period of the system is similar to that of the bare frame, which is usually long enough.

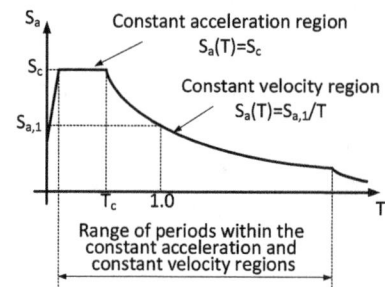

FIGURE 2 | Psuedo-acceleration response spectrum adopted in the study.

Seismic Environment

The seismic environment, in this study, is represented using a typical 5% damping pseudo-acceleration response-spectrum. It is assumed that the periods of the systems considered are within the period range of the constant acceleration and constant velocity regions of the spectrum (see, e.g., **Figure 2**). Thus, two parameters are controlling the shape of the spectrum. In this study, the elastic 5% damping spectral acceleration at some given period (with no behavior factor), $S_a(T')$, and the corner period, between the constant acceleration and the constant velocity regions, T_c, are adopted. The period at which $S_a(T')$ is evaluated will be chosen in the next sections.

Responses of Interest

In this paper, the frame is assumed to be the lateral load resisting system of the existing old building as well as its gravity carrying system. Thus, the maximum peak inter-story drift angle of all stories of the frame, ID_f, is of much importance. Here, the inter-story drift angle is the horizontal displacement of the story ceiling relative to the story floor normalized by the story height. Based on this parameter, one could assess if yielding occurs and to what extent. Inter-story drifts also indicate the extent of damage to some non-structural components (e.g., partition walls). Furthermore, the maximum peak elastic shear in the frame could be computed by $GA \cdot ID_f$ where GA is the smeared frame stiffness as will be explained in section "Controlling Parameters". Another important response related to the frame is its base over-turning moment, OTM_f, that affects the axial forces in its columns and its foundations. Recently, focus was also drawn to the level of absolute accelerations. These are especially important in cases where the structure is expected to remain operational after the seismic event. As most of the important functions in the considered buildings are expected to be supported by the frame, the maximum peak frame absolute acceleration, A_f, is considered as a response of interest.

The walls are the new lateral load resisting system. In general, no partition walls or gravity frames are rigidly connected to these walls; hence, their inter-story drifts are of no particular importance. The design of these walls and the design of their foundations are dictated by the wall base shear, BS_w, and base moment, OTM_w. Thus, these are the responses of interest related to the walls.

Finally, the responses of interest related to the connection between the frame and the wall, and to the EDDs are identified. Viscous dampers and springs are designed and sized based on the peak force they are expected to experience, F_d. Furthermore, this force is to be locally taken by the existing slabs at the region of the connection. Thus, the connection force serves as the first response of interest related to the connection. To avoid pounding between the frame and the wall, it is important to insert seismic gaps. Furthermore, another design parameter of EDDs is the stroke they are expected to experience. Thus, the second response parameter related to the EDDs is the maximum peak relative displacement between the frame and the wall, or the damper elongation/contraction, D_d.

The Parametric Study Design

Controlling Parameters

In view of the previous section, the parameters controlling the behavior of the problem are M_f, K_S, M_w, EI, H, h, C_d, K_d, $S_a(T')$, and T_c. For the sake of generality, "smeared" properties per unit height will be used as parameters rather than discrete properties. It is assumed that two systems with similar smeared properties but with different numbers of stories will have similar smeared responses. The error associated with this approximation is also assessed later on in Section "Analysis Approach." The use of smeared properties to transfer from discrete models to continuous ones is not new and has even been used in the seismic analysis of structures with EDDs (Lavan, 2012; Tubaldi, 2015). It is known to lead to very good approximations with accuracy growing with the number of stories. The shear stiffness of the frame is therefore $GA = K_s \cdot h$. The seismic mass per unit height taken by the frame is $m_f = M_f/h$ while that taken by the wall is $m_w = M_w/h$. The damping coefficient of the EDDs and their stiffness per unit height are $c_d = C_d/h$ and $k_d = K_d/h$, respectively. This reduces the group of controlling parameters to the following nine parameters: m_f, GA, m_w, EI, H, c_d, k_d, $S_a(T')$, and T_c. The responses of interest are, as before: ID_f, OTM_f, A_f, BS_w, OTM_w, F_d, and D_d.

Most of the responses of interest are either smeared measures to begin with (e.g., ID_f), or local responses at a specific point (e.g., OTM_f, A_f, BS_w, OTM_w, and D_d). The only response of interest to be smeared is therefore the damping force that is taken as $f_d = F_d/h$.

Physical Considerations in the Reduction of the Number of Controlling Parameters

Conducting a parametric study with as many as nine controlling parameters is not realistic. Although with today's computation capabilities it may be computationally feasible, it may not easily enable gaining insight. A clear way for presenting the study results

is also infeasible. To enable a more computationally efficient study, that would allow a clear presentation of the results, and enable gaining insight, advantage is taken of the physics of the problem. In this section, $S_a(T')$ and T_c will be eliminated from the parametric study as follows.

The problem at hand is a linear problem. It is well known that in such problems there is a linear relation between the magnitudes of the input loading and the magnitudes of the responses. That is, if the input spectrum is multiplied by a given factor, the responses of interest will change by the same factor. Thus, modified responses of interest will be adopted as the original responses of interest, normalized by $S_a(T')$. With this set of parameters adopted, the value of $S_a(T')$ has no effect on the parametric study results. Hence, it is eliminated.

In general, the effect of T_c on the responses of interest is a result of its effect on the spectral accelerations of the higher modes. The contribution of these modes to some responses of interest (e.g., base moment and shear) may be significant when the natural period is relatively large. There are two specific cases, however, where the actual value of T_c is insignificant. The first is when the periods of all relevant modes of the retrofitted system fall within the constant acceleration region. This is illustrated in **Figure 3A**. The second is when the periods of all relevant modes fall within the constant velocity region (**Figure 3B**). Thus, this study is partitioned to these two cases. In the first case, a constant acceleration spectrum is assumed and the value of $S_a(T')$ by which the responses are normalized is taken as the spectral acceleration at the fundamental period of the retrofitted system, i.e., the wall–EDD–frame system. If indeed, the periods of all relevant modes of the retrofitted system fall within the constant acceleration region, then $S_a(T') = S_c$. If the spectral acceleration at the fundamental period of the bare frame falls within the constant acceleration region, than the normalization could be done based on that value, which is also equal to S_c. In the second case, a constant velocity spectrum is assumed and the value of $S_a(T')$ is taken as the spectral acceleration at the fundamental period of the bare frame, $S_a(T_1)$.

Characteristic values for T_c in Israel usually range from 0.2 s to about 0.8 s in extreme cases. Most of the structures retrofitted are four stories and more, while two additional stories could be built and sold to fund the retrofitting expenses. Thus, the structures analyzed are usually six stories and more. Frame structures of that height, in old buildings that were not seismically designed, usually possess fundamental periods of about 1.2–1.8 s. Thus, although in many cases, most relevant modes will fall within the constant

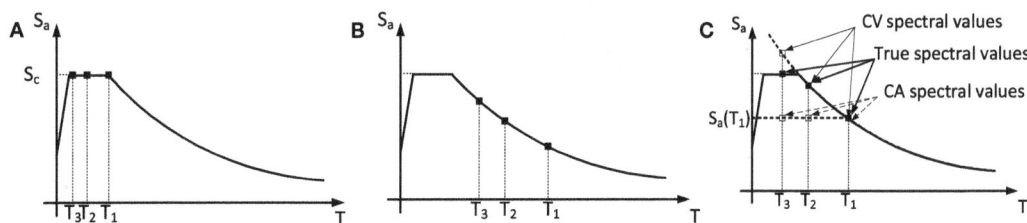

FIGURE 3 | Location of the first natural periods in the response spectrum: (A) all in the CA region, (B) all in the CV region, and (C) some in the CV region while others in the CA region.

velocity region, it is not uncommon to find cases where periods of some relevant modes fall within the constant velocity region while those of other modes fall within the constant acceleration region. In such cases, the responses of interest attained assuming that the periods of all relevant modes fall within the constant velocity region will serve as upper bounds. This is because the spectral accelerations of the higher modes, computed based on a constant velocity spectrum, are higher than the actual ones (**Figure 3C**). Similarly, the responses of interest attained assuming that the periods of all relevant modes fall within the constant acceleration region will serve as lower bounds. This is because the spectral accelerations of the higher modes, computed based on a constant acceleration spectrum, are lower than the actual ones (**Figure 3C**). Note that, in this case, $S_a(T')$ is taken as the spectral acceleration at the fundamental period of the retrofitted system, that is different from S_c. It should be noted that, for the frame, responses the upper bound and lower bound were found to be very close. That is, the frame responses are not considerably affected by higher modes contributions. However, if viscous dampers alone are used as connections, the upper bound and lower bound for wall responses were not as close. Nonetheless, the values of these responses, as will be shown later on, were much smaller than those attained with a rigid connection. Thus, for the sake of comparing alternatives, and for initial design, this approach leads to reasonable results.

Dimensional Analysis

Taking advantage of the physics of the problem reduced the number of controlling parameters from nine to seven. While this presents a significant simplification of the problem, a parametric study with seven controlling parameters is still complex to manage, present, and analyze. A further reduction of the number of controlling parameters is thus desired. This is done by means of dimensional analysis [see, e.g., Barenblatt (2003)]. The origins of dimensional analysis are probably motivated by experimental research where the number of required experiments may be drastically decreased. The use of such tools in numerical parametric studies has also been taken advantage of [e.g., Makris and Black (2004a,b) and Lavan (2012)]. Here, the π theorem is adopted.

The first stage of the π theorem is choosing the repeating variables. As the problem at hand involves three independent dimensions, e.g., length, mass, and time, the number of repeating variables may be up to three. Those are chosen as m_f, GA, and H. This choice is made for several reasons: for one, it is convenient that all non-dimensional parameters and responses are normalized by known parameters. As in the retrofitting problem stated above the frame parameters are given, this seems a natural choice. This will later allow a simple comparison between various retrofitting alternatives as it could be directly done based on the non-dimensional parameters and responses (assuming that the periods of all relevant modes of all retrofitting alternatives fall within one region of the spectrum). Note also that this choice will lead, later on, to some non-dimensional parameters that are well known from the behavior of wall-frame systems in static loadings. This is important as some engineers already developed some intuition to the order of magnitude of reasonable values of such parameters for given cases. Furthermore, with this choice, a clear rational could be used to determine the range of values for

the non-dimensional controlling parameters. This may prevent the parametric study, and the conclusions drawn from that, from considering cases that are not practical.

With m_f, GA, and H as the repeating variables, using dimensional analysis theory, the following non-dimensional controlling parameters are attained:

$$\pi_m = \frac{m_w}{m_f} \quad \pi_s = \sqrt{\frac{GA}{EI}}H \quad \pi_c = \frac{c_d \cdot H}{\sqrt{m_f \cdot GA \cdot \pi^2}} \quad \pi_k = \frac{k_d \cdot H^2}{GA} \tag{1}$$

The first of these parameters is the ratio of wall mass to frame mass. The second is the relative frame stiffness to wall stiffness. This is very well known as $\alpha \cdot H$ in wall–frame systems in static loads (Stafford Smith and Coull, 1991). Finally, the third and fourth parameters represent the relative EDD damping and stiffness, respectively.

From the dimensional analysis, the non-dimensional responses of interest are attained as well. Those are

$$\pi_{ID_f} = \frac{ID_f \cdot GA}{S_a(T') \cdot m_f \cdot H} \quad \pi_{OTM_w} = \frac{OTM_w}{S_a(T') \cdot m_f \cdot H^2}$$

$$\pi_{OTM_f} = \frac{OTM_f}{S_a(T') \cdot m_f \cdot H^2} \quad \pi_{BS_w} = \frac{BS_w}{S_a(T') \cdot m_f \cdot H}$$

$$\pi_{A_f} = \frac{A_f}{S_a(T')} \quad \pi_{f_d} = \frac{f_d}{S_a(T') \cdot m_f} \quad \pi_{D_d} = \frac{D_d \cdot GA}{S_a(T') \cdot m_f \cdot H^2}$$

$$\pi_T = T \cdot \sqrt{\frac{GA}{m_f \cdot H^2}} \quad \pi_\xi = \xi \tag{2}$$

Range of Values for the Controlling Parameters

The application considered herein focuses on an existing frame structure retrofitted with new walls and dampers. These walls would usually carry their self-weight only. Thus, practically, the seismic mass to be carried by the frame is larger than that taken by the wall. The range of values for the seismic mass carried by the wall to that carried by the frame is thus taken as $\pi_m = \frac{m_w}{m_f}$: $\frac{1}{10}, \frac{1}{4}, \frac{1}{2}$.

In general, the range of values for the parameter $\alpha \cdot H$ extends from 0, when the wall is infinitely stiff compared to the frame, to infinity, when the frame is infinitely stiff compared to the wall. For the application considered herein, the practical range of values is closer to the first case. This is both because the existing frames are usually flexible and because the considered buildings are low to medium-rise. Thus, values of $\pi_s = \sqrt{\frac{GA}{EI}}H$: $\frac{1}{10}, \frac{1}{2}, 1$ were considered. It is important to emphasize that values smaller than 0.1 were also analyzed and led to responses practically identical to those attained with 0.1.

Theoretically, the range of values for the damping coefficient of the dampers extends from 0 to infinity. Although the damping is very well utilized in the considered structural system, it is unlikely that practical use will be made of damping that would lead to very large damping ratios in the first modes, or even over-damped first modes. Thus, a preliminary study was conducted to tune the upper bound for the non-dimensional parameter to try to avoid such excessive damping as much as possible. The values of π_c: 0, 0.1, 0.2, 0.3, 0.4, 0.5, 0.6, 0.7, and 0.8 were therefore adopted.

The range of values for the stiffness of the dampers can also, theoretically, span from 0 to infinity. The latter represents a rigid connection between the frame and the wall. As both 0 and a relatively stiff damper are practical, the values for this parameter were chosen so as to lead to a good distribution in terms of the responses of interest. These are π_k: 0, 4.8, 13.6, 8 × 10⁸.

Analysis Approach

As indicated earlier, a discretized model is used for the analysis. The number of stories in this model was taken as 6, as this is believed to be the most common number of stories in such systems in Israel. Estimation of the potential errors in adopting the "smeared" properties and responses attained with a 6 story model, for structures with a different number of stories, was then carried out. This was done via analyses of models with different number of stories ranging from 4 to 50. Those showed errors of ±3% in most responses of interest. In some extreme cases, a few responses were off by −8% to +10%.

The equations of motion of the discretized model subjected to an input ground motion are given as follows:

$$\mathbf{M}\ddot{x}(t) + \mathbf{C}\dot{x}(t) + \mathbf{K}x(t) = \mathbf{f}(t) \tag{3}$$

where $\mathbf{M}$, $\mathbf{C}$, and $\mathbf{K}$ are the mass, damping, and stiffness matrices, respectively; $x(t)$ is the vector of coordinate displacements; an overdot represents a derivative with respect to time, t; and $\mathbf{f}(t)$ is a vector of loads. It should be emphasized that the considered systems include mechanical dampers. In general, and in the cases considered here, in particular, such systems may not qualify the Caughey criterion (Caughey and O'Kelly, 1965). Thus, for the purpose of analyzing such systems while modeling the seismic hazard via a response spectrum, use is made of a complex modal spectral analysis. Out of the procedures available for complex modal spectral analysis [e.g., Singh (1980), Takewaki (2004), and Song et al. (2008)], the one proposed by Song et al. (2008) was adopted and programed in Matlab. Using this approach, the equations are first brought to their following state-space form [see, e.g., Soong (1990)]:

$$\mathbf{A}\dot{y}(t) + \mathbf{B}y(t) = \mathbf{f}_S(t) \tag{4}$$

Here,

$$\mathbf{A} = \begin{bmatrix} \mathbf{0} & \mathbf{M} \\ \mathbf{M} & \mathbf{C} \end{bmatrix}; \quad \mathbf{B} = \begin{bmatrix} -\mathbf{M} & \mathbf{0} \\ \mathbf{0} & \mathbf{K} \end{bmatrix}; \quad \mathbf{f}_S(t) = \begin{Bmatrix} \mathbf{0} \\ \mathbf{f}(t) \end{Bmatrix};$$

$$\mathbf{y}(t) = \begin{Bmatrix} \dot{x}(t) \\ x(t) \end{Bmatrix} \tag{5}$$

Solving the eigenvalue problem corresponding to the homogeneous counterpart of Eq. 4 leads to the eigenvalues, λ_i, in the complex numbers domain from which the natural frequencies and damping ratios could be evaluated as follows:

$$\omega_i = |\lambda_i|\,; \quad \xi_i = \frac{\text{real}(\lambda_i)}{\omega_i} \tag{6}$$

where $|\cdot|$ and real$(\cdot)$ represent the absolute and the real part of a complex number, respectively. For each complex eigenvalue, a corresponding complex eigenvector (mode shape) could be

computed. From the complex eigenvalues and mode shapes, the following contribution of the complex mode i and its conjugate to a response of interest could be assessed:

$$(\omega_i^2 A_{0i}^2 + B_{0i}^2)\,|q_i(t)|_{\max}^2 \tag{7}$$

where A_{0i} and B_{0i} are computed based on the eigenvalue and mode shape i and its conjugate, and $|q_i(t)|_{\max}$ is the spectral displacement corresponding to the natural period and the damping ratio of these modes. Similarly, if real (overdamped) modes are attained, the contribution of the overdamped mode i could be assessed as follows:

$$(A_{0i}^P)^2\,\left|q_i^P(t)\right|_{\max}^2 \tag{8}$$

where A_{0i}^P is computed based on the eigenvalue and mode shape i, and $\left|q_i^P(t)\right|_{\max}$ is the spectral displacement corresponding to the natural period of that mode. Finally, the contributions of the various modes could be combined using various combination rules. One of those rules, which was adopted here, is the GCQC (Song et al., 2008).

Equations 7 and 8 require the spectral displacements of underdamped and overdamped SDOF systems, respectively. Those of the underdamped systems were evaluated by modifying the 5% damping spectral accelerations using the following factor:

$$R_\xi = \sqrt{\frac{0.1}{0.05 + \xi}} \tag{9}$$

where ξ is the damping ratio as a value (not in percent). This factor is given in Eurocode 8 (CEN-Comité Européen de Normalisation, 2003) with a lower bound of 0.55. This factor, without the lower bound, was compared with the results presented by Ramirez et al. (2001) up to 99.99% damping with good agreement. Thus, it was adopted as is, with no lower bound. The spectral displacements of the overdamped modes, where applicable, were evaluated using the approach presented by Song et al. (2008).

In order to verify the validity of the analysis approach and the Matlab code, a comparison of the responses attained using this code with those attained using time history analyses was made with good agreement. In this comparison, the displacements of the SDOF systems representing the modes were evaluated using time history analyses.

Once a verification of the analysis engine was made, a comprehensive parametric study that includes the range of parameters indicated in Section "Range of Values for the Controlling Parameters" was carried out. The results of the parametric study, that present the attained non-dimensional responses of interest as a function of the non-dimensional controlling parameters, are presented in section "Parametric study results and discussion".

Retrieving the Dimensional Responses of Interest

The non-dimensional responses are assessed in this study as part of the parametric study. The parametric study results are presented in section "Parametric study results and discussion". Once a solution for the non-dimensional system is at hand, with known non-dimensional responses of interest (the π values in Eq. 2), the dimensional responses of interest are retrieved using Eq. 2.

FIGURE 4 | General preview of a typical response presentation: (A) the matrix of graphs, (B) a cell in the matrix representing given wall and frame.

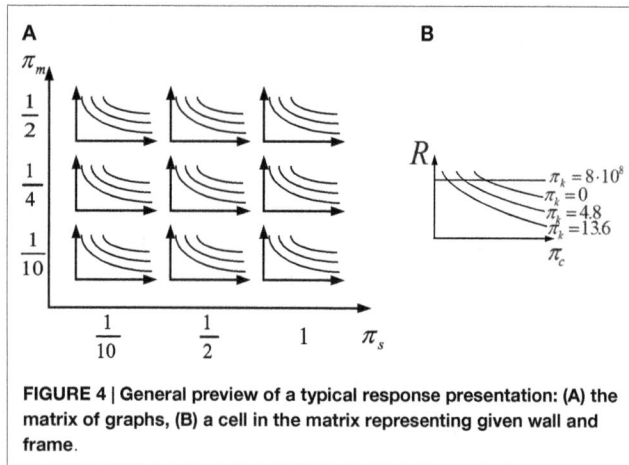

Parametric Study Results and Discussion

Presentation of the Parametric Study Results

Although great efforts were made to reduce the number of controlling parameters, their number amounts to four, for each region of the spectrum. Thus, the results are presented in separate graphs for each region of the spectrum. In addition, each response of interest is presented in a dedicated matrix of graphs. In this matrix, each column has a different value of π_s while each row has a different π_m. This is depicted in **Figure 4A**. Furthermore, in each "cell" of this matrix lines are plotted to present the response of interest as a function of π_c. Each of these lines is plotted for a different value of π_k. This is depicted in **Figure 4B**. These represent connections by viscous dampers (continuous blue); a flexible spring ($\pi_k = 4.8$) and a viscous damper (dashed green); a stiffer spring ($\pi_k = 13.6$) and a viscous damper (dash-dotted red), and; a stiff connection (dotted black). Some results of the parametric study will be now presented and discussed. Results related to other responses of interest could be found in Supplementary Material.

Natural Periods

The non-dimensional fundamental period of the system is presented in **Figure 5**. As can be seen, a connection with viscous dampers only retains the fundamental period of the frame (estimated using $\pi_c = 0$). In some systems having values of $\pi_s = 1$ a large damping leads to an increase in the fundamental period. In these cases, the first mode was found to be overdamped. When either springs are added, or rigid connections are used, the natural period decreases drastically. This may have a huge effect on the forces attracted by the system, as will be discussed later on.

Responses of Interest Related to the Frame

The first frame response of interest is its inter-story drift. This is presented in **Figures 6A,B** for the constant velocity and constant acceleration regions, respectively. As can be seen, a rigid connection between the frame and the wall, results in small inter-story drifts in the frame. This is more pronounced for small π_s values, which indicate a large stiffness of the wall. This is because a rigid connection leads to the same displacements in the frame and the wall. When the connection is made with viscous damping only,

small inter-story drifts can also be attained with a sufficiently large damping coefficient, especially in the constant velocity region. These drifts may be much smaller than those experienced by the bare frame (when a 0 value of the viscous damper is used). When the wall is not "infinitely" rigid, the drifts of the damped frame can sometimes be even as small as the ones attained with a rigid connection. With the decrease in wall stiffness, this is attained with a smaller amount of damping. When, in addition to damping, stiffness is added to the connection, the frame inter-story drift reduces considerably. This is true even if the amount of stiffness is relatively small. Another point that is worth noting is that, when a non-rigid connection is used, this response seems relatively insensitive to π_m and π_s. This is important in case there is some uncertainty with respect to structural properties. It should be noted that, while this discussion focused on the frame inter-story drift, similar trends were observed in some other frame responses (e.g., base shear, top displacement, and over-turning moment).

Another response of interest of the frame is its absolute acceleration. This is presented in **Figures 7A,B** for the constant velocity and acceleration regions, respectively. As can be seen, a rigid connection between the frame and the wall, results in extremely large absolute accelerations in the frame, especially in the constant velocity region. This is attributed to the fact that, with a rigid connection the floor mass is directly connected to the walls, which are relatively rigid. Thus, the periods of the modes that contribute to the frame displacements become very short, indicating large accelerations. When the connection is made with viscous damping only, the smallest absolute acceleration is usually attained. When, in addition to damping, stiffness is added to the connection, the absolute acceleration tends to somewhat increase.

Large absolute accelerations indicate damage to some non-structural components. This may be very important in structures that are expected to remain operational after the earthquake. Nevertheless, this is not their only importance. Absolute accelerations also indicate the level of inertia forces acting on the mass. These forces are to be transmitted from their point of action, through the existing slabs of the frame, to the lateral load resisting systems. These slabs are usually thin, and may sometimes be rib slabs. In addition, due to their year of construction, they sometimes are made of concrete of low quality. Thus, their ability to transfer these forces is limited, and is usually known with a large uncertainty. In addition, as new lateral load resisting systems exist, the path of the inertia forces may change considerably and be much different from the original one.

Responses of Interest Related to the Connection

The most important connection response of interest is the maximum peak connection force. It is strongly related to the feasibility of the retrofitting approach. While new dampers or connection elements can be designed to take these forces, they are connected to existing elements that are of limited given capacity. The maximum peak connection force is presented in **Figure 8**.

As can be seen, in the case of a rigid connection, the connection force is extremely large, much larger than any other alternative. This is more pronounced in the constant velocity region. When the connection is made with viscous damping only, the smallest

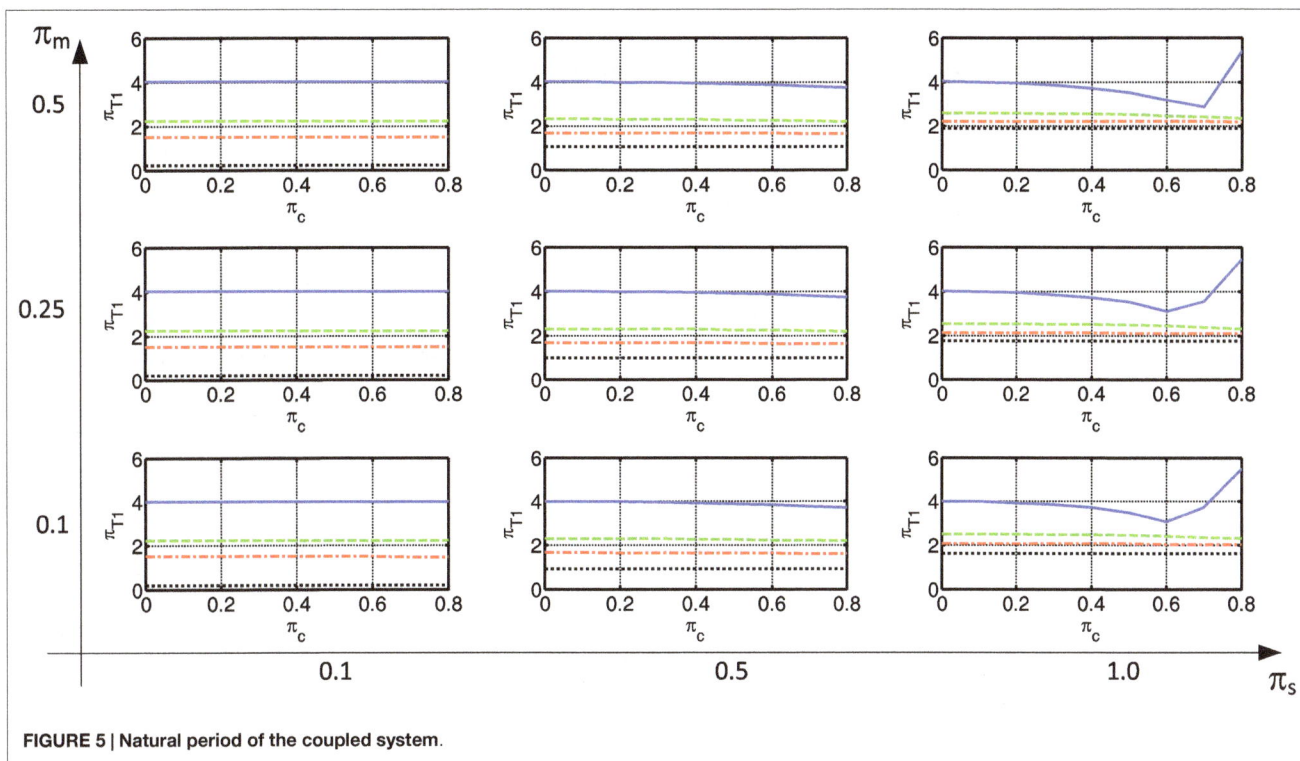

FIGURE 5 | Natural period of the coupled system.

connection force is usually attained. When, in addition to damping, stiffness is added to the connection, the connection force increases. This increase is more pronounced with a small damping. For larger values of damping, the increase of connection force due to the addition of some stiffness is much less pronounced, and sometimes even a decrease is attained.

As indicated in the Section "Introduction," large forces in the connecting elements are locally transmitted through the existing slabs of the building. These are limited in their given capacity. Furthermore, the large forces in the connections indicate that a large portion of the inertia forces acting on the slabs is transmitted to the new lateral load resisting system. As the path that these forces take from their point of action to the connections is much different from their original path to the old lateral load resisting system (frames), the slabs may not be able to withstand these forces. This is especially true if their magnitude is large. Thus, this response may lead to infeasibility of using a given connection type.

Responses of Interest Related to the Wall

The wall response of interest is the peak base over-turning moment (**Figure 9**). As can be seen, in the case of a rigid connection, the wall moment is larger than any other alternative. As in other responses, this is more pronounced in the constant velocity region. When the connection is made with viscous damping only, the smallest moment is usually attained. When, in addition to damping, stiffness is added to the connection, the moment increases. This increase is more pronounced with a small damping. For larger values of damping, the increase of moment due to the addition of some stiffness is less pronounced, and sometimes even a decrease is attained. It should be noted that, while this

discussion focused on the wall base moment, similar trends were observed in some other wall responses (e.g., base shear).

As indicated in the Section "Introduction," large over-turning moments in the wall, with the small gravity forces acting on them, may lead to large tension in their foundations. Depending on the soil type, this may lead to very large foundations. In turn, this may lead to high retrofitting costs and may sometimes fail a retrofitting project. This becomes more and more of an issue with a growing number of stories.

Summary of the Parametric Study

In view of the results presented above, frame inter-story drifts are the smallest when a rigid connection is used. Connections with viscous dampers and springs could also considerably reduce the drifts with respect to the bare frame, in some cases, even to the same levels as those attained with a rigid connection. When it comes to other responses such as frame absolute accelerations, connection force, and wall responses, a connection with dampers is much superior. Thus, it is suggested to use a connection with viscous dampers, possibly with springs as well, in any case that it can reduce the frame drifts to allowable values. This is especially true in the constant velocity region. It should be emphasized that a damped connection showed better reduction in all responses of interest, including frame drifts, with larger values of π_s. While these may be realistic for other purposes (e.g., new buildings), for the retrofitting projects discussed those were not considered.

Example

The example considers the seismic retrofitting of an existing six story frame structure. Each story area is $460\,\text{m}^2$ and its mass

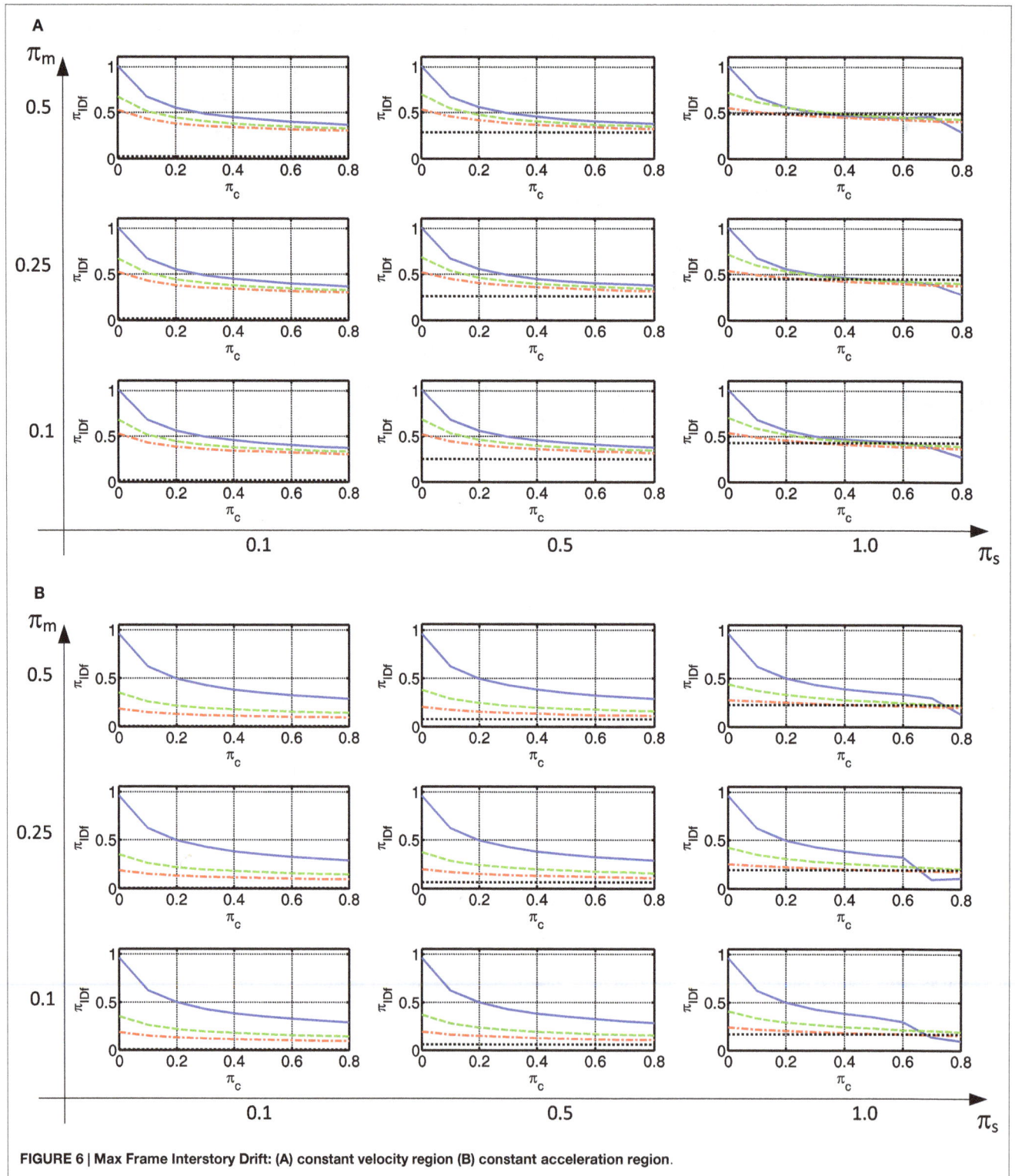

FIGURE 6 | Max Frame Interstory Drift: (A) constant velocity region (B) constant acceleration region.

is 600 ton (300 ton at the roof). The typical story height is 3 m and the total height of the building is 18 m. Thus, the mass per unit height is 200 ton/m. The typical story stiffness of the frames assuming cracked cross sections is 106,300 kN/m leading to a natural period of the building of 1.8 s. This period compares well with the rule of thumb proposed by Crowley and Pinho (2004)

for the yield period of existing European RC structures. This rule of thumb was originally proposed for better displacement estimation and, although thoroughly fundamented, it has not yet found its way to seismic codes. It should also be noted that if some inaccuracy in the natural period of the frame, or in its stiffness, exists, it affects the parameter π_s. As can be seen from

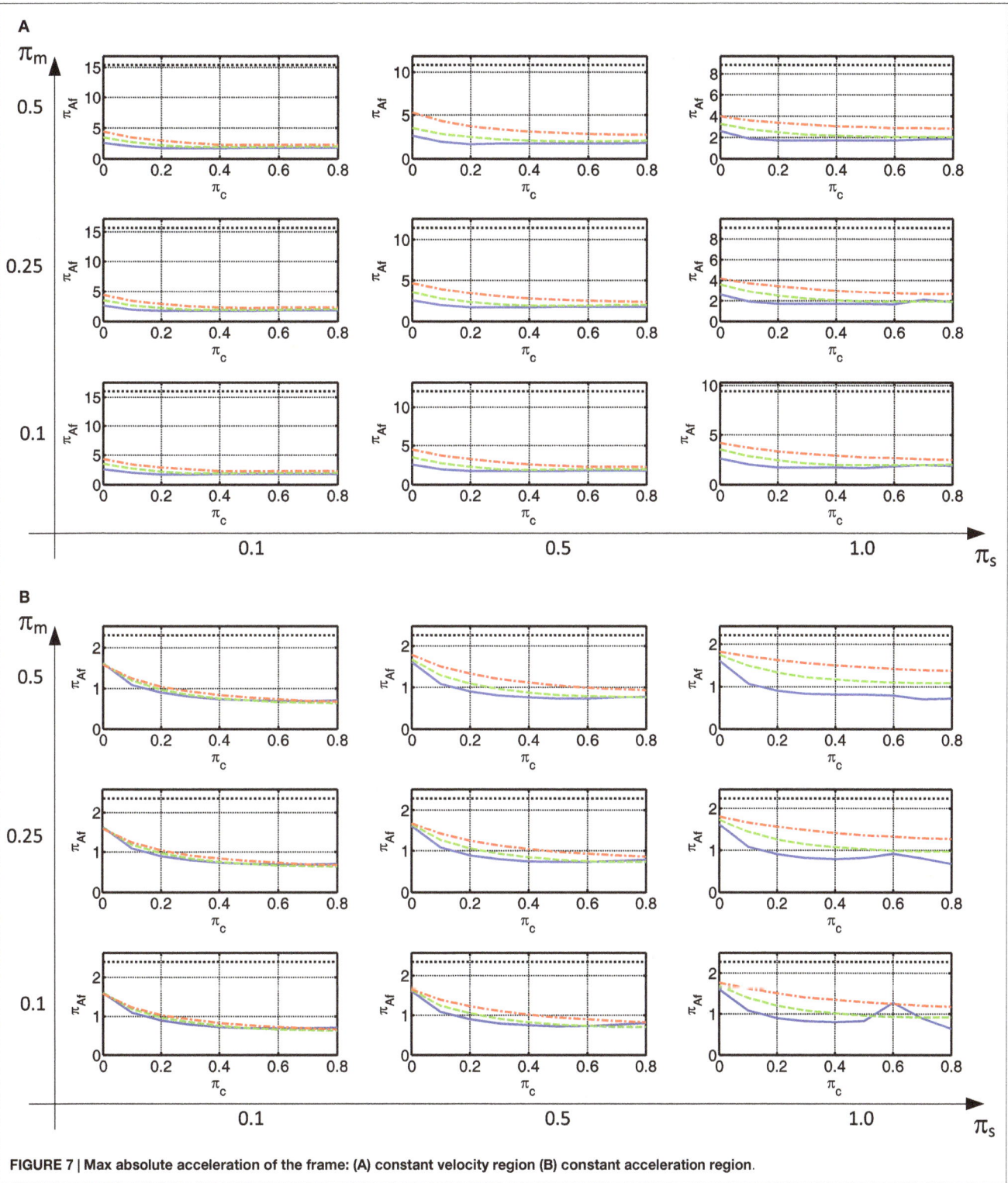

FIGURE 7 | Max absolute acceleration of the frame: (A) constant velocity region (B) constant acceleration region.

Figures 6–9, in this structure ($\pi_s \approx 0.5$, as will be calculated later) some inaccuracy in π_s is not expected to considerably affect the values of the estimated responses (wall responses and frame displacements). It would, however, affect the frame force related responses. As those are computed based on the displacements and the frame stiffness at a later stage, conservative assumptions could

be then made. The shear stiffness of the frame was computed as $GA = K_s/h = 318{,}900$ kN. The corner period of the representing spectrum is $T_c = 0.4$ s and its spectral pseudo-acceleration at 1.0 s is $S_1 = 2.5$ m/s^2.

Four cores were added to the building to supply shelter rooms. Their total area is 55 m^2 per floor. Their total cracked moment

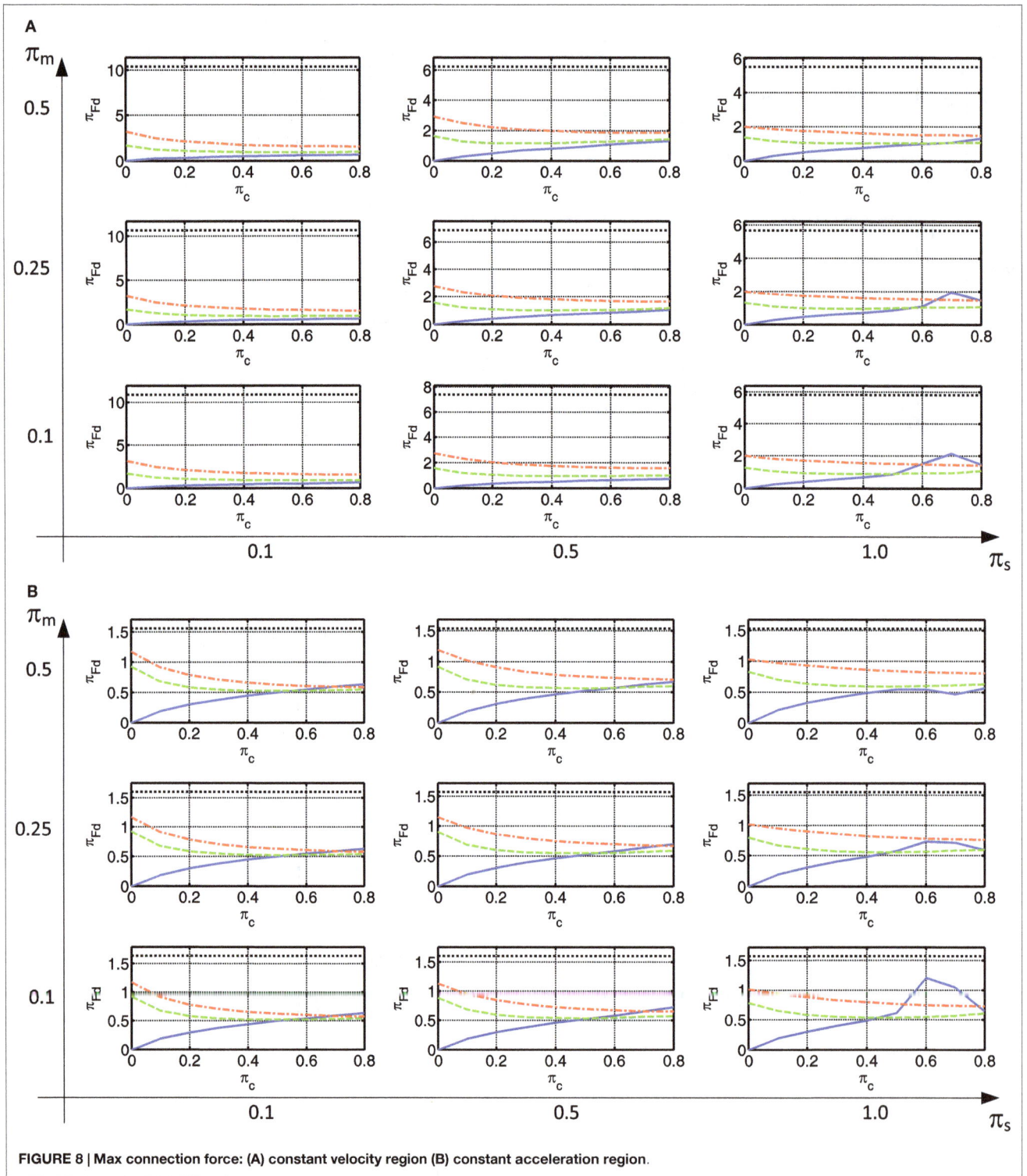

FIGURE 8 | Max connection force: (A) constant velocity region (B) constant acceleration region.

of inertia is 387×10^6 kN·m^2 and their mass per unit height is 52 ton/m. It is now desired to find the best connection type between the existing building and the new cores, as well as attain responses of interest for initial design.

The two known non-dimensional parameters of the system are π_m and π_s. Those are attained as $\pi_m = \frac{52}{200} = 0.26$ and

$\pi_s = \sqrt{\frac{318900}{387\cdot10^6}} \cdot 18 = 0.51$ and rounded to 0.25 and 0.5, respectively. Thus, the center graph in each graph matrix represents the structure at hand. In order to attain the non-dimensional responses from the graphs, one needs to determine what spectrum region is relevant. Thus, the periods of the first three modes are evaluated for each retrofitting alternative. Using damping only,

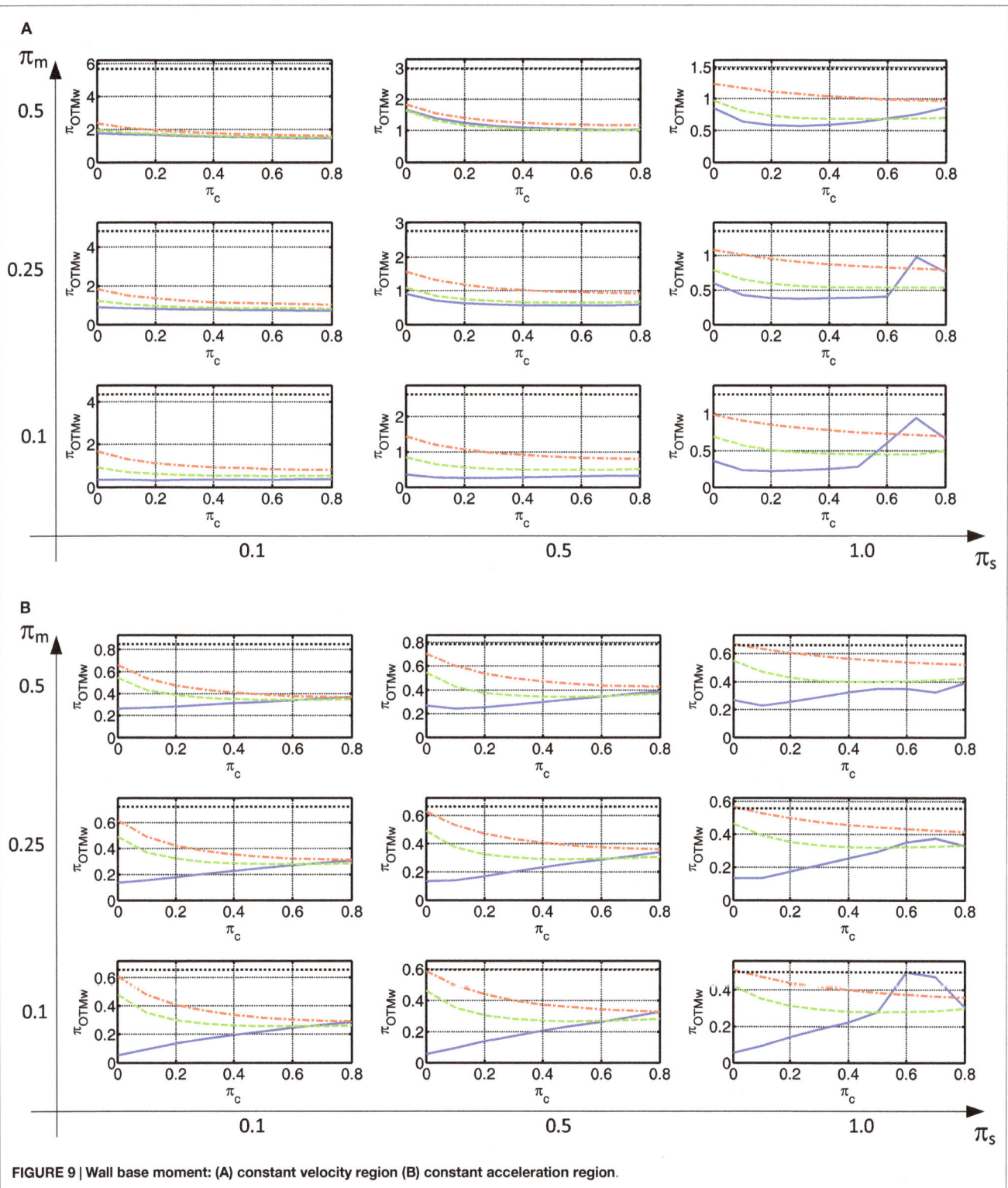

FIGURE 9 | Wall base moment: (A) constant velocity region (B) constant acceleration region.

the periods are 1.74, 0.63, and 0.37 s. With a small stiffness added, the periods became 0.95, 0.54, and 0.36 s. With a medium stiffness added to the damper, values of 0.82, 0.48, and 0.33 s are attained. Finally, with a rigid connection, the periods drop to 0.4, 0.09, and 0.04 s. Thus, when a non-rigid connection is used, the constant velocity graphs will be adopted with $S_a(T') = 1.38 \, \text{m/s}^2$ that

is evaluated at the fundamental period of the bare frame, 1.8 s. This is slightly conservative as the third mode has a period smaller then T_c. When a rigid connection is used, the constant acceleration graphs will be adopted with $S_a(T') = 6.25 \, \text{m/s}^2$ that is evaluated at the fundamental period of the rigidly connected system, 0.4 s.

TABLE 1 | Responses of interest using the various connection options.

Response	Bare frame	FVD connection	FVD + low stiffness spring connection	FVD + high stiffness spring connection	Rigid connection
		CV region	CV region	CV region	CA region
IDF [%]	1.57	0.70	0.62	0.56	0.44
BSW [kN]	–	4409	5215	8025	22,410
OTMF [kN*m]	55310	21,220	16,210	14,950	18,480
OTMW [k*m]	–	50,420	58,600	90,770	266,000
AF [m/s²]	3.45	2.39	2.63	3.87	14.26
Fd [kN/m]	–	180	284	502	1941
Dd [m]	–	0.0672	0.0453	0.0350	0.0000

IDF, inter-story drift; BSW, base shear of the wall; OTMF, over-turning moment at the frame; OTMW, over-turning moment at the wall; AF, acceleration at the frame; Fd, force at the damper; Dd, displacement at the damper-stroke.

The amount of damping that was adopted is $\pi_c = 0.4$. Thus, the damping coefficient per unit height was taken as $c_d = 556.53 \, \text{kN·s/m}^2$. The responses of interest, which were evaluated using the non-dimensional graphs, are summarized in **Table 1**. For example, the inter-story drift angle of the frame with a viscous damping connection was computed as follows:

$$ID_f = \frac{\pi_{ID_f} \cdot S_a(T') \cdot m_f \cdot H}{GA} = \frac{0.453 \cdot 1.38 \cdot 200 \cdot 18}{318,900} = 0.705\% \tag{10}$$

which translates to an inter-story drift of $0.00705 \cdot 300 = 2.11 \, \text{cm}$.

As can be seen, with damping only, the inter-story drift was reduced by 55% (72% with a rigid connection). If such a drift could be accommodated by the frame, a large decrease could be attained in other responses compared to those of the rigid connection option: the wall over-turning moment is 19% of that of the rigid connection; The frame acceleration is 17% of that of the rigid connection and 70% of that of the bare frame, and; the connection force is 12% of that of the rigid connection. These huge differences are attributed to three factors: a viscous damper as a connection dissipates energy thus reduces the response; the forces in viscous dampers are out-of-phase with forces due to displacements and with inertia forces, and; the period of the damped system is much longer than that of the rigidly connected system. Smaller drifts than those attained with viscous damping only are attained when some stiffness is added to the damping. Those are accompanied, however, with some increase in other responses.

Discussion

This paper studied the seismic behavior of wall–EDD–frame systems in the context of retrofitting existing frame structures. The

controlling non-dimensional parameters of such systems were identified and a rigorous parametric study was performed.

It was found that, frame inter-story drifts, as well as other frame responses, could efficiently be reduced when connecting the existing frame building to new walls with viscous dampers, with or without springs in parallel. Those responses were sometimes comparable to those attained with a rigid connection between the frame and the new wall. Other responses such as frame absolute accelerations, connection force, and wall responses, were much smaller when a connection with dampers was utilized in comparison to those attained with a rigid connection. These huge differences are attributed to three factors: a viscous damper as a connection dissipates energy thus reduces the response; the forces in viscous dampers are out-of-phase with forces due to displacements and with inertia forces, and; the period of the damped system is much longer than that of the rigidly connected system. Thus, it is suggested to use a connection with viscous dampers, possibly with springs as well, in any case that it can reduce the frame drifts to allowable values. This is especially true if the periods of the modes contributing to the bare frame response are within the constant velocity region. In addition, the sensitivity of the damped system to uncertainty in wall stiffness is much smaller than that of the system with a rigid connection.

Finally, tools were given for initial design of such retrofitting schemes. These enable both choosing the most appropriate retrofitting alternative and selecting initial values for its parameters, as demonstrated by the example.

References

Adachi, F., Yoshitomi, S., Tsuji, M., and Takewaki, I. (2013). Nonlinear optimal oil damper design in seismically controlled multi-story building frame. *Soil Dynam. Earthquake Eng.* 44, 1–13. doi:10.1016/j.soildyn.2012.08.010

Aguirre, J. J., Almazán, J. L., and Paul, C. J. (2013). Optimal control of linear and nonlinear asymmetric structures by means of passive energy dampers. *Earthquake Eng. Struct. Dynam.* 42, 377–395. doi:10.1002/eqe.2211

Aida, T., Aso, T., Takeshita, K., Takiuchi, T., and Fujii, T. (2001). Improvement of the structure damping performance by interconnection. *J. Sound Vib.* 242, 333–353. doi:10.1006/jsvi.2000.3349

Aydin, E., Boduroglu, M. H., and Guney, D. (2007). Optimal damper distribution for seismic rehabilitation of planar building structures. *Eng. Struct.* 29, 176–185. doi:10.1016/j.engstruct.2006.04.016

Barenblatt, G. I. (2003). *Scaling.* Cambridge: Cambridge University Press.

Bhaskararao, A. V., and Jangid, R. S. (2007). Optimum viscous damper for connecting adjacent SDOF structures for harmonic and stationary white-noise random excitations. *Earthquake Eng. Struct. Dynam.* 36, 563–571. doi:10.1002/eqe.636

Bigdeli, K., Hare, W., Nutini, J., and Tesfamariam, S. (2015). Optimizing damper connectors for adjacent buildings. *Optim. Eng.*

Caughey, T. K., and O'Kelly, M. E. J. (1965). Classical normal modes in damped linear dynamic systems. *J. Appl. Mech.* 32, 583–588. doi:10.1115/1.3627262

CEN-Comité Européen de Normalisation. (2003). *Eurocode8: Design of Structures for Earthquake Resistance. Part1: General Rules, Seismic Actions and Rules for Buildings.* Brussels: European Committee for Standardization.

Christopoulos, C., and Filiatrault, A. (2006). *Principles of Passive Supplemental Damping and Seismic Isolation.* Pavia: IUSS Press.

Cimellaro, G. P., and Lopez-Garcia, D. (2011). Algorithm for design of controlled motion of adjacent structures. *Struct. Contr. Health Monit.* 18, 140–148. doi:10.1002/stc.357

Constantinou, M.C., and Symans, M.D. (1992). *Experimental and Analytical Investigation of Seismic Response of Structures with Supplemental Fluid Viscous Dampers.* Report No. NCEER-92-0032. New York, NY: National Center for Earthquake Engineering Research, State University of New York at Buffalo.

Crowley, H., and Pinho, R. (2004). Period-height relationship for existing European reinforced concrete buildings. *J. Earthquake Eng.* 8, 93–119. doi:10.1080/13632460409350522

Dargush, G. F., and Sant, R. S. (2005). Evolutionary aseismic design and retrofit of structures with passive energy dissipation. *Earthquake Eng. Struct. Dynam.* 34, 1601–1626. doi:10.1002/eqe.497

Enrique Luco, J., and De Barros, F. C. P. (1998). Optimal damping between two adjacent elastic structures. *Earthquake Eng. Struct. Dynam.* 27, 649–659. doi:10.1002/(SICI)1096-9845(199807)27:7<649::AID-EQE748>3.0.CO;2-5

FEMA 356. (2000). *Prestandard and Commentary for the Seismic Rehabilitation of Buildings.* Washington, DC: Federal Emergency Management Agency.

Gidaris, I., and Taflanidis, A. A. (2015). Performance assessment and optimization of fluid viscous dampers through life-cycle cost criteria and comparison to alternative design approaches. *Bull. Earthquake Eng.* 13, 1003–1028.

Gluck, N., Reinhorn, A. M., Gluck, J., and Levy, R. (1996). Design of supplemental dampers for control of structures. *J. Struct. Eng.* 122, 1394–1399. doi:10.1061/(ASCE)0733-9445(1996)122:12(1394)

Hatzigeorgiou, G. D., and Pnevmatikos, N. G. (2014). Maximum damping forces for structures with viscous dampers under near-source earthquakes. *Eng. Struct.* 68, 1–13. doi:10.1016/j.engstruct.2014.02.036

Huang, X., and Zhu, H. P. (2013). Optimal arrangement of viscoelastic dampers for seismic control of adjacent shear-type structures. *J. Zhejiang Univ.* 14, 47–60. doi:10.1631/jzus.A1200181

Kanno, Y. (2013). Damper placement optimization in a shear building model with discrete design variables: a mixed-integer second-order cone programming approach. *Earthquake Eng. Struct. Dynam.* 42, 1657–1676. doi:10.1002/eqe.2292

Kim, J., Ryu, J., and Chung, L. (2006). Seismic performance of structures connected by viscoelastic dampers. *Eng. Struct.* 28, 183–195. doi:10.1016/j.engstruct.2005.05.014

Landi, L., Lucchi, S., and Diotallevi, P. P. (2014). A procedure for the direct determination of the required supplemental damping for the seismic retrofit with viscous dampers. *Eng. Struct.* 71, 137–149. doi:10.1016/j.engstruct.2014.04.025

Lavan, O. (2012). On the efficiency of viscous dampers in reducing various seismic responses of wall structures. *Earthquake Eng. Struct. Dynam.* 41, 1673–1692. doi:10.1002/eqe.1197

Lavan, O. (2015). A methodology for the integrated seismic design of nonlinear buildings with supplemental damping. *Struct. Contr. Health Monit.* 22, 484–499. doi:10.1002/stc.1688

Lavan, O., and Amir, O. (2014). Simultaneous topology and sizing optimization of viscous dampers in seismic retrofitting of 3D irregular frame structures. *Earthquake Eng. Struct. Dynam.* 43, 1325–1342. doi:10.1002/eqe.2399

Lavan, O., and Avishur, M. (2013). Seismic behavior of viscously damped yielding frames under structural and damping uncertainties. *Bull. Earthquake Eng.* 11, 2309–2332. doi:10.1007/s10518-013-9479-7

Lavan, O., Cimellaro, G. P., and Reinhorn, A. M. (2008). Noniterative optimization procedure for seismic weakening and damping of inelastic structures. *J. Struct. Eng.* 134, 1638–1648. doi:10.1061/(ASCE)0733-9445(2008)134:10(1638)

Lavan, O., and Dargush, G. F. (2009). Multi-objective evolutionary seismic design with passive energy dissipation systems. *J. Earthquake Eng.* 13, 758–790. doi:10.1080/13632460802598545

Lavan, O., and Levy, R. (2005). Optimal design of supplemental viscous dampers for irregular shear-frames in the presence of yielding. *Earthquake Eng. Struct. Dynam.* 34, 889–907. doi:10.1002/eqe.458

Lavan, O., and Levy, R. (2006). Optimal peripheral drift control of 3D irregular framed structures using supplemental viscous dampers. *J. Earthquake Eng.* 10, 903–923. doi:10.1142/S1363246906002931

Lavan, O., and Levy, R. (2009). Simple iterative use of Lyapunov's solution for the linear optimal seismic design of passive devices in framed buildings. *J. Earthquake Eng.* 13, 650–666. doi:10.1080/13632460902837736

Lavan, O., and Levy, R. (2010). Performance based optimal seismic retrofitting of yielding plane frames using added viscous damping. *Earthquake Struct.* 1, 307–326. doi:10.12989/eas.2010.1.3.307

Lin, J. L., Bui, M. T., and Tsai, K. C. (2014). An energy-based approach to the generalized optimal locations of viscous dampers in two-way asymmetrical buildings. *Earthquake Spectra* 30, 867–889. doi:10.1193/052312EQS196M

Lopez-Garcia, D., and Soong, T. T. (2002). Efficiency of a simple approach to damper allocation in MDOF structures. *J. Struct. Contr.* 9, 19–30. doi:10.1002/stc.3

Makris, N., and Black, C. J. (2004a). Dimensional analysis of rigid-plastic and elasto-plastic structures under pulse-type excitations. *J. Eng. Mech.* 130, 1006–1018. doi:10.1061/(ASCE)0733-9399(2004)130:9(1006)

Makris, N., and Black, C. J. (2004b). Dimensional analysis of bilinear oscillators under pulse-type excitations. *J. Eng. Mech.* 130, 1019–1031. doi:10.1061/(ASCE)0733-9399(2004)130:9(1006)

Martínez, C. A., Curadelli, O., and Compagnoni, M. E. (2013). Optimal design of passive viscous damping systems for buildings under seismic excitation. *J. Construct. Steel Res.* 90, 253–264. doi:10.1016/j.jcsr.2013.08.005

Murase, M., Tsuji, M., and Takewaki, I. (2013). Smart passive control of buildings with higher redundancy and robustness using base-isolation and interconnection. *Earthquakes Struct.* 4, 649–670. doi:10.12989/eas.2013.4.6.649

Nakashima, M., Lavan, O., Kurata, M., and Luo, Y. (2014). Earthquake engineering research needs in light of lessons learned from the 2011 Tohoku earthquake. *Earthquake Eng. Eng. Vib.* 13, 141–149. doi:10.1007/s11803-014-0244-y

Palermo, M., Muscio, S., Silvestri, S., Landi, L., and Trombetti, T. (2013). On the dimensioning of viscous dampers for the mitigation of the earthquake-induced effects in moment-resisting frame structures. *Bull. Earthquake Eng.* 11, 2429–2446. doi:10.1007/s10518-013-9474-z

Patel, C. C., and Jangid, R. S. (2010). Seismic response of dynamically similar adjacent structures connected with viscous dampers. *IES J. A Civ. Struct. Eng.* 3, 1–13.

Peng, Y., Mei, Z., and Li, J. (2014). Stochastic seismic response analysis and reliability assessment of passively damped structures. *J. Vib. Contr.* 20, 2352–2365. doi:10.1177/1077546313486910

Rama Raju, K., Ansu, M., and Iyer, N. R. (2014). A methodology of design for seismic performance enhancement of buildings using viscous fluid dampers. *Struct. Contr. Health Monit.* 21, 342–355. doi:10.1002/stc.1568

Ramirez, O. M., Constantinou, M. C., Kircher, C. A., Whittaker, A. S., Johnson, M. W., Gomez, J. D., and Chrysostomou, C. Z. (2001). *Development and Evaluation of Simplified Procedures for Analysis and Design of Buildings with Passive Energy Dissipation Systems-Revision 01.* Technical Report MCEER-00-0010. Buffalo, NY.

Richardson, A., Walsh, K., and Abdullah, M. (2013). Closed-form equations for coupling linear structures using stiffness and damping elements. *Struct. Contr. Health Monit.* 20, 259–281. doi:10.1002/stc.490

Silvestri, S., and Trombetti, T. (2007). Physical and numerical approaches for the optimal insertion of seismic viscous dampers in shear-type structures. *J. Earthquake Eng.* 11, 787–828. doi:10.1080/13632460601034155

Singh, M. P. (1980). Seismic response by SRSS for nonproportional damping. *J. Eng. Mech. Div.* 106, 1405–1419.

Singh, M. P., and Moreschi, L. M. (2001). Optimal seismic response control with dampers. *Earthquake Eng. Struct. Dynam.* 30, 553–572. doi:10.1002/eqe.23

Song, J., Chu, Y. L., Liang, Z., and Lee, G. C. (2008). *Modal Analysis of Generally Damped Linear Structures Subjected to Seismic Excitations.* Report No. MCEER-08-0005. New York, NY: Multidisciplinary Center for Earthquake Engineering Research, State University of New York at Buffalo.

Sonmez, M., Aydin, E., and Karabork, T. (2013). Using an artificial bee colony algorithm for the optimal placement of viscous dampers in planar building frames. *Struct. Multidiscip. Optim.* 48, 395–409. doi:10.1007/s00158-013-0892-y

Soong, T. T. (1990). *Active Structural Control*. Harlow: Longman Scientific & Technical.

Soong, T. T., and Dargush, G. F. (1997). *Passive Energy Dissipation Systems in Structural Engineering*. Chichester: John Wiley & Sons Ltd.

Stafford Smith, B., and Coull, A. (1991). *Tall Building Structures: Analysis and Design*. New York: John Wiley & Sons.

Sun, H. S., Liu, M. H., and Zhu, H. P. (2014). Connecting parameters optimization on unsymmetrical twin-tower structure linked by sky-bridge. *J. Cent. S. Univ.* 21, 2460–2468. doi:10.1007/s11771-014-2200-4

Takewaki, I. (1997). Optimal damper placement for minimum transfer functions. *Earthquake Eng. Struct. Dynam.* 26, 1113–1124. doi:10.1002/(SICI)1096-9845(199711)26:11<1113::AID-EQE696>3.0.CO;2-X

Takewaki, I. (2004). Frequency domain modal analysis of earthquake input energy to highly damped passive control structures. *Earthquake Eng. Struct. Dynam.* 33, 575–590. doi:10.1002/eqe.361

Takewaki, I. (2007). Earthquake input energy to two buildings connected by viscous dampers. *J. Struct. Eng.* 133, 620–628. doi:10.1061/(ASCE)0733-9445(2007)133:5(620)

Takewaki, I. (2009). *Building Control with Passive Dampers: Optimal Performance-based Design for Earthquakes*. Singapore: John Wiley & Sons Ltd. (Asia).

Trombetti, T., and Silvestri, S. (2004). Added viscous dampers in shear-type structures: the effectiveness of mass proportional damping. *J. Earthquake Eng.* 8, 275–313. doi:10.1080/13632460409350490

Trombetti, T., and Silvestri, S. (2006). On the modal damping ratios of shear-type structures equipped with Rayleigh damping systems. *J. Sound Vib.* 292, 21–58. doi:10.1016/j.jsv.2005.07.023

Trombetti, T., and Silvestri, S. (2007). Novel schemes for inserting seismic dampers in shear-type systems based upon the mass proportional component of the Rayleigh damping matrix. *J. Sound Vib.* 302, 486–526. doi:10.1016/j.jsv.2006.11.030

Tubaldi, E. (2015). Dynamic behavior of adjacent buildings connected by linear viscous/viscoelastic dampers. *Struct. Contr. Health Monit.* doi:10.1002/stc.1734

Tubaldi, E., Ragni, L., and Dall'Asta, A. (2015). Probabilistic seismic response assessment of linear systems equipped with nonlinear viscous dampers. *Earthquake Eng. Struct. Dynam.* 44, 101–120. doi:10.1002/eqe.2461

Zhang, R. H., and Soong, T. T. (1992). Seismic design of viscoelastic dampers for structural applications. *J. Struct. Eng.* 118, 1375–1392. doi:10.1061/(ASCE)0733-9445(1992)118:5(1375)

Zhang, W. S., and Xu, Y. L. (1999). Dynamic characteristics and seismic response of adjacent buildings linked by discrete dampers. *Earthquake Eng. Struct. Dynam.* 28, 1163–1185. doi:10.1002/(SICI)1096-9845(199910)28:10<1163::AID-EQE860>3.3.CO;2-S

Zhang, W. S., and Xu, Y. L. (2000). Vibration analysis of two buildings linked by Maxwell model-defined fluid dampers. *J. Sound Vib.* 233, 775–796. doi:10.1006/jsvi.1999.2735

Conflict of Interest Statement: The authors declare that the research was conducted in the absence of any commercial or financial relationships that could be construed as a potential conflict of interest.

Appendix

List of symbols

h	Typical story height
c_d	Damping coefficient of the EDDs per unit height
$\boldsymbol{f}(t)$	Load vector
f_d	Peak force experience by the EDDs per unit height
k_d	Stiffness of the EDDs per unit height
m_f	Seismic mass per unit height taken by the frame
m_w	Seismic mass per unit height taken by the wall
$\left.\lvert q_i(t)\rvert\right._{\max}$	Spectral displacement corresponding to the natural period and the damping ratio of an underdamped mode
$\left.\lvert q_i^p(t)\rvert\right._{\max}$	Is the spectral displacement corresponding to the natural period of an overdamped mode
$\boldsymbol{x}(t)$	Vector of coordinate displacements
t	Time
A_{0i}, B_{0i}, and A_{0i}^p	Coefficients
A_f	The maximum peak frame absolute acceleration
BS_w	Peak wall base shear
$\mathbf{C}$	Damping matrix
C_d	The damping coefficient of each connecting viscous damper
D_d	Maximum peak relative displacement between the frame and the wall, or the EDD's elongation/contraction
EI	The bending stiffness of the wall
F_d	Peak force experience by the EDDs
GA	Shear stiffness of the frame
H	The total height of building
ID_f	The maximum peak inter-story drift angle of all stories of the frame
$\mathbf{K}$	Stiffness matrix
K_d	The stiffness of each connecting spring
K_s	The story stiffness of the frame (shear force required to result in a unit inter-story drift)
$\mathbf{M}$	Mass matrix
M_f	The seismic mass per floor taken by the frame (equal for all floors)
M_w	The seismic mass per floor taken by the wall
N	Number of stories in the building
OTM_f	The peak base over-turning moment of the frame
OTM_w	Peak wall base moment
R_ξ	A correction factor to account for damping ratios different than 0.05
$S_a(T')$	The elastic 5% damping spectral acceleration
T_c	The corner period between the constant acceleration and the constant velocity regions
S_c	The elastic 5% damping spectral acceleration at a period of T_c
T_1	Fundamental period of the bare frame,
λ_l	Eigenvalues
ξ	Damping ratio (as a value, not in per-cent)
ξ_i	Damping ratio of the mode i
π_c	Relative EDD damping
π_{f_d}	Non-dimensional peak force experience by the EDDs per unit height
π_k	Relative EDD stiffness
π_m	Ratio of wall to frame mass
π_s	Relative frame stiffness to wall stiffness
π_{A_f}	Non-dimensional maximum peak frame absolute acceleration
π_{BS_w}	Non-dimensional peak wall base shear
π_{D_d}	Non-dimensional maximum peak relative displacement between the frame and the wall, or the EDD's elongation/contraction
π_{ID_f}	Non-dimensional maximum peak inter-story drift angle of all stories of the frame
π_{OTM_f}	Non-dimensional base over-turning moment of the frame
π_{OTM_w}	Non-dimensional base over-turning moment of the wall
π_T	Non-dimensional period of the system
π_ξ	Non-dimensional damping ratio of the system
ω_i	Natural frequency of the mode i
$\lvert \cdot \rvert$	Absolute of a complex number
real$(\cdot)$	Real part of a complex number

10

The 2015 Gorkha Nepal earthquake: insights from earthquake damage survey

Katsuichiro Goda[1]*, Takashi Kiyota[2], Rama Mohan Pokhrel[2], Gabriele Chiaro[2], Toshihiko Katagiri[2], Keshab Sharma[3] and Sean Wilkinson[4]

[1] Department of Civil Engineering, University of Bristol, Bristol, UK, [2] Institute of Industrial Science, University of Tokyo, Tokyo, Japan, [3] Department of Civil and Environmental Engineering, University of Alberta, Edmonton, AB, Canada, [4] School of Civil Engineering and Geosciences, Newcastle University, Newcastle upon Tyne, UK

The 2015 Gorkha Nepal earthquake caused tremendous damage and loss. To gain valuable lessons from this tragic event, an earthquake damage investigation team was dispatched to Nepal from 1 May 2015 to 7 May 2015. A unique aspect of the earthquake damage investigation is that first-hand earthquake damage data were obtained 6–11 days after the mainshock. To gain deeper understanding of the observed earthquake damage in Nepal, the paper reviews the seismotectonic setting and regional seismicity in Nepal and analyzes available aftershock data and ground motion data. The earthquake damage observations indicate that the majority of the damaged buildings were stone/brick masonry structures with no seismic detailing, whereas the most of RC buildings were undamaged. This indicates that adequate structural design is the key to reduce the earthquake risk in Nepal. To share the gathered damage data widely, the collected damage data (geo-tagged photos and observation comments) are organized using Google Earth and the kmz file is made publicly available.

Keywords: 2015 Nepal earthquake, earthquake damage survey, building damage, ground motion, aftershocks

Edited by:
Solomon Tesfamariam,
The University of British Columbia,
Canada

Reviewed by:
Vladimir Sokolov,
Karlsruhe Institute of Technology,
Germany
Takeshi Koike,
Kyoto University, Japan

*Correspondence:
Katsuichiro Goda,
Department of Civil Engineering,
University of Bristol, Queen's
Building, University Walk, Bristol BS8
1TR, UK
katsu.goda@bristol.ac.uk

Introduction

An intense ground shaking struck Central Nepal on 25 April 2015 (local time 11:56 a.m.). The moment magnitude of the earthquake was $M_w 7.8$ with its hypocenter located in the Gorkha region (about 80 km north–west of Kathmandu). The earthquake occurred at the subduction interface along the Himalayan arc between the Indian plate and the Eurasian plate (Avouac, 2003; Ader et al., 2012). The earthquake rupture propagated from west to east and from deep to shallow parts of the shallowly dipping fault plane [United States Geological Survey (USGS), (2015)], and consequently, strong shaking was experienced in Kathmandu and the surrounding municipalities. This was the largest event since 1934, $M_w 8.1$ Bihar–Nepal earthquake (Ambraseys and Douglas, 2004; Bilham, 2004). The 2015 mainshock destroyed a large number of buildings and infrastructure in urban and rural areas, and triggered numerous landslides and rock/boulder falls in the mountain areas, blocking roads, and hampering rescue and recovery activities. Moreover, aftershock occurrence has been active since the mainshock; several major aftershocks (e.g., $M_w 6.7$ and $M_w 7.3$ earthquakes in the Kodari region, north–east of Kathmandu) caused additional damage to rural towns and villages in the northern part of Central Nepal. As of 26 May 2015, the earthquake damage statistics for Nepal from the 25 April 2015 mainshock stand at the total number of 8,510 deaths and 199 missing[1]. In addition, the major aftershock that occurred on 12 May 2015 caused 163 deaths/missing. Center for Disaster Management and Risk Reduction Technology (CEDIM), (2015) reports that the total economic loss is in the order of 10 billion U.S. dollars, which is about a half of Nepal's gross domestic product. The 2015 earthquakes will have grave long-term socioeconomic impact on people and communities in Nepal [United Nations Office for the Coordination of Humanitarian Affairs (UN-OCHA), (2015)].

Earthquake field observations provide raw damage data of existing built environments and are useful for developing empirical correlation between ground motion intensity and damage severity for earthquake impact assessment of future events. To gain valuable lessons from this tragic event, an earthquake damage investigation team was jointly organized by the Japan Society of Civil Engineers and the Japan Geotechnical Society, and was dispatched to Nepal from 1 May 2015 to 7 May 2015. The survey trip was planned in such away that relatively large geographical areas that were affected by the earthquakes were covered to grasp spatial features of the damage in the earthquake-hit regions. A unique aspect of this damage investigation is that the data were collected at the early stage of disaster response and recovery (6–11 days after the mainshock), and thus first-hand earthquake damage observations were obtained before major repair work. The collected damage data, in the form of geo-tagged photos and some measurements (e.g., size of a landslide), are useful for other earthquake damage reconnaissance teams who visit Nepal several weeks after the mainshock, and serve as a starting point of longitudinal research of a recovery process from the earthquakes. To achieve this goal, damage photos that were taken during the survey trip are organized using Google Earth and are made publicly available; the kmz file is provided as supplementary resource of this paper. Viewers can download the photos directly and can use them for research and educational purposes; all photos are geo-tagged and are accompanied by brief comments.

This paper summarizes key findings of ground shaking damage in Nepal, and is organized as follows. To link building damage observations with available seismological data, seismotectonic setting of Nepal is reviewed, and earthquake rupture process and aftershock data, which are available from the U.S. Geological Survey (USGS), are analyzed to gain scientific insights into ground motions that were experienced during the mainshock and major aftershocks. It is important to note that strong motion observation networks in Nepal are not well developed and data are not publicly accessible. This means that the estimation of observed ground motions at building damage sites is highly uncertain. Currently, recorded time-history data of strong motion are only available at the KATNP station, which is located in the city center of Kathmandu. In this study, strong motion data at KATNP are analyzed and the results, in the form of elastic response spectra, are discussed by comparing with relevant ground motion prediction models [e.g., Kanno et al. (2006) and Boore and Atkinson (2008)] and with well-recorded strong motion data from the 2008 M_w7.9 Wenchuan China earthquake

(Lu et al., 2010), seismological features of which are broadly similar to the 2015 Nepal earthquake. Furthermore, issues related to ground motion estimation for prompt earthquake impact assessment [e.g., Jaiswal and Wald (2010) and Center for Disaster Management and Risk Reduction Technology (CEDIM), (2015)] are discussed by examining how the way source-to-site distance measures, as in ground motion prediction equations, are evaluated affects the scenario shake map of a large subduction event within a fault rupture zone (note: size of the fault rupture zone can be in the order of a few hundred kilometers for M_w8.0+ earthquakes). Such investigations provide new insights for improvements in producing more reliable scenario shake maps and prompt earthquake impact assessments (Goda and Atkinson, 2014). Subsequently, building typology in Nepal is reviewed briefly, followed by earthquake damage observations in Kathmandu, Melamchi, Trishuli, and Baluwa. Finally, key lessons from the 2015 Nepal earthquake are summarized.

Regional Seismicity and Ground Motion

This section aims at providing with relevant seismological information for interpreting earthquake damage survey observations in Nepal (which are discussed in the following section). First, seismotectonic and seismological aspects of the on-going mainshock–aftershock sequence are reviewed by analyzing available earthquake catalog data and source rupture models of the mainshock. Strong ground motion recordings at KATNP are analyzed to estimate the observed ground motion intensity in Kathmandu. Subsequently, scenario shake maps are generated by considering different source-to-site distance measures to highlight the influence of finite-fault source representation for a large earthquake in applications to prompt earthquake impact assessment.

Seismotectonic Setting and Seismic Hazard in Nepal

Nepal is located along the active Main Himalayan Thrust arc, where the subducting Indian plate and the overriding Eurasian plate interact. This region accommodates approximately a half of the tectonic convergence between these two plates, i.e., about 20 mm/year (Avouac, 2003; Ader et al., 2012). The locked part of the subduction interface has a low-dip angle (about 10°) and is located at depths of 4–18 km (Bilham, 2004), and has potential to generate M_w8+ earthquakes (Gupta, 2006).

Historically, Nepal hosted several large earthquakes (Ambraseys and Douglas, 2004; Bilham, 2004). A map of Nepal and locations of major historical seismic events are shown in **Figure 1**. Western Nepal experienced a M_w8.2 event in 1505. This event occurred west of the rupture zone of the 2015 earthquake and accumulated strain in this seismic gap region has not been released since then; thus, there is high potential for future large earthquakes in the western region. In Eastern Nepal, two known major earthquakes occurred in 1833 and 1934. In particular, the 1934 M_w8.1 Bihar–Nepal earthquake was destructive and caused many fatalities (+10,000 deaths). The 2015 Gorkha–Kodari earthquakes have ruptured a fault section that overlaps with the fault rupture plane of the 1934 earthquake (see **Figure 1**). It is noted that the rupture planes of

---

[1]http://earthquake-report.com/

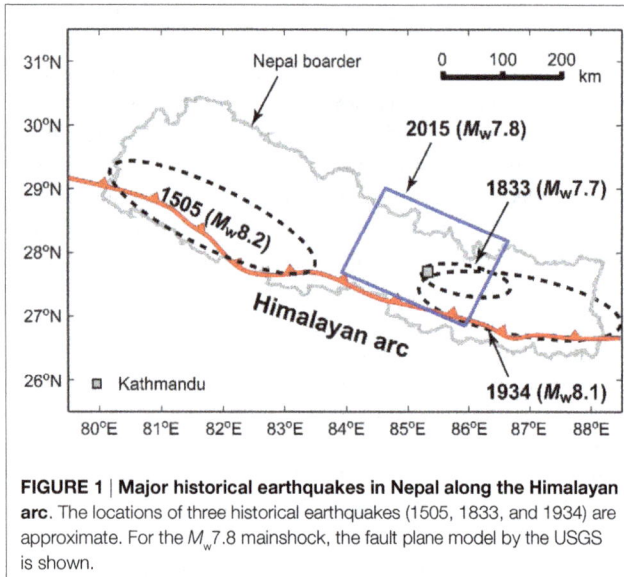

FIGURE 1 | Major historical earthquakes in Nepal along the Himalayan arc. The locations of three historical earthquakes (1505, 1833, and 1934) are approximate. For the M_w7.8 mainshock, the fault plane model by the USGS is shown.

the 1934 and 2015 earthquakes are directly beneath Kathmandu, although the locations of their hypocenters are east and west of Kathmandu, respectively.

Recently, several probabilistic seismic hazard studies have been conducted for Nepal by employing updated seismic source zone models based on improved earthquake catalogs and modern ground motion models [e.g., Nath and Thingbaijam (2012) and Ram and Wang (2013)]. The estimated peak ground acceleration (PGA) with 10% probability of exceedance in 50 years (i.e., return period of 475 years) in Western Nepal ranges between 0.5 and 0.6 g, whereas that in Eastern Nepal ranges between 0.3 and 0.6 g. These hazard estimates are obtained for rock sites, therefore, when typical soil sites are considered (e.g., Kathmandu Valley), they need to be increased. An important observation is that the ground motion shaking in Kathmandu during the 2015 mainshock (which is discussed in detail in the following) was less than the PGA estimates with 10% probability of exceedance in 50 years, which may be considered as a basis for seismic design in Nepal.

Fault Rupture Model of the 25 April 2015 Mainshock

Several earthquake rupture models for the 2015 mainshock have been developed [e.g., United States Geological Survey (USGS) (2015); Yagi (2015)]. A common feature of the estimated slip distributions is that large slips occurred north and north–east of Kathmandu, and the rupture propagated from the hypocenter (north–west of Kathmandu) toward east as well as south (deeper to shallower depth). The slip distribution of the USGS model is illustrated in **Figure 2A**. The fault length and width of the rupture plane are 220 and 165 km, respectively, and its strike and dip are 295° and 10°, respectively. **Figure 3** overlays the route of the survey trip over the USGS source model to put visited locations (i.e., Melamchi, Trishuli, and Baluwa) into perspective with respect to the earthquake slip distribution. The USGS source model has its maximum slip of 3.11 m (north of Kathmandu). It is also interesting to observe that the estimated slip near the hypocenter is 1.29 m,

which is about 40% of the maximum slip, and its distance from the maximum slip sub-fault (i.e., asperity) is about 70 km. By analyzing numerous earthquake rupture models statistically, Mai et al. (2005) found that the rupture often nucleates in the regions of low-to-moderate slip (sub-faults with slip <2/3 of the maximum slip) and close to the maximum slip sub-fault. The rupture nucleation of the 2015 mainshock (i.e., slip and location at the hypocenter) is in good agreement with these empirical rules suggested by Mai et al. (2005).

Aftershocks

In post-earthquake situations, one of the major concerns for evacuees and emergency response teams is the occurrence of major aftershocks, triggering secondary hazards. Generally, a larger earthquake is followed by more aftershocks, and returning to a background level of seismic activities takes longer. **Figure 2A** shows the spatial distribution of aftershocks that occurred before 25 May 2015 (30 days since the mainshock). The aftershock data are obtained from the USGS NEIC catalog[2]. Immediately after the mainshock, a moderate (M_w6.6) aftershock occurred near the hypocenter. On the other hand, the majority of aftershocks occurred in the Kodari region (north–east of Kathmandu); a notable event was the 12 May 2015 M_w7.3 aftershock, which caused additional damage and casualties. Comparison of the aftershock distribution with respect to the slip distribution of the mainshock indicates that the major aftershocks do not occur very near to the mainshock asperity (with large slip) but they occur in the surrounding areas of the mainshock asperity. This is because the spatial and temporal characteristics of aftershocks are the manifestation of internal crustal dynamics involving the redistribution of stress and displacement fields (Stern, 2002; Heuret et al., 2011).

To gain further insights into the aftershock occurrence process of the 2015 mainshock–aftershock sequence, statistical analysis of aftershock data is carried out by applying the Gutenberg–Richter law and the modified Omori law (Shcherbakov et al., 2005); the completeness magnitude is set to 4.5 for the analyses. The Gutenberg–Richter law describes the frequency–magnitude characteristics of an aftershock sequence, whereas the modified Omori law models a temporal decay of an aftershock occurrence rate. The fitting of the 2015 Nepal aftershock data to the Gutenberg–Richter relationship is satisfactory (**Figure 2B**); the estimated slope parameter (i.e., b-value) is −0.862. This slope is slightly gentler (i.e., more productive for larger aftershocks) than the typical b-value for global subduction earthquakes but within the expected range (Shcherbakov et al., 2013). **Figure 2C** shows that the modified Omori's law fits well with the aftershock data. The obtained parameters are typical for global subduction earthquakes (Shcherbakov et al., 2013). For example, the temporal decay parameter (i.e., p value, power parameter in the equation shown in **Figure 2C**) is 1.049, which is close to the global average of about 1.2 (by taking into account inherent variability of this parameter). The above results support the applicability of well-established empirical laws for characterizing the 2015 Nepal aftershock data. This is a useful confirmation from seismic risk management viewpoints because initial estimates of aftershock-related hazard can be obtained from

---

[2]http://earthquake.usgs.gov/earthquakes/search/

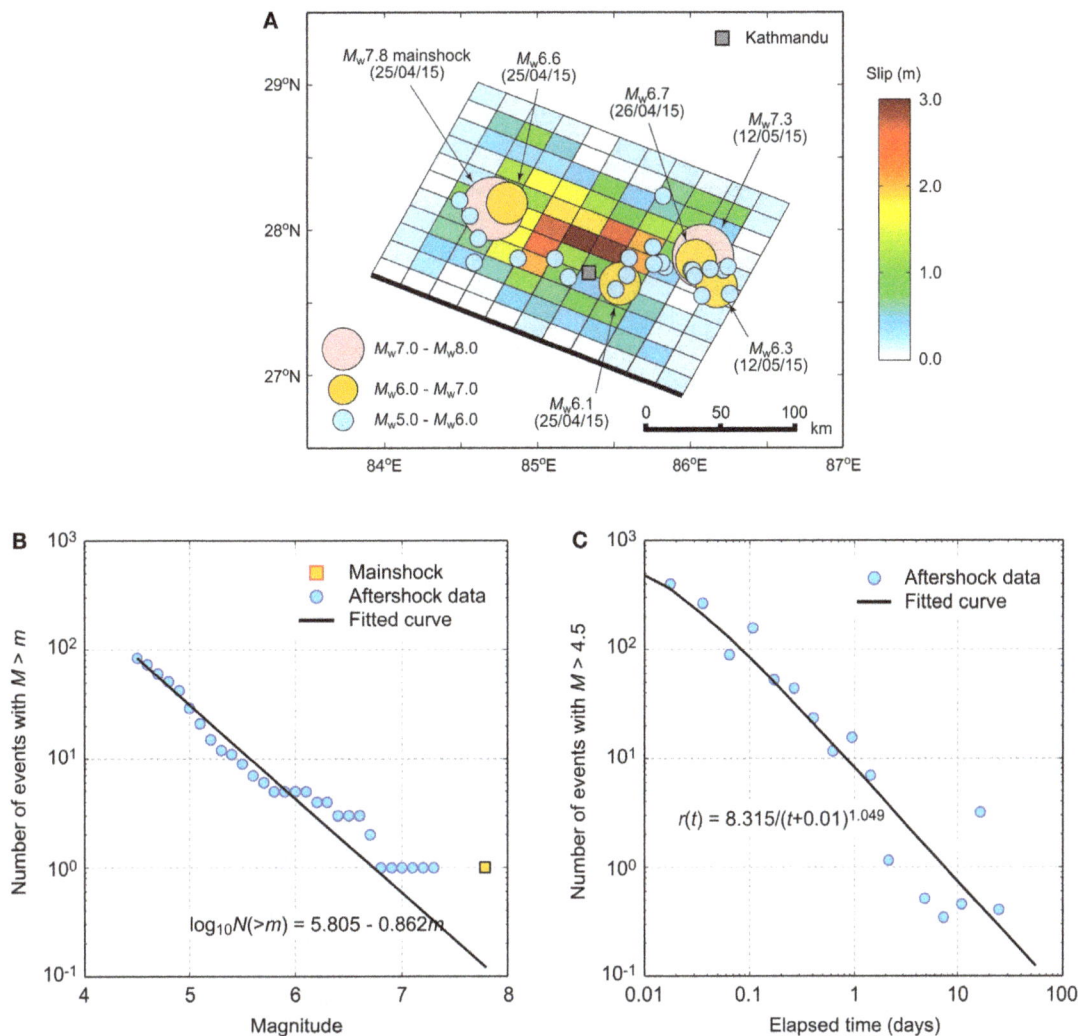

FIGURE 2 | (A) Aftershock distribution of the 2015 earthquake sequence; an earthquake source model by the USGS is shown. **(B)** Gutenberg–Richter relationship of the 2015 earthquake sequence. **(C)** Modified Omori law of the 2015 earthquake sequence.

the empirical aftershock models immediately after the mainshock (before real time data are collected and analyzed).

Ground Motion in Kathmandu

The accelerograms recorded at KATNP are publicly available[3]. In light of poor strong motion network in Nepal, the recorded ground motion data at KATNP are invaluable and serve as a benchmark in estimating ground motion intensity at unobserved locations in Kathmandu. **Figure 4** shows the location of the KATNP station; the map also shows the locations of the earthquake damage survey sites in Kathmandu. The KATNP station is located near the historical district in the city center (e.g., Durbar Square), where severe damage and collapse of old historical buildings occurred.

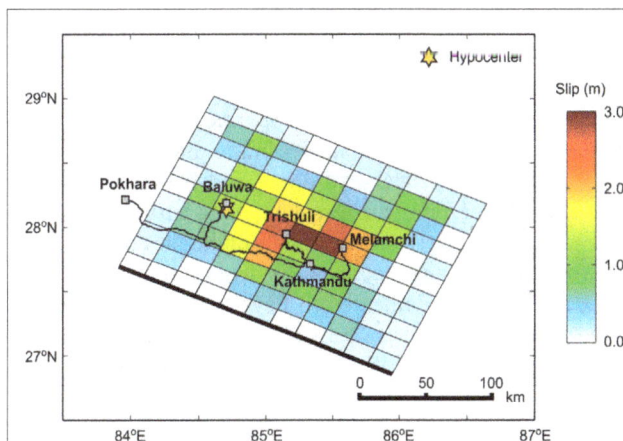

FIGURE 3 | Earthquake damage survey locations.

---

[3] http://www.strongmotioncenter.org/

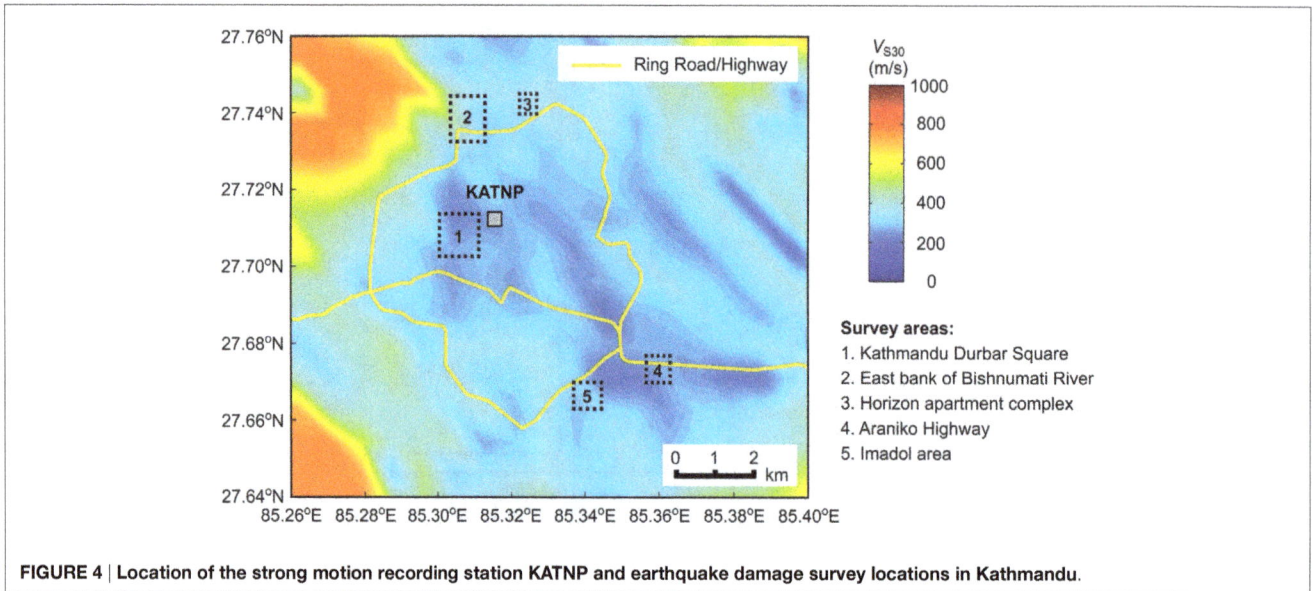

FIGURE 4 | Location of the strong motion recording station KATNP and earthquake damage survey locations in Kathmandu.

Prior to ground motion data analysis and estimation, it is important to review typical site conditions in Kathmandu, as they affect ground motion intensity significantly. Kathmandu is located in the Kathmandu Basin, where thick lacustrine and fluvio-lacustrine sediments are deposited (Sakai et al., 2002). The thickness of sediments (i.e., depth to bedrock) is in the range of 550–650 m. The setting of the Kathmandu Valley is similar to Mexico City (Paudyal et al., 2012), noting that during the 1985 Michoacán earthquake, long-period ground motions were significantly amplified in Mexico City due to soft lakebed deposits and caused catastrophic damage to mid-to-high-rise buildings. A seismic microzonation study in Kathmandu, conducted by Paudyal et al. (2012), indicates that the dominant periods of the ground at sites inside the Ring Road (see **Figure 4**) are between 1.0 and 2.0 s (i.e., high potential for resonating with long-period ground motions), and that the dominant period is correlated with the thickness of Pliocene and Quaternary deposits. The KATNP station is located within the long-dominant-period zone.

Another useful source of information in assessing site amplification potential of near-surface soil deposits in Kathmandu is the USGS global V_{S30} server (Wald and Allen, 2007)[4]. V_{S30} is the average shear-wave velocity in the uppermost 30 m and is often employed as a proxy site parameter in ground motion models [e.g., Kanno et al. (2006) and Boore and Atkinson (2008)]. Wald and Allen (2007) correlated V_{S30} data with topographic slope to derive the first-order estimate of the site amplification for two tectonic regimes, active and stable continental regions. The database is implemented to develop USGS ShakeMaps (Wald et al., 2005)[5], which are used for rapid earthquake impact assessment (Jaiswal and Wald, 2010)[6]. **Figure 4** shows the V_{S30}

contour map in Kathmandu. The map indicates that the central part of Kathmandu has soft surface deposits (typically NEHRP site class D, V_{S30} between 180 and 360 m/s). The V_{S30} value at the KATNP station is 250 m/s. It is noteworthy that V_{S30} is applicable to near-surface site amplification only; amplification of long-period seismic waves due to a large-scale geological structure (e.g., basin) should be taken into account separately.

Figures 5A,B show recorded accelerograms (three components) at KATNP for the $M_w7.8$ mainshock and for the $M_w7.3$ aftershock, respectively (note: among other recorded aftershock ground motions at KATNP, the $M_w7.3$ aftershock records show the most significant effects). An inspection of the time-history data indicates that the PGA of the recorded ground motions is about 150–170 and 70–80 cm/s² for the $M_w7.8$ mainshock and the $M_w7.3$ aftershock, respectively. These are significantly smaller than the PGA estimates with 10% probability of exceedance in 50 years from the recent regional seismic hazard studies (Nath and Thingbaijam, 2012; Ram and Wang, 2013). It is also observed that long-period components are present in the $M_w7.8$ mainshock records (**Figure 5A**). To further investigate the extent of ground shaking at KATNP, 5%-damped response spectra of the recorded accelerograms for the $M_w7.8$ mainshock and the $M_w7.3$ aftershock are calculated and compared in **Figure 5C**. The results suggest that the amplitudes of response spectra for the mainshock are greater than those for the major aftershock (also applicable to other aftershocks). For the $M_w7.8$ mainshock, two large peaks of response spectra are present at vibration periods around 0.2–0.6 s (N–S component only) and around 4.0–6.0 s (both N–S and E–W components). The former is attributed to direct shaking due to near-source ruptures, whereas the latter is caused by the combination of rich long-period content of seismic waves at source (because of large moment magnitude) and site amplification due to the basin effects. Given the existing building stock in Kathmandu/Nepal (the majority of buildings are low-to-mid rise and thus are likely to have vibration periods <1.0 s; Chaulagain et al., 2015), the main causes of severe structural damage and collapse of buildings in

---

[4]http://earthquake.usgs.gov/hazards/apps/vs30/
[5]http://earthquake.usgs.gov/earthquakes/shakemap/
[6]http://earthquake.usgs.gov/earthquakes/pager/

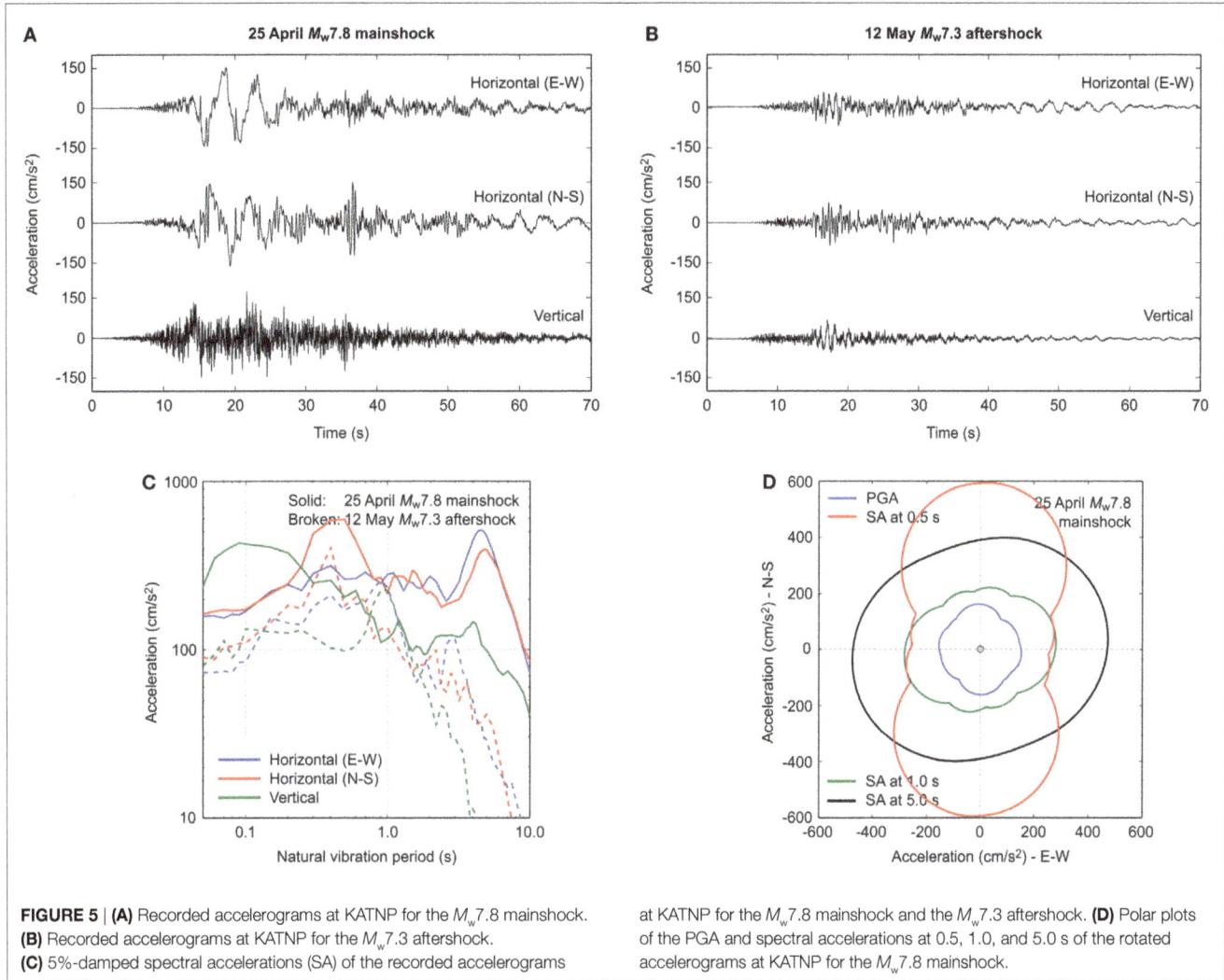

FIGURE 5 | (A) Recorded accelerograms at KATNP for the M_w7.8 mainshock. **(B)** Recorded accelerograms at KATNP for the M_w7.3 aftershock. **(C)** 5%-damped spectral accelerations (SA) of the recorded accelerograms at KATNP for the M_w7.8 mainshock and the M_w7.3 aftershock. **(D)** Polar plots of the PGA and spectral accelerations at 0.5, 1.0, and 5.0 s of the rotated accelerograms at KATNP for the M_w7.8 mainshock.

Kathmandu are due to the large peak in the short vibration period range. It is important to point out that buildings in Kathmandu were largely unaffected by the long-period ground motions in the Kathmandu Valley because of non-resonance. This was fortunate in the context of the current disaster. However, earthquake engineers should pay careful attention to long-period ground motions (Takewaki et al., 2011), when tall buildings are constructed in the central part of the Kathmandu Valley.

To further examine the orientation of ground motion parameters at KATNP for the M_w7.8 mainshock, PGA and 5%-damped spectral accelerations are computed by rotating accelerograms recorded at KATNP from 0° to 360° (Hong and Goda, 2007). The polar plots of PGA and spectral accelerations at 0.5, 1.0, and 5.0 s are shown in **Figure 5D**. The results are useful for understanding the orientation dependency of the peak seismic demand in the near-fault region (Huang et al., 2008). The results indicate that the spectral acceleration at 0.5 s (i.e., large response spectral peak in the short-vibration period range) is highly polarized; the ratio of the maximum-to-minimum response is about 2.5, while the degree of such polarization of the response spectra is much less pronounced

at other vibration periods. Although it is beyond the scope of this study, a further insight can be gained by investigating the effects of the orientation of ground motion with regard to the structural axis of damaged versus non-damaged buildings near the KATNP station.

Comparison of Observed Ground Motion in Kathmandu with Ground Motion from the 2008 Wenchuan Earthquake and Predicted Ground Motion

Due to the limited availability of recorded ground motions in Central Nepal, ground motion estimation may need to rely on: (1) ground motion data from other seismic regions having broad similarity with the target region [e.g., Sharma et al. (2009)], (2) empirical ground motion prediction models [e.g., Nath and Thingbaijam (2011)], or (3) ground motion simulations [e.g., Harbindu et al. (2014)]. In this study, the first two options are explored to gain insights into actual ground motions for the M_w7.8 mainshock.

For Option 1, ground motion data from the 2008 M_w7.9 Wenchuan earthquake (Lu et al., 2010) are analyzed. This earthquake is chosen because seismotectonic settings in Nepal and Tibet (i.e., southern and eastern sides of the Tibetan Plateau) are broadly similar and their earthquake magnitudes are comparable. The Wenchuan earthquake occurred along the Longmenshan fault Sichuan, China. The amplitude–distance plots of PGA and spectral accelerations at 0.5 and 5.0 s are shown in **Figure 6**; only records at soft soil sites (V_{S30} <400 m/s) are considered. The rupture distance (R_{rup}, shortest distance from a site of interest to the fault rupture plane) for the Wenchuan data is calculated using the fault plane model by Ji (2008).

For Option 2, a ground motion model by Kanno et al. (2006) (hereafter Kanno06) is adopted. This prediction equation was developed by using ground motion records from Japanese earthquakes and from worldwide shallow crustal earthquakes (i.e., Next Generation Attenuation database). The Kanno06 equation is selected among other applicable models [e.g., Boore and Atkinson (2008), Sharma et al. (2009), and Harbindu et al. (2014)] for three reasons. The first reason is that the performance test of various ground motion models conducted by Nath and Thingbaijam (2011) indicates that the Kanno06 equation is superior to other candidate models in predicting PGA at rock sites in Northern India and Nepal. Second, the applicable moment magnitude range of the Kanno06 equation covers the moment magnitude of the 2015 Nepal earthquake; for instance, regional equations by Sharma et al. (2009) and Harbindu et al. (2014) are not applicable to M_w8-class earthquakes. Third, the Kanno06 equation adopts R_{rup} as a representative distance measure, while the equation by Boore and Atkinson (2008) (hereafter BA08) adopts the Joyner–Boore distance (R_{jb}, shortest distance from a site of interest to the projected fault rupture plane on Earth's surface). The use of R_{jb} can be problematic because ground motion intensity for the locations above the fault rupture plane is evaluated using a uniform value of $R_{jb} = 0$ km (which results in significant bias of predicted ground motion intensity). This issue is revisited in the next subsection.

Figure 6 compares observed ground motions at KATNP (i.e., **Figure 5**) with the ground motion data from the M_w7.9 Wenchuan earthquake as well as the Kanno06 model. The rupture distance for KATNP (=11.1 km) is calculated using the USGS finite-fault plane model (i.e., **Figure 2A**). For the Kanno06 model, 16th and

84th percentile curves are also shown to indicate a typical range of predicted ground motion variability. **Figure 6A** indicates that the observed PGA at KATNP is significantly smaller than the Wenchuan data in the similar distance range and the predicted PGA based on the Kanno06 equation (below the 16th percentile curve). The below-average trend of the observed ground motion intensity, in comparison with the Wenchuan data and the Kanno06 model, persists for spectral accelerations at vibration periods <2.0 s (**Figure 6B**). These comparisons indicate that the level of short-period ground motion near KATNP during the 2015 mainshock was smaller than expected ground motion levels based on empirical data/models for similar scenarios. On the other hand, **Figure 6C** shows an opposite trend: the long-period spectral acceleration at KATNP is significantly greater than the counterparts based on the Wenchuan data and the Kanno06 model. The large spectral acceleration in the long vibration period range is attributed to the basin effects. It is also interesting to note that the recent ground motion prediction model, such as Boore et al. (2014), can take into account the basin effects using a depth-to-bedrock parameter [note: in **Figure 6**, the equation by Boore et al. (2014) is not considered because it is based on R_{jb}]. Using the empirical model by Boore et al. (2014), the expected site amplification due to the basin effects is a factor of two for vibration periods longer than 2.0 s; the observed long-period spectral acceleration can be better explained. Therefore, it is important to adopt advanced ground motion models that can account for major systematic components (e.g., faulting mechanism and basin amplification) in predicting ground motion intensity for future earthquakes.

Scenario Shake Map

Rapid earthquake impact reports [e.g., Center for Disaster Management and Risk Reduction Technology (CEDIM), (2015)] are useful because emergency officers and international aiding agencies can appreciate the expected level of destruction due to an earthquake at the very early stage of a disaster. In producing rapid earthquake impact assessment, scenario shake maps are the essential input. In these applications, shake maps are generated by using a suitable ground motion model together with observed

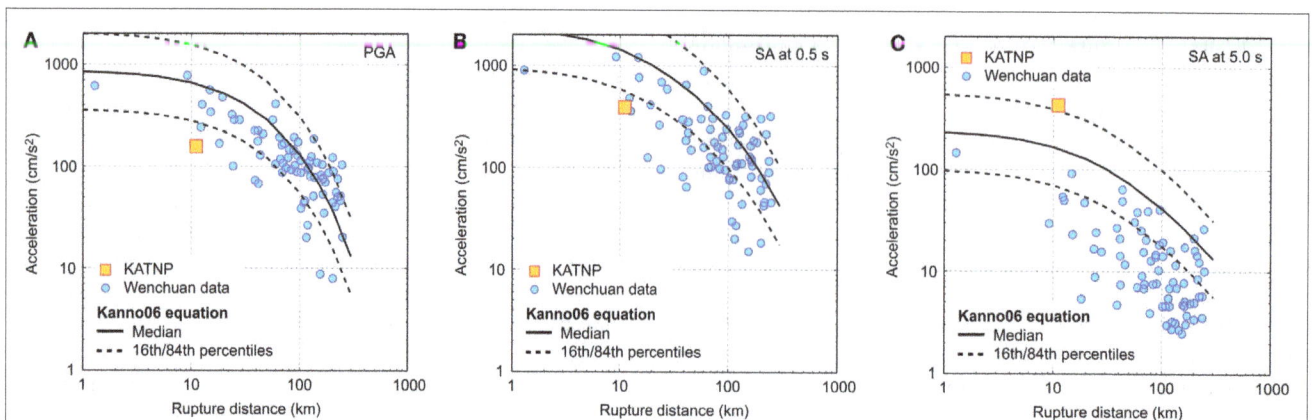

FIGURE 6 | Comparison of the observed PGA (A) and spectral accelerations [0.5 s in (B) and 5.0 s in (C)] at KATNP with the M_w7.9 Wenchuan ground motion data and with the ground motion model by Kanno et al. (2006).

instrumental data and seismic intensity information (e.g., DYFI; Atkinson and Wald, 2007)[7]. In seismic regions with limited monitoring capability of strong motion, shake maps are more dependent on the accuracy of an adopted ground motion model as well as on initial estimates of the seismic event (e.g., moment magnitude). This is because there will not be many real-time observations to constrain the shake map predictions.

Modern ground motion models adopt extended-source-based distance measures, such as R_{rup} and R_{jb} (i.e., calculation of these distance measures requires a fault plane model). A simpler representation of an earthquake source is a point source model; in this case, hypocentral and epicentral distances, R_{hypo} and R_{epi}, are often used. When a slip distribution is available, another useful distance measure is the shortest distance to the asperity R_{asp} (Goda and Atkinson, 2014). For large subduction events having large fault plane dimensions, the calculated distance measures can vary significantly, depending on how a fault plane model is defined and which distance measure is adopted. For instance, for the $M_w7.8$ mainshock, distance measures at KATNP are evaluated as: R_{rup} = 11.1 km, R_{jb} = 0.1 km (numerical lower bound), R_{hypo} = 85.3 km, R_{epi} = 76.8 km, and R_{asp} = 29.4 km. The influence of distance measures is particularly significant for large magnitude events.

The above-mentioned problem has an important implication on shake map generation for a large earthquake. To demonstrate this for the $M_w7.8$ mainshock, four scenario PGA shake maps are developed by considering different distance measures and ground motion models. The results are shown in **Figure 7**. **Figures 7–C** are based on the Kanno06 model together with R_{rup}, R_{hypo}, and R_{asp}, respectively, whereas **Figure 7D** is based on the BA08 model with R_{jb}. For all shake maps, V_{S30} information at individual sites is taken into account. Strictly, R_{hypo} and R_{asp} should not be used in the Kanno06 model (as the distance measures and the model development process are incompatible); this is for illustration only. **Figure 7A** shows the predicted PGAs at sites above the fault plane are large (0.5–0.7 g) and predicted PGA values gradually decrease toward north (i.e., the fault plane becomes deeper). **Figures 7B,C** show different patterns from **Figure 7A** because the distance measures are essentially defined for point source but with different source locations (i.e., hypocenter versus asperity). The predicted PGA values in **Figures 7B,C** are less than those in **Figure 7A** and are in more agreement with observed ground motion intensity in Kathmandu. **Figure 7D** shows the most significant difference from the observed ground motion intensity in Kathmandu because for all sites above the fault plane, the distance measure is set to R_{jb} = 0.1 km. Indeed, the USGS ShakeMap is similar to **Figure 7D** in terms of amplitude and spatial pattern of the shake map. Importantly, bias in estimated ground motions propagates into rapid earthquake impact assessment. The key issue here is that the current ground motion model together with a finite-fault plane can result in biased predictions of overall earthquake impact (which may affect subsequent decisions for emergency response actions). From practical viewpoints, this issue needs to be resolved in the near future.

[7]http://earthquake.usgs.gov/earthquakes/dyfi/

Earthquake Damage Survey

This section presents main observations and findings from the earthquake damage survey in Nepal. The building typology in Nepal is briefly reviewed, and then, field observations in Kathmandu, Melamchi, Trishuli, and Baluwa are discussed. The regional map of the visited locations is shown in **Figure 3**, and the main survey locations in Kathmandu are indicated in **Figure 4**. The cases discussed in the following are selected to highlight main observations from the survey trip. Numerous photos are available through the Google Earth file as supplementary material to this paper.

Building Typology in Nepal

Buildings in Nepal are vulnerable to seismic actions. The majority of houses and buildings are not seismically designed and constructed, lacking ductile behavior. Due to poor seismic performance, many buildings were damaged/collapsed and these structural failures caused many fatalities during the 2015 earthquake sequences. This subsection briefly summarizes general characteristics of building typology in Nepal. More complete information (e.g., statistics of building characteristics) is available in Chaulagain et al. (2015). According to the 2011 National Population and Housing Census, the total number of individual households in Nepal is 5,423,297, while the population is 26,494,504. The census data indicate that mud-bonded brick/stone masonry buildings are the most common in all geographical regions of Nepal (44.2%), followed by wooden buildings (24.9%). In urban areas (e.g., Kathmandu Valley), buildings with cement-bounded brick/stone (17.6%) and cement concrete (9.9%) are popular.

In Nepal, many masonry buildings are constructed with walls made of sun-dried/fired bricks or stone with mud mortar, and the building frame is made of wood. These types of buildings generally have flexible floors and roof, and are prevalent in rural areas. The masonry materials are of low strength and thus are seismically vulnerable. Recently, with the advancement of the cement in Nepal, brick/stone buildings are constructed with cement mortar. The wooden buildings are popular near the forest areas in Nepal. In these buildings, wooden pillars are made out of tree trunks and walls are constructed with wooden planks or bamboo net cement/mud mortar plaster. The reinforced concrete (RC) building is a modern form of construction in Nepal, which began in late 1970s. The RC moment resisting frame assembly is comprised of cast-in-place concrete beams and columns with cast-in-place concrete slabs for floor and roof. Most of the conventional RC constructions are non-engineered (i.e., not structurally designed) and thus lack sufficient seismic resistance. Engineered RC buildings, which are relatively new, often adopt the Indian standard code with seismic provisions.

Survey Results in Kathmandu

Many historical buildings in the Kathmandu Durbar Square (in front of the Old Royal Palace of the former Kathmandu Kingdom and is a UNESCO World Heritage site) were devastated (area 1 in **Figure 4**). **Figure 8A** shows the collapse of the Basantapur Tower. The complete destruction in the Durbar Square was in sharp contrast with undamaged buildings surrounding the Durbar Square (**Figure 8B**; several wall cracks can be found on these buildings;

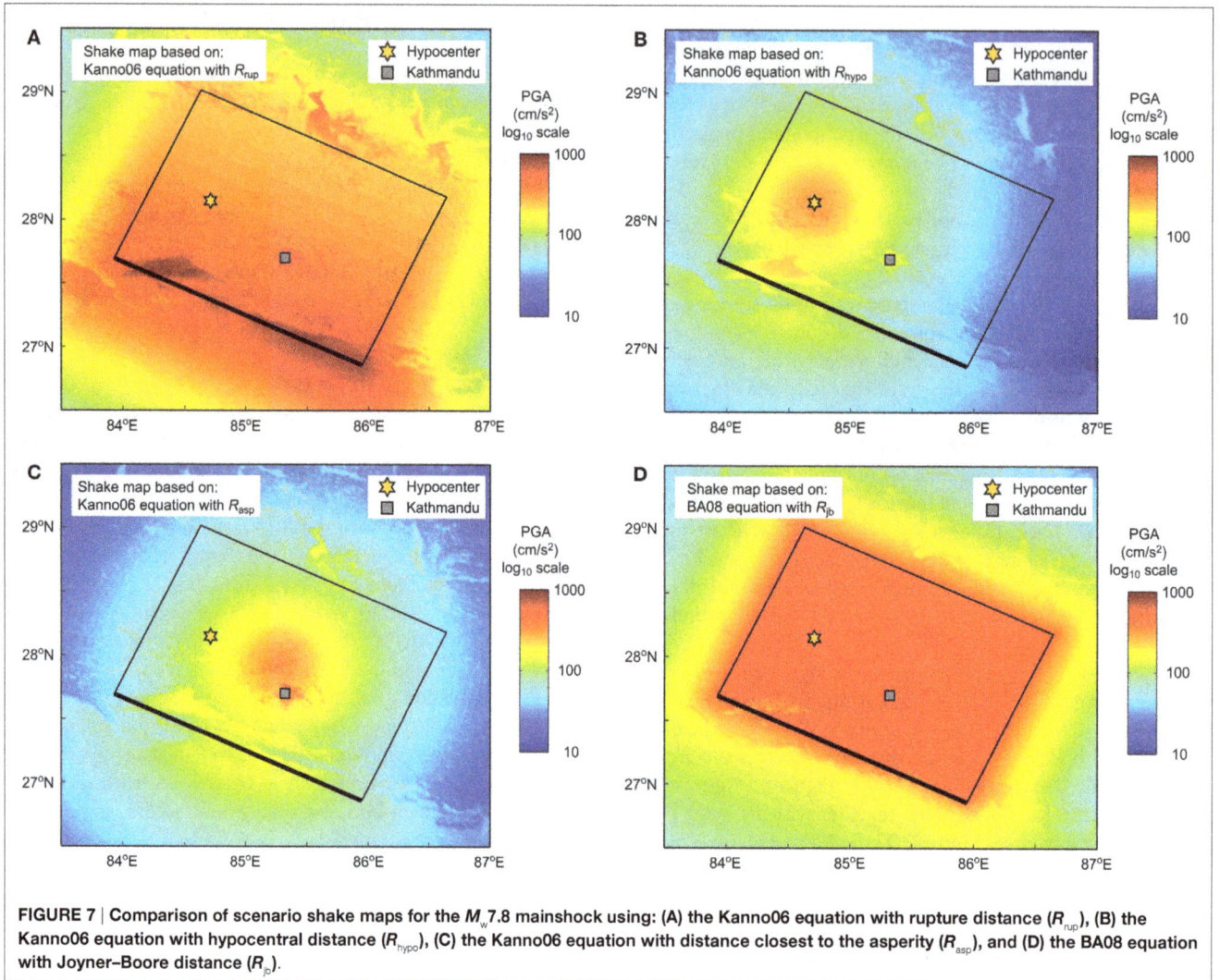

FIGURE 7 | Comparison of scenario shake maps for the M_w 7.8 mainshock using: (A) the Kanno06 equation with rupture distance (R_{rup}), (B) the Kanno06 equation with hypocentral distance (R_{hypo}), (C) the Kanno06 equation with distance closest to the asperity (R_{asp}), and (D) the BA08 equation with Joyner–Boore distance (R_{jb}).

however, the majority of the masonry buildings are structurally stable). This indicates that the ground shaking experienced in this area (note: this is relatively close to the KATNP station; see **Figure 4**) was sufficient to cause the collapses of the old historical buildings but was not to cause severe damage to the surrounding buildings. This observation was confirmed by walking through the Indra Chowk area (market squares near the Old Palace), where many old masonry buildings (three to six stories) were densely constructed. Nevertheless, there were several buildings that collapsed completely and some search and rescue activities were undertaken (**Figures 8C,D**).

There were numerous building collapses in the north–west section of the Ring Road along the Bishnumati River (area 2 in **Figure 4**). According to the local geomorphological map, sites within about 300 m from the river are alluvial (Holocene) soil deposits, whereas sites farther east are Pleistocene soil deposits. Therefore, site amplification effects due to different soil conditions may be expected in this area. A walk-through survey was carried out to investigate the spatial distribution of collapsed and severely damaged buildings in this area. Out of

28 collapsed or severely damaged buildings, 19 buildings were in the alluvial deposit area (**Figure 9A**), whereas 9 buildings were in the Pleistocene deposit area but nearer to the boundary (**Figure 9B**). This qualitatively confirms the effects of local site conditions on the building damage and collapse.

In area 3, there was a 16-story high-rise apartment complex (Park View Horizon). The walls of this building suffered from many major cracks along its height (**Figure 9C**). Currently, the apartments are unfit for living and residents have evacuated. The causes of the major damage of the Horizon apartments (and similar high-rise buildings in Kathmandu) may be attributed to the long-period ground motions (**Figure 5**). In addition, local topological features may have contributed to extensive damage there (the complex is on a hill).

Along the Araniko Highway between Kathmandu and Bhaktapur (area 4 in **Figure 4**), a section of the highway (about 200 m in length) built upon embankments was damaged due to the ground settlement. The amount of settlements was about 0.5–2.0 m, depending on locations (**Figure 9D**). The central section of the highway was constructed using reinforced soil retaining wall and

FIGURE 8 | Damage in Kathmandu (area 1 in Figure 4). (A) Collapse of the Basantapur Tower in the Kathmandu Durbar Square. **(B)** Undamaged buildings opposite of the Basantapur Tower in the Kathmandu Durbar Square. **(C)** Collapse of four 5- or 6-story old masonry buildings. **(D)** Collapse of a 4-story masonry building.

gravity-type retaining wall (2–3 m high and 100 m wide). The retaining walls were structurally intact and suffered from minor cracks and outward deformation only, whereas the natural slopes at both ends of the highway embankments experienced noticeable settlements (**Figure 9E**). Several buildings along the highway were tilted due to the settlements. A pedestrian footbridge crossing the highway suffered from the differential settlement of foundation, resulting in a gap of 45 cm between the bridge girder and the stair steps.

In area 5 (**Figure 4**), minor liquefaction, which was evidenced by sand boils and did not cause any structural damage, was observed in a small open land near a canal. In the surveyed area, a church was collapsed due to the ground shaking (**Figure 9F**). According to local residents, the church building was standing after the $M_w7.8$ mainshock but was collapsed due to the $M_w6.7$ aftershock on the following day. The extent of structural damage before the $M_w6.7$ aftershock is unknown. There were several houses that settled and tilted in this area. However, the degree of destruction in this area was minor.

Overall, earthquake damage in Kathmandu was not widespread but more localized. This may suggest that overall strong shaking experienced in Kathmandu was not extremely large. The areas that suffered from major destruction tend to have some local characteristics, such as soft soil conditions and structural deficiencies.

Survey Results in Melamchi

The survey was conducted along the road to Melamchi (about 30 km north–east of Kathmandu; **Figure 3**). Melamchi and the surrounding areas were close to the locations of major aftershocks (i.e., 26 April $M_w6.7$ aftershock and 12 May $M_w7.3$ aftershock; **Figures 2A** and **3**), and suffered from devastation due to these earthquakes. On the way to Melamchi, there were many small villages that suffered from earthquake damage. During interviews with local residents, they expressed serious concerns about incessant aftershocks and urgent need of repairs of the damaged houses before the arrival of rainy season. Proceeding north toward Melamchi, the occurrence of earthquake damage becomes more frequent.

Melamchi is a small town along the Indrawati River, and residents in the town have been involved with a major Melamchi Water Supply project[8], which diverts the river and channels its water to Kathmandu through tunnels. There were several factories along the road, which make water main pipes. Overall, the earthquake damage in Melamchi was severe, mostly affecting vulnerable masonry buildings, whereas the damage to RC buildings (4- to 5-story) was limited. For instance, the main

[8]http://www.melamchiwater.org/home/

FIGURE 9 | Damage in Kathmandu (areas 2–5 in Figure 4). (A) Collapsed building along the Bishnumati River (alluvial soil deposit area; area 2 in **Figure 4**). **(B)** Collapsed building (soft story collapse) near the Bishnumati River (boundary between alluvial and Pleistocene soil deposit areas; area 2 in **Figure 4**). **(C)** Horizon apartment buildings (area 3 in **Figure 4**). **(D)** Settlement of the Araniko Highway (area 4 in **Figure 4**). **(E)** Damage to the Araniko Highway (area 4 in **Figure 4**). **(F)** Collapsed church in the Imadol area (area 5 in **Figure 4**).

street of Melamchi was not completely destroyed (**Figure 10A**); most buildings looked undamaged based on their appearances, although several buildings were collapsed. On the other hand, buildings along a side street were devastated by the earthquakes (**Figures 10B,C**). The majority of the damaged buildings were made of brick and stone. Along the road, several sections of the slope suffered from shallow landsides (**Figure 10D**), their debris blocked the road at one time but was removed. There was a steel truss bridge with RC deck for vehicle crossing; the bridge was not damaged (inspected from backside). It has been reported that

further damage occurred in Melamchi due to the 12 May M_w7.3 aftershock. A further damage survey in Melamchi is required to investigate the effects of the aftershock with respect to the incurred damage prior to the aftershock (although it is beyond the scope of this study).

Survey Results in Trishuli

The survey was conducted along the road to Trishuli (about 30 km north–west of Kathmandu; **Figure 3**). One of the purposes of the trip was to investigate the earthquake damage near the Trishuli

FIGURE 10 | Damage in Melamchi (see Figure 3). **(A)** Main street in Melamchi. **(B)** Damaged stone masonry house. **(C)** Devastated street in Melamchi. **(D)** Shallow landslide along the main road.

hydroelectric station. Trishuli was closer to the hypocenter of the M_w7.8 mainshock, and thus severer damage, in comparison with Kathmandu, was expected. Along the way to Trishuli, earthquake damage in Ranipauwa (about 15 km north–west of Kathmandu) appeared relatively minor. Proceeding further north–west, earthquake damage to houses and landslides along the mountain slopes were observed more frequently. The rock fall, as secondary hazard, can be dangerous; a bus was hit by fallen boulder and several people were killed (**Figure 11A**). In Battar (about 25 km north–west of Kathmandu), a large number of brick/stone masonry buildings were collapsed (**Figure 11B**). The building materials of these damaged buildings were of poor quality; for example, two different types of the fragile bricks were used in one of the damaged houses (**Figure 11C**). According to local residents, many buildings were collapsed due to the 25 April M_w6.6 aftershock, which occurred 30 min after the mainshock.

In Trishuli, there was an earth fill dam for hydroelectric power generation. The main body of the dam was the excavated and compacted soil. The height of the dam was 12 m (upstream side) and 20 m (downstream side), and the crest width was about 4 m. Due to the earthquake, there were cracks at upstream side of the dam and fissures on the crest. Moreover, liquefaction (as evidenced by silt boils) and lateral spreading (**Figure 11D**) occurred inside of the dam reservoir due to the earthquake. The operation of the power generation had been suspended since the

following day of the mainshock; at the time of the visit, no power was available in nearby villages. Overall, the earthquake damage to the Trishuli dam will not cause severe problems immediately. However, the extent of cracking along the dam axis may suggest a deterioration of the dam body, which may be accelerated into the dam failure by future earthquakes or penetration of rain water into the dam body through cracks. It is important to mention that in worst-case scenarios (note: this earthquake is not the extreme case in terms of ground shaking intensity), catastrophic dam failures could have been caused. As there are several major hydroelectric projects along the Trishuli River as well as in other major rivers in Nepal, ensuring dam safety against large earthquakes is important.

Survey Results in Baluwa

The survey team visited Baluwa (about 70 km north–west of Kathmandu; **Figure 3**) along the Daraudi River, which is close to the epicenter of the M_w7.8 mainshock. One of the aims for this visit was to investigate the earthquake damage very near to the epicenter. Along the Kathmandu–Pokhara highway (e.g., Abu Khaireni, a town located at an intersection between the main highway and the Daraudi link road; about 30 km from the epicenter), no major earthquake damage was observed. At distances of about 18 km from the epicenter, earthquake damage to houses was observed; proceeding further north toward Baluwa, the extent

FIGURE 11 | Damage in Trishuli (see Figure 3). (A) Destroyed bus due to boulder fall. (B) Damaged brick masonry house in Battar. (C) Different types of bricks used in the damaged masonry house in Battar. (D) Ground fissures in the Trishuli dam reservoir.

of earthquake damage to houses became severer. The first stone house that was collapsed due to the earthquakes was about 4.5 km from the epicenter. Similarly, many shallow landslides and rock falls were observed along the road to Baluwa (**Figure 12A**); the first middle-size landslide was observed at distances of about 15 km from the epicenter. At one location, the debris from a landslide blocked the road completely (**Figure 12B**; note: detour was possible). The spatial distribution of the collapsed houses and landslides was limited to the locations near the epicenter (within 10–15 km radius), and was in contrast with Melamchi and Trishuli (i.e., farther from the epicenter). This can be understood by referring to the slip distribution of the mainshock (**Figures 2A** and **3**).

A large slope failure was observed at the northern boundary of Baluwa (**Figure 12C**); the length and height of the slope failure were 300 and 100 m, respectively. The fallen boulders and debris blocked the road completely, disconnecting villages at the upstream of the Daraudi River (e.g., Barpak, 5 km north of Baluwa); people can reach these places on foot only. This hampered rescue and recovery activities by governments and international aid teams significantly, highlighting the importance of functional critical infrastructure during the natural disaster emergency. The houses in Baluwa were devastated by the earthquakes and many residents lived in tents (**Figure 12D**). Local residents mentioned that the number of fatalities in Baluwa was small because many of the residents were in the field for agricultural work at the time of the earthquake. Major concerns about the arrival of rainy season were expressed by the local residents.

Conclusion

The M_w7.8 subduction earthquake occurred along the Main Himalayan Thrust arc and triggered numerous major aftershocks. The earthquake damage was catastrophic, causing the fatalities of more than 8,500 and billions of dollars in economic loss. This paper presented important earthquake field observations in Nepal in the aftermath of the M_w7.8 mainshock. A unique aspect of the earthquake damage investigation is that the data were collected 6–11 days after the mainshock, and thus first-hand earthquake damage observations were obtained. To share the gathered damage data widely, geo-tagged photos with observation comments were organized using Google Earth and the kmz file was made publicly available. In the future, the updated version of the Google Earth file, containing more damage photos and measurements from follow-up investigations, will be available from http://www.gdm.iis.u-tokyo.ac.jp/index_e.html. Viewers can download the photos directly and can use them for research and educational purposes. To gain deeper understanding of the observed earthquake damage in Nepal, the seismotectonic setting and regional seismicity in Nepal were reviewed and available aftershock data and ground motion data were analyzed. In addition to ground motion data analysis, scenario shake maps were generated by trialing different combinations of applicable ground motion models and source-to-site distance measures to highlight the potential biases caused in estimated ground motion maps and prompt earthquake impact assessments for a large subduction earthquake.

FIGURE 12 | Damage in Baluwa (see Figure 3). (A) Fallen boulder. **(B)** Shallow landslide; debris blocked the road. **(C)** Large landslide (100 m high and 300 m wide); debris blocked the road and disconnected villages further north of Baluwa. **(D)** Devastated houses in Baluwa.

The main results from the earthquake damage surveys in Nepal are as follows:

1. In Kathmandu, earthquake damage to old historical buildings was severe, whereas damage to the surrounding buildings was limited. The damaged buildings were stone/brick masonry structures with wooden frames. The RC frame buildings performed well for this earthquake. This may indicate that ground motion intensity experienced in Kathmandu was not so intense, in comparison with those predicted from probabilistic seismic hazard studies for Nepal. Therefore, a caution is necessary related to future earthquakes in Nepal because the 2015 earthquake is not necessarily the worst-case scenario.

2. The Kathmandu Basin is deposited by thick soft sediments. This has led to the generation of long-period ground motions in the Kathmandu Valley. Although the majority of the existing buildings in Kathmandu were not directly affected by the long-period ground motions, such seismic waves can pose serious risks to high-rise buildings. Adequate earthquake engineering design considerations are essential for reducing potential seismic risk to these structures.

3. The building damage in Kathmandu was localized to specific areas. It appeared that the building collapse sites were affected by local soil characteristics and/or structural deficiencies. In this regard, microzonation studies provide valuable insights into earthquake damage occurrence.

4. Some buildings that were severely damaged by the mainshock were collapsed due to major aftershocks. The capability for aftershock forecasting, building evacuation procedure, building inspection and tagging, and building repairs and retrofitting (low-cost solutions) need to be improved to mitigate the earthquake damage potential.

5. In the mountain areas, numerous villages were devastated by the earthquake sequence and major landslides were triggered. On occasion, landslides blocked roads, disconnecting remote villages. The redundancy of the local transportation network in Nepal needs to be improved for enhancing the resilience of rural communities.

Acknowledgments

The authors thank Pradeep Pokhrel for his great assistance during the field survey. The financial support by the JSPS KAKENHI (15H02631) is greatly acknowledged. The work was also funded by the EPSRC grant (EP/I01778X/1) for the Earthquake Engineering Field Investigation Team (EEFIT).

References

Ader, T., Avouac, J. P., Liu-Zeng, J., Lyon-Caen, H., Bollinger, L., Galetzka, J., et al. (2012). Convergence rate across the Nepal Himalaya and interseismic coupling on the Main Himalayan fault: implications for seismic hazard. *J. Geophys. Res.* 117, B04403. doi:10.1029/2011JB009071

Ambraseys, N. N., and Douglas, J. (2004). Magnitude calibration of north Indian earthquakes. *Geophys. J. Int.* 159, 165–206. doi:10.1111/j.1365-246X.2004.02323.x

Atkinson, G. M., and Wald, D. J. (2007). Did you feel it? intensity data: a surprisingly good measure of earthquake ground motion. *Seismol. Res. Lett.* 78, 362–368. doi:10.1785/gssrl.78.3.362

Avouac, J. P. (2003). Mountain building, erosion and the seismic cycle in the Nepal Himalaya. *Adv. Geophys.* 46, 1–80. doi:10.1016/S0065-2687(03)46001-9

Bilham, R. (2004). Earthquakes in India and the Himalaya: tectonics, geodesy and history. *Ann. Geophys.* 47, 839–858. doi:10.4401/ag-3338

Boore, D. M., and Atkinson, G. M. (2008). Ground motion prediction equations for the average horizontal component of PGA, PGV, and 5% damped PSA at spectral periods between 0.01 s and 10.0 s. *Earthquake Spectra* 24, 99–138. doi:10.1193/1.2830434

Boore, D. M., Stewart, J. P., Seyhan, E., and Atkinson, G. M. (2014). NGA-West2 equations for predicting PGA, PGV, and 5% damped PSA for shallow crustal earthquakes. *Earthquake Spectra* 30, 1057–1085. doi:10.1193/070113EQS184M

Center for Disaster Management and Risk Reduction Technology (CEDIM). (2015). *Nepal Earthquakes – Report #3.* Available at: https://www.cedim.de/english/index.php

Chaulagain, H., Rodrigues, H., Spacone, E., and Varum, H. (2015). Seismic response of current RC buildings in Kathmandu Valley. *Struct. Eng. Mech.* 53, 791–818. doi:10.12989/sem.2015.53.4.791

Goda, K., and Atkinson, G. M. (2014). Variation of source-to-site distance for megathrust subduction earthquakes: effects on ground motion prediction equations. *Earthquake Spectra* 30, 845–866. doi:10.1193/080512EQS254M

Gupta, I. D. (2006). Delineation of probable seismic sources in India and neighbourhood by a comparative analysis of seismotectonic characteristics of the region. *Soil Dynam. Earthquake Eng.* 26, 766–790. doi:10.1016/j.soildyn.2005.12.007

Harbindu, A., Gupta, S., and Sharma, M. L. (2014). Earthquake ground motion predictive equations for Garhwal Himalaya, India. *Soil Dynam. Earthquake Eng.* 66, 135–148. doi:10.1016/j.soildyn.2014.06.018

Heuret, A., Lallemand, S., Funiciello, F., Piromallo, C., and Faccenna, C. (2011). Physical characteristics of subduction interface type seismogenic zones revisited. *Geochem. Geophys. Geosyst.* 12, Q01004. doi:10.1029/2010GC003230

Hong, H. P., and Goda, K. (2007). Orientation-dependent ground-motion measure for seismic-hazard assessment. *Bull. Seismol. Soc. Am.* 97, 1525–1538. doi:10.1785/0120060194

Huang, Y. N., Whittaker, A. S., and Luco, N. (2008). Maximum spectral demands in the near-fault region. *Earthquake Spectra* 24, 319–341. doi:10.1193/1.2830435

Jaiswal, K. S., and Wald, D. J. (2010). An empirical model for global earthquake fatality estimation. *Earthquake Spectra* 26, 1017–1037. doi:10.1193/1.3480331

Ji, C. (2008). *Preliminary Result of the May 12, 2008 M_w 7.97 Sichuan Earthquake.* Available at: http://www.geol.ucsb.edu/faculty/ji/big_earthquakes/2008/05/12/ShiChuan.html

Kanno, T., Narita, A., Morikawa, N., Fujiwara, H., and Fukushima, Y. (2006). A new attenuation relation for strong ground motion in Japan based on recorded data. *Bull. Seismol. Soc. Am.* 96, 879–897. doi:10.1785/0120050138

Lu, M., Li, X. J., An, X. W., and Zhao, J. X. (2010). A comparison of recorded response spectra from the 2008 Wenchuan, China, earthquake with modern ground-motion prediction models. *Bull. Seismol. Soc. Am.* 100, 2357–2380. doi:10.1785/0120090303

Mai, P. M., Spudich, P., and Boatwright, J. (2005). Hypocenter locations in finite-source rupture models. *Bull. Seismol. Soc. Am.* 95, 965–980. doi:10.1785/0120040111

Nath, S. K., and Thingbaijam, K. K. S. (2011). Peak ground motion predictions in India: an appraisal for rock sites. *J. Seismol.* 15, 295–315. doi:10.1007/s10950-010-9224-5

Nath, S. K., and Thingbaijam, K. K. S. (2012). Probabilistic seismic hazard assessment of India. *Seismol. Res. Lett.* 83, 135–149. doi:10.1785/gssrl.83.1.135

Paudyal, Y. R., Bhandary, N. P., and Yatabe, R. (2012). Seismic microzonation of densely populated area of Kathmandu Valley of Nepal using microtremor observations. *J. Earthquake Eng.* 16, 1208–1229. doi:10.1080/13632469.2012.693242

Ram, T. D., and Wang, G. (2013). Probabilistic seismic hazard analysis in Nepal. *Earthquake Eng. Eng. Vib.* 12, 577–586. doi:10.1007/s11803-013-0191-z

Sakai, H., Fujii, R., and Kuwahara, Y. (2002). Changes in the depositional system of the Paleo-Kathmandu Lake caused by uplift of the Nepal Lesser Himalayas. *J. Asian Earth Sci.* 20, 267–276. doi:10.1016/S1367-9120(01)00046-3

Sharma, M. L., Douglas, J., Bungum, H., and Kotadia, J. (2009). Ground-motion prediction equations based on data from Himalayan and Zagros regions. *J. Earthquake Eng.* 13, 1191–1210. doi:10.1080/13632460902859151

Shcherbakov, R., Goda, K., Ivanian, A., and Atkinson, G. M. (2013). Aftershock statistics of major subduction earthquakes. *Bull. Seismol. Soc. Am.* 103, 3222–3234. doi:10.1785/0120120337

Shcherbakov, R., Turcotte, D. L., and Rundle, J. B. (2005). Aftershock statistics. *Pure Appl. Geophys.* 162, 1051–1076. doi:10.1007/s00024-004-2661-8

Stern, R. J. (2002). Subduction zones. *Rev. Geophys.* 40, 1012. doi:10.1029/2001RG000108

Takewaki, I., Murakami, S., Fujita, K., Yoshitomi, S., and Tsuji, M. (2011). The 2011 off the Pacific coast of Tohoku earthquake and response of high-rise buildings under long-period ground motions. *Soil Dynam. Earthquake Eng.* 31, 1511–1528. doi:10.1016/j.soildyn.2011.06.001

United Nations Office for the Coordination of Humanitarian Affairs (UN-OCHA). (2015). Available at: http://www.unocha.org/nepal

United States Geological Survey (USGS). (2015). Available at: http://earthquake.usgs.gov/earthquakes/eventpage/us20002926#scientific_finitefault

Wald, D. J., and Allen, T. I. (2007). Topographic slope as a proxy for seismic site conditions and amplification. *Bull. Seismol. Soc. Am.* 97, 1379–1395. doi:10.1785/0120060267

Wald, D. J., Worden, B. C., Quitoriano, V., and Pankow, K. L. (2005). *ShakeMap Manual: Technical Manual, User's Guide, and Software Guide.* Golden, CO: U.S. Geological Survey, 132.

Yagi, T. (2015). Available at: http://www.geol.tsukuba.ac.jp/~yagi-y/EQ/20150425/index.html

Conflict of Interest Statement: The authors declare that the research was conducted in the absence of any commercial or financial relationships that could be construed as a potential conflict of interest.

11

Displacement-based seismic design of symmetric single-storey wood-frame buildings with the aid of N2 method

Panagiotis Mergos[1]* and Katrin Beyer[2]

[1] Research Centre for Civil Engineering Structures, Department of Civil Engineering, City University London, London, UK,
[2] Earthquake Engineering and Structural Dynamics Laboratory, Department of Civil Engineering, École Polytechnique Fédérale de Lausanne, Lausanne, Switzerland

Edited by:
Solomon Tesfamariam,
The University of British Columbia,
Canada

Reviewed by:
P. Rajeev,
Swinburne University of Technology,
Australia
Hossein Mostafaei,
FM Global, USA

*Correspondence:
Panagiotis Mergos,
Research Centre for Civil Engineering
Structures, Department of Civil
Engineering, City University London,
Northampton Square,
London EC1V 0HB, UK
panagiotis.mergos.1@city.ac.uk

This paper presents a new methodology for the displacement-based seismic design of symmetric single-storey wood-frame buildings. Previous displacement-based design efforts were based on the direct displacement-based design approach, which uses a substitute linear system with an appropriate stiffness and viscous damping combination. Despite the fact that this method has shown to produce promising results for wood structures, it does not fit into the framework of the Eurocode 8 (EC8) provisions. The methodology presented herein is based on the N2 method, which is incorporated in EC8 and combines the non-linear pushover analysis with the response spectrum method. The N2 method has been mostly applied to reinforced concrete and steel structures. In order to properly implement the N2 method for the case of wood-frame buildings, new behavior factor–displacement ductility relationships are proposed. These relationships were derived from inelastic time history analyses of 35 SDOF systems subjected to 80 different ground motion records. Furthermore, the validity of the N2 method is examined for the case of a timber shear wall tested on a shake table and satisfactory predictions are obtained. Last, the proposed design methodology is applied to the displacement-based seismic design of a realistic symmetric single-storey wood-frame building in order to meet the performance objectives of EC8. It is concluded that the simplicity and computational efficiency of the adopted methodology make it a valuable tool for the seismic design of this category of wood-frame buildings, while the need for extending the method to more complex wood-frame buildings is also highlighted.

Keywords: performance, seismic, design, wood-frame, structures, N2

Introduction

In Eurocode 8 – Part 1 (CEN, 2004), the performance-based seismic design of structures is based on a force-based approach. Force-based seismic design is adopted by the codes since engineers are more familiar with this methodology as it resembles the design for other load cases, such as gravity loads or wind loads. However, nowadays, it is widely recognized (Priestley et al., 2007; Fardis, 2009) that force-based design is not a rational way for implementing performance-based seismic design. This is the case because structural and non-structural damage is directly related to imposed displacements and/or deformations. In force-based design, displacements and deformations are only checked at the end of the design procedure in order to establish that they are below some predefined limits.

For wood-frame buildings, correct application of force-based design is further undermined by several drawbacks, such as (i) the requirement of the definition of an initial ("elastic") stiffness and period (timber shear walls behave non-linearly even in the very early stages of their lateral response), (ii) a lack of appropriate behavior factor q–displacement ductility μ_Δ relationships in the literature and therefore in design codes. It is recalled herein that q is the ratio of the peak force F_{el} that would have developed if the system behaved linearly elastically to the yield strength F_y of the actual system. The displacement ductility μ_Δ is the ratio of the maximum displacement response to the yield displacement d_y. Moreover, capacity design approaches have not yet been fully developed for the case of wood-frame buildings.

To overcome the short-comings of a force-based design approach, several displacement-based design approaches have been developed (Sullivan et al., 2003). The fundamental concept of displacement-based design is to design a structure in order to achieve, rather be bounded by, a performance level for a given seismic action. One of the best-known procedures that falls within this category is the direct displacement-based design (DDBD) method, which was initially developed by Priestley (1993) and Priestley and Kowalsky (2000). DDBD methodologies assume a substitute linear system (Shibata and Sozen, 1976) with an appropriate stiffness and viscous damping combination that best reproduces the response of the inelastic system at the performance level under investigation.

Several researchers have applied the DDBD approach to wood-frame buildings. Filiatrault and Folz (2002) developed a performance-based seismic design methodology for wood-frame buildings, which is based on the DDBD. Pang and Rosowsky (2009) developed a new DDBD procedure for performance-based seismic design of mid-rise wood-frame buildings. Pang et al. (2010) simplified the methodology by Pang and Rosowsky (2009) and applied their approach to a six-storey wood-frame building, which was tested at full scale as part of the NEESWood project (van de Lindt and Liu, 2006). Furthermore, Wang et al. (2010) developed a set of factors for use in the methodology by Pang et al. (2010) in order to meet pre-specified performance levels with certain probabilities of non-exceedance.

Although DDBD methods have shown to produce promising results for wood structures, they do not fit into the framework of the Eurocode 8 (EC8) provisions, which do not adopt the linear substitute structure approach for assessing seismic response. Instead, they use a non-linear static assessment method the so-called N2 method (Fajfar and Gaspersic, 1996; Fajfar, 1999, 2000). This method is a capacity spectrum method, which combines the non-linear static (pushover) analysis and the response spectrum approach. The method has been widely applied to reinforced concrete, steel, and unreinforced masonry structures. For timber buildings, however, only few applications have been carried out.

Fragiacomo et al. (2011) used the N2 method to assess the seismic response of a multi-storey crosslam massive wooden building. They indicate that the lack of appropriate behavior factor q–displacement ductility μ_Δ relationships for systems with

significant pinching and stiffness degradation is the main drawback for the application of the method to timber structures.

The main objective of this paper is the application of the N2 method to the direct performance-based seismic design of symmetric single-storey wood-frame buildings. To meet this goal, a new design methodology is developed and applied to a realistic wood-frame building case study.

Furthermore, to apply the N2 method to wood-frame buildings in a consistent manner, new behavior factor q–displacement ductility μ_Δ relationships of SDOF systems representative of the hysteretic response of wood-frame buildings are developed. Then, the validity of the N2 method with the new q–μ_Δ relationships is verified against experimental results of wood-frame buildings.

Description of the N2 Method for SDOF Systems

In this section, the basic steps of the N2 method for SDOF systems, which can be assumed representative of symmetric in plan single-storey wood-frame buildings, are outlined. Furthermore, the limitations of this method when applied to wood-frame buildings are highlighted and discussed. The N2 method can also be applied to MDOF systems by transforming the MDOF system into a SDOF system that represents the first mode behavior.

The N2 method for SDOF systems comprises the following basic steps.

Step 1

The data required for the application of the method are obtained. For computing the structure's capacity, the structural configuration of the SDOF system with mass m needs to be determined (**Figure 1A**). Seismic demand is represented by a pseudo-acceleration response spectrum (**Figure 1B**).

Step 2

The elastic acceleration–displacement response spectrum (ADRS) is determined by the following relationship, which is valid for a constant viscous damping ratio (e.g., 5%), and S_{de} and S_{ae} represent elastic displacement and acceleration spectra, respectively.

$$S_{de} = \frac{T^2}{4 \cdot \pi^2} S_{ae}. \tag{1}$$

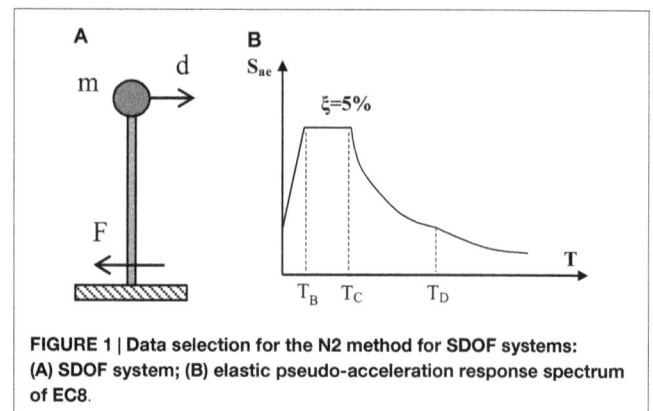

FIGURE 1 | Data selection for the N2 method for SDOF systems: (A) SDOF system; (B) elastic pseudo-acceleration response spectrum of EC8.

The inelastic acceleration S_a and inelastic displacement S_d spectra for a constant displacement ductility level μ_Δ are then determined by the following relationships, where q is the behavior factor due to ductility (i.e., hysteretic response).

$$S_a = \frac{S_{ae}}{q}. \tag{2}$$

$$S_d = \frac{S_{de}}{q} \cdot \mu_\Delta. \tag{3}$$

For the derivation of the inelastic spectra, it is necessary to determine the relationship between q and μ_Δ. In the next section, new q–μ_Δ relationships are developed to account for the special hysteretic characteristics of wood-frame buildings.

Step 3

A pushover analysis is conducted and the relationship between top displacement d versus base shear F is established.

Step 4

Having established the actual force F versus displacement d relationship of the SDOF system, an approximate elastoplastic envelope (**Figure 2**) is obtained, by using an appropriate bilinearisation method (FEMA-273, 1997; Fajfar, 2000), with yielding point (d_y, F_y).

The elastic period T of the equivalent SDOF system is calculated as

$$T = 2 \cdot \pi \sqrt{\frac{m \cdot d_y}{F_y}}. \tag{4}$$

The elastoplastic capacity curve is finally transformed into acceleration versus displacement format by the following relationship

$$S_a = \frac{F}{m}. \tag{5}$$

Step 5

The performance (target) point with displacement d_t is determined graphically by the intersection of the demand (inelastic spectrum) and capacity curve when both are plotted in the displacement–acceleration format (**Figure 2**). Alternatively, d_t can be calculated by simple closed-form expressions as explained in Annex B of EC8-Part 1.

Step 6

At this last step, local demands developed at the level of target displacement are compared (if required) with the corresponding acceptable limits (capacities) for the performance level under examination.

From the description of the basic steps of the N2 method for SDOF systems, the following flaws related to the application of this method to symmetric single-storey wood-frame buildings can be identified:

- The method uses inelastic response spectra that have not been developed for structural systems representative of wood-frame buildings, which are characterized by hysteretic behavior with significant stiffness and strength deterioration and important pinching effect.
- The method uses displacement ductility μ_Δ, which is a function of the displacement of the structure at yielding. Yield displacement is used also for the determination of the elastic stiffness and elastic period. However, wood-frame buildings exhibit a non-linear behavior right when subjected to horizontal loads. Hence, the definition of the yield displacement is ambiguous and may not be representative of the actual non-linear response.

Relationship Between Behavior Factor q and Ductility Demand μ_Δ for Wood-Frame Buildings

As explained in the previous section, the relationship between q-factor and μ_Δ is a key feature of the N2 method. It is recalled that EC8 adopts the inelastic response spectra derived by Vidic et al. (1994) for the determination of seismic demand. These spectra were derived using hysteretic relationships representative of the flexural response of reinforced concrete and steel members: the elastoplastic model and the Q-model Vidic et al. (1994); for both models a positive post-yield slope of 10% was assumed. These models may therefore not be representative and accurate for timber structures, whose hysteretic behavior is characterized by significant stiffness and strength deterioration and important pinching effect.

To investigate herein the validity of the relationship between q-factor and μ_Δ, a number of inelastic time history analyses are conducted for SDOF systems representative of the timber shear walls' hysteretic response for several ground motion records from regions of low to moderate and high seismicity. The same set of

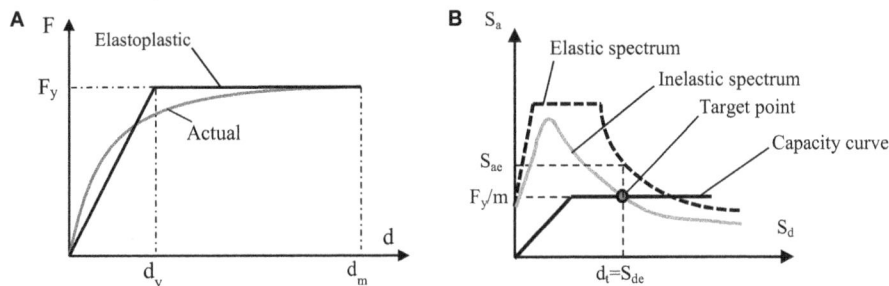

FIGURE 2 | N2 method: (A) equivalent bilinear capacity curve; (B) determination of target point (case where $T \geq T_C$).

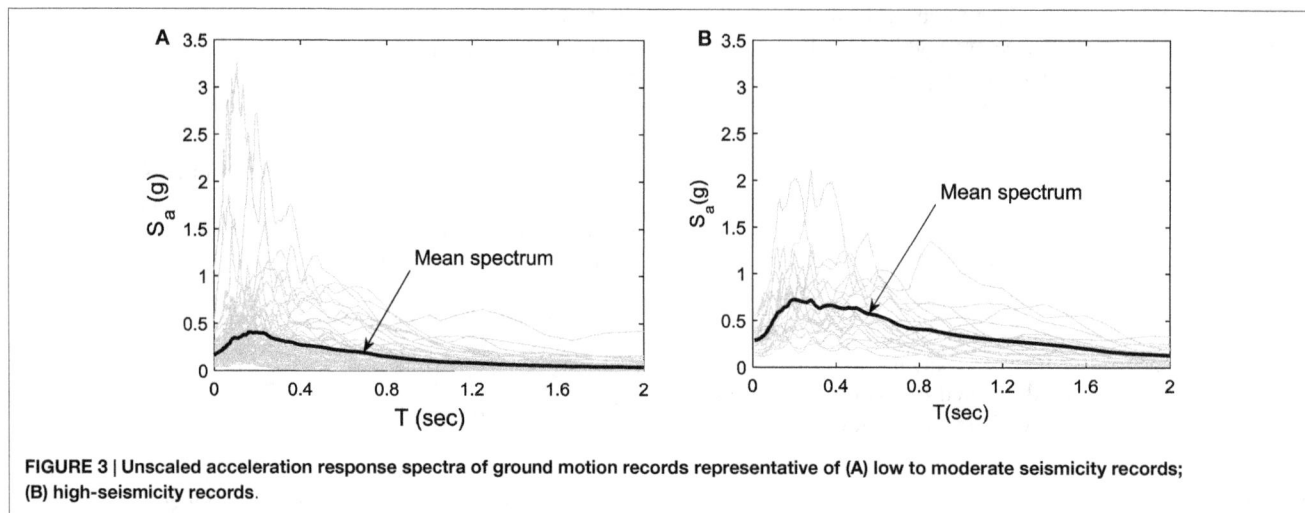

FIGURE 3 | Unscaled acceleration response spectra of ground motion records representative of (A) low to moderate seismicity records;
(B) high-seismicity records.

ground motions has been used in a study on loading protocols (Mergos and Beyer, 2014).

More particularly, 60 records were taken to be representative of a low to moderate seismicity region like the city of Sion in Switzerland. The city of Sion is situated in the Rhone Valley and has a design PGA of 0.16 g for 10% in 50 years (10/50) hazard level. For this site, the de-aggregation of hazard results is readily available (Giardini et al., 2004). All ground motions have a moment magnitude within the range of $4.3 \leq M_w \leq 6.6$ and an epicentral distance within the range $5\,km \leq R \leq 33\,km$. The unscaled acceleration response spectra of the ground motion records selected for low to moderate seismicity regions are shown in **Figure 3A**.

In addition to the 60 ground motion records representative of low to moderate seismicity regions, a set of 20 ground motion records used by Krawinkler et al. (2001) as representative of high-seismicity regions is also considered herein. All ground motions of this set have a moment magnitude within the range of $6.7 \leq M_w \leq 7.3$ and an epicentral distance within the range $14\,km \leq R \leq 26\,km$. The unscaled acceleration response spectra of the ground motion records used for high-seismicity regions are shown in **Figure 3B**.

The selected ground motion records are scaled one by one in order to match the spectral acceleration of the horizontal elastic spectrum (Krawinkler et al., 2001) of EC8-Part 1 for the 10/50 seismic hazard level at the fundamental period of the structure. The target EC8 elastic spectrum is derived for soil class C. The PGA for the 10/50 seismic hazard level and the site of Sion is taken equal to 0.16 g, while for the high-seismicity earthquakes it is taken equal to 0.40 g.

In order for a SDOF system to be representative of a structural system, an appropriate force-displacement hysteretic model has to be selected. For the timber shear walls, the Wayne–Stewart hysteretic model is adopted herein with the hysteretic parameter values proposed by Stewart (1987) for plywood sheathed timber walls (**Figure 4**). It is worth noticing that similar hysteretic parameters have been proposed by Filiatrault et al. (2003) for different types of timber walls [i.e., oriented strandboard (OSB), stucco, and gypsum]. The obtained results are therefore applicable to a wide range of timber wall buildings.

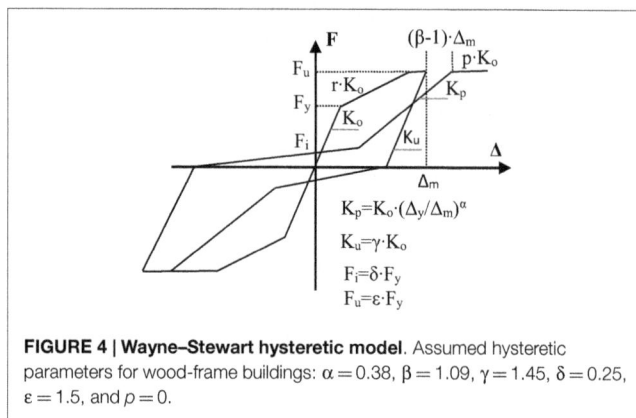

FIGURE 4 | **Wayne–Stewart hysteretic model**. Assumed hysteretic parameters for wood-frame buildings: $\alpha = 0.38$, $\beta = 1.09$, $\gamma = 1.45$, $\delta = 0.25$, $\varepsilon = 1.5$, and $p = 0$.

The following values of elastic periods of the SDOF systems are assumed in this study: $T = 0.2, 0.3, 0.45, 0.6, 0.8, 1.0$, and $1.5\,s$. Moreover, the following values of behavior factors q are examined in accordance with EC8 provisions: $q = 1, 2, 3, 4$, and 5. The yield strength F_y of each SDOF systems is calculated from the ordinate of the design EC8 spectrum for the 10/50 seismic hazard level for the period T and the q-factor under investigation. The viscous damping ratio ζ is assumed equal to 5%. The post-yield stiffness ratio r (ratio of post-yield to elastic stiffness) of the SDOF systems is assumed equal to 10%. The same value has been adopted by Vidic et al. (1994) for the derivation of the q–μ_Δ relationships of EC8. Furthermore, it is close to the values proposed by Filiatrault et al. (2003) for different types of timber shear walls (i.e., OSB, Stucco, and Gypsum).

In total, 35 different SDOF systems are examined and $35 \times 80 = 2800$ time history analyses are conducted. For each SDOF system, the median values of the maximum displacement ductility demands μ_{max} from the 60 low to moderate seismicity records and the 20 high-seismicity records are calculated. The main difference between high seismicity and low to moderate seismicity records relates to the number of cycles the structure is exposed to (Mergos and Beyer, 2014).

Figure 5A compares the calculated maximum displacement ductility demands μ_{max} with the analytical predictions μ_{EC8} of the

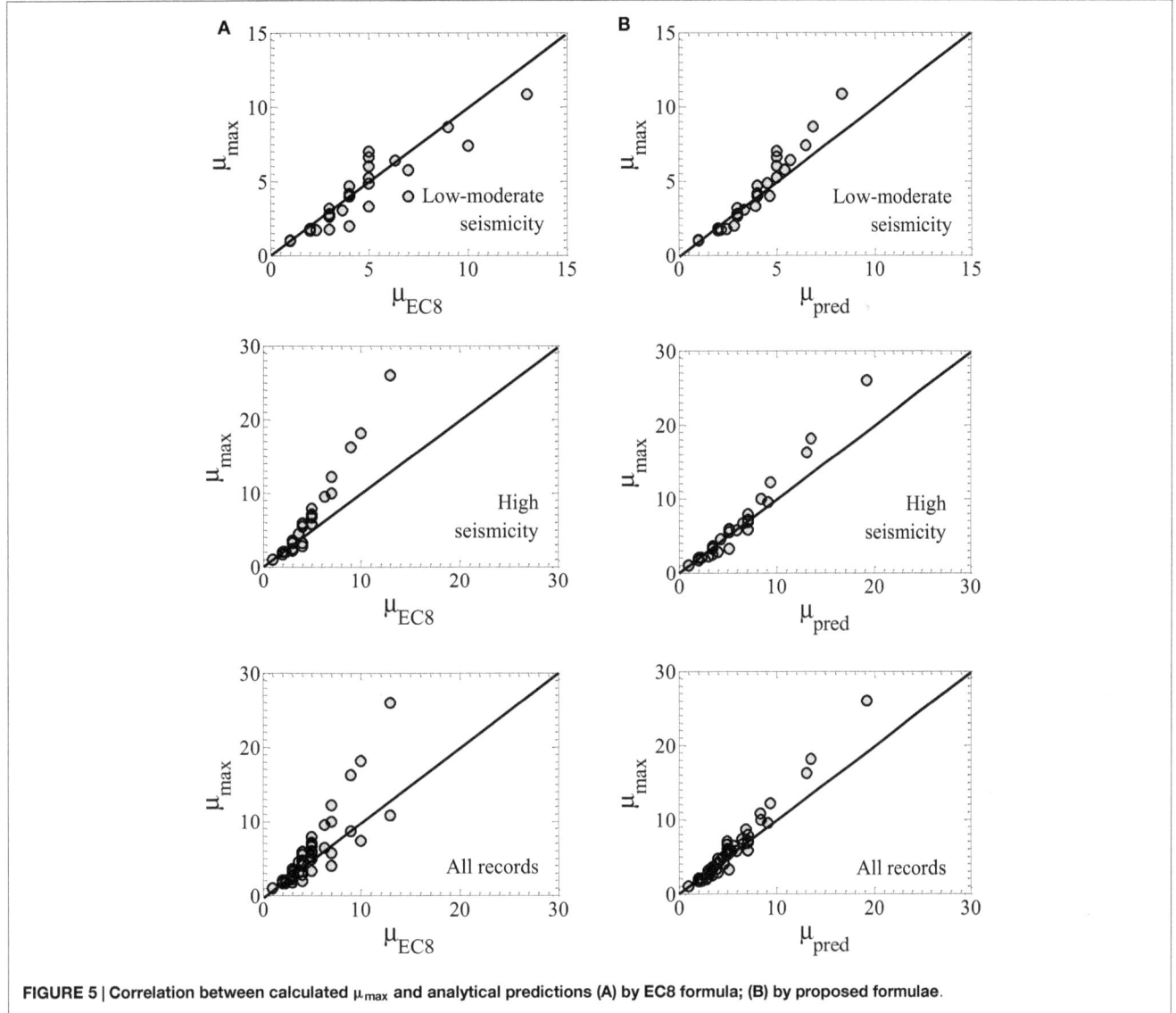

FIGURE 5 | Correlation between calculated μ_{max} and analytical predictions (A) by EC8 formula; (B) by proposed formulae.

formula adopted in EC8 for correlating displacement ductility and q-factor. The correlations between μ_{EC8} and μ_{max} are summarized in **Table 1**. It shows that EC8 formula tends to overpredict μ_{max} for the low to moderate seismicity regions and to underpredict maximum demands for the high-seismicity regions. When considering both seismicity levels as a single data set, the coefficient of determination R^2 is 0.69, the mean μ_{max}/μ_{EC8} ratio is 1.06, the median 1.00, and the coefficient of variation 0.29.

To improve the predictions of μ_{max}, a new formula is proposed herein, which builds on the existing formula of EC8, but it is more general by the introduction of two empirical parameters, such as α and β. The general form of this equation is the following:

$$T \geq T_C \rightarrow \mu_\Delta = 1 + (q-1)^\alpha \rightarrow q = 1 + \sqrt[\alpha]{\mu_\Delta - 1}$$

$$T < T_C \rightarrow \mu_\Delta = 1 + (q-1)^\alpha \cdot \left(\frac{T_C}{T}\right)^\beta$$

$$\rightarrow q = 1 + \sqrt[\alpha]{(\mu_\Delta - 1) \cdot \left(\frac{T}{T_C}\right)^\beta}. \tag{6}$$

TABLE 1 | Statistics of the calculated maximum ductility demands μ_{max} versus analytical predictions μ_{EC8} of EC8.

Seismicity	Mean of μ_{max}/μ_{EC8}	Median of μ_{max}/μ_{EC8}	Coefficient of variation of μ_{max}/μ_{EC8}	Coefficient of determination R^2
Low to moderate	0.93	0.97	0.21	0.81
High	1.19	1.14	0.28	0.64
All	1.06	1.00	0.29	0.69

Equation 6 becomes EC8 formula for $\alpha = \beta = 1$. Furthermore, Eq. 6 gives for $q = 1$ always $\mu_{max} = 1$. Furthermore, for $T = T_C$ both relationships (for $T \geq T_C$ and $T < T_C$) yield the same predictions. Hence, it is always a continuous equation as a function of T. Parameters, such as α and β, are evaluated in order to yield the best correlation of the analytical predictions μ_{pred} with the calculated maximum demands μ_{max}.

TABLE 2 | Parameter values and statistics of calculated maximum ductility demands μ_{max} versus analytical predictions μ_{pred} of the proposed equation.

Seismicity	α	β	Mean of μ_{max}/μ_{pred}	Median of μ_{max}/μ_{pred}	Coefficient of variation of μ_{max}/μ_{pred}	Coefficient of variation of R^2
Low to moderate	1.00	0.55	1.00	1.00	0.17	0.89
High	1.30	1.00	1.00	1.00	0.17	0.90
All	1.10	1.00	1.00	1.00	0.25	0.69

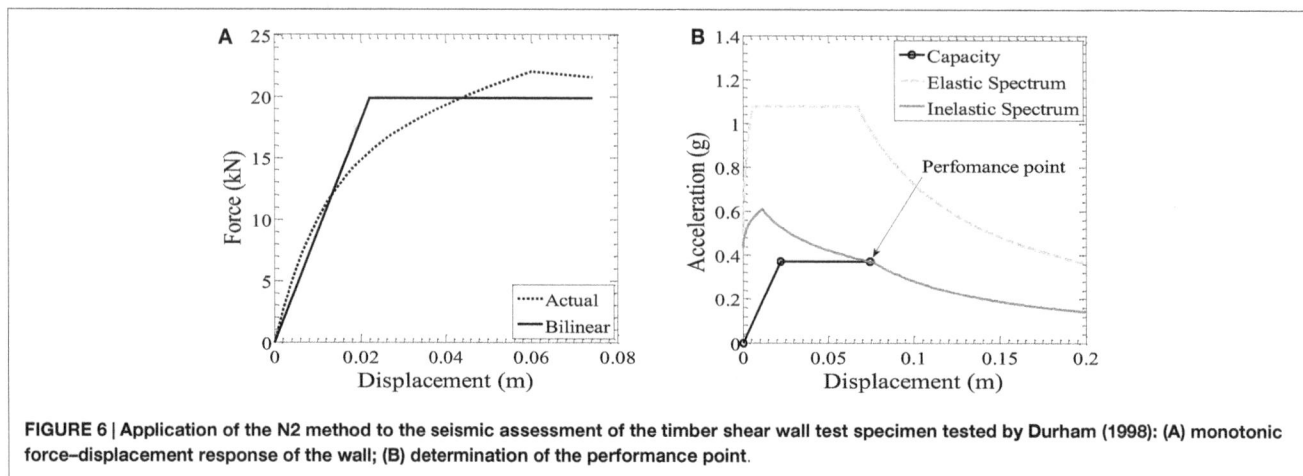

FIGURE 6 | Application of the N2 method to the seismic assessment of the timber shear wall test specimen tested by Durham (1998): (A) monotonic force–displacement response of the wall; (B) determination of the performance point.

In the very short-period range ($0 \leq T \leq T_B$), and in order to assure smooth inelastic spectra for all structural periods, it is proposed that parameter β is replaced by β_s given by:

$$\beta_s = 1 - (1 - \beta) \cdot \frac{T}{T_B}. \qquad (7)$$

Table 2 summarizes the proposed parameters values as well as the correlations between μ_{pred} and μ_{max}. In addition, **Figure 5B** illustrates the correlations between the predictions of the proposed formula and μ_{max} for the case of low to moderate seismicity, the case of high seismicity and for both levels of seismicity. For all records, **Figure 5B** shows the predictions of the combination of equations proposed for the two different seismicity levels and not the prediction of the equation developed when considering records from both seismicity levels as a single data set.

The proposed formulae yields always better results than the EC8 formula. The mean and median ratios μ_{max}/μ_{pred} are equal to unity. The formulae dealing with either the low to moderate seismicity level or the high-seismicity level provide better coefficients of variation (0.17 instead of 0.25) and coefficients of determination (0.89 and 0.90 instead of 0.69) than the formula involving ground motion records for both seismicity levels. This clearly advocates the adoption of different formulas for correlating μ_{max} and q for low to moderate and high-seismicity regions in the case of wood-frame buildings.

Validation of the N2 Method Against Experimental Results for Wood-Frame Buildings

To the best of the authors' knowledge, the N2 method has not yet been validated against experimental results of wood-frame

buildings. To investigate the validity of the method, the N2 method is applied to the assessment of the seismic behavior of a timber shear wall tested on a seismic table by Durham (1998). The dimensions of the wall were 2.4 m × 2.4 m. Studs were placed at every 400 mm. Sheathing panels were 9.5 mm thick oriented strand boards with 1.5 GPa elastic shear modulus. Three panels were used to sheath the panel. Sheathing to framing connectors were 50 mm long-spiral nails.

The applied vertical load on the wall was 50 kN and the specimen was subjected to the E–W component of the 1992 California Earthquake as recorded at Joshua Tree station. The ground motion was scaled to have a peak ground acceleration of 0.36 g. The displacement time history at the top of the specimen was recorded and the maximum displacement was found to be approximately 60 mm.

Using the computer software CASHEW, Folz and Filiatrault (2000, 2001) found that the monotonic force F–displacement d response of the timber wall can be approximated by the "actual" envelope curve that is shown in **Figure 6A**. The part of the curve up to 20% loss of maximum strength is only examined herein.

For the representation of seismic demand, the elastic acceleration response spectrum of EC8-Part 1 (Type 1) is adopted with PGA = 0.36 g and soil type B (soil factor 1.2). This soil type is chosen because the average shear wave velocity $V_{s,30}$ of the site where the accelerogram was recorded is approximately 380 m/s.

Figure 6 presents the application of the N2 method to the seismic assessment of the timber wall. **Figure 6A** shows the actual and equivalent bilinear curves up to the calculated performance point. The bilinear curve is derived by assuming that the elastic branch passes through the point of the actual curve with an ordinate equal to 0.6 F_y and that the areas beneath the actual and bilinear curve are the same. The same procedure is adopted in FEMA-273 (1997) and by Fajfar (2000).

It is important to mention here that the methodology adopted for the derivation of the bilinear envelope curve may affect significantly the accuracy of the results. Hence, it is generally recommended that the methodology for deriving the idealized bilinear curves is calibrated against experimental results of timber shear walls. Alternatively, different approaches can be applied and the most conservative results should be adopted.

Figure 6B illustrates the determination of the performance point of the structure as the intersection of the inelastic response spectrum and the capacity curve in the ADRS format. For the determination of the inelastic spectrum, Eq. 6 was applied. The predictions of the N2 method are the following: yield displacement $d_y = 22$ mm, elastic period $T = 0.49$ s, displacement ductility demand $\mu_\Delta = 3.40$, and maximum developed displacement 74 mm. The estimated maximum displacement is 23% higher than the actual one developed during the experimental procedure (60 mm). Furthermore, it is on the side of safety since the predicted displacement is larger than the one measured in the experiment. This occurs because the behavior factor is slightly underestimated for the actual ductility demand. Lower q values drive to higher required design strengths. It is worth pointing out that application of EC8 q–μ_Δ relationships leads to 66 mm maximum displacement prediction, which is only 10% higher than the actual displacement demand. However, overall, the new q–μ_Δ relationships provide better estimations of maximum displacements and ductilities as shown in **Figure 5**.

Displacement-Based Design with the N2 Method

Method Description

In this section, a displacement-based seismic design methodology for wood-frame buildings is proposed with the aid of the N2 method. The procedure is displacement driven since it starts with a target displacement that the structure is allowed to develop for a given level of seismicity and calculates the required strength of the wood-frame building. In particular, the design aims at determining the required nailing patterns of the wood-frame building timber walls. The procedure assumes that other structural components of timber walls like the hold-down or tie-down systems are properly designed. Furthermore, the methodology is only valid for symmetric single-storey wood-frame buildings that can be sufficiently represented by SDOF systems.

The analytical steps of the proposed methodology are the following.

Step 1

For the force capacity and start the sentence directly with: the structural configuration of timber walls apart from the nailing pattern (i.e., dimensions of walls, thickness of sheathing panels) is determined. The seismic demand is represented by an elastic pseudo-acceleration response spectrum (**Figure 1**). The target displacement d_t is established by the performance level of the building (EC8-Part 3, FEMA-273).

Step 2

The main aim of this step is to determine the yield displacement d_y of the equivalent bilinear force F–displacement d response of the SDOF system. Conveniently, determination of d_y can be done by using simplifying expressions as the case for other structural types (i.e., reinforced concrete buildings). However, to the best of the authors' knowledge, these relationships do not yet exist for timber shear walls and consequently for wood-frame buildings.

Wang et al. (2010) conducted pushover analyses, using the computer software CASHEW (Folz and Filiatrault, 2000) of timber shear walls with the same structural configuration, but different nailing spacings. After normalizing the backbone curves of the walls F_i–d_i by the maximum force F_{ui}, they observed that a single normalized (average) backbone curve F_i/F_{ui} versus displacement d_i can be assumed for all nailing spacings with adequate accuracy (see next section).

Making use of this observation, it is suggested herein that for each timber shear wall structural configuration, a pushover analysis is conducted by assuming a nailing spacing. The obtained F_i–d_i is then normalized with respect to maximum strength F_{ui}. More accurately, pushover analyses for different nailing spacings are conducted and normalized and then an average normalized F_i/F_{ui}–d_i is obtained. This normalized backbone curve is assumed as representative of all nailing spacings. Hence, it is assumed that nailing spacing affects only the maximum shear wall strength F_{ui}.

For the case that the walls are deformed in pure racking mode (without uplift) and consequently they undergo the same lateral top displacement the force capacity of the wood-frame building can be taken as the sum of the timber shear walls force capacities for the same displacement level. Hence, if the building comprises walls with the same structural configuration, the normalized force F/F_u–d displacement relationship of the building is the same as the normalized backbone curves of the single walls. If the building is composed of different shear walls then the normalized force F/F_u–d displacement of the building should be obtained by dividing the total force F–d displacement relationship of the building by the force capacity of the building F_u.

After establishing, the normalized force F/F_u–d displacement relationship of the building an equivalent normalized elastoplastic bilinear curve is established with yield point (F_y/F_u, d_y). It is important to note that no iterations are required for the determination of the equivalent normalized bilinear curve. This is the case because the displacement of the final point of the equivalent bilinear curve (F_y/F_u, d_m) (**Figure 2**) is constant and equal to d_t.

Step 3

Here, the elastic and inelastic ADRS are computed. First, the elastic ADRS is defined and then by calculating target displacement ductility $\mu_\Delta = d_t/d_y$ and using Eqs 2, 3, and 6 the inelastic ADRS is established.

Step 4

From the target displacement d_t and the inelastic spectrum, the required strength F_{yt}/m is obtained (**Figure 7**). Hence, required yield F_{yt} and ultimate strength F_{ut} are directly calculated.

Step 5

A nailing spacing is selected yielding strength capacity F_u for the building higher than F_{ut}.

Case Study

Introduction

Wang et al. (2010) examined the monotonic F_i–d_i relationship of timber shear walls with the following structural characteristics: 1.2 m width B and 3.1 m height H, sheathing panels attached to the framing members vertically with 11.1 mm thick OSB attached with $d_n = 3.3$ mm diameter (8 d_n length) nails. Interior nail spacing is 305 mm. Four different cases of constant edge (perimeter) nail spacing were examined: 51, 76, 102, or 152 mm. The obtained F_i–d_i relationships are shown in **Figure 8A**. As mentioned in the previous section, Wang et al. (2010) observed that when normalizing with respect to F_{ui} a single (average) backbone curve can be applied for all the different values of edge nail spacing (**Figure 8B**). Furthermore, it was found that the ultimate strength F_{ui} is equal to 32.4, 22.3, 17.1, and 11.5 kN for 51, 76, 102, and 152 mm edge nail spacing, respectively. The displacement d_{ui} (corresponding approximately to 20% loss of maximum strength) is 0.124 m (4% drift).

As a design example in this section, it is assumed that a symmetric in plan single-storey wood-frame building (**Figure 9**) is composed by eight of these shear walls in each direction. It is also assumed that the walls are deformed in pure racking mode (without uplift) and consequently they undergo the same lateral top displacement and their force capacities can be added.

The seismic mass of the building is m = 90 ton and the design PGA for the 10/50 seismic hazard level is equal to 0.16 g. The building is constructed on soil type C according to the categorization of EC8. The building will be designed to comply with the performance objectives of **Table 3**. The design aims at determining the appropriate edge nailing spacing for the construction of the walls.

Design Objectives

The adopted performance levels are similar to the ones used in EC8-Part 3 (CEN, 2005) for other types of structures (i.e., reinforced concrete, steel) apart from the damage limitation (DL) limit state, which is taken herein to coincide with the provisions of EC8-Part 1 for buildings having ductile non-structural elements. For these non-structural elements, the interstorey drift is limited to 0.75% (**Table 3**). The target displacement for the near collapse (NC) limit state d_u is taken equal to the displacement corresponding to 20% loss of maximum strength of a single timber shear wall $d_{ui} = 0.124$ m (4% drift). The target displacement at the SD limit state is taken equal to $0.75 \cdot d_u = 0.093$ m (3% drift).

The seismic hazard levels for the significant damage (SD) and near collapse (NC) limit states are taken in accordance with the recommendations of EC8-Part 1. The DL limit state is checked indirectly by the multiplication of the target normalized drift d_{SD} for the SD limit state by the reduction factor $v = 0.5$ for importance classes I and II, which accounts for the lower return period of the seismic action associated with this performance level following the approach of EC8-Part 1.

As mentioned previously, the design drifts are approximately $d_{NC} = 4\%$ for the NC limit state and $d_{SD} = 0.75 \cdot 4 = 3\%$ for the SD limit state. However, in order for the DL design objective to be achieved, the following should hold:

$$d_{SD} \cdot v \leq 0.75\% \cdot H \rightarrow d_{SD}/H \leq 0.75\%/0.5 = 1.5\% \quad (8)$$

Consequently, the design should be performed for the following drift limits: $d_{NC} = 4\%$ (0.124 m displacement) for the 2/50 (2% is probability of exceedance in 50 years) seismic hazard level, and $d_{SD} = 1.5\%$ (0.046m displacement) for the 10/50 (10% is probability of exceedance in 50 years) hazard level.

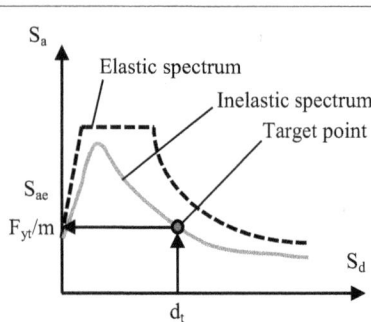

FIGURE 7 | Determination of required strength F_{yt}/m for a given target displacement d_t with the N2 method.

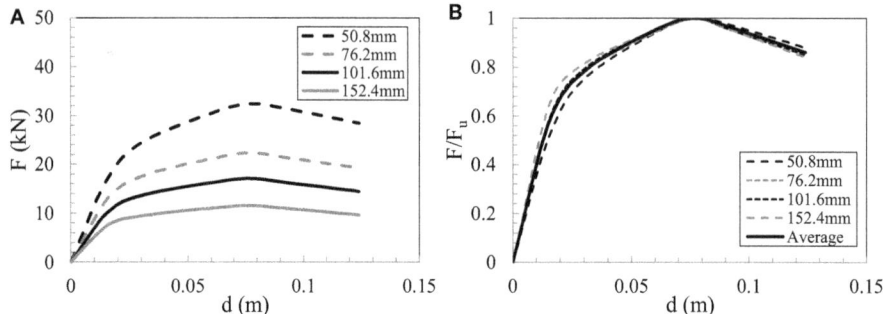

FIGURE 8 | Backbone curves of timber shear wall with different nailing patterns (Wang et al., 2010) (A) force F–displacement d (B) normalized force F/F_u–displacement d.

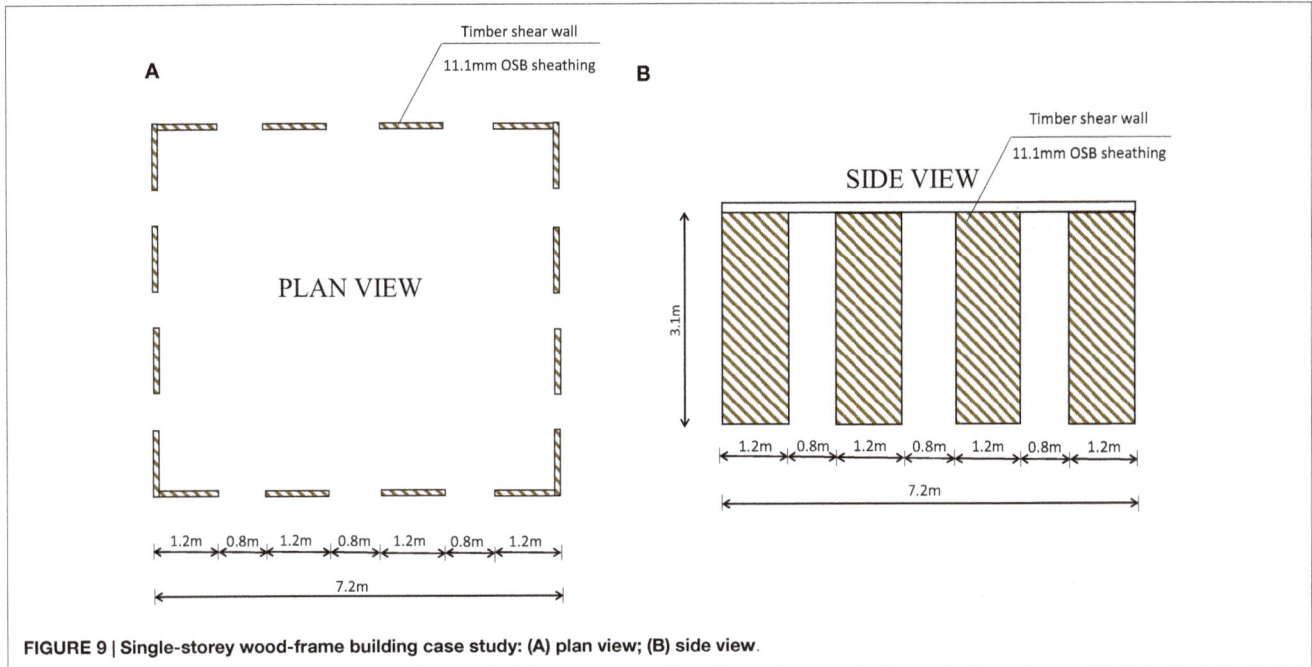

FIGURE 9 | Single-storey wood-frame building case study: (A) plan view; (B) side view.

TABLE 3 | Performance objectives for seismic design of case-study wood-frame building.

	Seismic design performance level		
	Damage limitation (DL)	Significant damage (SD)	Near collapse (NC)
		Seismic hazard level	
	–	10% probability of exceedance in 50 years	2% probability of exceedance in 50 years
Limit state	$d_{SD} \cdot v \le 0.0075 \cdot H$	$d_{SD} \le 0.75 \cdot d_u$	$d_{NC} \le d_u$

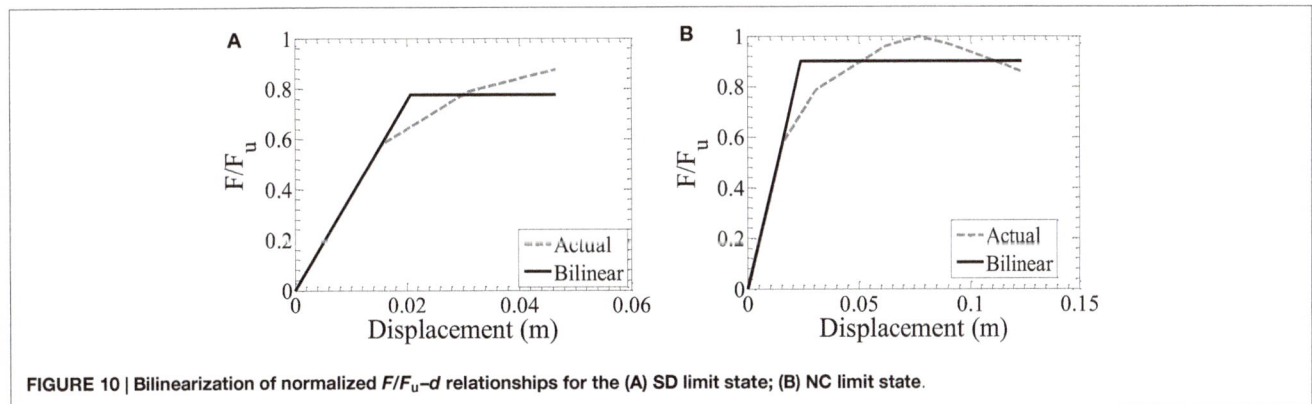

FIGURE 10 | Bilinearization of normalized F/F_u–d relationships for the (A) SD limit state; (B) NC limit state.

In order to define the design PGA value for the NC limit state (2/50 seismic hazard level), the following equation, proposed in EC8-Part 1, is used:

$$a_{g,NC} = a_{g,SD} \cdot \left(\frac{P_{NC}}{P_{SD}} \right)^{-1/3}$$

$$= 0.16g \cdot \left(\frac{2}{10} \right)^{-1/3} = 0.16g \cdot 1.71 = 0.27g \quad (9)$$

where $a_{g,NC}$ is the design PGA for the NC limit state, $a_{g,SD}$ is the design PGA for the SD limit state, $P_{SD} = 10\%$ is probability of exceedance in 50 years for the SD limit state, and $P_{NC} = 2\%$ is probability of exceedance in 50 years for the NC limit state.

Bilinearization of Normalized Force–Displacement Relationship

The second step for the seismic design with the proposed N2 method is the bilinearization of the normalized force–displacement envelopes (**Figure 10**). It is pointed out that since the building is composed by identical timber shear walls the normalized F/F_u–d behavior of the building matches the F_i/F_u–d_i of the single walls.

The 0.6 F_y rule for the bilinearization of the F/F_u–d curve is used as described in the validation of N2 method section. The target displacements are set equal to the design displacements for each performance level. Hence, an iterative procedure is not required for the bilinearizations.

Figure 10 presents the equivalent bilinear elastoplastic curves obtained for the two different limit states. Furthermore, **Table 4** presents the characteristic values of the equivalent bilinear curves as well as the respective displacement ductility demands μ_Δ.

Inelastic Displacement Spectra

The next step of the design procedure involves the derivation of the elastic and then the inelastic displacement spectra in ADRS format for the displacement ductility demands of the equivalent bilinear curves (**Figure 11**). For the derivation of the inelastic ADRS spectra, the new q–μ_Δ equations proposed in this study (Eqs 6 and 7) have been applied.

Calculation of Required Strength

From the inelastic ADRS spectra and the target displacements, the required F_{yt}/m strengths can readily be obtained as shown in **Figure 11**. In this figure, it can be seen that the F_{yt}/m for the 10/50 seismic hazard level is 0.184 g and for the 2/50 seismic hazard level is 0.084 g. Hence, the 10/50 seismic hazard level (for DL limit state) governs the seismic design of this wood-frame building.

The design base shear for the wood-frame building is calculated as

$$F_{b,tot} = m \cdot S_a = 90 \cdot 9.81 \cdot 0.184 = 162kN \qquad (10)$$

TABLE 4 | Characteristic values of equivalent bilinear F/F_u–d curves and displacement ductility demands for the two design seismic hazard levels.

Seismic hazard level	d_y (m)	d_t (m)	μ_Δ	F_y/F_u
10/50	0.021	0.046	2.21	0.775
2/50	0.024	0.124	5.17	0.900

This base shear will be carried by the eight shear walls. If F_{uti} is the required maximum strength capacity of each of these walls then it becomes

$$F_{b,tot} = 8 \cdot 0.775 \cdot F_{uti} \rightarrow F_{uti} = \frac{F_{b,tot}}{8 \cdot 0.775} = \frac{162.45}{8 \cdot 0.775} = 26.2kN \qquad (11)$$

In the previous equation, the factor 0.775 is taken from the equivalent bilinear curve for the critical 10/50 seismic hazard level (see **Table 4**).

Selection of Nailing Pattern

To complete the seismic design, the edge nail spacing that yields ultimate strength F_{ui} for each timber shear wall higher than 26.2 kN should be selected. According to the analyses of Wang et al. (2010), the required ultimate strength F_{uti} lies between the strengths provided by 50.8 mm edge nail spacing and 76.2 mm edge nail spacing. Hence, 50.8 mm edge nail spacing should be assigned for undertaking the design base shear with safety.

Discussion

Traditional seismic design of wood-frame buildings is force based. However, force-based seismic design does not directly control structural and non-structural damage, which are deformation and/or displacement related. Hence, displacement-based design is a more rational design approach.

Earlier displacement-based design efforts focused on the DDBD approach. Despite the fact that this method has shown to produce promising results for wood structures, it does not fit into the framework of the Eurocode 8 (EC8) provisions, which use as non-linear static assessment method the N2 method.

This study investigates the application of the N2 method to the seismic design of wood-frame buildings. The N2 method requires proper behavior factor q versus displacement ductility μ_Δ relationships for the derivation of the inelastic spectra, which do not exist for hysteretic systems representative of wood-frame buildings.

To tackle this limitation, new q–μ_Δ relationships are proposed herein based on numerous inelastic time history analyses of SDOF systems representative of wood-frame buildings. The new formulae can be considered as extensions of the q–μ_Δ relationships proposed by Vidic et al. (1994). It is found that different formulae

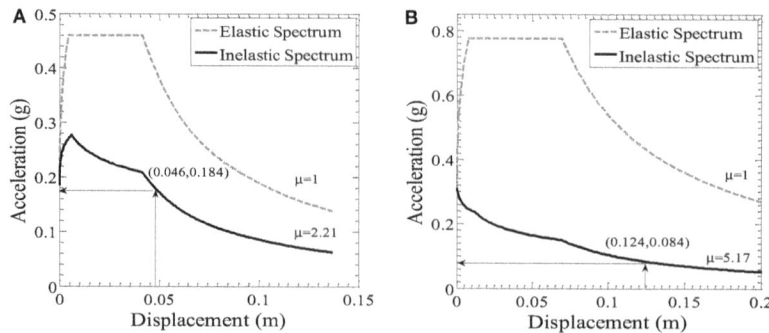

FIGURE 11 | Determination of required F_{yt}/m strengths for the (A) 10/50 seismic hazard level and (B) 2/50 seismic hazard level.

should be applied for regions of low to moderate seismicity and for regions of high seismicity since the hysteretic response of wood-frame systems depends strongly on the number of cycles the structure is exposed to (Mergos and Beyer, 2014).

Another limitation of the N2 method relative to its application to wood-frame buildings is the fact that it requires the definition of displacement ductility and subsequently of the yield displacement. The definition of yield displacement is ambiguous for wood-frame buildings that behave strongly non-linearly from the early stages of their lateral response. Hence, validation of the N2 method is required before its application to the seismic design of new buildings.

Since the N2 method has not yet been verified against experimental results of wood-frame buildings, a verification case study is examined herein for a single timber shear wall tested on a shake table. It is found that the method overestimates maximum displacement demand by 23%.

A new displacement-based seismic design methodology for symmetric, single-storey wood-frame buildings with the aid of the N2 method is also developed in this paper. The procedure is displacement driven since it starts with a target displacement the structure is allowed to develop for a given level of seismicity

and calculates the required nailing spacing of timber shear walls. The procedure assumes that other structural components of timber walls like the hold-down or tie-down systems are properly designed and remain elastic.

The proposed design approach is applied to the design of a realistic wood-frame building. The design methodology is computationally efficient since it does not require an iterative procedure. Furthermore, it can be directly incorporated in the framework of performance-based design. However, in its present form, it is limited to SDOF systems. Hence, the extension of the method to multi-storey and/or asymmetric in plan wood-frame buildings that behave as MDOF systems is required.

Acknowledgments

The authors thank the reviewers for their comments that helped to improve the manuscript. Financial support for this research was provided by the Swiss National Science Foundation within project NRP-66 (Resource Wood, Project Nr. 406640-136900). The opinions, findings, and conclusions expressed in this paper are those of the authors and do not necessarily reflect those of the sponsoring organization.

References

CEN. (2004). *Eurocode 8: Design of Structures for Earthquake Resistance, Part 1: General Rules, Seismic Actions and Rules for Buildings*. Brussels: European Standard EN 1998-1.

CEN. (2005). *Eurocode 8: Design of Structures for Earthquake Resistance, Part 3: Assessment and Retrofitting of Buildings*. Brussels: European Standard EN 1998-3.

Durham, J. P. (1998). *Seismic Response of Wood Shearwalls with Oversized Oriented Strand Board Panels*. MSc Thesis, University of British Columbia, Vancouver, BC.

Fajfar, P. (1999). Capacity spectrum method based on inelastic demand spectra. *Earthq. Eng. Struct. Dyn.* 28, 979–993. doi:10.1002/(SICI)1096-9845(199909)28: 9<979::AID-EQE850>3.0.CO;2-1

Fajfar, P. (2000). A nonlinear analysis method for performance-based seismic design. *Earthq. Spectra* 16, 573–592. doi:10.1193/1.1586128

Fajfar, P., and Gaspersic, P. (1996). The N2 method for the seismic damage analysis of RC buildings. *Earthq. Eng. Struct. Dyn.* 25, 31–46. doi:10.1002/(SICI) 1096-9845(199601)25:1<31::AID-EQE534>3.0.CO;2-V

Fardis, M. (2009). *Seismic Design, Assessment and Retrofitting of Concrete Buildings*. Dordrecht, NY: Springer.

FEMA-273. (1997). *NEHRP Guidelines for the Seismic Rehabilitation of Buildings*. Washington, DC: Federal Emergency Management Agency.

Filiatrault, A., and Folz, B. (2002). Performance-based seismic design of wood-framed buildings. *ASCE J. Struct. Eng.* 128, 39–47. doi:10.1155/2014/240952

Filiatrault, A., Isoda, H., and Folz, B. (2003). Hysteretic damping of wood framed buildings. *Eng. Struct.* 25, 461–471. doi:10.1016/S0141-0296(02) 00187-6

Folz, B., and Filiatrault, A. (2000). *CASHEW – Version 1.0: A Computer Program for Cyclic Analysis of Wood Shear Walls*. Rep. No. SSRP-2000/10, Structural Systems Research Project. San Diego, CA: Dept. of Structural Engineering, Univ. of California.

Folz, B., and Filiatrault, A. (2001). Cyclic analysis of wood shear walls. *ASCE J. Struct. Eng.* 127, 433–441. doi:10.1061/(ASCE)0733-9445(2001)127:4(433)

Fragiacomo, M., Dujic, B., and Sustersic, I. (2011). Elastic and ductile design of multi-storey crosslam massive wooden buildings under seismic actions. *Eng. Struct.* 33, 3043–3053. doi:10.1016/j.engstruct.2011.05.020

Giardini, D., Wiemer, S., Fäh, D., and Deichmann, N. (2004). *Seismic Hazard Assessment of Switzerland*. Zurich: Swiss Seismological Service.

Krawinkler, H., Parisi, F., Ibarra, L., Ayoub, A., and Medina, R. (2001). *Development of a Testing Protocol for Woodframe Structures*. Richmond, CA: CUREE Publication No. W-02.

Mergos, P. E., and Beyer, K. (2014). Loading protocols for European regions of low to moderate seismicity. *Bull. Earthq. Eng.* 12, 2507–2530. doi:10.1007/ s10518-014-9603-3

Pang, W., and Rosowsky, D. V. (2009). Direct displacement procedure for performance-based seismic design of mid-rise wood-framed structures. *Earthq. Spectra* 25, 583–605. doi:10.1193/1.3158932

Pang, W., Rosowsky, D. V., Pei, S., and Van de Lindt, J. W. (2010). Simplified direct displacement design of six-storey wood-framed building and pretest seismic performance assessment. *ASCE J. Struct. Eng.* 136, 813–825. doi:10.1061/(ASCE) ST.1943-541X.0000181

Priestley, M. J. N. (1993). Myths and fallacies in earthquake engineering – conflicts between design and reality. *Bull. NZSEE* 26, 329–341.

Priestley, M. J. N., Calvi, G. M., and Kowalsky, M. J. (2007). *Direct Displacement-Based Seismic Design of Structures*. Pavia: IUSS Press.

Priestley, M. J. N., and Kowalsky, M. J. (2000). Direct-displacement-based seismic design of concrete buildings. *Bull. NZSEE* 33, 421–444.

Shibata, A., and Sozen, M. (1976). Substitute structure method for seismic design in reinforced concrete. *ASCE J. Struct. Eng.* 102, 1–18.

Stewart, W. G. (1987). *The Seismic Design of Plywood Sheathed Shear Walls*. Ph.D. Thesis, University of Canterbury. Christchurch, New Zealand.

Sullivan, T., Calvi, G. M., Priestley, M. J. N., and Kowalsky, M. J. (2003). The limitations and performances of different displacement-based design methods. *J. Earthq. Eng.* 7, 201–241. doi:10.1142/S1363246903001012

van de Lindt, J. W., and Liu, H. (2006). "Correlation of observed damage and FEMA 356 drift limits: results from one-story wood-frame house shake table tests," in *Proc., 2006 Structures Congress* (Reston, VA: ASCE).

Vidic, T., Fajfar, P., and Fischinger, M. (1994). Consistent inelastic design spectra: strength and displacement. *Earthq. Eng. Struct. Dyn.* 23, 502–521. doi:10.1002/ eqe.4290230504

Wang, Y., Rosowsky, D. V., and Weichiang, P. (2010). Performance-based procedure for direct displacement-based design of engineered wood-frame buildings. *ASCE J. Struct. Eng.* 136, 978–988. doi:10.1061/(ASCE)ST.1943-541X.0000188

Conflict of Interest Statement: The authors declare that the research was conducted in the absence of any commercial or financial relationships that could be construed as a potential conflict of interest.

12

Performance of rocking systems on shallow improved sand: shaking table testing

Angelos Tsatsis[1] and Ioannis Anastasopoulos[2]*

[1] *Laboratory of Soil Mechanics, School of Civil Engineering, National Technical University of Athens, Athens, Greece,* [2] *Division of Civil Engineering, School of Engineering Physics and Mathematics, University of Dundee, Dundee, UK*

Edited by:
*Panagiotis Mergos,
City University London, UK*

Reviewed by:
*Marios Panagiotou,
University of California Berkeley, USA
Michalis F. Vassiliou,
ETH Zürich, Switzerland*

*Correspondence:
*Ioannis Anastasopoulos,
Division of Civil Engineering, School of
Engineering Physics and
Mathematics, University of Dundee,
Nethergate, Dundee DD14HN, UK
i.anastasopoulos@dundee.ac.uk*

Recent studies have highlighted the potential benefits of inelastic foundation response during seismic shaking. According to an emerging seismic design scheme, termed *rocking isolation*, the foundation is intentionally *under-designed* to promote rocking and limit the inertia transmitted to the structure. Such reversal of capacity design may improve the seismic performance, drastically increasing the safety margins. However, the benefit comes at the expense of permanent settlement and rotation, which may threaten post-earthquake functionality. Such undesired deformation can be maintained within tolerable limits, provided that the safety factor against vertical loading FS_V is adequately large. In such a case, the response is uplifting dominated and the accumulation of settlement can be limited. However, this is not always feasible as the soil properties may not be ideal. Shallow soil improvement may offer a viable solution and is therefore worth investigating. Its efficiency is related to the nature of rocking, which tends to mobilize a shallow stress bulb. To this end, a series of shaking table tests are conducted, using an idealized slender bridge pier as conceptual prototype. Two systems are studied, both lying on a square foundation of width B. The first corresponds to a lightly loaded and the second to a heavily loaded structure. The two systems are first tested on poor and ideal soil conditions to demonstrate the necessity for soil improvement. Then, the efficiency of shallow soil improvement is studied by investigating their performance on soil crusts of depth $z/B = 0.5$ and 1. It is shown that a $z/B = 1$ dense sand crust is enough to achieve practically the same performance with the ideal case of dense sand. A shallower $z/B = 0.5$ improvement layer may also be considered, depending on design requirements. The efficiency of the soil improvement is ameliorated with the increase of rotation amplitude, and with the number of the cycles of the seismic motion.

Keywords: rocking, seismic performance, soil improvement, physical modeling, shaking table

Introduction

According to current seismic codes, the foundation soil is not allowed to fully mobilize its strength, and plastic deformation is restricted to above-ground structural members. "Capacity" design is applied to the foundation *guiding* failure to the superstructure, thus, prohibiting mobilization of soil bearing capacity, uplifting and/or sliding, or any relevant combination. However, a significant body of pragmatic evidence provides robust justification that allowing strongly non-linear foundation response is not only unavoidable but may also be advantageous (Housner, 1963; Paolucci, 1997;

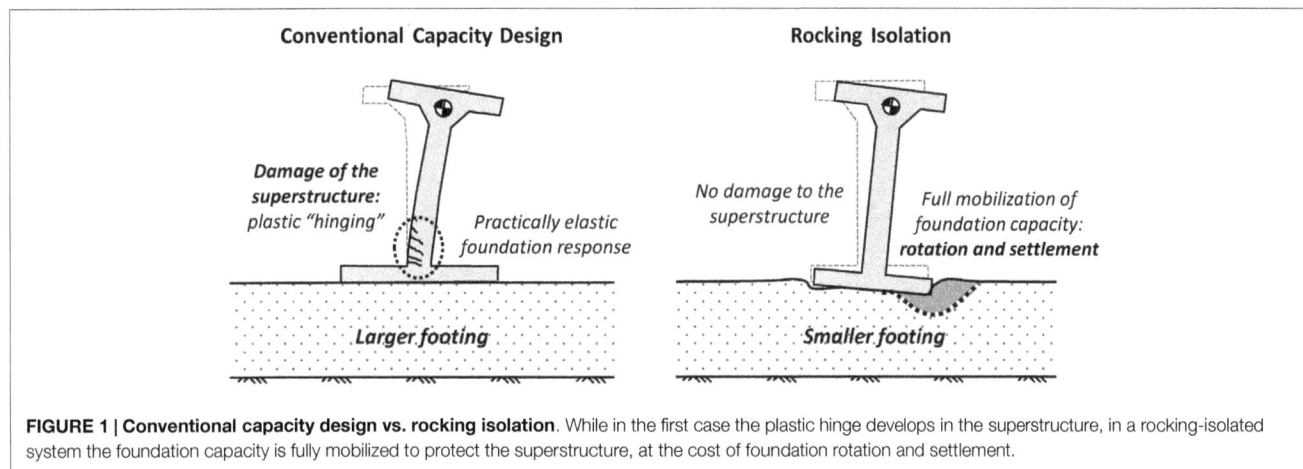

FIGURE 1 | Conventional capacity design vs. rocking isolation. While in the first case the plastic hinge develops in the superstructure, in a rocking-isolated system the foundation capacity is fully mobilized to protect the superstructure, at the cost of foundation rotation and settlement.

Pecker, 1998, 2003; Gazetas et al., 2003; Gajan et al., 2005; Kawashima et al., 2007; Anastasopoulos et al., 2010a).

Non-linear soil–foundation–structure response is simulated by means of (a) Winkler-based models that capture the settlement–rotation response of the footing (Yim and Chopra, 1985; Nakaki and Hart, 1987; Allotey and El Naggar, 2003, 2007; Chen and Lai, 2003; Houlsby et al., 2005; Harden and Hutchinson, 2006; Raychowdhury and Hutchinson, 2009); (b) sophisticated macro-element models, where the entire soil–foundation system is replaced by a single element that describes the generalized force–displacement behavior of the foundation (Nova and Montrasio, 1991; Paolucci, 1997; Pedretti, 1998; Crémer, 2001; Crémer et al., 2001; Le Pape and Sieffert, 2001; Grange et al., 2008; Chatzigogos et al., 2009, 2011); and (c) finite elements (or finite differences), modeling the superstructure, the foundation, and the soil in detail (Tan, 1990; Butterfield and Gottardi, 1995; Taiebat and Carter, 2000; Gourvenec, 2007; Anastastasopoulos et al., 2010b; Anastasopoulos et al., 2011). Physical modeling has also been applied to experimentally simulate non-linear soil–foundation–structure response, by means of (a) large-scale dynamic and cyclic pushover testing, focusing on non-linear soil–foundation response (Negro et al., 2000; Faccioli et al., 2001; Antonellis et al., 2015); (b) centrifuge model testing, also taking account of non-linear superstructure response (Kutter et al., 2003; Gajan et al., 2005; Gajan and Kutter, 2008, 2009); and (c) reduced-scale cyclic pushover and shaking table testing (Paolucci et al., 2008; Shirato et al., 2008; Drosos et al., 2012).

In this framework, recent studies have investigated that the idea of exploiting inelastic foundation response in order to limit the stresses transmitted onto the superstructure during strong shaking. As schematically illustrated in **Figure 1**, in contrast to conventional capacity design the foundation is deliberately "*under-designed*" to promote rocking, limiting the inertia forces transmitted onto the superstructure. The effectiveness of such an alternative seismic design philosophy, termed as "*rocking isolation*" (Mergos and Kawashima, 2005), has been explored analytically and experimentally for bridge piers (Anastasopoulos et al., 2010a, 2013a) and frames (Gelagoti et al., 2012; Anastasopoulos et al., 2013b). Such "reversal" of capacity design may

substantially improve the performance, drastically increasing the safety margins.

Yet, this benefit comes at the expense of permanent settlement and rotation, which could threaten the serviceability of the structure. Such undesired deformation can be maintained within tolerable limits, provided that the safety factor against static (vertical) loads FS_V is adequately large (Gajan et al., 2005). In such a case, the response of the foundation is uplifting dominated, and there is no substantial accumulation of cyclic settlement. The response is markedly different for lower values of FS_V, becoming sinking dominated: excessive soil yielding takes place underneath the foundation, leading to accumulation of substantial settlement and permanent rotation. Evidently, ensuring an adequately large FS_V in order to promote uplifting-dominated response greatly depends on the exact soil properties, which may not always be the ones that are desired. Shallow soil improvement (a concept that is commonly applicable in geotechnical engineering) may offer a viable solution to this problem.

Such a remediation technique has been introduced and experimentally investigated in Anastasopoulos et al. (2012). Based on the results of reduced-scale monotonic and cyclic pushover tests, the concept of shallow soil improvement was proven to be quite effective. Its efficiency is directly related to the nature of foundation rocking, which tends to mobilize a shallow stress bulb underneath the foundation. Although cyclic pushover testing has offered valuable evidence, the nature of seismic shaking is undeniably different. To this end, this paper goes one step further, exploring the efficiency of shallow soil improvement through reduced-scale shaking table testing.

Problem Definition and Experimental Setup

A series of reduced-scale shake table tests were conducted at the Laboratory of Soil Mechanics of the National Technical University of Athens (NTUA) to explore the efficiency of shallow soil improvement under dynamic loading. Based on the work presented in Anastasopoulos et al. (2012), a slender rocking-isolated bridge pier of height $h = 9$ m supported on a surface square footing of width $B = 3$ m is used as a conceptual prototype. Taking account of the capacity of the NTUA shaking table, a linear geometric

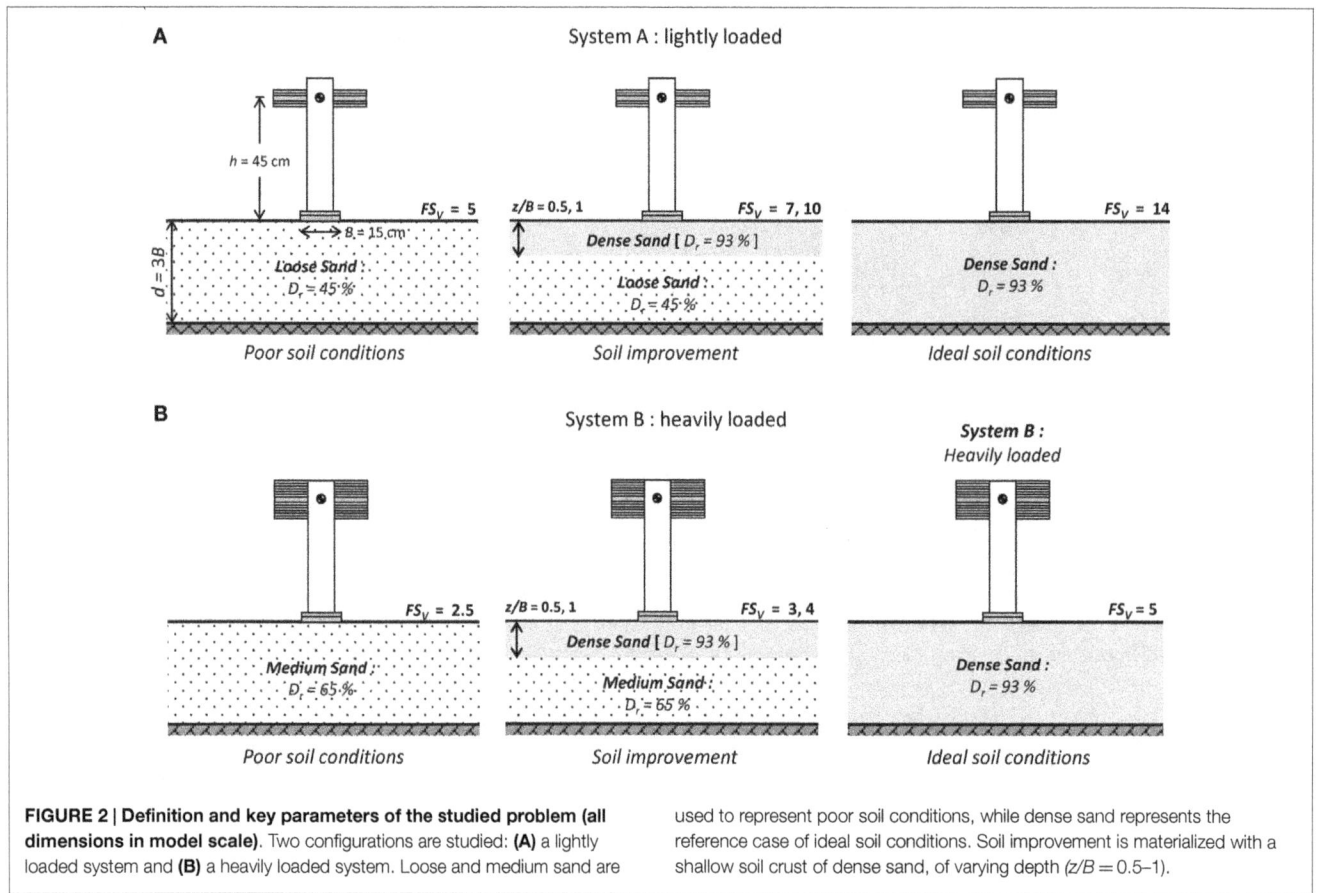

FIGURE 2 | Definition and key parameters of the studied problem (all dimensions in model scale). Two configurations are studied: **(A)** a lightly loaded system and **(B)** a heavily loaded system. Loose and medium sand are used to represent poor soil conditions, while dense sand represents the reference case of ideal soil conditions. Soil improvement is materialized with a shallow soil crust of dense sand, of varying depth ($z/B = 0.5$–1).

scale of 1:20 ($n = 20$) was selected for the experiments, and model properties were scaled down according to relevant scaling laws (Muir Wood, 2004). In all cases examined, to focus on foundation performance, the superstructure is assumed rigid and elastic.

As schematically illustrated in **Figure 2**, two superstructure systems are studied: (a) *System A*, which is representative of a lightly loaded structure having a relatively large FS_V (**Figures 2A,B**); and *System B*, being representative of a heavily loaded structure, characterized by a relatively low FS_V. These two systems were selected to model distinctly different foundation performance, from uplifting-dominated (*System A*) to sinking-dominated response (*System B*). Three different soil profiles were simulated in the experiments: (a) medium ($D_r = 65\%$) and loose ($D_r = 45\%$) sand for *System A* and *System B*, respectively, representing poor soil conditions; (b) soil improvement by means of a shallow "crust" of dense sand, of varying depth $z/B = 0.5$–1; and (c) the reference case of dense ($D_r = 93\%$) sand, representing ideal soil conditions.

Physical Modeling

The physical model consists of a square $B = 15$ cm aluminum footing, rigidly connected to a pair of rigid steel columns. The latter support a rigid aluminum slab, positioned at height $h = 45$ cm above the foundation level. The superstructure mass is composed of a number of steel plates, installed symmetrically above and below the aluminum slab, so as to maintain the center of mass at the same level. The mass of the model was adjusted by adding or removing steel plates. Sandpaper was placed underneath the foundation to achieve a realistically rough foundation–soil interface (corresponding to a coefficient of friction ≈ 0.7). The model was placed inside a rigid soil container, lying on an adequately deep sand stratum of depth $d = 3B = 45$ cm, and at an adequately large distance ($L \approx 5B = 75$ cm) from the container walls.

The soil consists of dry-pluviated *Longstone* sand. The sand, characterized by a uniformity coefficient $C_u = 1.42$ and mean grain sized diameter $D_{50} = 0.15$ mm and, was pluviated using an automated sand raining system. The latter is a custom-built system, capable of producing sand specimens of controllable relative density D_r, with exceptional repeatability. The properties of the sand have been measured through laboratory tests, also conducted at NTUA, and are documented in Anastastasopoulos et al. (2010b). In reduced-scale testing, the stress field in the soil cannot be reproduced correctly. Given the fact that the strength of the sand is stress dependent, the reduced effective stresses in the model unavoidably lead to an increase of the mobilized friction angle (compared to the prototype). Such scale effects can only be avoided through centrifuge modeling, and should be carefully contemplated when interpreting 1 g test results. Even though, in the case of the investigated problem, the stresses due to the superstructure dead load are prevailing, minimizing the adverse role of scale effects.

Even though, aiming to avoid scaling-related misinterpretations, a series of vertical push tests were conducted to measure the bearing capacity of the $B = 15$ cm square foundation for all soil

TABLE 1 | Summary of monotonic vertical loading.

	System A			System B		
	Configuration	N_{ult} (kN)	FS_V	Configuration	N_{ult} (kN)	FS_V
Poor soil conditions	Loose sand ($D_r = 45\%$)	1.7	5	Medium sand ($D_r = 65\%$)	2.5	2.5
Soil improvement	$z/B = 0.5$ on top of loose sand	2.4	7	$z/B = 0.5$ on top of medium sand	3	3
	$z/B = 1$ on top of loose sand	3.4	10	$z/B = 1$ on top of medium sand	3.9	4
Ideal soil conditions	Dense sand ($D_r = 93\%$)	4.8	14	Dense sand ($D_r = 93\%$)	4.8	5

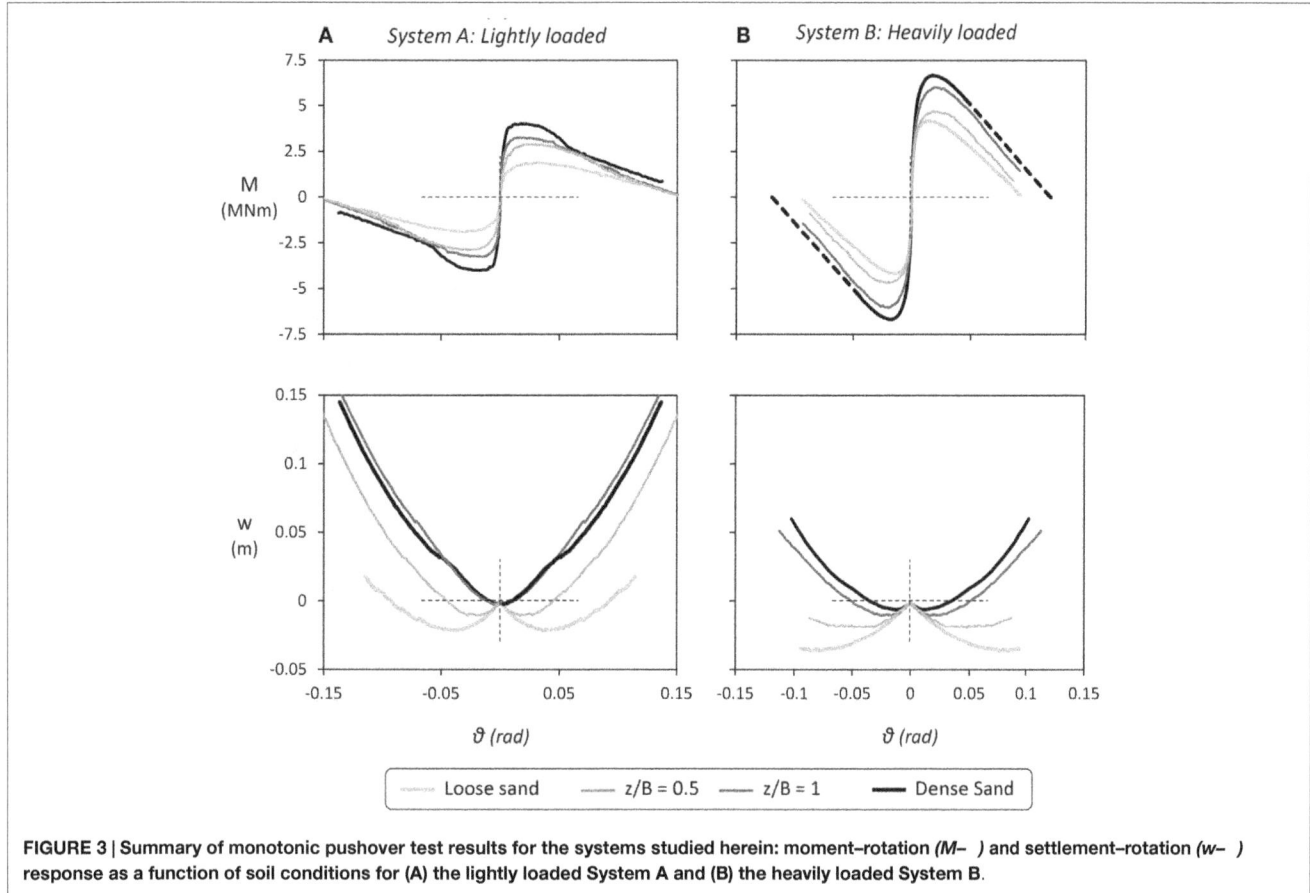

FIGURE 3 | Summary of monotonic pushover test results for the systems studied herein: moment–rotation (M–) and settlement–rotation (w–) response as a function of soil conditions for (A) the lightly loaded System A and (B) the heavily loaded System B.

conditions examined. Based on the results of these tests, which are summarized in **Table 1**, the mass of the two superstructure models was adjusted to produce the desired FS_V for the reference case of ideal soil conditions (i.e., for dense sand): $FS_V = 14$ for the lightly loaded *System A* (materialized using a mass of 35 kg), and $FS_V = 5$ for the heavily loaded *System B* (materialized using a mass of 100 kg).

Summary of Monotonic Response

Before proceeding to the testing sequence, a brief discussion of the monotonic response of the studied systems is necessary. A detailed description of the pushover tests and their key results can be found in Anastasopoulos et al. (2012). **Figure 3** summarizes the moment–rotation (–) and settlement–rotation (w–) response of the two systems founded on the four different soil profiles.

In the case of the lightly loaded System A (**Figure 3A**), when founded on poor soil conditions (i.e., loose sand) its maximum moment capacity reaches $M_{max} = 1.8$ MN/m (unless otherwise

stated, all results are discussed in prototype scale). Considering the dynamic response, a critical acceleration ac can be defined as the maximum acceleration that can develop at the mass of the oscillator (representing the bridge deck). For rocking-isolated systems, such as those examined herein, a_c is bounded by the moment capacity of the foundation, and can be defined as follows:

$$a_c = M_{max}/mgh \qquad (1)$$

Based on the above, *System A* founded on poor soil is characterized by a critical acceleration $a_c = 0.072$ g. The moment capacity is substantially increased for the case of shallow soil improvement, leading to a proportional increase of the critical acceleration to $a_c = 0.117$ g for $z/B = 0.5$ and $a_c = 0.130$ g for a deeper $z/B = 1$ dense sand crust. The latter is still lower than the one for ideal soil conditions, $a_c = 0.162$ g, but the efficiency of the improvement is evident. Likewise, for the heavily loaded *System B* (**Figure 3B**), the footing on poor soil conditions has a moment capacity

TABLE 2 | Summary of monotonic pushover loading.

	FS_V	M_{max} (MN/m)	a_c (g)	$K_{initial}$ (MN/m)	$T_{initial}$ (s)
System A: Lightly Loaded					
Loose sand ($D_r = 45\%$)	5	1.8	0.072	7	1.21
$z/B = 0.5$	7	2.9	0.117	13	0.92
$z/B = 1$	10	3.2	0.130	20	0.75
Dense sand	14	4	0.162	26	0.66
System B: Heavily Loaded					
Medium sand ($D_r = 65\%$)	2.5	4.2	0.060	30	1.03
$z/B = 0.5$	3	4.5	0.065	33	0.98
$z/B = 1$	4	5.9	0.085	40	0.89
Dense sand	5	6.7	0.095	45	0.84

$M_{max} = 4.2$ MN/m, which translates to $a_c = 0.065$ g. Applying a layer of improved soil of depth $z/B = 0.5$, the critical acceleration increases only slightly to $a_c = 0.065$ g ($M_{max} = 4.5$ MN/m). A deeper $z/B = 1$ soil crust is required to attain a substantial increase of the critical acceleration to $a_c = 0.085$ g. The latter is only slightly lower than for ideal soil conditions ($a_c = 0.095$ g), confirming the efficiency of shallow soil improvement in this case also. It is worth mentioning that an excitation with acceleration exceeding the critical acceleration of the systems does not necessarily mean collapse; the structure will topple only when the imposed rotation of the footing is larger than the maximum rotation as shown in the $M-$ curves.

Most importantly, in both cases shallow soil improvement leads to a substantial reduction of the tendency for accumulation of settlement. As evidenced by the $w-$ response, in both cases, the application of shallow soil improvement tends to suppress the sinking behavior of the foundation. For the lightly loaded *System A* in particular, a $z/B = 1$ soil crust is sufficient to achieve almost the same $w-$ response with the reference case of ideal soil condition (dense sand), while an even shallower $z/B = 0.5$ crust is also quite effective. This has been confirmed by the slow-cyclic tests, which can be found in Anastasopoulos et al. (2012). On the other hand, for the heavily loaded *System B*, a $z/B = 0.5$ soil crust is clearly not enough: the response is improved but remains sinking-dominated. A deeper $z/B = 1$ soil improvement is needed to ensure uplifting-dominated response. Indeed, in this case, the $w-$ response becomes almost identical to the case of ideal soil conditions (dense sand). The performance of the tested configurations is summarized in **Table 2**.

Instrumentation and Testing Sequence

The model was instrumented to allow direct recording of translational and rotational deformations, and lateral accelerations. As shown in the photos of **Figure 4**, wire (WDT) and laser displacement transducers (LDTs) were utilized to measure horizontal and vertical displacements.

Two LDTs were used to measure the sliding displacement of the footing, while two WDTs were used to record the horizontal displacement of the oscillator mass. Four additional WDTs were used to measure the vertical displacement at the four corners of the mass, in order to measure the rotation and the settlement of the physical model in both directions (the direction of loading and the transverse). Accelerometers were installed at characteristic

FIGURE 4 | Photos showing the instrumentation.

locations, on the model (on the foundation and at mass level), and embedded within the soil at 5 cm depth: one directly underneath the foundation and another one at a distance, where free-field conditions are restored. In addition, visual data were obtained using high-definition cameras, recording both the response of the entire system, and the response of the foundation (settlement, uplift, sliding) from a closer view from behind.

The different configurations were subjected to shaking table testing, using a variety of real seismic records and artificial motions as base excitation (**Figure 5**). The selected seismic motions were scaled down in time (divided by $\sqrt{n}$) according to the relevant scaling laws. Three different seismic shaking sequences were imposed, as summarized in **Table 3**. They consist of artificial motions (sinusoidal excitations) or real records, or a combination of the two. In all cases examined, the PGA of the seismic excitations was well beyond the critical acceleration of the tested systems (progressively increasing). This was done to force the systems to behave strongly non-linear in order to explore their response in the plastic (after yielding) as well as in their metaplastic regime (after peak conditions are reached and the resistance of the footing decreases due to P–δ effects).

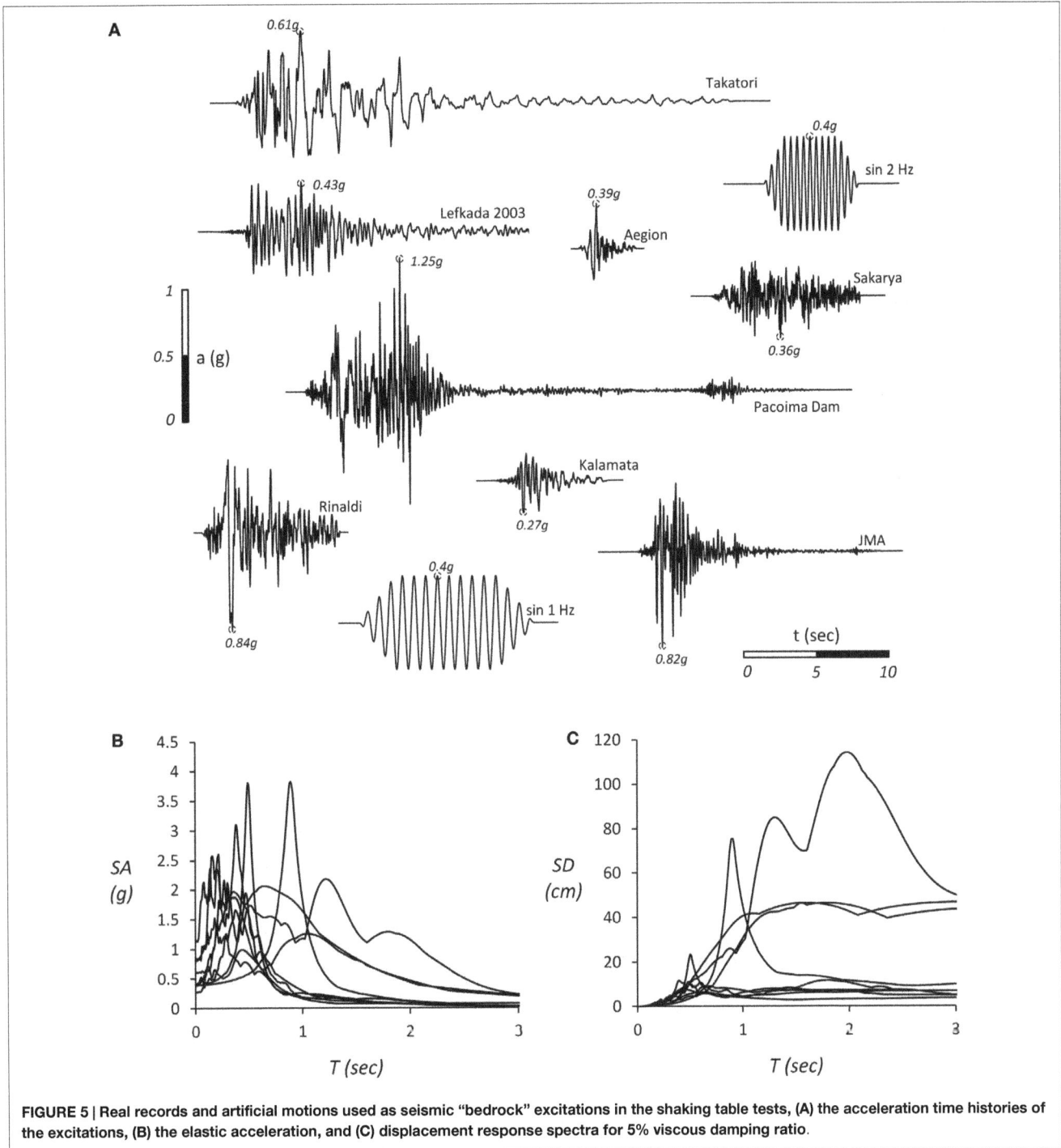

FIGURE 5 | Real records and artificial motions used as seismic "bedrock" excitations in the shaking table tests, (A) the acceleration time histories of the excitations, (B) the elastic acceleration, and (C) displacement response spectra for 5% viscous damping ratio.

Seismic Performance of the Lightly-Loaded System A

System A was subjected to the seismic excitation sequences I and II. As discussed later on, sequence III was used for *System B* only. In this section, selected results are presented for the lightly loaded *System A*, aiming to gain insight of the main features affecting its seismic performance and to assess the effectiveness of shallow soil improvement under truly dynamic conditions.

Truly Dynamic vs. Slow-Cyclic Response

In terms of monotonic and slow-cyclic pushover response, shallow soil improvement was proven to be quite effective (Anastasopoulos et al., 2012). **Figure 6** summarizes the results of such testing, depicting the settlement per cycle w_c as a function of the imposed cyclic rotation amplitude $_c$. Although this section focuses on the lightly loaded *System A*, the results for *System B* are also presented for completeness. For both systems, a $z/B = 1$ dense sand crust is proven enough to achieve practically the same performance as the ideal case of dense sand. For the shallower $z/B = 0.5$ soil improvement layer, the cyclic settlement reduction is quite evident, but the response differs substantially from the ideal case of dense sand for both systems. Even though, a shallower $z/B = 0.5$ soil improvement layer may be adequate, depending on design requirements. The efficiency of shallow soil improvement is ameliorated with the increase in the cyclic rotation amplitude, especially in the case of the lightly loaded *system A*.

Figure 7 summarizes the performance of the lightly loaded system subjected to the Aegion seismic excitation. Although the record of the 1995 Ms 6.2 Aegion (Greece) earthquake is considered as a moderate intensity seismic excitation, it does contain a single strong motion pulse of 0.39 g, which is well above the critical acceleration ac of *System* A for all cases considered (as previously

discussed, a_c ranges from 0.072 g for poor soil to 0.15 g for ideal conditions). To make things worse, due to soil amplification the maximum acceleration measured at the free field reaches 0.58 g (**Figure 7A**). As a result, the response of the system is highly non-linear for all cases examined. As shown in **Figure 7B**, the strong motion pulse of the Aegion record leads to a maximum rotation $_{max}$ of roughly 0.01 rad, which is not particularly sensitive to the soil conditions. Naturally, $_{max}$ is slightly larger for poor soil conditions (loose sand), but the differences are practically negligible. A simplified explanation for this can be derived using an equivalent linear approach: the initial natural period of the four systems that ranges between 0.66 s for the system lying on dense sand and 1.21 s for the one lying on loose sand increases substantially with the non-linear response of the rocking footing, yielding effecting periods that correspond to substantially diminished spectral accelerations in the area of the spectrum where the differences in the stiffness of the soil deposit or in the effective damping ratio do not alter the response significantly. The imposed rotation is irrecoverable when the system is founded on loose sand, with the residual rotation $_{res}$ being almost the same as $_{max}$. This is not the case for ideal soil conditions (dense sand), where $_{res}$ is practically 0: the system returns to each original position. Shallow soil improvement proves quite effective in reducing the residual rotation, with the deeper $z/B = 1$ crust being advantageous: $_{res} = 0.002$ rad as opposed to 0.0025 rad of the shallower $z/B = 0.5$ crust.

As expected, under such strongly non-linear foundation response the rocking system accumulates dynamic settlement. As

TABLE 3 | The tree seismic shaking sequences of the shaking table tests.

Sequence I		Sequence II		Sequence III	
Excitation	PGA (g)	Excitation	PGA (g)	Excitation	PGA (g)
sin 2 Hz	0.2	Aegion	0.39	sin 2 Hz	0.1
sin 2 Hz	0.4	Kalamata	0.4	sin 2 Hz	0.15
sin 1 Hz	0.2	Lefkada 2003	0.2	sin 2 Hz	0.2
sin 1 Hz	0.4	JMA	0.4	sin 2 Hz	0.25
Pacoima Dam	1.25	Rinaldi	1.14	sin 2 Hz	0.3
Sakarya	0.36	Takatori	0.36	sin 2 Hz	0.35
Lefkada 2003	0.43			sin 2 Hz	0.4

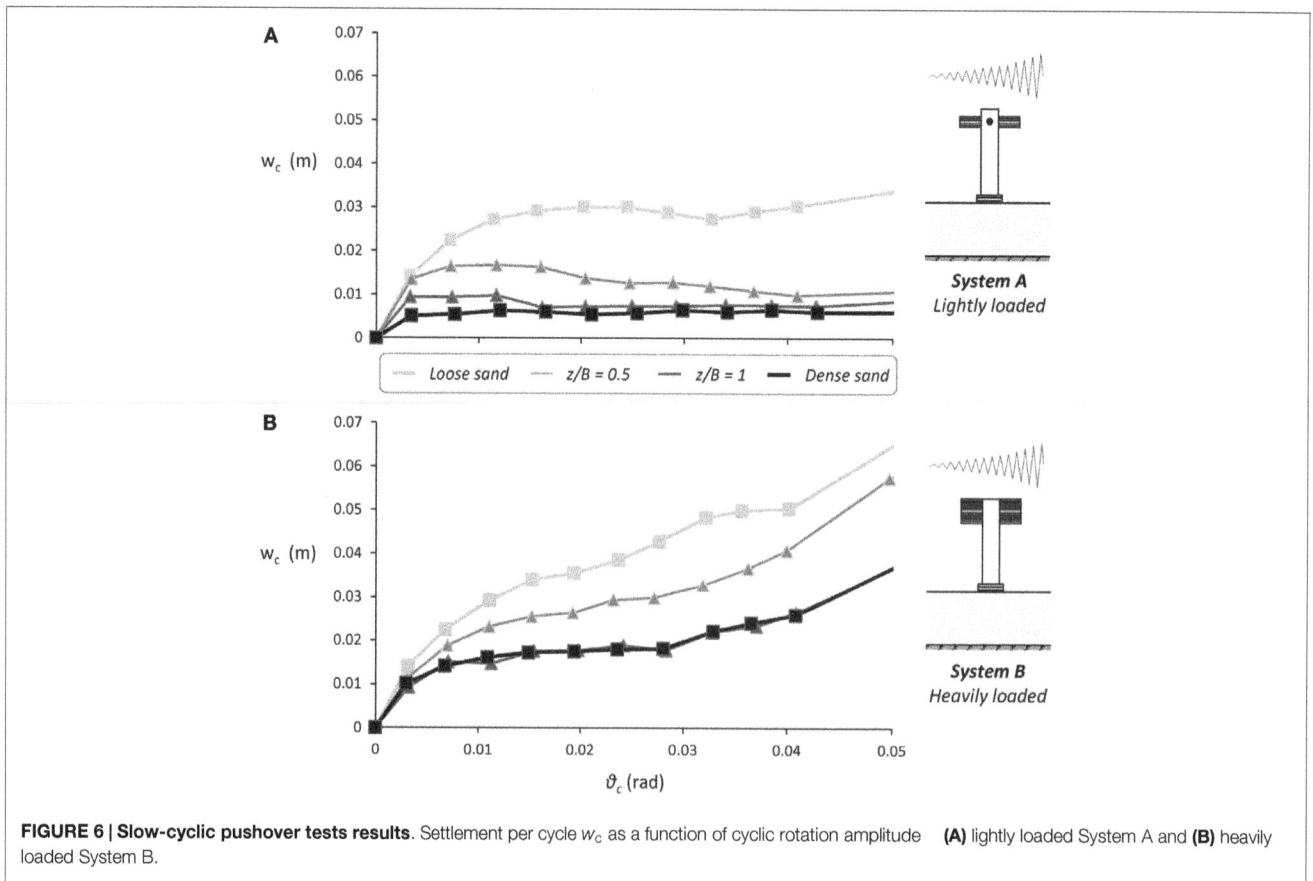

FIGURE 6 | Slow-cyclic pushover tests results. Settlement per cycle w_c as a function of cyclic rotation amplitude **(A)** lightly loaded System A and **(B)** heavily loaded System B.

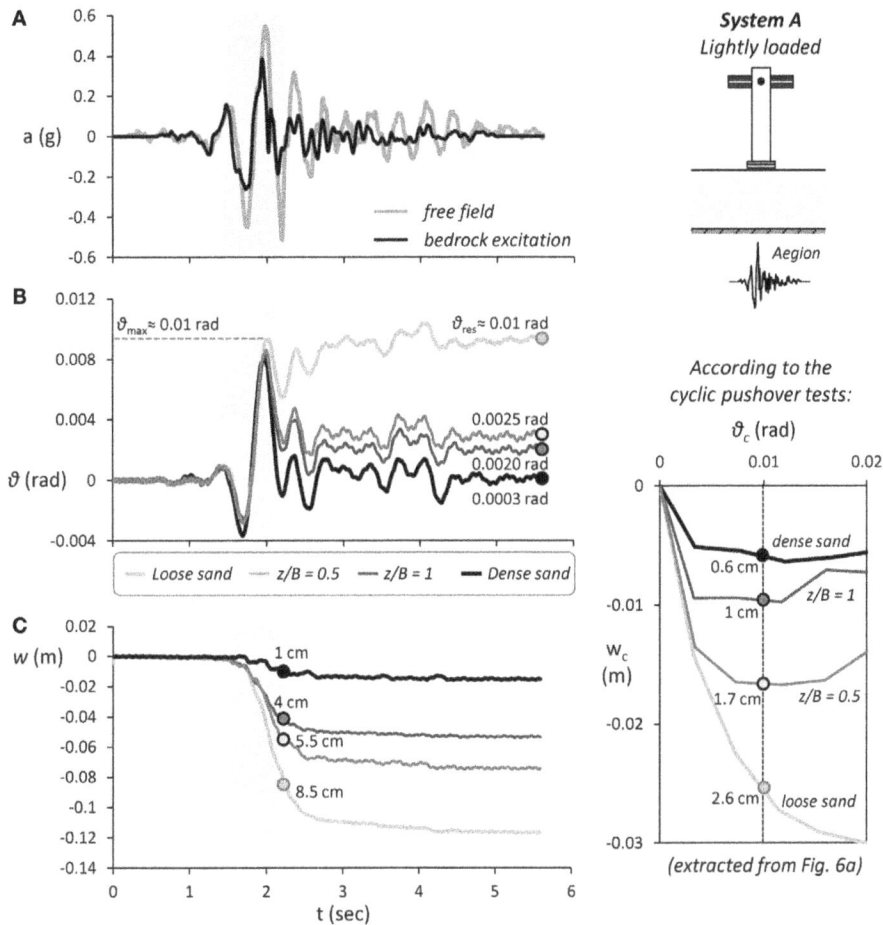

FIGURE 7 | Seismic performance of the lightly loaded System A subjected to moderate seismic shaking using the Aegion record as excitation–comparison of shallow soil improvement with poor (loose sand) and ideal (dense sand) soil conditions. Time histories of **(A)** free field and base excitation, **(B)** foundation rotation, and **(C)** settlement. An extract from **Figure 6A** (bottom right) is also shown to allow direct comparison with the cyclic pushover tests.

shown in **Figure 7C**, shallow soil improvement is quite effective in reducing the accumulated settlement. It should be noted that the settlement shown in this figure as well as in all the respective figures in the ensuing, corresponds to the total settlement of the footing, while the settlement of the free field is omitted. This is done because the instrument used to measure the settlement of the free field malfunctioned during some of the tests, and a meaningful comparison would not be possible. However, for the cases where it was measured, the free field settlement proved to be minor compared to the settlement of the footing. For example, for the lightly loaded system lying on the shallower crust and subjected to a sinusoidal excitation of frequency 2 Hz and amplitude 0.4 g, one of the most adverse motions, the measured free field settlement was 6.6 cm compared to the 39.2 cm for the footing. When founded on poor soil conditions (loose sand), the settlement reaches 12 cm. Applying shallow soil improvement of depth $z/B = 0.5$, the accumulated settlement is reduced to 7 cm, while a deeper $z/B = 1$ soil crust leads to further reduction of the settlement to 5 cm. Although such a value is substantially larger than the dynamic settlement under ideal soil conditions (merely

1.5 cm), the efficiency of shallow soil improvement is undeniable. Observe that the larger part of the accumulated settlement is due to the single strong motion pulse of the Aegion record. In the case of loose sand, 8.5 out of 12 cm of the total accumulated settlement take place during this pulse. The same applies to the remaining configurations, with most of the settlement taking place during that single pulse: 5.5 out of 7 cm for the $z/B = 0.5$ crust; 4 out of 5 cm for the $z/B = 1$ crust; and 1 out of 1.5 cm for the case of ideal conditions.

With the response being so straight forward, it is interesting to compare the results of the shake table tests to what would be expected on the basis of the cyclic pushover tests. Going back to **Figure 6A** (an extract of which is reproduced in **Figure 7** to allow direct comparisons), for cyclic rotation $\vartheta_c = \vartheta_{max} \approx 0.01$ rad, the system on loose sand would accumulate cyclic settlement $w_c = 2.6$ cm, while as we saw under truly dynamic loading it actually accumulates $w_{dyn} = 8.5$ cm. Similarly, in the case of the $z/B = 0.5$ shallow soil improvement, the system settles $w_{dyn} = 5.5$ cm as opposed to the $w_c = 1.7$ cm, and the same applies to the $z/B = 1$ ($w_{dyn} = 4$ cm as opposed to $w_c = 1$ cm)

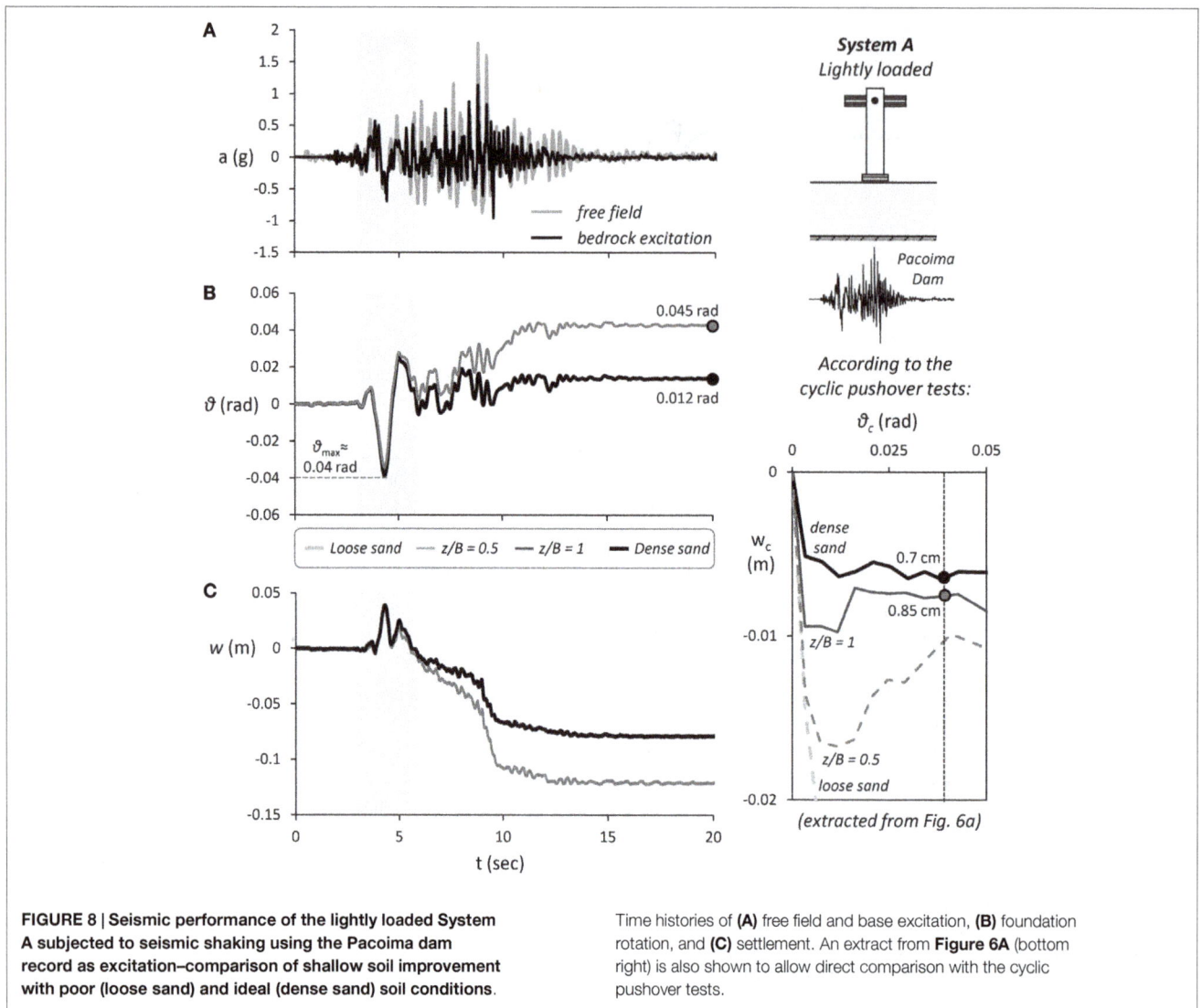

FIGURE 8 | Seismic performance of the lightly loaded System A subjected to seismic shaking using the Pacoima dam record as excitation–comparison of shallow soil improvement with poor (loose sand) and ideal (dense sand) soil conditions.

Time histories of **(A)** free field and base excitation, **(B)** foundation rotation, and **(C)** settlement. An extract from **Figure 6A** (bottom right) is also shown to allow direct comparison with the cyclic pushover tests.

and to the ideal case of dense sand ($w_{dyn} = 1$ cm as opposed to $w_c = 0.6$ cm). Hence, although qualitatively the shake table tests confirm the findings of the slow-cyclic pushover tests, from a quantitative point of view, there are very substantial differences. These differences can only be attributed to the dynamic response of the soil, which cannot possibly be captured through cyclic loading. Under dynamic loading, the deformation of the soil underneath the footing is not only due to the stresses imposed by the rocking foundation (inertia loading) but is also affected by the shear stresses that develop within the soil due to the seismic shaking itself (kinematic loading). Even in the absence of a rocking foundation, due to the developing shear stresses within the soil (kinematic loading), the sand would settle: dynamic compaction. However, compared to the free field where the soil is compacted under 0 normal stress at the surface, the soil underneath the footing is compacted under the weight of the footing, leading thus to increased settlements.

Such dynamic compaction proves to be quite intense for the case of loose ($D_r = 45\%$) sand: $w_{dyn} - w_c = 5.9$ cm. In the case of dense ($D_r = 93\%$) sand, such effects are suppressed and the

differences are much lower: $w_{dyn} - w_c = 0.4$ cm. Things are slightly more complicated in the case of shallow soil improvement. While the dense sand crust should not be prone to such effects, underneath there is still loose sand, which will settle due to dynamic compaction. Naturally, the depth of the loose sand layer is reduced with the increase of the depth of the improvement crust: from $3B$ in the case of loose sand, to $2.5B$ for $z/B = 0.5$, and to $2B$ for $z/B = 1$. Since the amount of soil compaction is proportionate to the depth of the loose sand layer, there should be an analogy here as well. Starting from the previously mentioned values for loose sand ($w_{dyn} - w_c = 5.9$ cm), the expected values for $z/B = 0.5$ and $z/B = 1$ should be 5 cm ($= 6$ cm $\times 2.5B/3B$) and 4 cm ($= 6$ cm $\times 2B/3B$), respectively. The experimental results verify the above simplified approach with minor divergence: the difference between the dynamic settlement and the respective predicted by the slow-cyclic pushover tests is 3.8 cm for the case of $z/B = 0.5$ and 3 cm for the $z/B = 1$, with the small difference attributed to the fact that the two mechanisms that lead to footing settlement (rocking of the footing and dynamic compaction of the soil deposit under the weight of the footing) act simultaneously and therefore

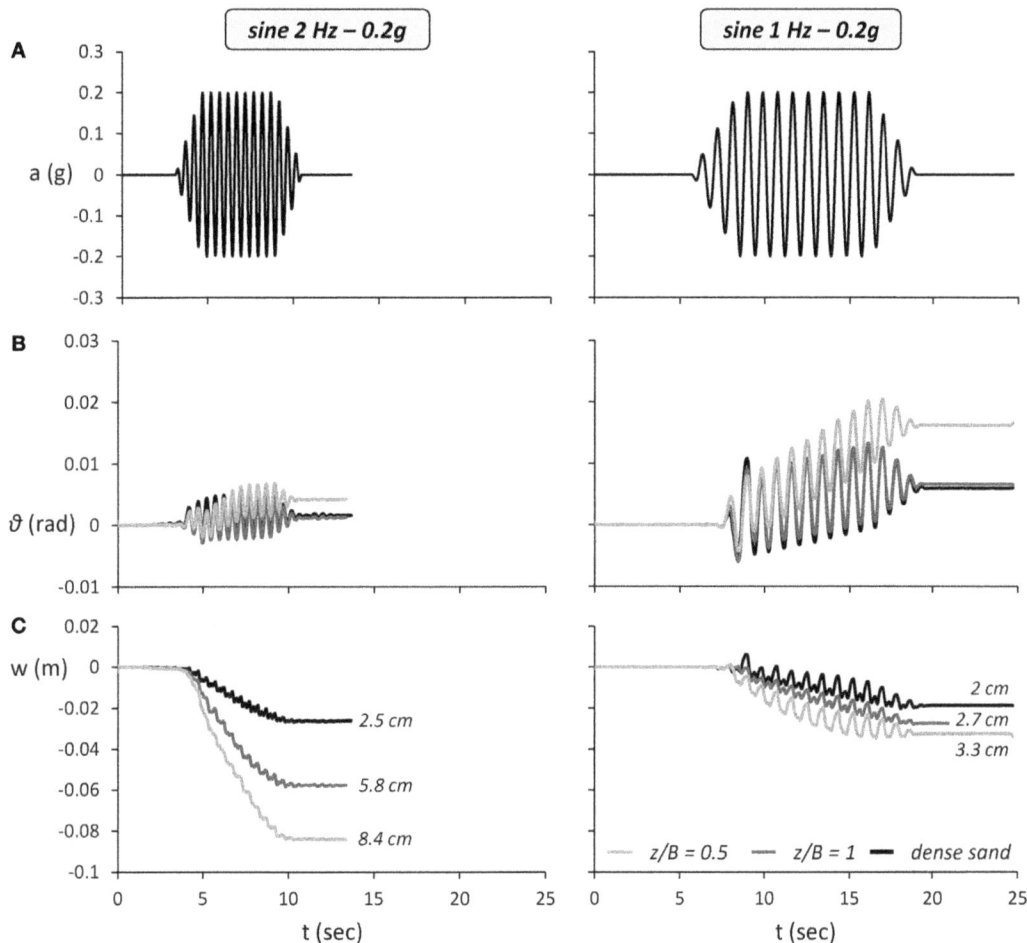

FIGURE 9 | The effect of excitation frequency on the efficiency of shallow soil improvement. Comparison of shallow soil improvement with ideal soil conditions for the lightly loaded System A subjected to seismic shaking using two 15-cylce sinusoidal excitations of frequency $f = 2$ Hz (left) and $f = 1$ Hz (right). Time histories of **(A)** free field and base excitation, **(B)** foundation rotation, and **(C)** settlement.

are coupled resulting in less settlement than if they were acting separately. Nonetheless, the results confirm that the differences in the response (compared to cyclic loading) are due to dynamic compaction of the underlying loose sand.

The Effect of Excitation Frequency

The performance of *System A* subjected to more intense seismic shaking, using as seismic excitation the Pacoima Dam record from the 1971 San Fernando earthquake, is summarized in **Figure 8**. In contrast to previous seismic excitation (Aegion), the Pacoima Dam record is definitely a very strong seismic record. Apart from its impressive *PGA* of 1.25 g, this record is characterized by a long duration (and hence, low frequency) directivity pulse, having an amplitude of 0.6 g (see the shaded area in **Figure 8A**). Given the previously discussed critical acceleration of the system on loose sand, merely $a_c = 0.072$ g collapse should have been expected, and this is exactly what happened. The same applies to the $z/B = 0.5$ crust. The system managed to survive such strong shaking only when founded on dense sand, or in the case of the deeper $z/B = 1$ soil improvement. This alone is a very important conclusion,

confirming the efficiency of the $z/B = 1$ shallow soil improvement in terms of survivability. The remaining discussion will focus on the two systems that did not collapse.

The response can roughly be divided in two phases. In the first phase, which approximately lasts from $t = 3$ to 6 s, the previously mentioned strong directivity pulse dominates the response. With a very large period $T \approx 1.2$ s, this pulse drives both systems well within their metaplastic regime, developing a maximum rotation ≈ 0.04 rad (**Figure 8B**). The second phase (for $t > 6$ s) is characterized by a multitude of strong motion cycles of even larger amplitude (up to 1.25 g) but of substantially smaller period, ranging from 0.1 to 0.4 s. As revealed by the free field acceleration measurements (**Figure 8A**), due to soil amplification, there are three acceleration peaks in excess of 1 g, with one of them reaching a *PGA* of 1.8 g. Under such unrealistically extreme seismic excitation, either on dense sand or on $z/B = 1$ soil improvement, the rocking system survives. Although the differences in $_{max}$ are again negligible, there is a substantial difference in $_{res} = 0.045$ rad for the case of $z/B = 1$ soil improvement, compared to 0.012 rad for ideal soil conditions.

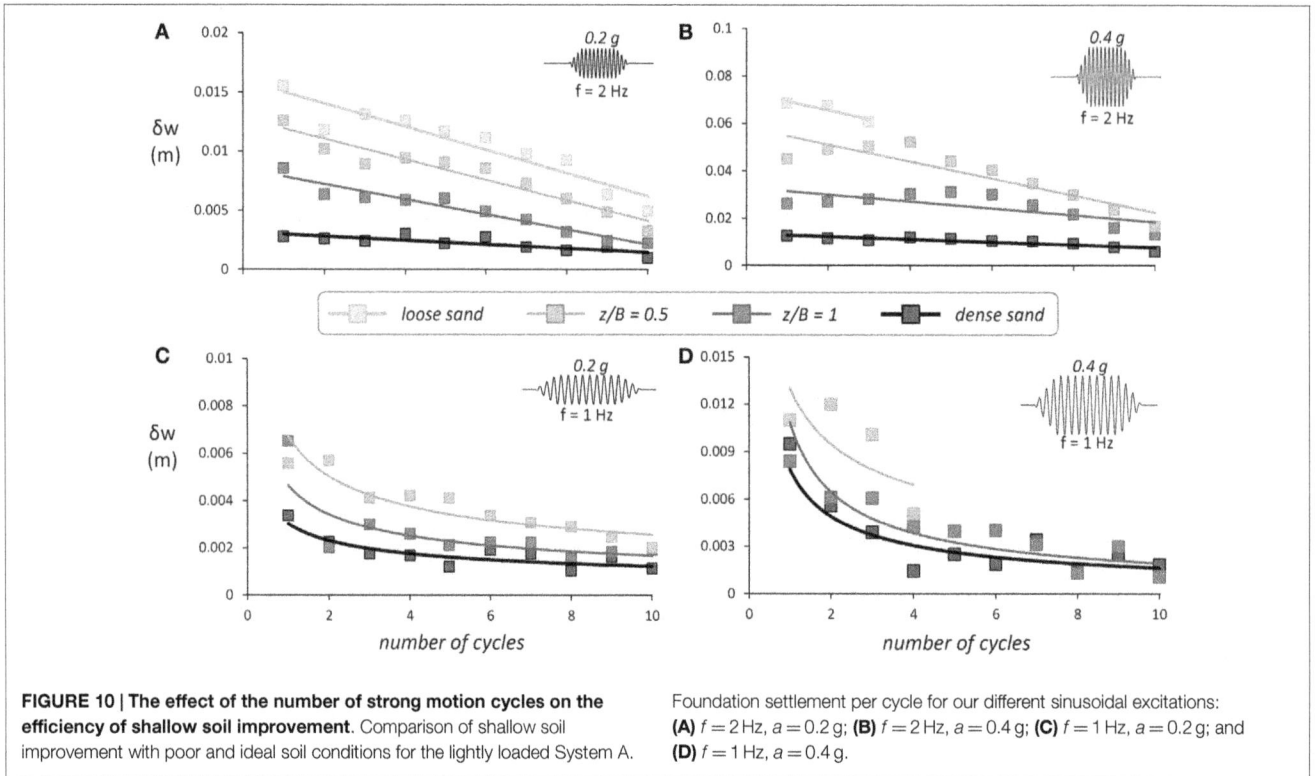

FIGURE 10 | The effect of the number of strong motion cycles on the efficiency of shallow soil improvement. Comparison of shallow soil improvement with poor and ideal soil conditions for the lightly loaded System A. Foundation settlement per cycle for our different sinusoidal excitations: **(A)** $f = 2\,Hz$, $a = 0.2\,g$; **(B)** $f = 2\,Hz$, $a = 0.4\,g$; **(C)** $f = 1\,Hz$, $a = 0.2\,g$; and **(D)** $f = 1\,Hz$, $a = 0.4\,g$.

In terms of settlement accumulation, the response is distinctly different during the two phases of the response. As illustrated in **Figure 8C**, during the long period directivity pulse (phase 1), the settlement is minimal in both cases. In fact, the rocking system is mainly subjected to uplifting and the accumulated settlement at $t = 6\,s$ does not exceed 1 cm. During this phase of response, the $z/B = 1$ crust is proven very effective, exhibiting practically identical behavior to that of the ideal case of dense sand. The performance is markedly different during the second phase, which is characterized by a multitude of strong motion cycles of larger amplitude but of much higher frequency. The settlement mainly takes place during this second phase, with the accumulated settlement reaching 8 cm for dense sand and 12 cm in the case of the $z/B = 1$ soil crust. It is worth reminding that the above values refer to the prototype structure. The correct scaling of both the structure (by 20 times) and of the frequency of the excitation (by 4.5 times) assure the similitude in the settlement between the prototype problem and the model. Although such scaling is not perfect in 1 g testing, it is the best that can be done and this is generally accepted in such procedures. Even though, the settlement may be affected by scale effects, as recently shown in Kokkali et al. (2015), where we compared 1 g with centrifuge model testing. The comparison indicates that the cyclic foundation settlement in 1 g testing is over-estimated. Due to the incorrect scaling of geostatic stresses, the strength and the dilative behavior of the sand are over-estimated, but the shear stiffness is under-estimated. And under cyclic loading, this leads to larger foundation settlement. This is a limitation of the presented work that should be clearly spelled out.

Let us now compare the results of the shake table tests to the "prediction" on the basis of the cyclic pushover tests. As for the previous case, an extract from **Figure 6A** is included in **Figure 8** to facilitate the comparison. Focusing on the first phase of response, for a cyclic rotation $\vartheta_c = \vartheta_{max} \approx 0.04\,rad$, the system would accumulate cyclic settlement $w_c = 0.7\,cm$ and 0.85 cm, in the case dense sand and $z/B = 1$ soil improvement, respectively. In the shaking table tests, the dynamic settlement during the first phase is quite similar, not exceeding $w_{dyn} \approx 1\,cm$. The differences are much larger during the second phase of response, with the final accumulated dynamic settlement reaching 8 cm under ideal soil conditions and almost 12 cm in the case of shallow soil improvement. In both cases, the magnitude of the accumulated settlement is larger than what would have been predicted on the basis of the cyclic pushover test results. The reasons are the same with those previously discussed, but the accumulation of settlement and the efficiency of shallow soil improvement are clearly affected by the excitation frequency.

To further clarify the role of excitation frequency, two idealized 15-cylce sinusoidal motions with a *PGA* of 0.2 g are used, with the only difference being the frequency: $f = 1$ and 2 Hz. **Figure 9** compares the performance of the two cases of soil improvement with the ideal case of dense sand. Although the acceleration is exactly the same (**Figure 9A**), the "slower" $f = 1\,Hz$ excitation will develop larger (theoretically double) ground displacement compared to the "faster" $f = 2\,Hz$ sinus. And since the rotation of the rocking system depends largely on the ground displacement, the $f = 1\,Hz$ excitation should produce larger rotation. Indeed, as shown in the rotation time histories (**Figure 9B**), the rotation amplitude per cycle of the systems subjected to the $f = 1\,Hz$ sinusoidal excitation is approximately double the respective rotation when the systems are subjected to the $f = 2\,Hz$ excitation.

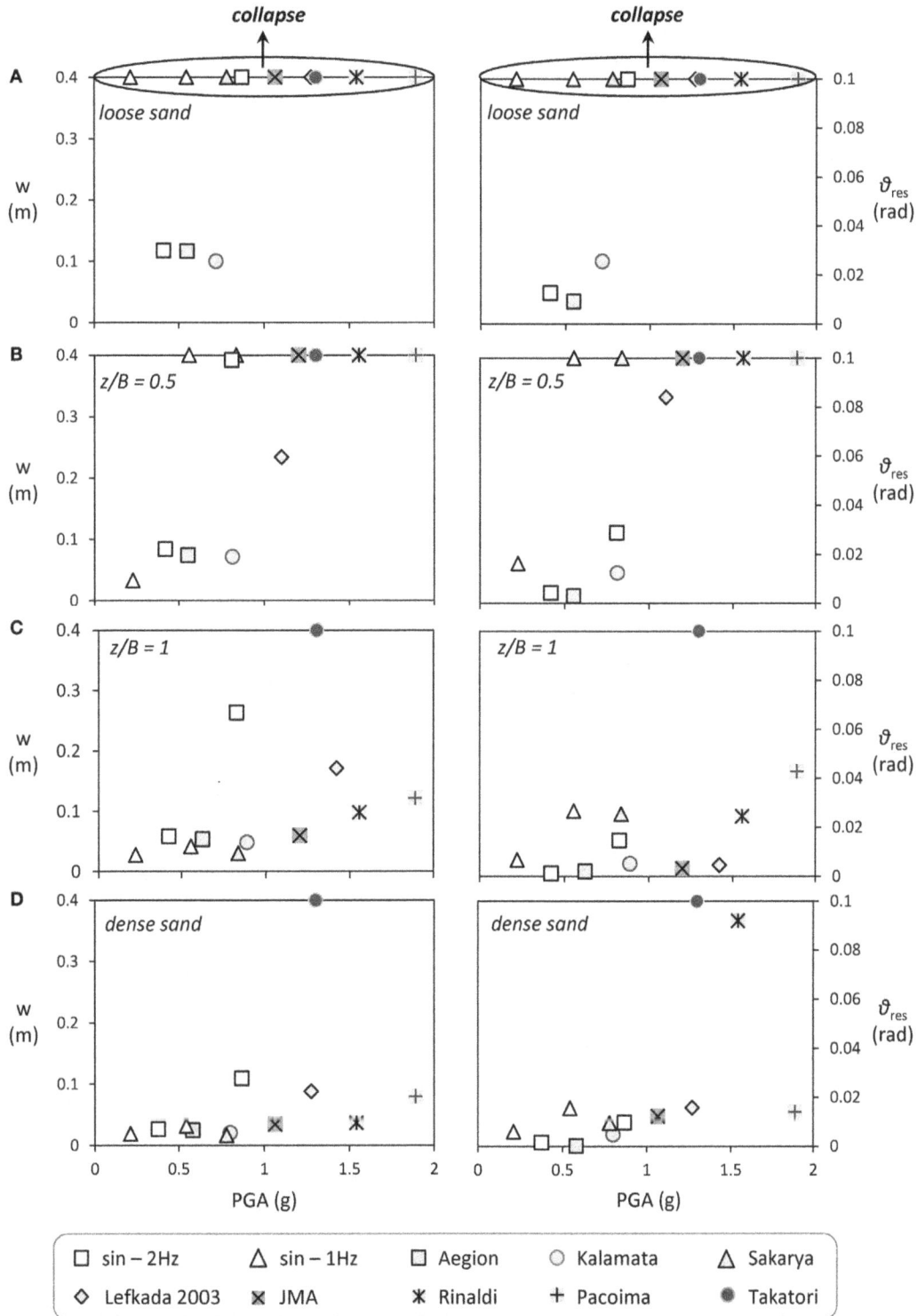

FIGURE 11 | Synopsis of the performance of the lightly loaded System A subjected to seismic shaking sequences I and II. Settlement w (left) and residual rotation $_{res}$ (right) as a function of PGA: **(A)** loose sand-poor soil conditions; **(B)** $z/B = 0.5$; **(C)** $z/B = 1$; and **(D)** dense sand-ideal soil conditions.

Figure 9C compares the settlement for the two excitation frequencies. In the case of dense sand, the settlement is not particularly sensitive to the excitation frequency. The settlement w reaches 2.5 cm for the high-frequency $f = 2$ Hz sine, being only slightly lower (2 cm) for the lower frequency $f = 1$ Hz excitation. This is in accord with the cyclic pushover test results, according to which the cyclic settlement of *System A* on dense sand remains practically constant for $0.003 < _c < 0.03$ rad. The small

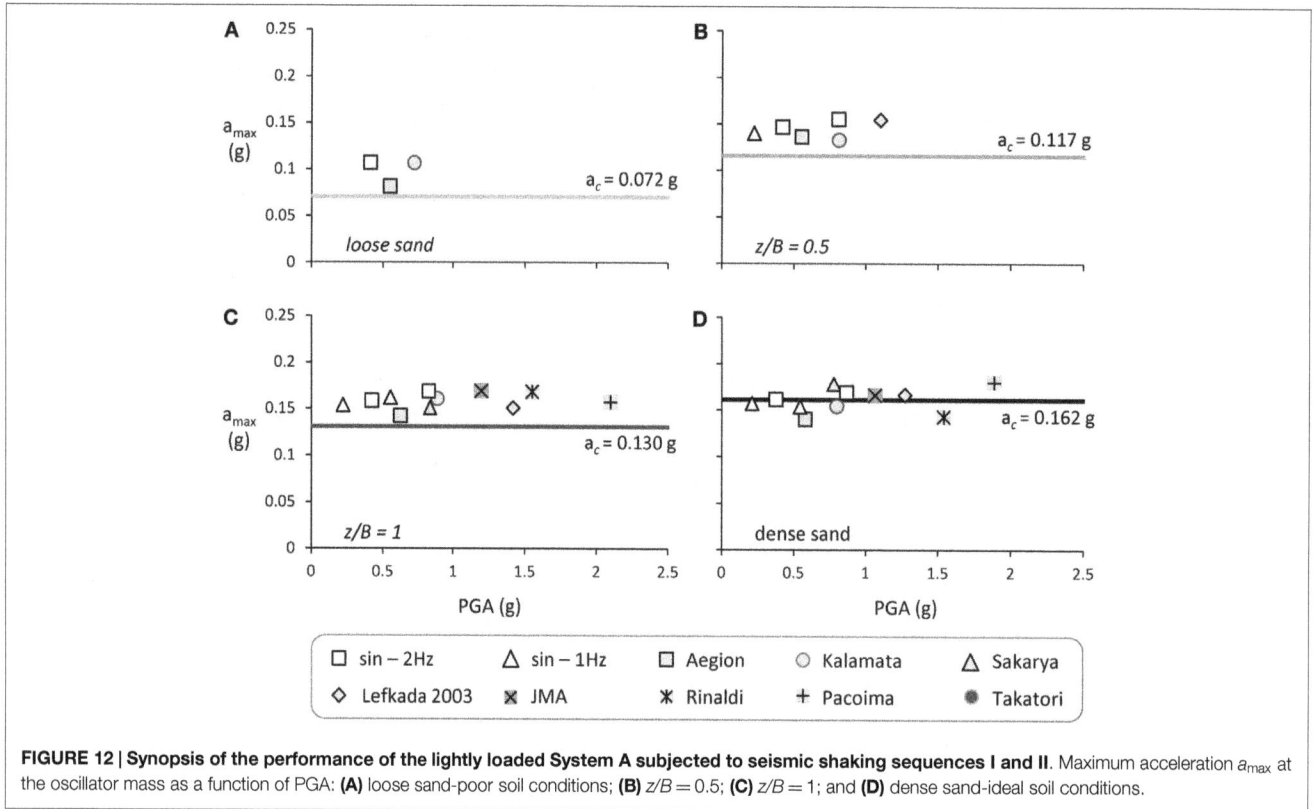

FIGURE 12 | Synopsis of the performance of the lightly loaded System A subjected to seismic shaking sequences I and II. Maximum acceleration a_{max} at the oscillator mass as a function of PGA: **(A)** loose sand-poor soil conditions; **(B)** $z/B = 0.5$; **(C)** $z/B = 1$; and **(D)** dense sand-ideal soil conditions.

difference is possibly related to a limited amount of dynamic compaction, which mainly affected the $f = 2$ Hz sine, which was the first excitation in this seismic shaking sequence (see also **Table 3**).

The differences are much more pronounced for the two cases of shallow soil improvement. In both cases (i.e., $z/B = 0.5$ and 1.0), the accumulated settlement is much larger for the high-frequency $f = 2$ Hz excitation: roughly two times larger than for the low-frequency $f = 1$ Hz sine. This very substantial difference can not only be solely attributed to dynamic compaction of the underlying loose sand but is also related to the dependence of the efficiency of shallow soil improvement on cyclic rotation. In agreement with the results of monotonic and cyclic pushover tests, the efficiency of the crust is found to increase with rotation (the amplitude of   for $f = 1$ Hz is almost twice as much for $f = 2$ Hz). While for smaller rotation the foundation is in full contact with the supporting soil, generating a deeper stress bulb, and hence, being affected by the underlying loose sand layer, when uplifting is initiated the effective foundation width is drastically decreased, reducing the depth of the generated stress bulb. Hence, a larger portion of the rocking-induced stresses are obtained by the "healthy" soil material of the crust, improving the performance of the system.

The Effect of the Number of the Cycles

Apart from rotation, the efficiency of shallow soil improvement is also ameliorated with the number of strong motion cycles. **Figure 10** summarizes the results of seismic shaking sequence I, presenting the dynamic settlement w per cycle of motion with respect to the number of cycles, and as a function of excitation frequency and amplitude. In all cases examined, irrespective of

excitation frequency or amplitude, the settlement per cycle of motion reduces with the number of cycles, thanks to soil densification underneath the footing.

For the two high-frequency ($f = 2$ Hz) excitations, the rate of settlement δw is reduced almost linearly with the number of strong motion cycles. Quite interestingly, the decrease of δw is much more intense when the frequency of excitation is lower ($f = 1$ Hz). In this case, the first two or three cycles are enough to cause substantial dynamic compaction of the soil, and as a result, the remaining strong motion cycles are not leading to any substantial additional settlement. As previously discussed, the oscillation of the lightly loaded system is of significantly larger amplitude for the low-frequency sinusoidal excitation, and therefore, the first two or three cycles of large rotational amplitude are enough to compact the sand under the footing. On the contrary, the sand is continuously compacted due to small vibrations of the footing when the system is subjected to the $f = 2$ Hz sinusoidal excitation. Moreover, the decreasing trend of settlement accumulation should also be attributed to the rotation accumulation of the systems; as the systems unavoidably accumulate rotation toward the one side, they do not execute a symmetric cycle of rotation rather, they tend to tilt even more. Since settlement development is correlated with the compaction of the sand as the structure returns to its initial position, it is reasonable to assume that settlement accumulation is also affected by the rotation accumulation.

Synopsis and Discussion

Figure 11 summarizes the performance of the lightly loaded *System A* subjected to shaking sequences I and II in terms of settlement w and residual rotation   res as a function of *PGA*. Although

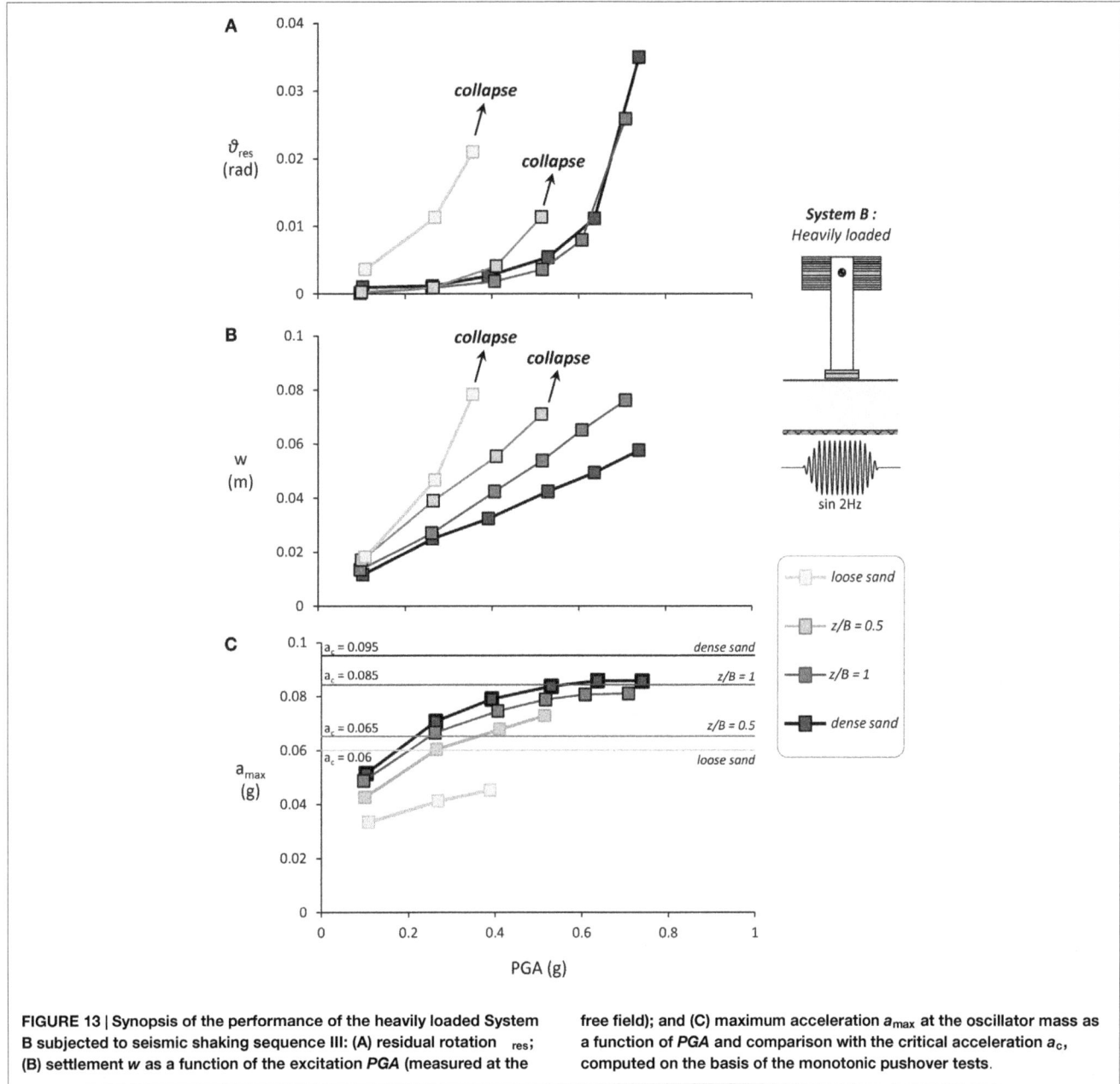

FIGURE 13 | Synopsis of the performance of the heavily loaded System B subjected to seismic shaking sequence III: (A) residual rotation ϑ_{res}; (B) settlement w as a function of the excitation *PGA* (measured at the free field); and (C) maximum acceleration a_{max} at the oscillator mass as a function of *PGA* and comparison with the critical acceleration a_c, computed on the basis of the monotonic pushover tests.

the excitations were imposed in a sequence (i.e., one after the other), the results presented herein refer to values recorded during each excitation (not the cumulative ones). Therefore, the results are not to be considered representative of the performance of the system subjected to each excitation separately, but rather in a comparative manner in order to assess the efficiency of shallow soil improvement.

As vividly shown in **Figure 11A**, when the system is founded on loose sand (representative of poor soil conditions), it can only sustain 3 out of 12 seismic excitations (considering both shaking sequences). Even for these three excitations, the settlement is quite substantial (in excess of 10 cm). The improvement is quite evident for shallow soil improvement of depth $z/B = 0.5$ (**Figure 11B**). The system is able to withstand seismic excitations

of *PGA* up to 1.1 g without toppling. Observe that the residual rotation ϑ_{res} is reduced by almost 50% compared to the untreated case of loose sand. However, the settlement w is not reduced to the same extent, and the system still topples in 7 out of 12 seismic excitations. A deeper $z/B = 1$ dense sand crust is required to decrease the settlement substantially (**Figure 11C**). In this case, the performance is practically the same with that of the ideal case of dense sand (**Figure 11D**), confirming the efficiency of shallow soil improvement with a $z/B = 1$ dense sand crust.

The main scope of rocking isolation is the reduction of the inertia transmitted onto the superstructure. It is therefore critical to ensure that shallow soil improvement is not canceling the isolation effect. This is confirmed **Figure 12**, where the acceleration a_{max} at

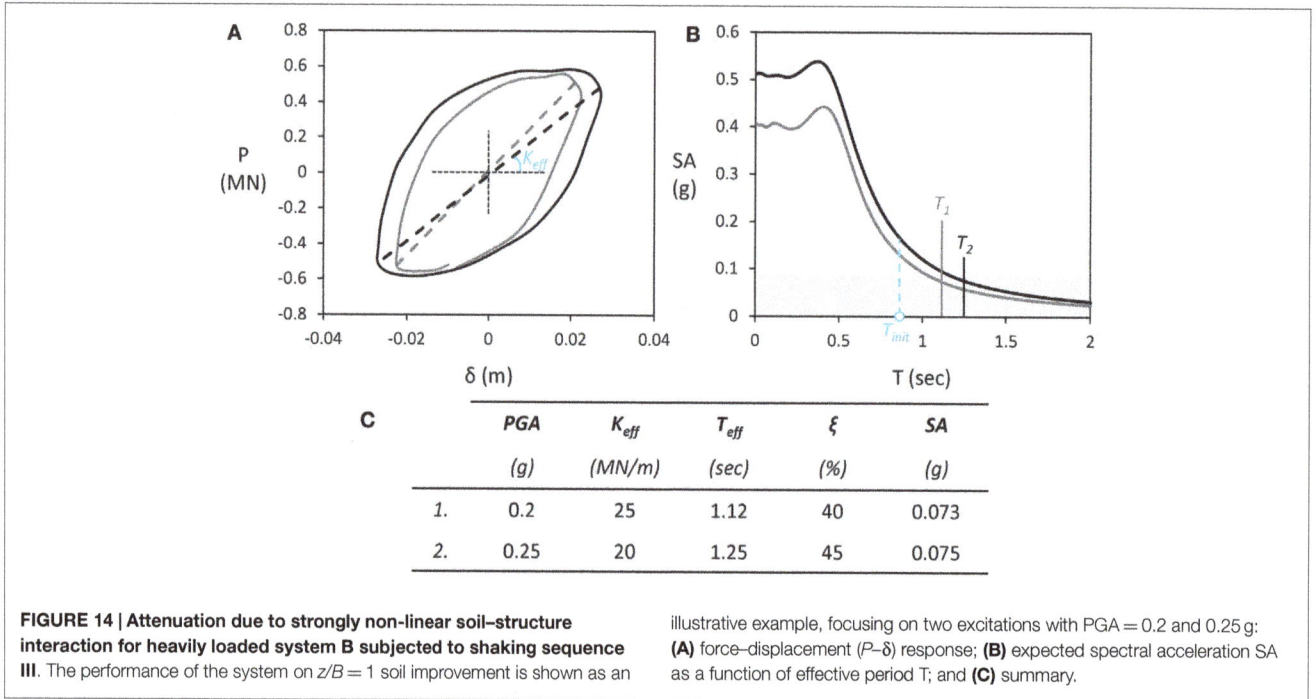

FIGURE 14 | Attenuation due to strongly non-linear soil–structure interaction for heavily loaded system B subjected to shaking sequence III. The performance of the system on $z/B = 1$ soil improvement is shown as an illustrative example, focusing on two excitations with PGA = 0.2 and 0.25 g: **(A)** force–displacement (P–δ) response; **(B)** expected spectral acceleration SA as a function of effective period T; and **(C)** summary.

the oscillator mass (representing the bridge deck) is plotted as a function of the PGA of the seismic excitation (measured in the free field). The results of the shake table tests are also compared with the previously discussed critical acceleration a_c, computed on the basis of the results of monotonic (static) pushover tests. In all cases examined, during the dynamic loading the footing exhibits a certain degree of overstrength ($a_{max} > a_c$). In fact, this overstrength is more significant as the FS_V value decreases. As a result, when the system is founded on loose sand the recorded maximum acceleration is $a_{max} = 0.1$ g (on average) as opposed to $a_c = 0.072$ g: an overstrength of roughly 30%. In the case of $z/B = 0.5$ shallow soil improvement, the overstrength is reduced but still quite substantial: $a_{max} = 0.14$ g as opposed to $a_c = 0.117$ g, 20% of overstrength. Further increase of the depth of the improvement layer to $z/B = 1$ leads to further reduction of the overstrength to 12% ($a_{max} = 0.155$ g while $a_c = 0.13$ g), with the ideal case of dense sand (in which case $FS_V = 14$) exhibiting minor, if any, overstrength. These conclusions are in full agreement with the results of slow-cyclic pushover tests (Anastasopoulos et al., 2012). The increased maximum measured accelerations could also be correlated with oscillations due to impact, a phenomenon discussed by various researchers [Chopra and Yim (1985) and Acikgoz and DeJong (2012, 2013) among others]. However, in the cases examined herein, the acceleration time histories recorded at the superstructure revealed no high-frequency oscillations that could be related to impact. This is expected since the systems rock on compliant soil rather than a rigid base.

Seismic Performance of the Heavily-Loaded System B

In this section, the seismic performance of the heavily loaded *System B* is discussed. In this case, the system is subjected to

shaking sequence III, which is composed of sinusoidal excitations of frequency $f = 2$ Hz of increasing PGA (from 0.1 to 0.4 g). This different shaking sequence was used as the much lower capacity of *System B* did not allow shaking with the much harsher excitations of the other two sequences.

Figure 13 summarizes the performance of *System B* founded on the four different soil profiles (loose sand, $z/B = 0.5$ and 1 soil improvement, and dense sand), focusing on the residual rotation res (**Figure 13A**) and settlement w (**Figure 13B**). Evidently, the heavily loaded system lying on loose sand is quite unstable, accumulating rather substantial rotation and settlement even for relatively low levels of PGA, and toppling for $PGA = 0.4$ g. In terms of residual rotation, the performance is improved rather spectacularly with shallow soil improvement, even for $z/B = 0.5$ (**Figure 13A**). For PGA < 0.4 g, the performance is almost identical to the ideal case of dense sand. However, in the case of the shallower $z/B = 0.5$ soil crust, *System B* topples for $PGA = 0.5$ g. Further increase of the improvement depth to $z/B = 1$ leads to a much more stable performance, and almost identical response with the ideal case of dense sand, even for the maximum imposed PGA.

In terms of settlement (**Figure 13B**), the deeper $z/B = 1$ soil improvement leads to a substantial improvement. A more shallow $z/B = 0.5$ crust is not as effective. Quite interestingly, up to a PGA of 0.2 g the system on $z/B = 0.5$ soil improvement settles almost the same as the one on loose sand. The efficiency of the crust starts improving for larger acceleration amplitudes, when uplifting starts to dominate the response, but soon after that the system collapses. Based on this result, it may safely be argued that such a shallow $z/B = 0.5$ improvement is not enough for such heavily loaded systems. On the other hand, the deeper $z/B = 1$ soil crust proves quite effective, with the settlement being roughly 25% larger compared to the ideal case of dense sand.

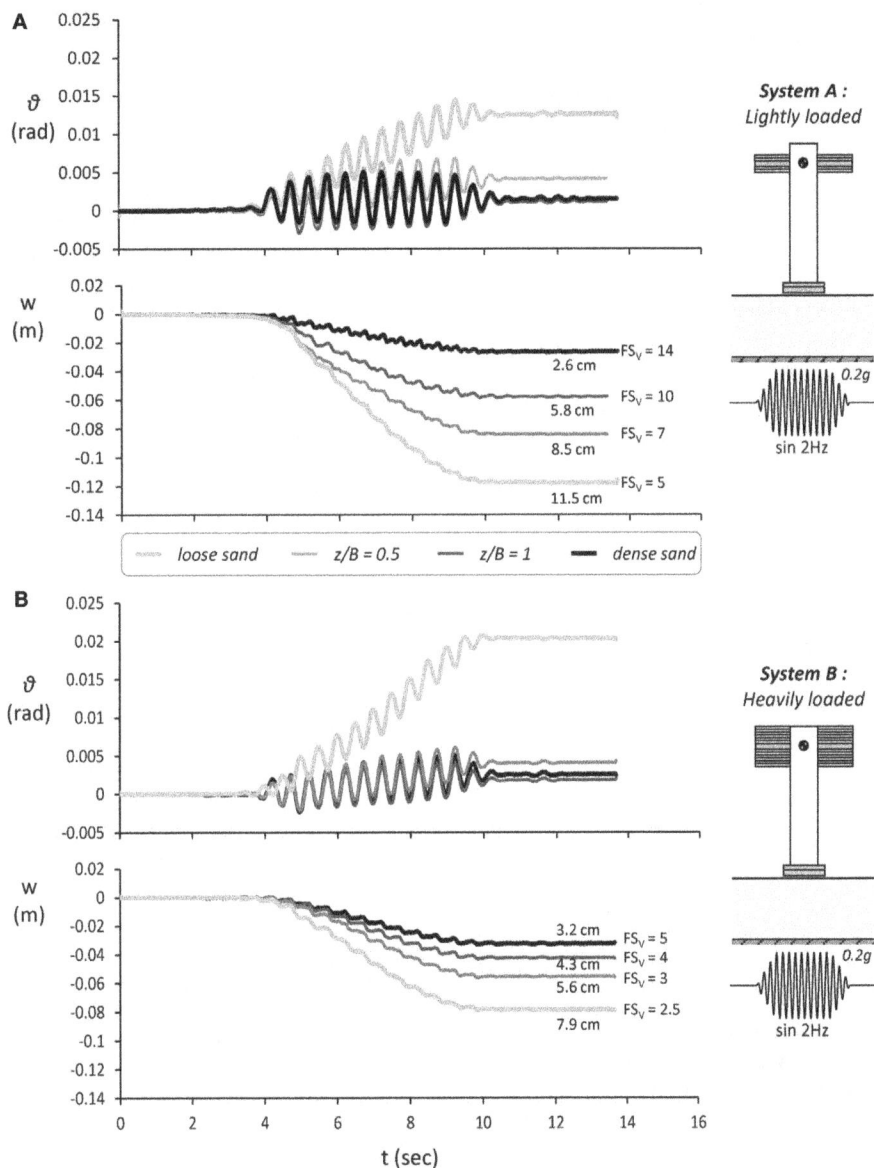

FIGURE 15 | The effect of "de-amplification" due to strongly non-linear SSI. Time histories of rotation and settlement w for all soil profiles examined, subjected to an f = 2 Hz sine of PGA = 0.2 g. Comparison of **(A)** lightly loaded System A; with **(B)** heavily loaded System B.

Figure 13C compares the performance in terms of maximum acceleration measured at the oscillator mass (representing the deck) as a function of PGA of the excitation (measured at the soil surface, in the free field). In stark contrast to the lightly loaded *System A*, where overstrength was apparent, the measured accelerations a_{max} are much lower than the corresponding critical acceleration a_c. The fact that for all four soil profiles a_{max} increases with the excitation PGA is attributed to soil densification and to the fact that the pier is gradually tilting toward the one side resulting to increased a_{max} on the opposite side. Quite interestingly, although *System B* is not reaching its ultimate moment capacity (since a_{max} is lower than the corresponding a_c), the response is profoundly non-linear as revealed by the accumulation of rotation in all cases examined (see **Figure 13B**). It may therefore be concluded that,

although the ultimate capacity is not reached, largely non-linear soil–structure interaction (SSI) takes place, leading to a rather intense attenuation of the seismic motion for all four soil profiles examined.

Figure 14 attempts to shed more light to such effects, using the case of $z/B = 1$ soil improvement as an illustrative example and focusing on two sinusoidal excitations having a PGA of 0.2 and 0.25 g. For these two seismic excitations, the maximum measured acceleration at the oscillator mass is 0.073 and 0.078 g, respectively – in both cases, lower than the critical acceleration $a_c = 0.085$ g (based on the results of monotonic pushover tests). **Figure 14A** illustrates the force–displacement (P–δ) response for the two cases examined, focusing on the steady state oscillation (i.e., after the first two to three cycles of motion). Although the ultimate capacity of the footing is not reached, the response is highly non-linear. Based on the illustrated loops, a very high hysteretic damping ratio can be calculated, ranging between 40 and 45% for $PGA = 0.2$ and $0.25\,g$, respectively. Such non-linear response unavoidably leads to a decrease of the effective stiffness K_{eff} of the system, and to an increase of its effective period T_{eff}.

Figure 14B presents the acceleration response spectra of the measured free field acceleration at the ground surface (which is considered as the input to the rocking system) for the two excitations under consideration, accounting for the damping ratios that have been calculated from the respective load–displacement loops. Based on the results of the monotonic pushover tests, the initial stiffness (i.e., for very small strains) of the system is $K_o = 40\,MN/m$. Hence, accounting for quasi-elastic SSI (there is always some non-linearity, even before initiation of seismic shaking), the initial natural period of the rocking system is $T_o = 0.89\,s$. Quite strikingly, when non-linear SSI is considered, the effective stiffness of the system drops to $K_{eff} = 24$ and $20\,MN/m$, resulting to $T_{eff} = 1.12$ and $1.25\,s$ for $PGA = 0.2$ and $0.25\,g$, respectively (**Figure 14C**). This substantial increase in the effective period of the system leads to rather intense "de-amplification" of the input motion, resulting to spectral accelerations SA lower than the critical acceleration $a_c = 0.085\,g$. Based on the above simplified rationale, the expected accelerations at the oscillator mass are 0.073 and $0.075\,g$ for $PGA = 0.2$ and $0.25\,g$, respectively, which are in very good agreement with the experimental measurements. Despite its simplicity and the fact that there are methods of non-linear analysis much more advanced, the above rationale provides an excellent prediction of the response at least for the specific cases examined herein.

Lightly vs. Heavily Loaded Systems: The Effect of De-Amplification

The previously discussed "de-amplification" proves to significantly affect response of the system. **Figure 15** compares the performance of the heavily loaded *System B* with the lightly loaded *system A*, further elucidating the effect of "de-amplification." The comparison is performed for seismic excitation with an $f = 2\,Hz$ sine having a PGA of $0.2\,g$.

Time histories of rotation and settlement w of the lightly loaded system (**Figure 15A**) are compared to the heavily loaded system (**Figure 15B**) for all four soil profiles. Given the much lower moment capacity of the heavily loaded system, it would have been reasonable to expect inferior performance. It is reminded that the factors of safety FS_V of the heavily loaded systems are much lower than those of the lightly loaded ones (ranging from 2.5 to 5 as opposed to 5 to 14), and the same applies to their critical accelerations a_c. The reality, however, is different. The lightly loaded systems are subjected to much more settlement. For example, the lightly loaded system founded on $z/B = 1$ soil improvement settles 6 cm, while the settlement of its heavily loaded counterpart does not exceed 4.5 cm. At this point, it should be noted that while for the lightly loaded *system A* this particular excitation is the first of shaking sequence I, for the heavily loaded *system B*, it is preceded by two other excitations, which means that some densification may have already taken place. Moreover, the loose sand representing poor soil conditions is of lower relative

density in the case of the lightly loaded system ($D_r = 45\%$ as opposed to 60%). Even though, these two factors alone cannot fully explain the difference in settlement.

This counter-intuitive behavior is easily explainable considering the effects of the previously discussed "de-amplification." Observe the rotation time histories of **Figure 15B** that the lightly loaded systems are rocking with maximum rotation $= 0.003\,rad$ (on average), while the heavily loaded systems are experiencing lower $= 0.002\,rad$ (on average). Although the seismic excitation is the same, being much stiffer, the lightly loaded system is excited much more: its lower natural period is closer to the dominant period of the seismic excitation. In stark contrast, the heavily loaded system is much more flexible to start with, and becomes even more flexible as soon as it starts responding non-linearly. As previously discussed, due to such non-linear SSI, significant degradation of the system's effective stiffness takes place, leading to a substantial increase of its effective period, which in turn leads to de-amplification. As a result, the heavily loaded systems prove more resilient to this particular seismic excitation than the lightly loaded ones. Naturally, overall the lightly loaded systems are much less vulnerable, as also revealed by the cases in which the heavily loaded systems toppled.

It is also interesting to compare the behavior of the two systems with the same FS_V: the lightly loaded system lying on loose sand and the heavily loaded system lying on dense sand. The two systems exhibit remarkably different behavior: not only does the heavily loaded system accumulate less settlement but also the rotation amplitude of its oscillation is notably smaller as well. This proves that the FS_V alone cannot describe the dynamic response of two systems subjected to the same excitation. As shown in other studies (Kourkoulis et al., 2012), the response of two rocking systems of the same FS_V and aspect ratio h/B subjected to the same excitation can be similar provided that there is the appropriate analogy in the stiffness of the foundation soil. In this case, the loose sand deposit proves to be relatively less stiff leading thus to increased rotation and settlement accumulation for the lightly loaded system.

Conclusion

Aiming to explore the efficiency of shallow soil improvement as a means to mitigate settlement accumulation due to non-linear response of the footing during an earthquake, this paper experimentally investigated the seismic response of two conceptual bridge piers represented by two relatively slender $h/B = 3$ systems, both lying on square foundation of width B. The first one corresponds to a lightly loaded structure (relatively large FS_V), while the second refers to a heavily loaded structure (relatively low FS_V), deliberately chosen to model distinctly different foundation performance, from uplifting-dominated to sinking-dominated response. The two systems were subjected to reduced-scale shaking table testing at the Laboratory of Soil Mechanics of the NTUA. They were first tested on poor soil conditions in order to demonstrate the necessity for soil improvement. Then, the effectiveness of shallow soil improvement was studied by investigating the performance of the two systems on soil crusts of depth $z/B = 1$ and 0.5. Finally, the performance of the two systems lying on the improved soil profiles was compared to that considering ideal soil conditions.

The main conclusions of the presented research can be summarized as follows:

- Based on the conducted reduced-scale tests, and at least for the cases examined herein, the concept of shallow soil improvement is proven quite effective in reducing the dynamic settlement of the footing. For both systems, a $z/B = 1$ dense sand crust is enough to achieve practically the same performance (in terms of settlement) with the ideal case of dense sand. A shallower $z/B = 0.5$ soil improvement may also be considered effective, depending on design requirements.
- The results of the shaking table tests are in very good qualitative agreement with previously published (Anastasopoulos et al., 2012) experimental results from monotonic and slow-cyclic pushover tests. In quantitative terms, the differences are non-negligible with the shaking table tests yielding much larger settlement for all cases examined. The tests presented herein not only confirm the key conclusions of the static experiments but also reveal substantial differences,

which are attributed to kinematic soil response and dynamic compaction–mechanisms that cannot possibly be simulated through static pushover testing.

- As with the slow-cyclic pushover tests, the performance of shallow soil improvement is found to depend on the rotation amplitude. Real records and artificial motions of different frequency were examined, which forced the two systems to oscillate at various rotation amplitudes. It was shown that with the increase of the rotation amplitude the effectiveness of the soil crusts increases.
- The performance of shallow soil improvement is ameliorated with the number of cycles of the motion. The rate of settlement reduces with the increase of the number of cycles for all cases examined.

Acknowledgments

The financial support for this paper has been provided under the research project "DARE," which is funded through the European Research Council's (ERC) "IDEAS" Programme, in Support of Frontier Research-Advanced Grant, under contract/number ERC-2-9-AdG228254–DARE.

References

Acikgoz, S., and DeJong, M. J. (2012). The interaction of elasticity and rocking in flexible structures allowed to uplift. *Earthquake Eng. Struct. Dynam.* 45, 2177–2194. doi:10.1002/eqe.2181

Acikgoz, S., and DeJong, M. J. (2013). "Analytical and experimental observations on vibration modes of flexible rocking structures," in *SECED – Society for Earthquake and Civil Engineering Dynamics Young Engineers Conference.* Newcastle.

Allotey, N., and El Naggar, M. H. (2003). Analytical moment–rotation curves for rigid foundations based on a Winkler model. *Soil Dynam. Earthquake Eng.* 23, 367–381. doi:10.1016/S0267-7261(03)00034-4

Allotey, N., and El Naggar, M. H. (2007). An investigation into the winkler modeling of the cyclic response of rigid footings. *Soil Dynam. Earthquake Eng.* 28, 44–57. doi:10.1016/j.soildyn.2007.04.003

Anastasopoulos, I., Gazetas, G., Loli, M., Apostolou, M., and Gerolymos, N. (2010a). Soil failure can be used for earthquake protection of structures. *Bull. Earthquake Eng.* 8, 309–326. doi:10.1007/s10518-009-9145-2

Anastasopoulos, I., Gelagoti, F., Kourkoulis, R., and Gazetas, G. (2011). Simplified constitutive model for simulation of cyclic response of shallow foundations: validation against laboratory tests. *J. Geotech. Geoenviron. Eng.* 137, 1154–1168. doi:10.1061/(ASCE)GT.1943 5606.0000534

Anastasopoulos, I., Gelagoti, F., Spyridaki, A., Sideri, T. Z., and Gazetas, G. (2013a). Seismic rocking isolation of asymmetric frame on spread footings. *J. Geotech. Geoenviron. Eng.* 140, 133–151. doi:10.1061/(ASCE)GT.1943-5606.0001012

Anastasopoulos, I., Loli, M., Georgarakos, T., and Drosos, V. (2013b). Shaking table testing of rocking-isolated bridge pier. *J. Earthquake Eng.* 17, 1–32. doi:10.1080/13632469.2012.705225

Anastasopoulos, I., Kourkoulis, R., Gelagoti, F., and Papadopoulos, E. (2012). Rocking response of SDOF systems on shallow improved sand: an experimental study. *Soil Dynam. Earthquake Eng.* 40, 15–33. doi:10.1016/j.soildyn.2012.04.006

Anastastasopoulos, I., Georgarakos, P., Georgiannou, V., Drosos, V., and Kourkoulis, R. (2010b). Seismic performance of bar-mat reinforced-soil retaining wall: shaking table testing versus numerical analysis with modified kinematic hardening constitutive model. *Soil Dynam. Earthquake Eng.* 30, 1089–1105. doi:10.1016/j.soildyn.2010.04.002

Antonellis, G., Gavras, A. G., Panagiotou, M., Kutter, B. L., Guerrini, G., Sander, A. C., et al. (2015). Shake table test of large-scale bridge columns supported on rocking shallow foundations. *J. Geotech. Geoenviron. Eng.* 141, 04015009. doi:10.1061/(ASCE)GT.1943-5606.0001284

Butterfield, R., and Gottardi, G. (1995). "Simplifying transformations for the analysis of shallow foundations on sand," in *Proc. 5th Int. Offshore and Polar Eng Conf* (The Hague), 534–538.

Chatzigogos, C. T., Figini, R., Pecker, A., and Salençon, J. (2011). A macroelement formulation for shallow foundations on cohesive and frictional soils. *Int. J. Numer. Anal. Meth. Geomech.* 35, 902–931. doi:10.1002/nag.934

Chatzigogos, C. T., Pecker, A., and Salençon, J. (2009). Macroelement modeling of shallow foundations. *Soil Dynam. Earthquake Eng.* 29, 765–781. doi:10.1016/j.soildyn.2008.08.009

Chen, X. L., and Lai, Y. M. (2003). Seismic response of bridge piers on elastic-plastic Winkler foundation allowed to uplift. *J. Sound Vibrat.* 266, 957–965. doi:10.1016/S0022-460X(02)01382-2

Chopra, A. K., and Yim, S. C. S. (1985). Simplified earthquake analysis of structures with foundation uplift. *J. Struct. Eng.* 111, 906–930. doi:10.1061/(ASCE)0733-9445(1985)111:4(906)

Crémer, C. (2001). *Modélisation du Comportement Non Lineaire des Foundations Superficielles Sous Seisme.* Ph.D. Thesis, LMT-Cachan, France.

Crémer, C., Pecker, A., and Davenne, L. (2001). Cyclic macro-element for soil-structure interaction: material and geometrical nonlinearities. *Int. J. Numer. Anal. Meth. Geomech.* 25, 1257–1284. doi:10.1002/nag.175.abs

Drosos, V., Georgarakos, T., Loli, M., Anastasopoulos, I., Zarzouras, O., and Gazetas, G. (2012). Soil-foundation-structure interaction with mobilization of bearing capacity: an experimental study on sand. *J. Geotech. Geoenviron. Eng.* 138, 1369–1386. doi:10.1061/(ASCE)GT.1943-5606.0000705

Faccioli, E., Paolucci, R., and Vivero, G. (2001). "Investigation of seismic soil-footing interaction by large scale cyclic tests and analytical models," in *Proc. 4th International Conference on Recent Advances in Geotechnical Earthquake Engineering and Soil Dynamics* (San Diego, CA). Paper no. SPL-5.

Gajan, S., and Kutter, B. L. (2008). Capacity, settlement, and energy dissipation of shallow footings subjected to rocking. *J. Geotech. Geoenviron. Eng.* 134, 1129–1141. doi:10.1061/(ASCE)1090-0241(2008)134:8(1129)

Gajan, S., and Kutter, B. L. (2009). Effects of moment-to-shear ratio on combined cyclic load-displacement behavior of shallow foundations from centrifuge experiments. *J. Geotech. Geoenviron. Eng.* 135, 1044–1055. doi:10.1061/(ASCE)GT.1943-5606.0000034

Gajan, S., Kutter, B. L., Phalen, J. D., Hutchinson, T. C., and Martin, G. R. (2005). Centrifuge modeling of load-deformation behavior of rocking shallow foundations. *Soil Dynam. Earthquake Eng.* 25, 773–783. doi:10.1016/j.soildyn.2004.11.019

Gazetas, G., Apostolou, M., and Anastasopoulos, I. (2003). "Seismic uplifting of foundations on soft soil, with examples from Adapazari (Izmit 1999, Earthquake)," in *Proceedings of BGA International Conference on Foundations* (Dundee: University of Dundee), 37–50.

Gelagoti, F., Kourkoulis, R., Anastasopoulos, I., and Gazetas, G. (2012). Rocking isolation of low-rise frame structures founded on isolated footings. *Earthquake Eng. Struct. Dynam.* 41, 1177–1197. doi:10.1002/eqe.1182

Gourvenec, S. (2007). Shape effects on the capacity of rectangular footings under general loading. *Géotechnique* 57, 637–646. doi:10.1680/geot.2007.57.8.637

Grange, S., Kotronis, P., and Mazars, J. (2008). A macro-element for the circular foundation to simulate 3D soil-structure interaction. *Int. J. Numer. Anal. Meth. Geomech.* 32, 1205–1227. doi:10.1002/nag.664

Harden, C., and Hutchinson, T. (2006). Investigation into the effects of foundation uplift on simplified seismic design procedures. *Earthquake Spectra* 22, 663–692. doi:10.1193/1.2217757

Houlsby, G. T., Amorosi, A., and Rojas, E. (2005). Elastic moduli of soils dependent on pressure: a hyperelastic formulation. *Géotechnique* 55, 383–392. doi:10.1680/geot.2005.55.5.383

Housner, G. W. (1963). The behavior of inverted pendulum structures during earthquakes. *Bull. Seismol. Soc. Am.* 53, 403–417.

Kawashima, K., Nagai, T., and Sakellaraki, D. (2007). "Rocking seismic isolation of bridges supported by spread foundations," in *Proc. of 2nd Japan-Greece Workshop on Seismic Design, Observation, and Retrofit of Foundations* (Tokyo), 254–265.

Kokkali, P., Anastasopoulos, I., Abdoun, T., and Gazetas, G. (2015). Static and cyclic rocking on sand: centrifuge versus reduced scale 1g experiments. *Géotechnique* 64, 865–880. doi:10.1680/geot.14.P.064

Kourkoulis, R., Anastasopoulos, I., Gelagoti, F., and Kokkali, P. (2012). Dimensional analysis of SDOF systems rocking on inelastic soil. *J. Earthquake Eng.* 16, 995–1022. doi:10.1080/13632469.2012.691615

Kutter, B. L., Martin, G., Hutchinson, T. C., Harden, C., Gajan, S., and Phalen, J. D. (2003). *Status Report on Study of Modeling of Nonlinear Cyclic Load-Deformation Behavior of Shallow Foundations.* Davis, CA: University of California; PEER Workshop.

Le Pape, Y., and Sieffert, J. P. (2001). Application of thermodynamics to the global modelling of shallow foundations on frictional material. *Int. J. Numer. Anal. Meth. Geomech.* 25, 1377–1408. doi:10.1002/nag.186

Mergos, P. E., and Kawashima, K. (2005). Rocking isolation of a typical bridge pier on spread foundation. *J. Earthquake Eng.* 9, 395–414. doi:10.1142/S1363246905002456

Muir Wood, D. (2004). *Geotechnical Modelling.* London: Spon Press.

Nakaki, D. K., and Hart, G. C. (1987). "Uplifting response of structures subjected to earthquake motions," in *Report No. 2.1-3, U.S.-Japan Coordinated Program for Masonry Building Research* (Los Angeles, CA).

Negro, P., Paolucci, R., Pedretti, S., and Faccioli, E. (2000). "Large-scale soil-structure interaction experiments on sand under cyclic loading," in *Proceedings of the 12th World Conference on Earthquake Engineering* (Aukland), 1191.

Nova, R., and Montrasio, L. (1991). Settlement of shallow foundations on sand. *Geotechnique* 41, 243–256. doi:10.1680/geot.1991.41.2.243

Paolucci, R. (1997). Simplified evaluation of earthquake induced permanent displacements of shallow foundations. *J. Earthquake Eng.* 1, 563–579. doi:10.1142/S1363246997000210

Paolucci, R., Shirato, M., and Yilmaz, M. T. (2008). Seismic behaviour of shallow foundations: shaking table experiments vs numerical modeling. *Earthquake Eng. Struct. Dynam.* 37, 577–595. doi:10.1002/eqe.773

Pecker, A. (1998). "Capacity design principles for shallow foundations in seismic areas," in *Proc. 11th European Conference on Earthquake Engineering* (Paris: A.A. Balkema Publishing).

Pecker, A. (2003). "A seismic foundation design process, lessons learned from two major projects: the Vasco de Gama and the Rion Antirion bridges," in *ACI International Conference on Seismic Bridge Design and Retrofit* (San Diego, CA: University of California at San Diego).

Pedretti, S. (1998). *Non-Linear Seismic Soil-Foundation Interaction: Analysis and Modeling Methods.* Ph.D. thesis, Politechnico di Milano, Milan.

Raychowdhury, P., and Hutchinson, T. C. (2009). Performance evaluation of a nonlinear Winkler-based shallow foundation model using centrifuge test results. *Earthquake Eng. Struct. Dynam.* 38, 679–698. doi:10.1002/eqe.902

Shirato, M., Kouno, T., Asai, R., Nakani, N., Fukui, J., and Paolucci, R. (2008). Large-scale experiments on nonlinear behavior of shallow foundations subjected to strong earthquakes. *Soils Found.* 48, 673–692. doi:10.3208/sandf.48.673

Taiebat, H. A., and Carter, J. P. (2000). Numerical studies of the bearing capacity of shallow foundations on cohesive soil subjected to combined loading. *Géotechnique* 50, 409–418. doi:10.1680/geot.2000.50.4.409

Tan, F.S. (1990). *Centrifuge and Theoretical Modelling of Conical Footings on Sand.* Ph.D. thesis, University of Cambridge, Cambridge, UK.

Yim, S. C., and Chopra, A. K. (1985). Simplified earthquake analysis of structures with foundation uplift. *J. Struct. Eng.* 111, 906–930. doi:10.1061/(ASCE)0733-9445(1985)111:12(2708)

Conflict of Interest Statement: The authors declare that the research was conducted in the absence of any commercial or financial relationships that could be construed as a potential conflict of interest.

13

Seismic performance evaluation framework considering maximum and residual inter-story drift ratios: application to non-code conforming reinforced concrete buildings in Victoria, BC, Canada

Solomon Tesfamariam[1]* and Katsuichiro Goda[2]

[1] School of Engineering, The University of British Columbia, Kelowna, BC, Canada, [2] Department of Civil Engineering, University of Bristol, Bristol, UK

Edited by:
Panagiotis Mergos,
City University London, UK

Reviewed by:
Mohammad Mehdi Kashani,
University of Bristol, UK
Sameh Samir F. Mehanny,
Cairo University, Egypt
Anaxagoras Elenas,
Democritus University of Thrace,
Greece

*Correspondence:
Solomon Tesfamariam,
The University of British Columbia,
EME 4253 – 1137 Alumni Avenue,
Kelowna, BC V1V 1V7, Canada
solomon.tesfamariam@ubc.ca

This paper presents a seismic performance evaluation framework using two engineering demand parameters, i.e., maximum and residual inter-story drift ratios, and with consideration of mainshock–aftershock (MSAS) earthquake sequences. The evaluation is undertaken within a performance-based earthquake engineering framework in which seismic demand limits are defined with respect to the earthquake return period. A set of 2-, 4-, 8-, and 12-story non-ductile reinforced concrete (RC) buildings, located in Victoria, BC, Canada, is considered as a case study. Using 50 mainshock and MSAS earthquake records (2 horizontal components per record), incremental dynamic analysis is performed, and the joint probability distribution of maximum and residual inter-story drift ratios is modeled using a novel copula technique. The results are assessed both for collapse and non-collapse limit states. From the results, it can be shown that the collapse assessment of 4- to 12-story buildings is not sensitive to the consideration of MSAS seismic input, whereas for the 2-story building, a 13% difference in the median collapse capacity is caused by the MSAS. For unconditional probability of unsatisfactory seismic performance, which accounts for both collapse and non-collapse limit states, the life safety performance objective is achieved, but it fails to satisfy the collapse prevention performance objective. The results highlight the need for the consideration of seismic retrofitting for the non-ductile RC structures.

Keywords: seismic performance, maximum inter-story drift, residual inter-story drift, non-code conforming reinforced concrete building, mainshock–aftershock earthquake

Introduction

Motivation

The eastern and western provinces of Canada are subject to moderate to large magnitude earthquakes. As a result, Canadian buildings are prone to earthquake-induced damage (Bruneau and Lamontagne, 1994; Ventura et al., 2005). Since 1900, several destructive earthquakes have been reported (**Table 1**; **Figure 1**), including the 1918 and 1946 earthquakes in Vancouver Island and the 1949, 1965, and 2001 (Nisqually) deep earthquakes in Washington, DC, USA. The recurrence of the Cascadia subduction earthquakes (magnitudes of 8–9) can affect a vast region of the Pacific coast from Vancouver Island to Washington/Oregon (Hyndman and Rogers, 2010). For large interface events, intense long-period ground motions having long duration are anticipated. To assess seismic performance of structures and infrastructure more accurately, a novel seismic performance evaluation framework that accounts for probabilistic characteristics of multivariate engineering demand parameters caused by major earthquake ground motions as well as their aftershock ground motions is proposed. The developed methodology is applied to a set of non-ductile reinforced concrete (RC) structures that are located in Victoria, British Columbia (BC), Canada. In the framework, regional seismicity in southwestern BC is fully taken into account in defining seismic performance levels and in evaluating the non-linear structural responses via rigorous ground motion record selection.

Through lessons learned from performance of buildings during previous earthquakes and research over the last three decades, Canadian seismic design provisions have evolved (Mitchell et al., 2010). The first attempt for seismic hazard quantification in Japan and North America followed the 1923 Kanto (Tokyo) earthquake and the 1933 Long Beach (California) earthquake (Atkinson, 2004; Otani, 2004). Subsequently, the first edition of the National Building Code of Canada (NBCC) was published in 1941 (NRCC, 1941) and adopted provisions for seismic design based on the 1935 Uniform Building Code (UBC) in an appendix. Initially, the earthquake hazard quantification was introduced through seismic coefficients. Later, the provisions were incorporated into the main text of the 1953 NBCC and Canadian seismic zoning map was introduced. However, the seismic zones were introduced on a qualitative evaluation of hazard (Atkinson, 2004). The 1965 NBCC adopted the first seismic modification factor, as the construction factor, in the calculation of the minimum seismic base shear (NRCC, 1965). In late 1960s, the probabilistic quantification of seismic hazard has gained popularity. In 1970, the seismic code was changed to include the structural flexibility factor in addition to the construction factor (NRCC, 1970). To date, although the state of knowledge has improved, the same methods are still used in modern design codes; for engineering design purposes, these hazard factors in the newer code have been calibrated to a previous version (Atkinson, 2004). In the

1985, 1990, and 1995 NBCC, zonal velocity ratios (which have only four categories) are used to define seismic design loads at building locations, whereas since the 2005 NBCC, uniform hazard spectrum (UHS) is introduced to provide more site-specific seismic hazard values for calculating seismic design loads for buildings.

In BC, seismic provisions of the NBCC were not adopted by municipalities until 1973 (Ventura et al., 2005). Therefore, most of the pre-1970 buildings constructed in BC may have limited seismic capacity against severe earthquake forces (Onur et al., 2005). The cause–effect relationships of earthquake-induced damage on buildings designed without seismic capacity methods are summarized in **Table 2**. Most of these older buildings are currently operational and are required to be further assessed and upgraded to improve life safety (LS) and to mitigate potential economic consequences due to seismic damage.

In Canada, different building vulnerability assessment techniques have been proposed with different levels of complexity, ranging from a simple scoring to more detailed methods of non-linear structural analyses. The Institute for Research in Construction (IRC) of the National Research Council has developed a national seismic screening manual for buildings and different performance modifiers are taken into consideration (Rainer, 1992; Foo and Davenport, 2003). The methodology of the IRC manual follows the 1988 FEMA-154 screening guidelines (ATC, 2002). This seismic screening manual computes the seismic priority index (SPI), which is obtained as a summation of two indices, structural index (SI) and non-structural index (NSI). Saatcioglu et al. (2013) have updated the manual in accordance with the 2005 NBCC. Ventura et al. (2005) has developed building classification and fragility curves for southwestern BC to estimate the probability of damage at a given seismic intensity. The method

TABLE 1 | Damaging earthquakes in eastern and western Canada.

Damaging earthquakes in western Canada	Damaging earthquakes in eastern Canada
1949 M8.1 Queen Charlotte Islands earthquake	1988 M5.9 Saguenay earthquake
1946 M7.3 Vancouver Island earthquake	1944 M5.6 Cornwall-Massena earthquake
1918 M7.0 Vancouver Island earthquake	1935 M6.2 Timiskaming earthquake
1872 M7.4 North Cascades earthquake	1929 M7.2 Grand Banks earthquake
1700 M9.0 Cascadia earthquake	1925 M6.2 Charlevoix-Kamouraska earthquake

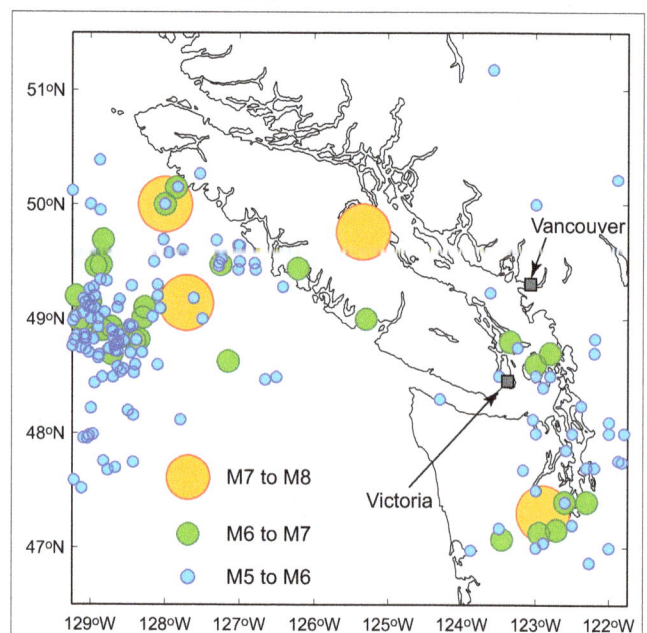

FIGURE 1 | Regional seismicity in southwestern British Columbia, Canada.

TABLE 2 | Cause–effect relationships for buildings designed without seismic capacity methods (Liel and Deierlein, 2008; Tesfamariam and Saatcioglu, 2008).

Cause	Effect
Inadequate anchorage of longitudinal reinforcement	• Yield strength of the reinforcement not being developed during the cyclic loading caused by earthquakes • Lap splices may fail if placed in potential plastic hinge regions
Inadequate anchorage of transverse reinforcement	• Transverse reinforcement will not be effective if not properly anchored and/or of insufficient quantity • 90° end hooks are inadequate for perimeter hoops, since spalling of cover concrete will result in loss of anchorage; end hooks should be bent through at least 135°
Inadequate quantities of transverse reinforcement	• Failure in shear

TABLE 3 | Vision 2000 recommended seismic performance objectives for buildings (SEAOC, 1995).

Earthquake design level (probability of exceedance)	Performance limit states			
	Immediate occupancy (IO)	Damage control (DC)	Life safety (LS)	Collapse prevention (CP)
Frequent (50% PE in 30 years)	▪	×	×	×
Occasional (50% PE in 50 years)	◆	▪	×	×
Rare (10% PE in 50 years)	◇	?	▪	×
Very rare (2% PE in 50 years)		◇	?	▪

▪ *Basic objective – proposed NBCC normal importance.*
◆ *Essential service objective – proposed NBCC high importance.*
◇ *Safety critical objective – not proposed NBCC category.*
× *Unacceptable performance for new construction.*
The color shades are provided to group the performance limit states.

was used for regional damage estimation and is not intended for individual buildings.

Performance-Based Seismic Evaluation of Buildings

Cornell and Krawinkler (2000) proposed a rational means of integrating the probabilistic performance-based earthquake engineering for seismic design and risk assessment. The analytical procedure probabilistically integrates seismic hazard analysis, structural analysis, damage assessment, and loss estimation. Performance-based design philosophy is adopted in the 2005 Canadian seismic design code (DeVall, 2003) following Structural Engineers Association of California (SEACO) Vision 2000 (SEAOC, 1995). The maximum inter-story drift ratio (MaxISDR) is used in Canadian and most building codes as the only performance metric. Relationships between different earthquake return periods and acceptable performance limit states in terms of MaxISDR are shown in **Table 3**. It can be highlighted that for frequent [50% probability of exceedance (PE) in 30 years], occasional (50% PE in 50 years), rare (10% PE in 50 years), and very rare (2% PE in 50 years) earthquake levels, the corresponding design performance limit states are immediate occupancy (IO), damage control (DC), LS, and collapse prevention (CP), respectively. Descriptions of the limit states are summarized in **Table 4**. In the Canadian code, the limit states for IO, DC, LS, and CP, in terms of MaxISDR are 0.2, 0.4, 1, and 2.5%, respectively. These limit state values are lower than values suggested in FEMA P-58-1 (2012). In this paper, limit state values similar to FEMA P-58-1 (2012) will be used.

Consideration of Maximum and Residual Drift Ratios in Seismic Risk Assessment of Structures

The seismic performance of a structure is often evaluated through MaxISDR. Recent post-earthquake functionality assessment of structures has highlighted that residual inter-story drift ratio (ResISDR) is an important factor in the post-earthquake safety of a building and economic feasibility of repair and reconstruction

(Kawashima et al., 1998; Ruiz-García and Miranda, 2006; Ramirez and Miranda, 2009; FEMA P-58-1, 2012). MacRae and Kawashima (1997) and Kawashima et al. (1998) implemented the first time risk assessment of bridges based on residual drift. **Table 4** summarizes the limits of MaxISDR and ResISDR for IO, DC, LS, and CP based on FEMA 356 (2000) and FEMA P-58-1 (2012). The ResISDR limits for CP are expressed in terms of the design shear force V_{design} normalized by the building weight W to consider cases where P-delta might be dominant at smaller drift ratios.

Christopoulos et al. (2003) and Pampanin et al. (2003) studied the effect of residual drift on single-degree-of-freedom (SDOF) and multi-degree-of-freedom (MDOF) systems, respectively. Christopoulos et al. (2003) proposed an assessment criterion as a weighted sum of structural and non-structural residual drifts. Pampanin et al. (2003) further extended this formulation into a MDOF system and proposed a seismic performance evaluation framework based on a MaxISDR–ResISDR matrix. In the absence of extensive data and information, in FEMA P-58-1 (2012), a simple relation between MaxISDR and ResISDR was provided for the four limit states. Erochko et al. (2011) have proposed a mechanistic equation to estimate residual drifts as a function of expected peak drift and elastic recoverable drift. For post-earthquake risk assessment of buildings, the residual drift can be easily measured, and as a result, the maximum drift is typically estimated as a function of residual drift (Hatzigeorgiou and Beskos, 2009; Erochko et al., 2011; Hatzigeorgiou et al., 2011; Christidis et al., 2013). Reported equations that relate MaxISDR and ResISDR are summarized in **Table 5**.

In the seismic performance assessment, the values for MaxISDR and ResISDR are subject to significant uncertainty and are dependent on each other. Uma et al. (2010) extended the performance-based seismic assessment framework by Pampanin et al. (2003) by taking into account the joint occurrence of MaxISDR and ResISDR of a SDOF system (modeled by a bivariate lognormal probability function). On the other hand, Goda and Tesfamariam (2015) have shown that MaxISDR and ResISDR of a MDOF system are statistically dependent, and that their marginal distributions can be represented by the Frechet and generalized

TABLE 4 | Limit states for maximum and residual inter-story drift ratios (FEMA 356, 2000; FEMA P-58-1, 2012).

Damage state	Description	Maximum inter-story drift ratio (MaxISDR) (%)	Residual inter-story drift ratio (ResISDR)
Immediate occupancy (IO)	No structural realignment is necessary for structural stability; however, the building may require adjustment and repairs to non-structural and mechanical components that are sensitive to the building alignments (e.g., elevator rails, curtain walls, and doors)	0.4	0.2% (equal to the maximum out-of-plumb tolerance typically permitted in new construction)
Damage control (DC)	Realignment of structural frame and related structural repairs required to maintain permissible drift limits for non-structural and mechanical components and to limit degradation in structural stability (i.e., collapse safety)	0.9	0.5%
Life safety (LS)	Major structural realignment is required to restore margin of safety for lateral stability; however, the required realignment and repair of the structure may not be economically and practically feasible (i.e., the structure might be at total economic loss)	2.5	1%
Collapse prevention (CP)	Residual drift is sufficiently large that the structure is in danger of collapse from earthquake aftershocks (note: this performance point might be considered as equivalent to collapse but with greater uncertainty)	4.5	• High ductility systems 4% $<0.5V_{design}/W$ • Moderate ductility systems 2% $<0.5V_{design}/W$ • Limited ductility systems 1% $<0.5V_{design}/W$

TABLE 5 | Equations to relate residual and maximum inter-story drift ratios.

Reference	Description	Comments
MacRae and Kawashima (1997)	$d_r = \begin{cases} d_l & \text{for } H_{yb} \leq 0 \leq H_{yt} \\ d_L = d_y\left(\dfrac{1-r}{r}\right) & \text{for } r > 0 \text{ and either } H_{yb} \geq 0 \text{ or } H_{yt} \leq 0 \end{cases}$	d_r = maximum residual drift; d_l = displacement where the elastic response line at the end of the analysis crosses the zero force line; d_L = displacement where the post-elastic line intersects the zero force line; d_y = yield displacement; r = bilinear factor = k_2/k_1; k_1 = initial elastic stiffness; k_2 = second post-yielding stiffness; H_{yt} = yield force at the top of the line; H_{yb} = yield force at the bottom of the line Structure type: bilinear SDOF systems
Kawashima et al. (1998)	$d_r = \begin{cases} (\mu-1)(1-r)d_y & \text{for } r(\mu-1) < 1 \\ [(1-r)/r]d_y & \text{for } r(\mu-1) \geq 1 \end{cases}$	d_r = maximum residual drift; μ = ductility factor; d_y = yield displacement; r = bilinear factor = k_2/k_1; k_1 = initial elastic stiffness; k_2 = second post-yielding stiffness Structure type: bilinear SDOF systems
Christopoulos et al. (2003)	$RDDI = \begin{cases} \phi \times RDDI_s + \chi \times RDDI_{NS} & \text{for } RDDI_s < 1 \\ 1 & \text{for } RDDI_s = 1 \end{cases}$	RDDI = residual deformation damage index [0, 1]; $RDDI_s$ = residual deformation structural damage index [0, 1]; $RDDI_{NS}$ = residual deformation non-structural damage index [0, 1]; φ and χ = relative importance factors for structural failure and non-structural failure, respectively Structure type: hysteretic SDOF systems
Hatzigeorgiou et al. (2011)	$d_m = (a_1 T + a_2 d_r + a_3 d_r^2 + a_4 T d_r) \times (1 + a_5 r + a_6 r^2)$	d_m = maximum drift; T = period (in seconds); d_r = maximum residual drift; $a_{1...6}$ = regression coefficients Structure type: Bilinear SDOF systems
Erochko et al. (2011)	$d_r = d - d_e$	d_r = maximum residual drift; d = drift; d_e = elastic recoverable drift = yield shear/elastic stiffness of a typical story Structure type: steel building; MDOF with 2 and 12 stories in height; special moment-resisting frames and buckling-restrained braced frames
FEMA P-58-1 (2012)	$d_r = \begin{cases} 0 & \text{for } d \leq d_y \\ d_r = 0.3(d - 3d_y) & \text{for } d_y < d < 4d_y \\ d_r = d - 3d_y & \text{for } d \geq 4d_y \end{cases}$	d_r = maximum residual drift; d = drift; d_y = yield drift Structure type: MODF systems
Christidis et al. (2013)	$|d_m| = (a_1 + a_2 \ln(N) + a_3 |d_r|) \times (1 + a_4 r)$	d_m = maximum drift; N = number of story; d_r = maximum residual drift; r = bilinear factor = k_2/k_1; k_1 = initial elastic stiffness; k_2 = second post-yielding stiffness; $a_{1...4}$ = regression coefficients Structure type: steel building; MDOF with 3, 6, 9, 12, 15 and 20 stories in height; moment resisting steel frames and concentrically X-braced steel frames

Pareto distributions, respectively, whereas their dependence can be characterized by different copulas (e.g., normal, *t*, Gumbel, Frank, Clayton, and asymmetrical Gumbel). Tesfamariam and Goda (2015) further developed the copula-based multivariate seismic demand model and applied it to seismic loss assessment of a non-code conforming RC building with consideration of mainshock–aftershock (MSAS) earthquake records.

Mainshock–Aftershock Earthquakes on RC Buildings

The 2011 M_w6.3 Christchurch earthquake in New Zealand (Elwood, 2013; Leite et al., 2013) and the 2011 M_w9.0 Tohoku earthquake in Japan (Goda et al., 2013, 2015) have highlighted vulnerability of buildings subject to MSAS earthquake sequences. There are an increasing number of studies on vulnerability assessment of RC buildings subject to MSAS sequences. Ryu et al. (2011) presented a methodology for developing fragilities for mainshock-damaged SDOF buildings by performing incremental dynamic analysis (IDA, Vamvatsikos and Cornell, 2002) with aftershock ground motions. The aftershock fragilities are computed conditional on the damage caused by the mainshock (MS) earthquake. Their analyses showed that the effect of aftershocks is not significant. Hatzigeorgiou and Liolios (2010) quantified vulnerability of non-code and code conforming RC frames with prevalent irregularity. The MSAS sequences were obtained from actual MSAS records and 40 artificial seismic sequences. They concluded that aftershocks have significant impact on drift demand of the non-code conforming and irregular buildings. Tesfamariam et al. (2015) investigated MSAS earthquakes on non-code conforming RC frames with vertical irregularity. A set of 50 MSAS earthquake sequences was selected for Vancouver with consideration of regional seismic hazard. For the irregular structures, the MSAS sequences caused higher drift values than MS records only. Tesfamariam and Goda (2015) investigated the effect of MSAS earthquake sequences on a 4-story non-code conforming RC building. Their results showed that the MSAS earthquake had no marked effect on collapse and loss assessment of the RC building. This study, with the consideration of seismicity in Victoria, BC, extends the 4-story RC building investigated in Tesfamariam and Goda (2015) to 2-, 8-, and 12-story RC buildings. The building vulnerability assessment is further undertaken for collapse and non-collapse damage limit states.

Research Objective and Methodology

The objective of this paper is to carry out probabilistic building vulnerability assessment with consideration of regional probabilistic seismic hazard. The novel aspects of the proposed building vulnerability assessment are as follows:

i.   Consideration of non-code conforming RC buildings having different story numbers, i.e., 2-, 4-, 8-, and 12-story, extending the work by Tesfamariam and Goda (2015) for the 4-story RC building;

ii. Consideration of three earthquake sources, i.e., crustal, inslab, and interface to reflect regional seismicity of southwestern BC in record selection (i.e., subduction environments and extensive ground motion datasets for the 2011 Tohoku

earthquake records, which can be regarded as closest proxy for the Cascadia subduction events);

iii. Consideration of MSAS sequences as seismic excitation;

iv. Multivariate seismic demand modeling, MaxISDR and ResISDR, for seismic performance evaluation; and

v.   Consideration of collapse and non-collapse limit states in the form of a bivariate seismic performance matrix.

Figure 2 illustrates a methodology for probabilistic building vulnerability assessment. It consists of five basic steps:

- Step 1: finite-element (FE) models of the 2-, 4-, 8-, and 12-story RC buildings are prepared to consider non-linear behavior of structural components and assembly. Modal analysis is performed to identify the three dominant fundamental periods (T_1, T_2, and T_3).

- Step 2: a suite of ground motions which corresponds to a target seismic hazard level is selected on the basis of T_1 by reflecting detailed characteristics of regional seismic hazard. Multiple conditional mean spectra (CMS) for different earthquake types are employed as target response spectra (Baker, 2011; Goda and Atkinson, 2011). Each ground motion consists of a MS record and a MSAS sequence.

- Step 3: a set of RC frames are analyzed through IDA to collapse limit states for the suite of MS records and MSAS sequences, and the performance parameters MaxISDR and ResISDR are recorded for each motion. The collapse fragilities are evaluated using the IDA results.

- Step 4: from non-collapsed results, marginal probability distributions of MaxISDR and ResISDR are derived, and corresponding dependency is characterized using copulas.

- Step 5: the performance matrix (**Table 3**) and limit states (**Table 4**) are used to carry out seismic performance evaluation. From the seismic performance evaluation, the probability unsatisfactory seismic performance with regard to the specified limit state criteria is derived for the MS and MSAS earthquake records.

Salient features of the key components of the framework are explained in the following.

Structural Model

Tesfamariam and Goda (2015) studied the effect of MSAS earthquake records on the loss assessment of a 4-story non-code conforming RC space frame structure. This study extends this investigation to archetypical structures with different story numbers reported in Liel and Deierlein (2008). The archetype structures are: 2-, 4-, 8-, and 12-story non-code conforming RC buildings; the structures were designed as a space frame, and all columns and beams were part of the lateral resisting system. The buildings were designed according to the 1967 UBC seismic provisions (ICBO, 1967). Beam and column elements have the same amount of over-strength; each element is 15% stronger than the code-minimum design level. The design is governed by strength and stiffness requirements, as the 1967 UBC had few requirements for special seismic design or ductile detailing.

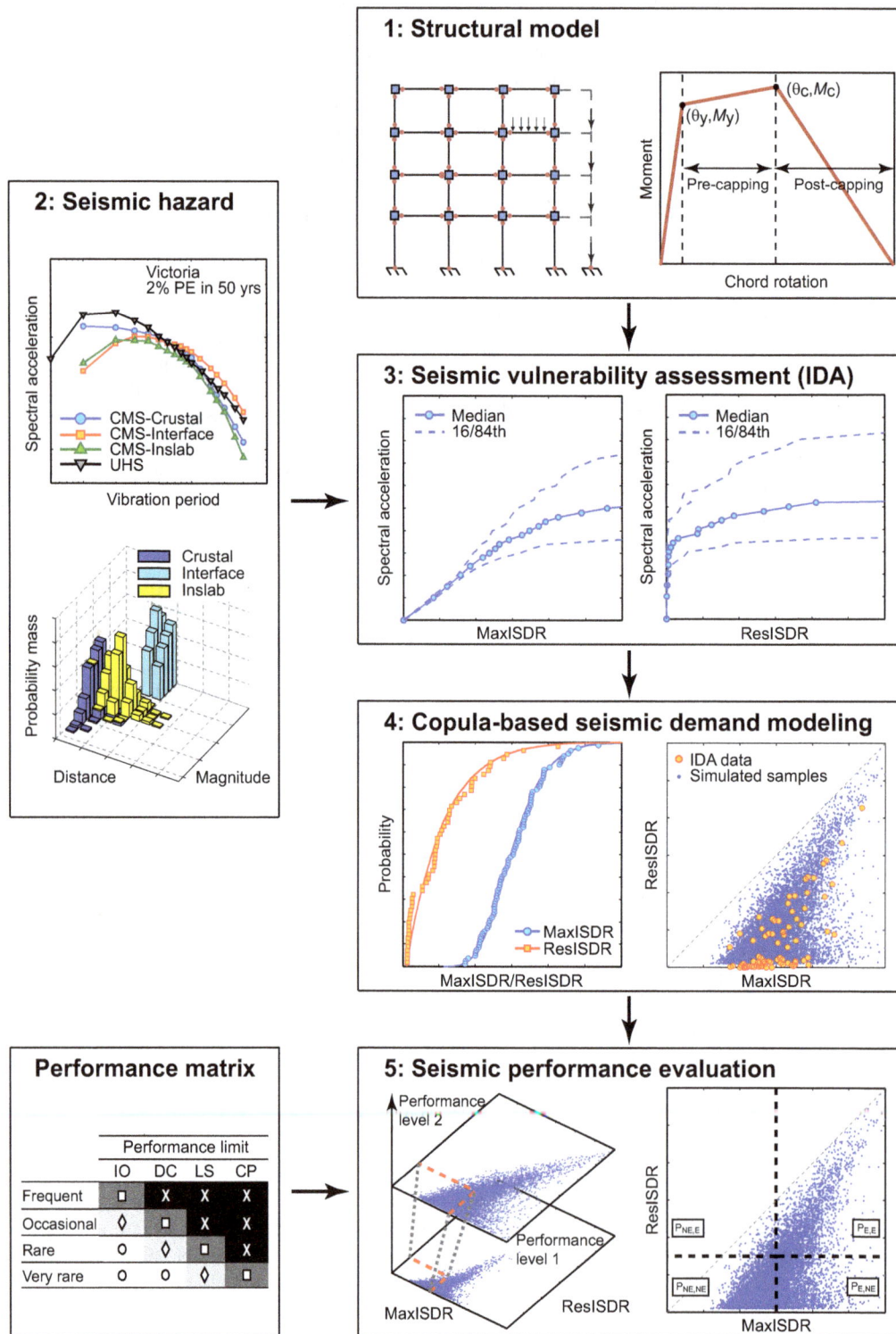

FIGURE 2 | Probabilistic seismic vulnerability assessment framework.

Finite-element modeling of structures can be achieved using a fiber or lumped plasticity model. In the fiber model, the element cross section is discretized and corresponding non-linear material properties of the core concrete, cover concrete, and reinforcing bars are assigned. On the other hand, in the lumped plasticity model, non-linearity of the beam-column element is introduced at the two ends (hinges), which are connected by an elastic element. Advantages and disadvantages of each approach are summarized in **Table 6**. Haselton et al. (2008) indicated that the lumped plasticity model, equipped with adequate hysteretic models for plastic hinges, can simulate global collapse behavior well (note: they observed that the fiber model may be numerically unstable when the responses become highly non-linear).

Figure 3A shows a schematic of the 4-story building. It has a floor area of 38.1 m (125 ft) by 53.3 m (175 ft); columns are spaced at 7.6 m (25 ft), and story heights are 4.6 m (15 ft) and 4.0 m (13 ft) at the ground floor and higher floor levels, respectively. The non-ductile RC models used in this paper are developed by Liel and Deierlein (2008). The models are based on a lumped plasticity approach in Open system for earthquake engineering simulation (OpenSees, McKenna et al., 2000). The lumped plasticity element models used to simulate plastic hinges in beam-column elements (**Figure 3B**) utilize a tri-linear non-linear spring model that is developed by Ibarra et al. (2005) and implemented in OpenSees by Altoontash (2004). **Figure 3B** shows the tri-linear backbone curve, coupled with the associated hysteretic rules, which is used to model the structures to post-peak response and near-collapse response. The post-peak response enables modeling of the strain hardening behavior associated with concrete crushing, rebar buckling and fracture, and bond failure (Haselton et al.,

2008; Liel and Deierlein, 2008). Liel and Deierlein (2008) and Haselton et al. (2008) reported that the Ibarra et al. model was calibrated with data from 255 RC column test results. Details of the calibration process and building details are provided in Liel and Deierlein (2008) and Haselton et al. (2008); for brevity, they are not repeated here. P-Δ effects are modeled using a leaning column. The vibration periods for the first three modes for the 2-, 4-, 8-, and 12-story buildings are summarized in **Table 7**.

Seismic Hazard for Victoria and Ground Motion Selection

Victoria is the provincial capital of BC and is located at the southern tip of Vancouver Island (**Figure 1**). Due to its geographical location, Victoria is affected by three types of earthquakes. The first type of the influential events is an earthquake at shallow depth in the crust; historically, the 1918 and 1946 earthquakes fall under this category. The other two types of the influential earthquakes are related to the movements of the Juan de Fuca Plate, Explorer Plate, Gorda Plate, and North American Plate in the Cascadia subduction zone. In the subducting slab, deep earthquakes occur (e.g., 2001 Nisqually earthquake), while at the plate interfaces, mega-thrust subduction earthquakes, as larger as M_w9.0, occur (e.g., 1700 Cascadia earthquake, Hyndman and Rogers, 2010). It is important to recognize that the three types of dominant earthquakes in southwestern BC have distinct characteristics in terms of recurrence interval, earthquake magnitude, location, and depth and thus should be treated differently.

The key features of the critical earthquake scenarios for a given location can be evaluated quantitatively via probabilistic seismic hazard analysis. Atkinson and Goda (2011) conducted seismic hazard studies for southwestern BC, by incorporating recent advancements in seismology. Typical outputs from probabilistic seismic hazard analysis, which are essential for seismic performance assessment of buildings and infrastructure, are the UHS and seismic deaggregation. Currently, the UHS at 2% PE in

TABLE 6 | Advantages and disadvantages of fiber and lumped plasticity models (Haselton et al., 2008).

Fiber model	Lumped plasticity model
Ability to consider shear flexibility by modeling shear DOF in the sections	Not able to capture shear flexibility
Used where cracking and tension-stiffening behavior governs	Used for collapse prediction
Inability to capture deterioration of the steel reinforcing bars due to rebar buckling and low-cycle fatigue	Captures deterioration of steel rebar due to buckling and low-cycle fatigue
Not able to capture strength and stiffness deterioration	Well captures strength and stiffness deterioration to assess global collapse

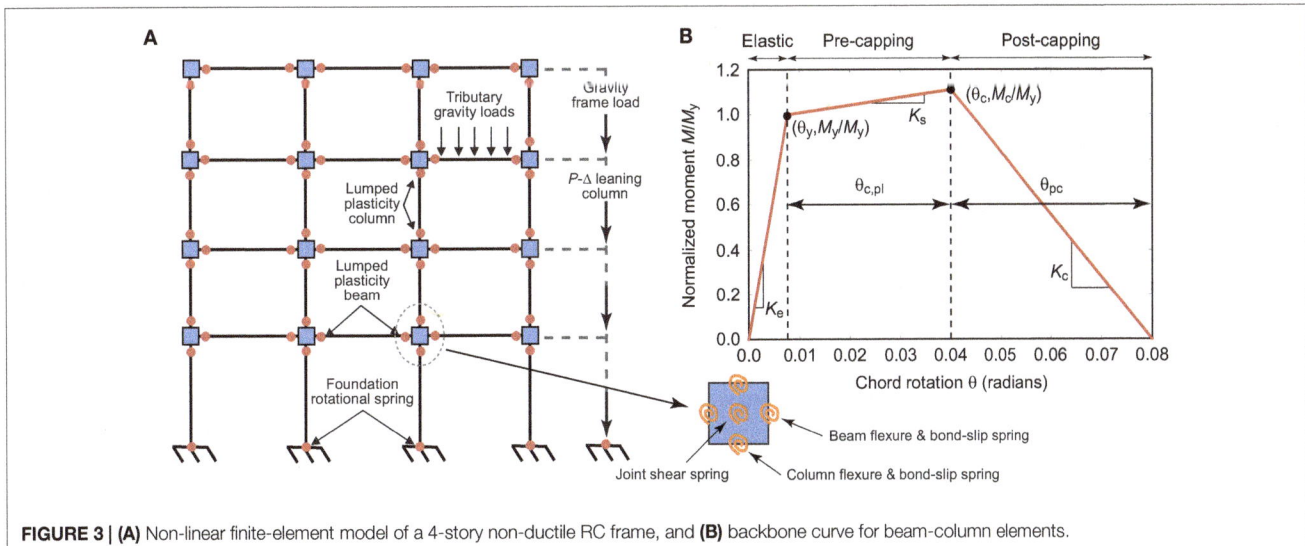

FIGURE 3 | (A) Non-linear finite-element model of a 4-story non-ductile RC frame, and **(B)** backbone curve for beam-column elements.

50 years (equivalent to the return period of 2500 years) is adopted as the basis for seismic design provisions for new construction in Canada. The seismic deaggregation identifies critical earthquake scenarios (for instance, in terms of magnitude, distance, and earthquake type) for a selected probability level. **Figure 4A** shows UHS for Victoria at 10, 5, and 2% PE in 50 years, where the site condition is set to site class C, which is represented by the average shear-wave velocity in the upper 30 m between 360 and 760 m/s. The three probability levels are relevant for assessing the seismic performance of structures in Canada. **Figure 5** shows the seismic deaggregation results for $T = 1.0$ and 2.0 s for 10, 5, and 2% PE in 50 years; the selected vibration periods correspond to the adopted seismic intensity measure (IM) for the 2-story building and the 4-, 8-, and 12-story buildings, respectively (**Table 7**). In **Figure 5**, relative contributions due to crustal, mega-thrust (Cascadia) interface, and deep inslab earthquakes are indicated. The seismic deaggregation results suggest that relative contributions due to the Cascadia subduction earthquakes increase with the probability level and the seismic hazard values for longer vibration periods are affected more significantly by the large subduction events. The variable characteristics of the dominant scenarios are important for seismic performance evaluations and thus should be taken into account in selecting ground motion records for non-linear dynamic analyses of structural models.

Careful record selection is of critical importance to produce unbiased estimates of seismic vulnerability. In particular, when record scaling is implemented to reach high seismic excitation levels, record selection needs to account for the spectral shape effects (Luco and Bazzurro, 2007). One practical method that is widely adopted for mitigating the record scaling bias is the CMS method (Baker, 2011). In the CMS-based record selection, the target response spectrum is modified based on dominant earthquake scenarios and relevant ground motion prediction equations at a selected performance level. Typically, the base target response spectrum for record selection is a UHS and is further modified based on the mean scenarios obtained from seismic deaggregation; several tens of ground motion records that match the modified target response spectrum (i.e., CMS) are selected as input motion. However, for the seismic environments in southwestern BC, it may be too simplistic to use a single target response spectrum for a given probability level because three dominant earthquakes with different characteristics are present (**Figure 5**). For this reason, in this study, the multiple CMS-based record selection method by Goda and Atkinson (2011) is adopted, which defines three different target spectra considering the different earthquake characteristics and ground motion prediction

models for these earthquake types. Examples of the CMS for crustal, interface, and inslab earthquakes are shown in **Figures 4B,C**; **Figure 4B** is for the 2-story building, whereas **Figure 4C** is for the 4-, 8-, and 12-story buildings. It is noted that the CMS for the interface events have richer spectral content with respect to other two earthquake types because of larger earthquake magnitudes and longer propagation paths.

Another important aspect for record selection is to prepare a suitable ground motion dataset for the seismic environments of interest. For southwestern BC, the base ground motion dataset should contain records from large mega-thrust subduction events. Moreover, the record database should contain as-recorded MSAS sequence records. To achieve these requirements, a new composite database of real MSAS sequences is compiled by combining the database that was constructed based on the Next Generation Attenuation database (Goda and Taylor, 2012) and the new database for Japanese earthquakes from the K-NET, KiK-nt, and SK-net (Goda et al., 2015). It is noteworthy that the new Japanese database includes records from the 2011 Tohoku earthquake, which may be considered as appropriate surrogate for the Cascadia subduction events. The composite dataset consists of 606 real MSAS sequence records; 75 sequences are from the NGA database and 531 sequences are from the Japanese database (each sequence has two horizontal components). This database is the largest dataset for as-recorded MSAS sequences and is sufficient to select a suitable set of record sequences by taking into account various requirements, such as earthquake type, magnitude, distance, and site class.

Incremental Dynamic Analysis

Incremental dynamic analysis implements a series of non-linear dynamic analyses by scaling a set of input ground motions based on an adopted IM, and develops prediction equations of engineering demand parameters (EDP, e.g., MaxISDR and ResISDR) at different IM levels. The IM is the spectral acceleration at the fundamental period of a structure. For the different building story numbers, the maximum scaling required in IDA can vary. For the 2-story building, the spectral acceleration at 1.0 s is selected as IM (**Table 7**) and the scaling range in IDA is varied from 0.05 to 1.4 g. For the 4-, 8-, and 12-story buildings, the spectral acceleration at 2.0 s (i.e., IM) ranges from 0.05 to 0.7 g. In general, numerical instability is encountered when the inter-story drift ratio of the frames exceeds 0.10. The first occurrence of such large deformation responses is treated as "*collapse*" (Vamvatsikos and Cornell, 2002). In characterizing the inelastic demand, non-linear responses that are in "collapse" and "non-collapse" states are distinguished. The collapse results are modeled by collapse fragility curves (see Collapse Fragility Assessment), whereas the non-collapse results are represented by multivariate seismic demand models (see Coupla-Based Seismic Demand Modeling). Eventually, the overall performance of the building is assessed by integrating collapse results and non-collapse results in the Section "Seismic Performance Evaluation."

Incremental dynamic analysis is carried out for the 2-, 4-, 8-, and 12-story RC frames using the set of 50 MS records as well

TABLE 7 | First three fundamental periods of 2-, 4-, 8-, and 12-story buildings.

Building story	Design ID	Period(s)		
		Mode-1	Mode-2	Mode-3
2	3001	1.10	0.20	0.03
4	3004	1.92	0.55	0.27
8	3016	2.23	0.80	0.41
12	3026	2.35	0.85	0.47

as a set of 50 MSAS sequences, which are selected based on the multiple CMS-based procedures. The IDA results for both MS records and MSAS sequences (i.e., EDP-IM plot) are shown in **Figures 6** and **7**; **Figure 6** is for MaxISDR, whereas **Figure 7** is for ResISDR. To present the uncertainty of the IDA results, 16th–84th percentile curves (corresponding to mean ± 1 SD), are included in the figures. The overall characteristics of the

IDA curves for MaxISDR and ResISDR are different; the former increases gradually with the seismic intensity level, whereas the latter increases rapidly when the seismic intensity level reaches in the range of 0.2–0.3 g for the 2-story building and 0.15–0.20 g for the 4-, 8-, and 12-story buildings; similar observations are also noted in FEMA P-58-1 (2012). It is noteworthy that the uncertainty of ResISDR is much greater than that of MaxISDR,

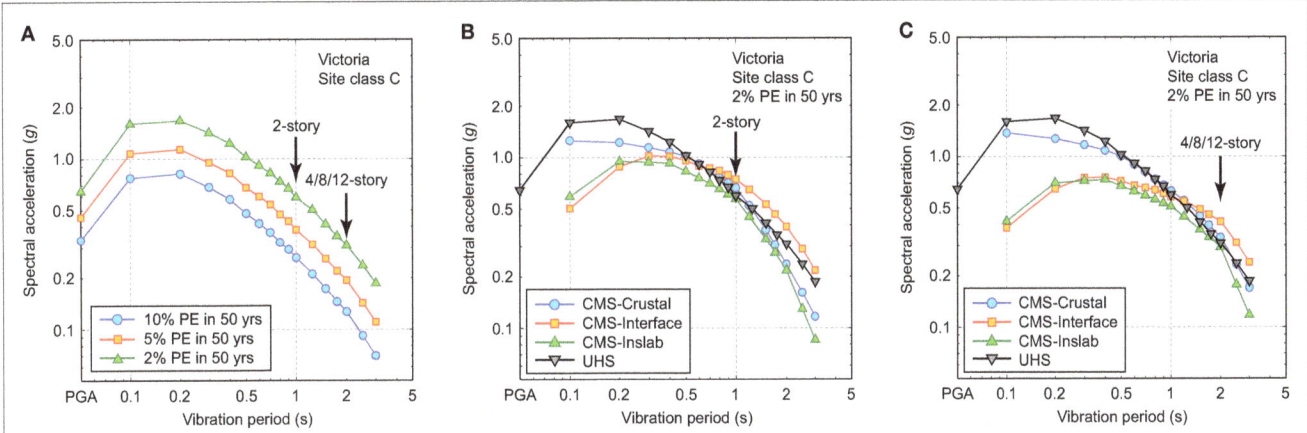

FIGURE 4 | (A) Uniform hazard spectra, (B) conditional mean spectra for the anchor vibration period of 1.0 s (for the 2-story RC frame), and (C) conditional mean spectra for the anchor vibration period of 2.0 s (for the 4-, 8-, and 12-story RC frames).

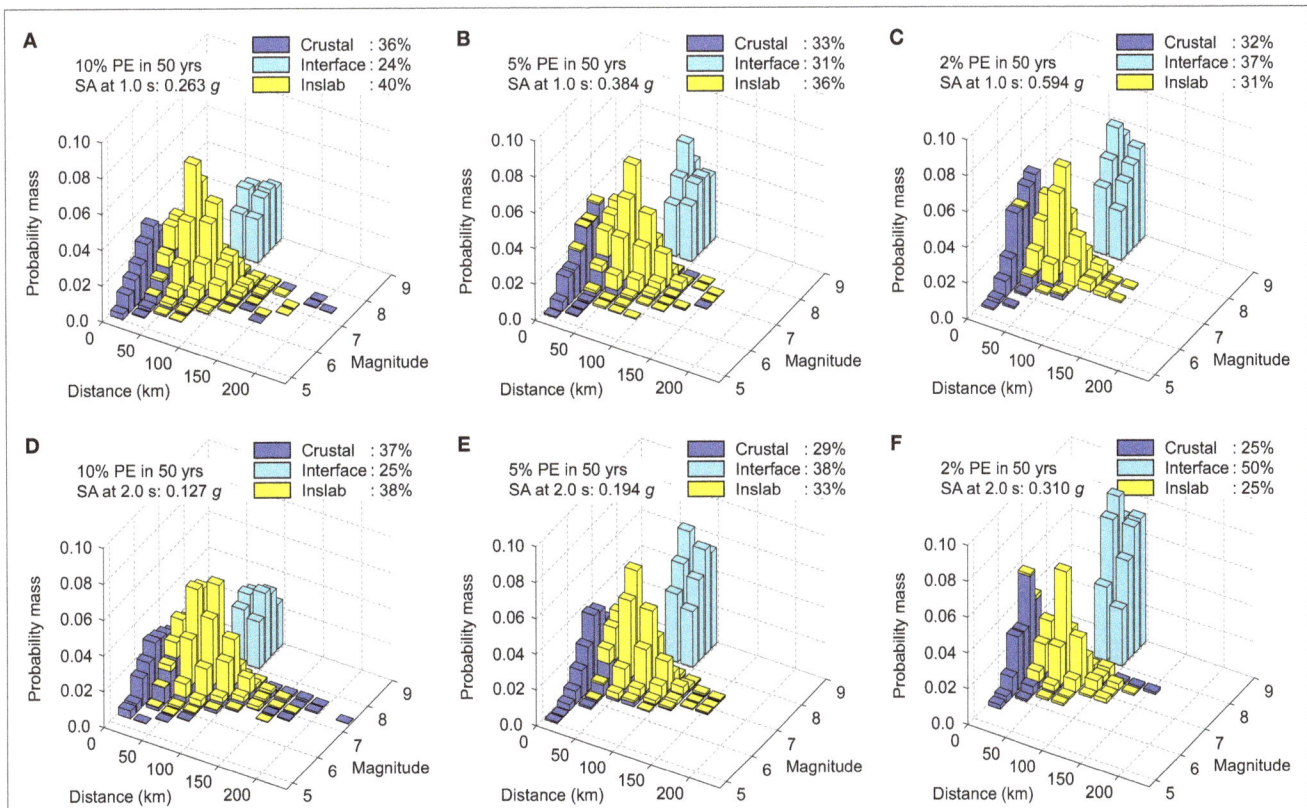

FIGURE 5 | Seismic deaggregation results for Victoria: (A–C) T = 1.0 s for 10, 5, and 2% probability of exceedance in 50 years, and (D–F) T = 2.0 s for 10, 5, and 2% probability of exceedance in 50 years.

as noted by Ruiz-García and Miranda (2006). To appreciate the differences of the IDA curves for the buildings with different story numbers, the 50th, 16th, and 84th percentile curves for the 4-, 8-, and 12-story buildings are overlaid together in **Figure 8**, noting that the same IM is adopted for these buildings (thus the IDA results can be compared directly). The results shown in **Figure 8** indicate that for a given seismic excitation level, both MaxISDR and ResISDR decrease with the story number; therefore, for the considered non-ductile RC frames, the 4-story building is more vulnerable than the other taller buildings.

Moreover, from the EDP-IM plots, it can be observed that the impact of aftershock records is significant for the 2-story building (**Figures 7A** and **8A**), whereas such marked effects diminish with increase in story number (**Figures 7B–D–8B–D**). One of the main reasons for the pronounced influence of aftershock records on MaxISDR and ResISDR for the 2-story building is related to its fundamental period (≈1.0 s; **Table 7**) and the dominant spectral content of the aftershock records; generally, aftershock records have richer spectral content in the short vibration period range (Goda et al., 2015). For all cases, the impact of MSAS earthquake sequence is more significant for ResISDR as compared with MaxISDR. For instance, for the 4-story building (**Figures 7B** and

8B), in terms of median, the consideration of MSAS sequences leads to 5–10% increase for MaxISDR and up to 100% increase for ResISDR with respect to the results for MS records.

Collapse Fragility Assessment

The collapse fragility can be represented by a lognormal cumulative distribution function (CDF):

$$P_{C} = \Phi\left(\frac{\ln\left(x / \theta\right)}{\beta}\right) \qquad (1)$$

where P_C is the probability that a ground motion with IM $= x$ will cause the structure to collapse, $\Phi(\bullet)$ is the standard normal CDF, θ is the median of the fragility function (the IM level with 50% probability of collapse), and β is the SD of lnIM (sometimes referred to as the dispersion parameter). **Figure 9** shows the collapse fragility results (raw data and fitted lognormal curve) for MS records and MSAS sequences. The estimated values of θ and β are also provided in the figure. The impact of aftershocks is pronounced for the 2-story building, where the median collapse capacity θ is reduced by 13% (i.e., the curve is shifted toward

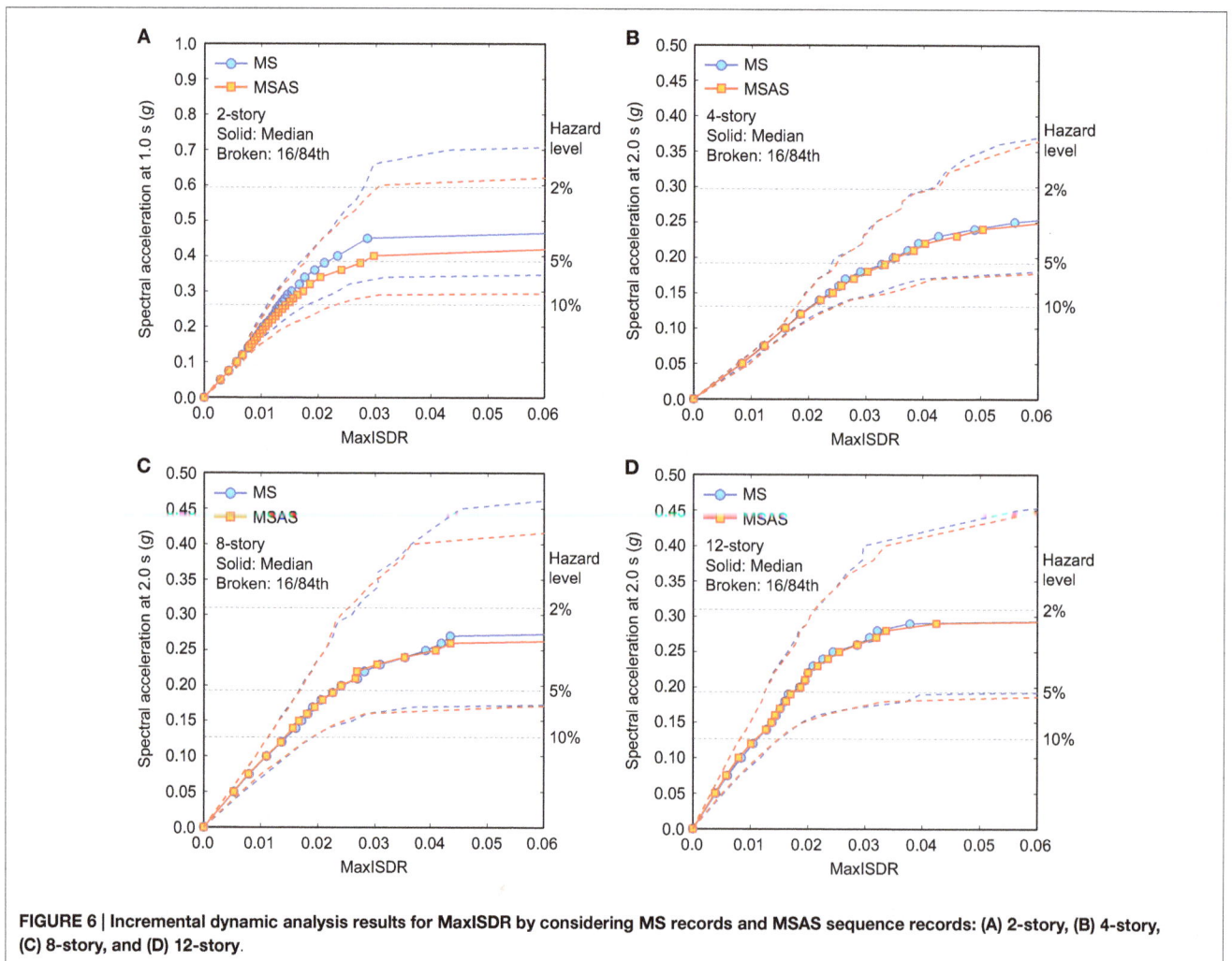

FIGURE 6 | Incremental dynamic analysis results for MaxISDR by considering MS records and MSAS sequence records: (A) 2-story, (B) 4-story, (C) 8-story, and (D) 12-story.

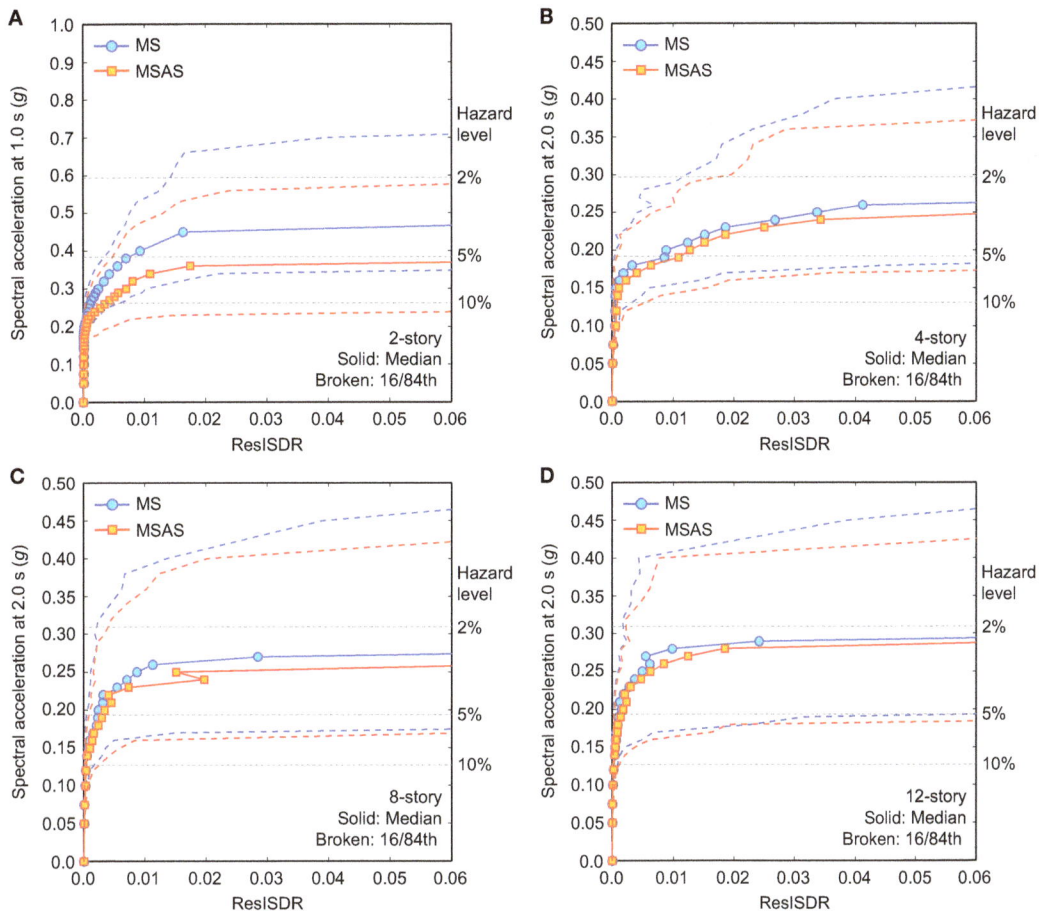

FIGURE 7 | Incremental dynamic analysis results for ResISDR by considering MS records and MSAS sequence records: (A) 2-story, (B) 4-story, (C) 8-story, and (D) 12-story.

left). On the other hand, the collapse fragility curves of the 4-, 8-, and 12-story buildings show no or slight differences. These results are consistent with the IDA curves shown in **Figure 6**. Furthermore, in **Figure 10**, the collapse fragility results for the 4-, 8-, and 12-story buildings are superimposed. The comparison shown in **Figure 10** indicates that the median collapse capacity θ as well as the dispersion β increases with the story number; the differences of the collapsed fragility curves are more pronounced at the greater seismic excitation levels.

Coupla-Based Seismic Demand Modeling

MaxISDR and ResISDR are statistically dependent (Goda and Tesfamariam, 2015) and thus this should be taken into account when these EDPs are characterized. For the seismic demand modeling, first, marginal probability distributions of MaxISDR and ResISDR should be developed, and second, corresponding dependence needs to be characterized. The probabilistic modeling of MaxISDR and ResISDR is performed at individual IM levels using non-collapse MaxISDR and ResISDR data (note: the number of available data points for seismic demand modeling decreases with the IM level because more data fall into collapse states; **Figure 9**).

Figure 11A shows the scatter plot for the 4-story building by considering MS records at 5% PE in 50 years level. In the figure, marginal distributions of MaxISDR and ResISDR are plotted along the horizontal axis and vertical axis, respectively. Note that ResISDR has a heavy right tail. Goda and Tesfamariam (2015) considered six probability distributions, i.e., lognormal, Gumbel, Frechet, Weibull, gamma, and generalized Pareto, for marginal probability distribution modeling of MaxISDR and ResISDR. For MS records and MSAS sequences, Goda and Tesfamariam (2015) showed that the Frechet distribution (Eq. 2) and generalized Pareto distribution (Eq. 3) are suitable for MaxISDR and ResISDR, respectively. The probability density functions of the Frechet and the generalized Pareto models are given by:

$$f(x) = \frac{\xi}{\sigma}\left(\frac{x-\mu}{\sigma}\right)^{-1-\xi} \exp\left[-\left(\frac{x-\mu}{\sigma}\right)^{-\xi}\right] \tag{2}$$

and,

$$f(x) = \frac{1}{\sigma}\left(1 + \xi\frac{x-\mu}{\sigma}\right)^{-(1/\xi+1)} \tag{3}$$

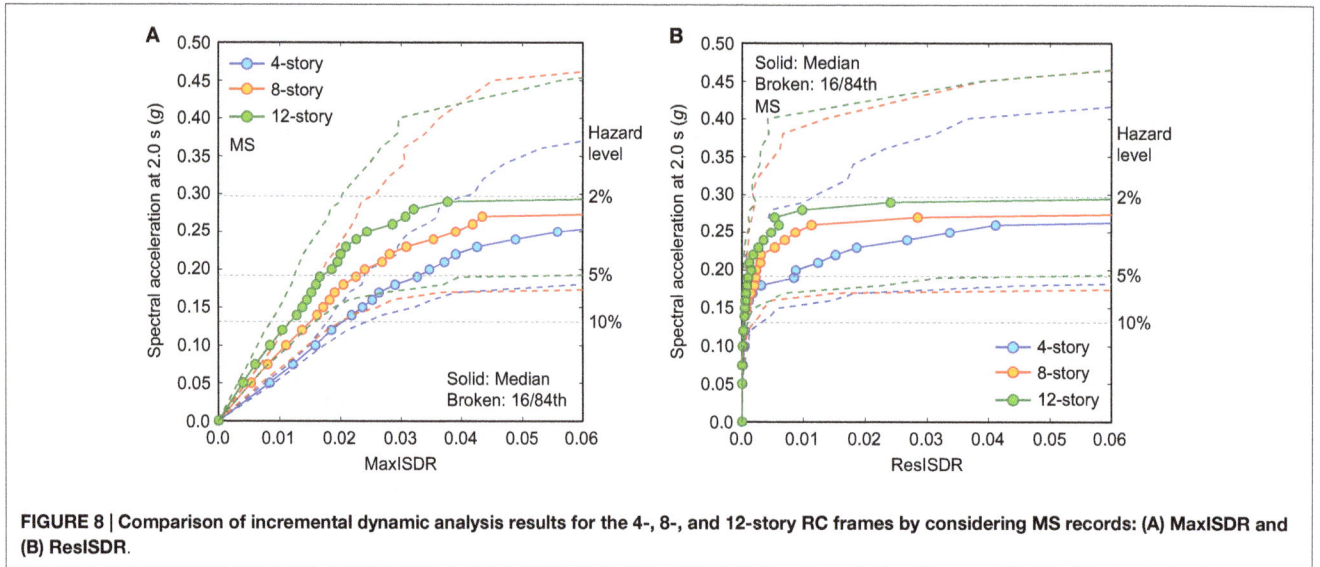

FIGURE 8 | Comparison of incremental dynamic analysis results for the 4-, 8-, and 12-story RC frames by considering MS records: (A) MaxISDR and (B) ResISDR.

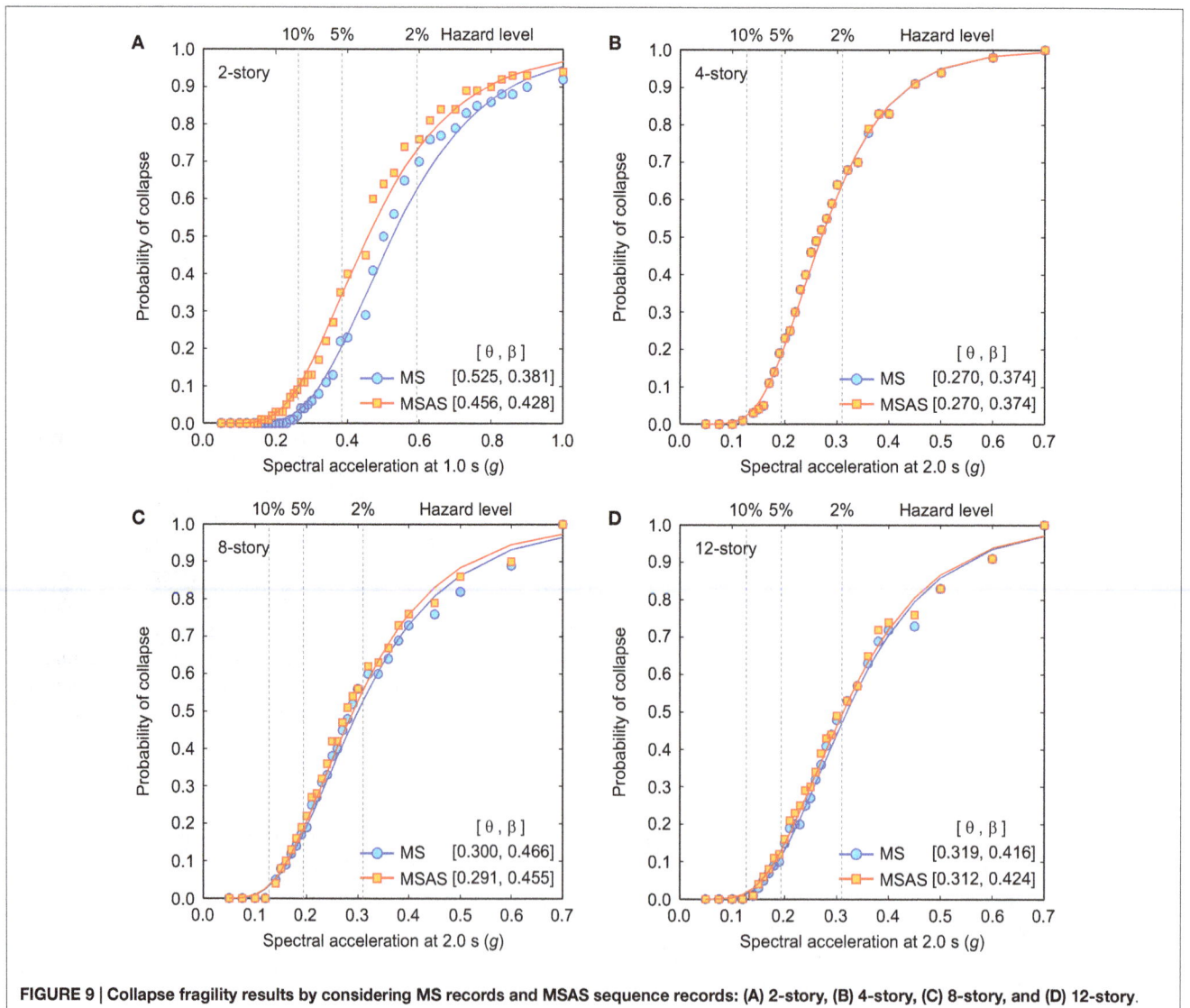

FIGURE 9 | Collapse fragility results by considering MS records and MSAS sequence records: (A) 2-story, (B) 4-story, (C) 8-story, and (D) 12-story.

where μ is the location parameter, and σ is the scale parameter, and ξ is the shape parameter. These marginal distributions are non-normal (in particular, ResISDR); in such cases, conventional multivariate normal (or lognormal) distribution modeling is not ideal, and a more elaborate approach is necessary.

The dependence of MaxISDR and ResISDR can be characterized by using elliptical copulas, such as normal and *t*, and Archimedean copulas, such as Gumbel, Frank, and Clayton (McNeil et al., 2005). The asymmetric Archimedean copula is a mixture of one of the Archimedean copulas and the independence copula; this copula class is useful in modeling data that exhibit uneven distribution of the data points along the upper-left-lower-right diagonal line in the transformed space. In the context of joint probability distribution modeling of MaxISDR and ResISDR, the uneven distribution of the data is related to the physical relationship between MaxISDR and ResISDR (i.e., MaxISDR ≥ ResISDR; Goda and Tesfamariam, 2015). To model the observed dependence of MaxISDR and ResISDR (e.g., scatter plot shown in **Figure 11A**), parametric copula functions are fitted to empirical copula samples using the maximum likelihood method (McNeil et al., 2005). The copula fitting of MaxISDR and ResISDR at various IM levels suggests that overall, the Gumbel (or asymmetrical Gumbel) copula (Eq. 4) is suitable for the majority of the cases examined in this study.

$$C_\delta(u_1, u_2) = \exp\left(-[(-\ln u_1)^\delta + (-\ln u_2)^\delta]^{1/\delta}\right), \delta > 1 \qquad (4)$$

where u_1 and u_2 are the uniform random variables, and δ is the model parameter.

The developed statistical seismic demand models of MaxISDR and ResISDR can be used for seismic performance evaluation of structures. For instance, considering the fitted dependence function for the 4-story building at 5% PE in 50 years, numerous copula samples are first generated; their marginal distributions are uniformly distributed with the specified dependence characteristics. Using the simulated copula samples and the fitted marginal distribution models for MaxISDR and ResISDR, pairs of MaxISDR and ResISDR samples can be obtained using the inverse transformation method. The results of 5,000,000 simulations are presented in **Figure 11B**. Indeed, similar figures can be generated for different building story numbers as well as seismic hazard levels, and can be used in the seismic performance evaluation.

Seismic Performance Evaluation

The collapse fragility curves and the joint probability model of non-collapse inelastic seismic demands outlined in the previous sections can now be used to carry out performance-based

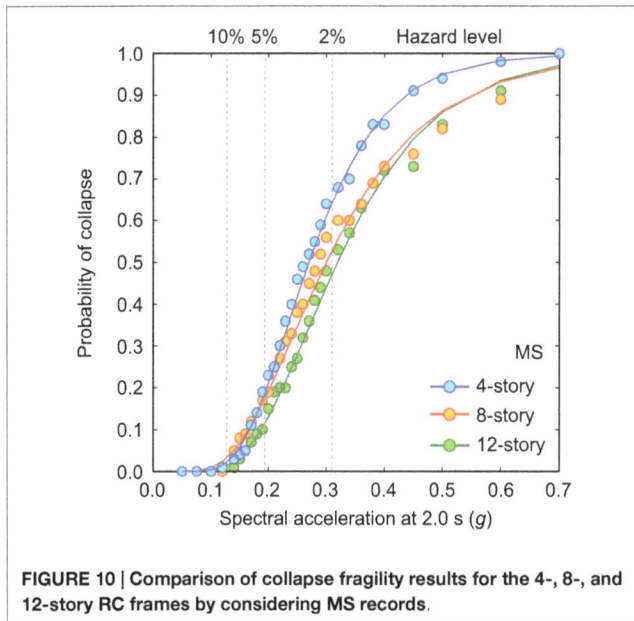

FIGURE 10 | Comparison of collapse fragility results for the 4-, 8-, and 12-story RC frames by considering MS records.

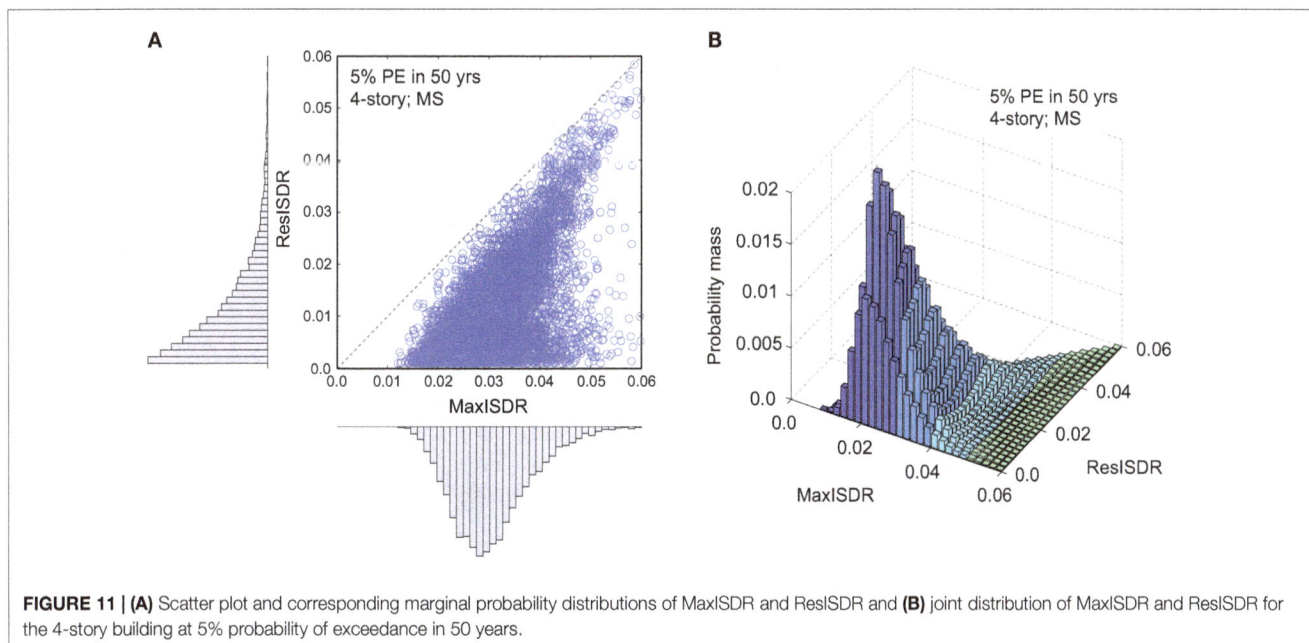

FIGURE 11 | (A) Scatter plot and corresponding marginal probability distributions of MaxISDR and ResISDR and (B) joint distribution of MaxISDR and ResISDR for the 4-story building at 5% probability of exceedance in 50 years.

evaluation of a building with the limit states provided in **Tables 3** and **4**. For example, for the NBCC normal importance buildings (*basic objective*), the acceptable limit states are LS and CP for 10 and 2% PE in 50 years, respectively. For the NBCC high importance buildings (*essential service objective*), the required limit states are more stringent and correspond to DC and LS for the same performance levels. For these cases, the corresponding values of [MaxISDR, ResISDR] are as follows (**Table 4**):

- *Basic objective:* LS = [2.5, 1.0%] for 10% PE in 50 years and CP = [4.5, 1.0%] for 2% PE in 50 years;
- *Essential service objective:* LS = [0.9, 0.5%] for 10% PE in 50 years and CP = [2.5, 1.0%] for 2% in 50 years.

For the structural models that are considered in this study (which should meet the *basic objective*), IO and DC are not applicable to evaluate their seismic performances based on bivariate structural responses. This is because the structures, when subjected to expected ground motions at IO and DC hazard levels, are essentially linear-elastic and residual responses are very small (near zero). In other words, the seismic hazard levels corresponding to 50% PE in 30 or 50 years are mainly related to the serviceability limit state and are too low to cause significant non-linear responses. As our focus in this paper is upon the non-linear responses, LS and CP are mainly concerned and an intermediate seismic performance level between LS and CP, i.e., 5% PE in 50 years (corresponding to the return period of 1000 years), is introduced.

To illustrate the proposed seismic performance evaluation method, three performance levels, i.e., 10, 5, and 2% PE in 50 years, are considered with the limit states of [MaxISDR, ResISDR] = [2.0, 1.0%], [3.0, 1.5%], and [5.0, 2.0%], respectively. These demand levels are similar to those presented in **Table 4**. **Figure 12A** shows the scatter plots of MaxISDR and ResISDR (for non-collapse cases) at the three performance levels, noting that the collapse cases are dealt with collapse fragility curves (**Figure 9**). The corresponding limit states are indicated with red broken lines. By connecting the limit state thresholds at different performance levels (gray broken lines) and plotting the seismic demands in bivariate space (blue dots), the evolution of the seismic performance evaluation of the structure can be visualized, facilitating the better understanding of the seismic performance of the structure at multiple seismic excitation levels.

The overall performance of the building is assessed through unconditional probability of unsatisfactory seismic performance (P_{NS}) (i.e., overall measure at a seismic performance level). The steps followed to compute P_{NS} are outlined below, with the results shown in **Figure 12B** as an example. **Figure 12B** illustrates the calculations of the probabilities of exceedance and non-exceedance of the specified limit state thresholds for the 5% PE in 50 years performance level for non-collapse cases. First, from **Figure 12B**, four probabilities of exceedance and non-exceedance can be derived:

- the lower-left quadrant corresponds to the probability of joint non-exceedance of the MaxISDR and ResISDR limits, $P_{NE,NE}$ (=0.494),

- the lower-right quadrant corresponds to the probability of exceedance of the MaxISDR limit and non-exceedance of the ResISDR limit, $P_{E,NE}$ (=0.271),
- the upper-left quadrant corresponds to the probability of non-exceedance of the MaxISDR limit and exceedance of the ResISDR limit, $P_{NE,E}$ (=0.025), and
- the upper-right quadrant corresponds to the probability of joint exceedance of the MaxISDR and ResISDR limits, $P_{E,E}$ (=0.210).

The four probabilities are useful for assessing the causes of unsatisfactory seismic performance for non-collapse cases. A large value of $P_{E,NE}$ tends to indicate that the unsatisfactory seismic performance is due to MaxISDR, whereas a large value of $P_{NE,E}$ suggests that the structure may need to be demolished after the earthquake. It is noteworthy that $P_{NE,NE}$, $P_{E,NE}$, $P_{NE,E}$, and $P_{E,E}$ are conditional probabilities upon non-collapse cases. Second, the collapse probability P_C and the non-collapse probability P_{NC}, i.e., $P_{NC} = (1 - P_C)$, need to be evaluated for the given seismic intensity level using the corresponding collapse fragility curve (**Figure 9**). Finally, once the different probability values are obtained as outlined above, the value of P_{NS} can be calculated by:

$$\begin{aligned} P_{NS} &= P_C + P_{NC} \times (P_{NE,E} + P_{E,NE} + P_{E,E}) \\ &= P_C + P_{NC} \times (1 - P_{NE,NE}) \end{aligned} \quad (5)$$

Figure 13 shows 4 by 3 panels (i.e., four buildings and three performance levels) of the bivariate MaxISDR–ResISDR data/performance limits for MS records; four conditional probabilities of exceedance and non-exceedance as well as collapse/non-collapse probabilities are indicated in the figure, whereas **Figure 14** shows the same set of results for MSAS sequences. To facilitate the comparison of the calculated probabilities for different cases, values of $P_{NE,NE}$, $P_{E,NE}$, $P_{NE,E}$, $P_{E,E}$, P_C, P_{NC}, and P_{NS} are summarized in **Tables 8** and **9** for MS records and MSAS sequences, respectively.

Figures 13 and **14** show that MaxISDR and ResISDR become severer with the increase in the seismic performance level; this can be inspected from the scatter of the data points as well as the increase of the collapse probability. The 2- and 4-story buildings are more vulnerable, in comparison with the 8- and 12-story buildings. The collapse probabilities for the 2-story building are generally greater than those for the 4-story building; however, for the non-collapse cases, MaxISDR and ResISDR data are more widely distributed and consequently, conditional probabilities of unsatisfactory seismic performance (e.g., $P_{E,NE}$, $P_{NE,E}$, and $P_{E,E}$) for the 4-story building are greater than those for the 2-story building. Overall, unconditional probabilities of unsatisfactory seismic performance for the 4-story building are greater than others (P_{NS} in **Tables 8** and **9**). Note that the causes of unsatisfactory seismic performance vary depending on building story numbers and performance levels for the non-collapse cases. For example, for the 2-story building, unsatisfactory performance is mainly due to large residual seismic demands; in this case, the damaged building may be demolished. On the other hand, for the 4-story building (e.g., 5% PE in 50 years), the unsatisfactory performance is mainly due to excessive peak transient seismic demands. These

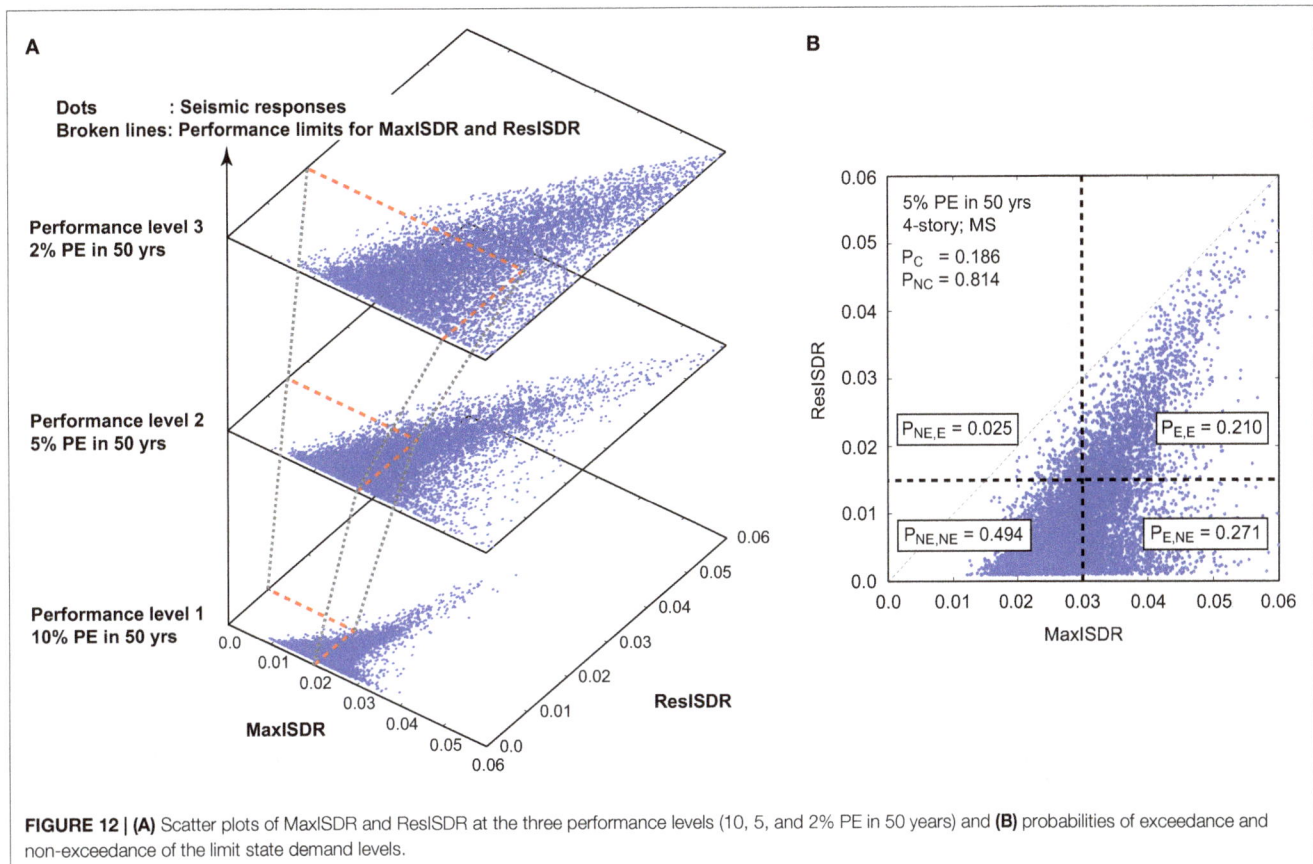

FIGURE 12 | (A) Scatter plots of MaxISDR and ResISDR at the three performance levels (10, 5, and 2% PE in 50 years) and **(B)** probabilities of exceedance and non-exceedance of the limit state demand levels.

results indicate that different counter measures may need to be implemented for different buildings as their damage mechanisms may be different.

The comparison of the results shown in **Figures 13** and **14** as well as **Tables 8** and **9** suggests that the observations made for MS records are generally applicable to MSAS sequences. However, additional seismic demands due to major aftershocks have noticeable influence on both MaxISDR and ResISDR for the 2-story building (**Figures 6A** and **9A**). Consequently, counter measures against aftershock risks should be specific to building types (i.e., dynamic structural characteristics and susceptible failure mode).

Importantly, the calculated values of PNS listed in **Tables 8** and **9** indicate that for all four non-ductile buildings, their seismic capacities may be judged as satisfactory (because PNS is relatively low) at the LS performance level (i.e., return periods of 500–1000 years), whereas they fail to meet the CP performance level required by the current standards suggested by FEMA P-58-1 (2012). Therefore, for this class of non-ductile RC buildings, seismic retrofitting should be implemented to improve the seismic performance.

Discussion and Conclusion

The primary objective of the building design code was LS. In developed countries, this has been met through improved seismic design provisions. Seismic vulnerability of existing buildings remains to be a major concern because of the use of older design codes and/or poor construction practices at the time of design and construction. Most of these older buildings are currently operational and are required to be further assessed and upgraded to improve potential economic consequences due to seismic damage. An accurate assessment of potential impact of future destructive earthquakes is essential for effective disaster risk reduction. Probabilistic seismic risk analysis entails comprehensive understanding of ground shaking information, such as fault rupture process, wave propagation, and site effects, as well as vulnerability of structures, such as structural damage accumulation, seismic loss generation, and societal/economic impact (Cornell and Krawinkler, 2000). Through probabilistic calculus, it evaluates the potential damage and loss that a certain group of structures is likely to experience due to various seismic events (Tesfamariam and Goda, 2015).

The current state of the art for seismic performance assessment of buildings in North America is FEMA P-58-1 (2012). It has been developed based on generic ground motions that are applicable to the seismicity of California, which might not be compatible with the seismicity in Canada. Furthermore, the damage observed from the MSAS sequence of the 2011 $M_w6.3$ Christchurch earthquake in New Zealand has highlighted the need for further study on the collapse risk of RC buildings in Canada (Elwood, 2013). The rigorous probabilistic seismic performance evaluation method can be used to aid in an informed decision-making by comparing performance metrics of alternative seismic risk mitigation measures quantitatively. Accurate representation

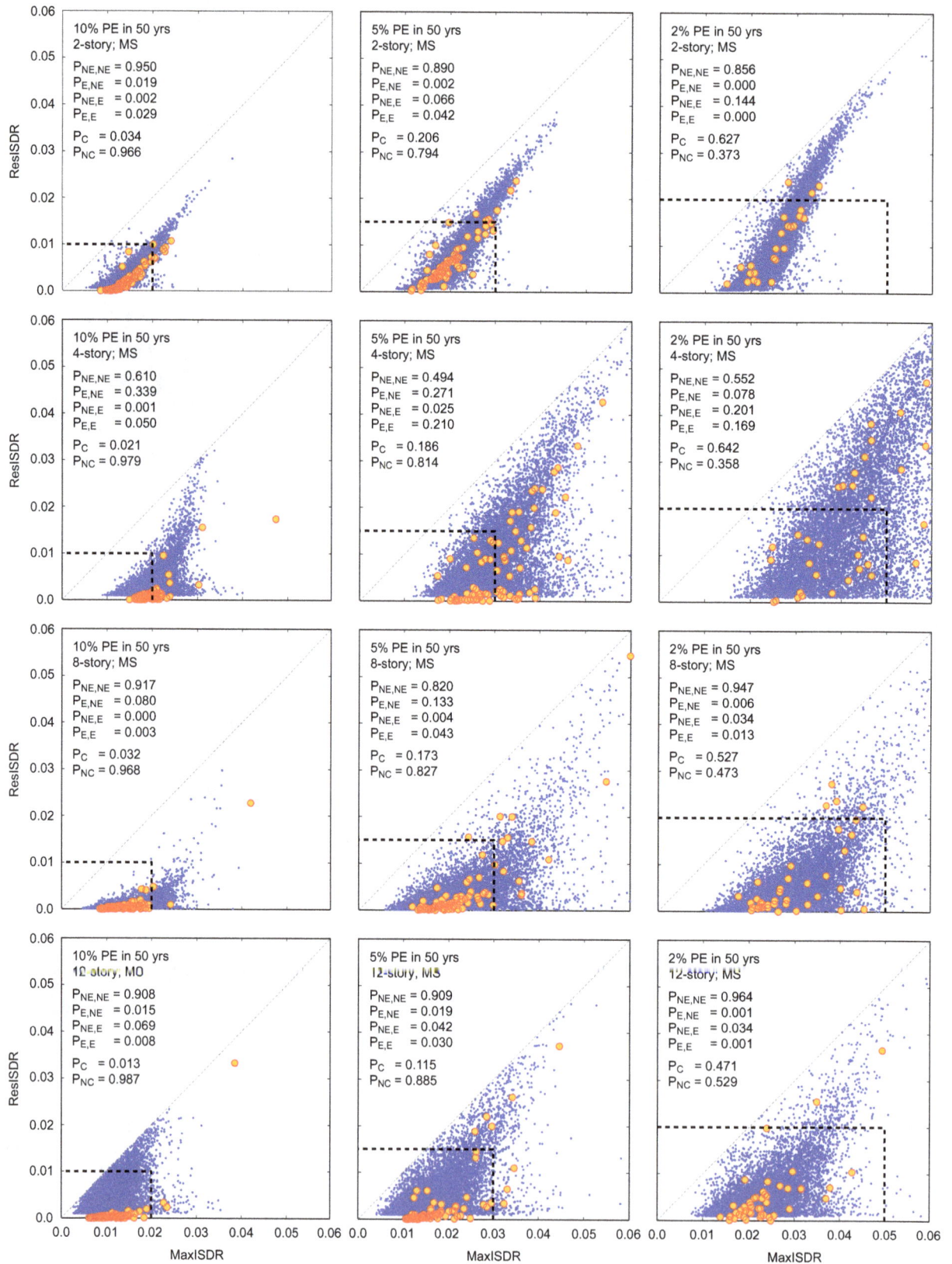

FIGURE 13 | Mainshock earthquake record – probabilities of exceedance and non-exceedance of the 2-, 4-, 8-, and 12-story building (rows) and 10, 5, and 2% PE in 50 years hazard levels (columns). Red circles are the IDA results.

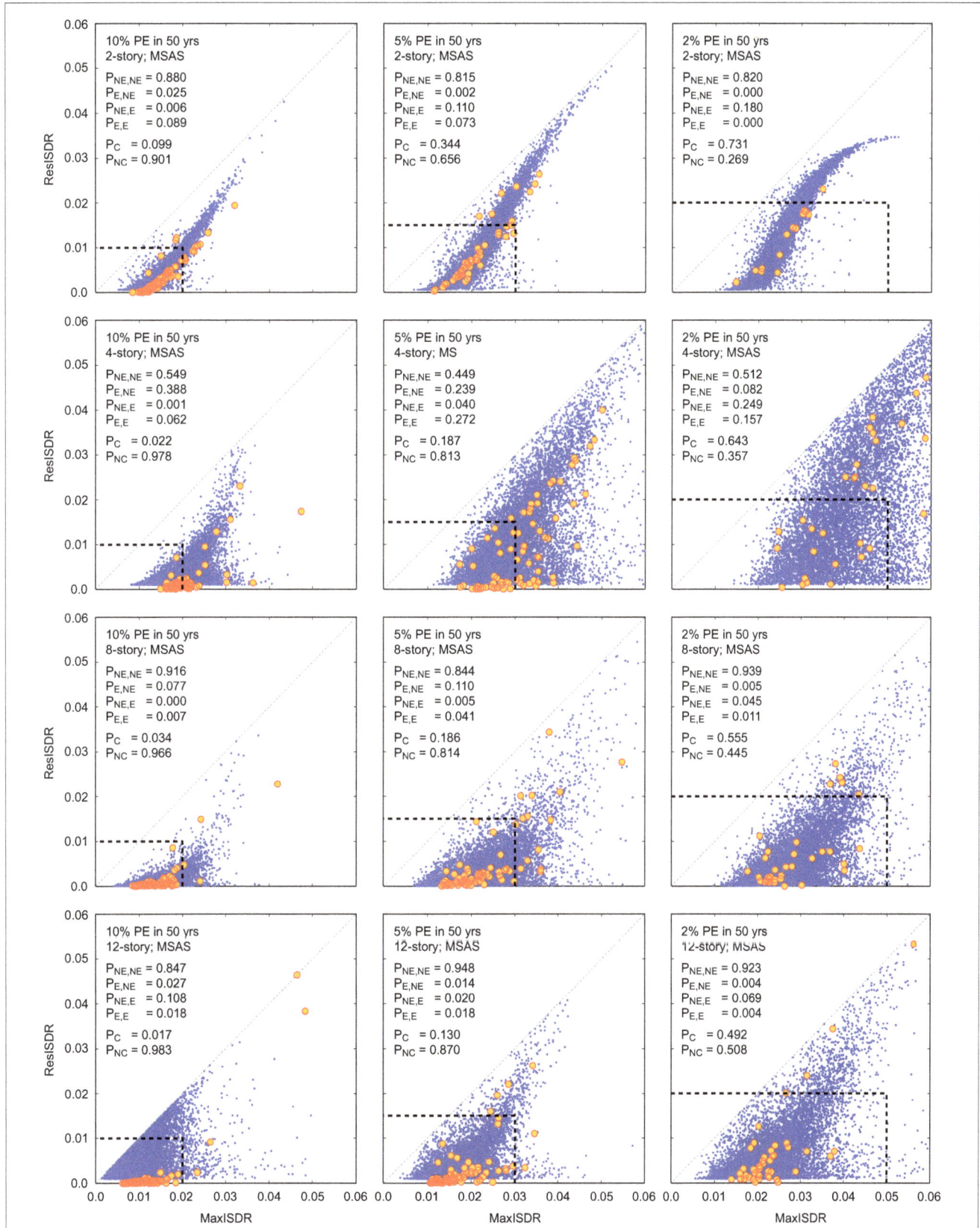

FIGURE 14 | Mainshock–aftershock earthquake sequence records – probabilities of exceedance and non-exceedance of the 2-, 4-, 8-, and 12-story building (rows) and 10, 5, and 2% PE in 50 years hazard levels (columns). Red circles are the IDA results.

TABLE 8 | Collapse and non-collapse probabilities and probabilities of exceedance and non-exceedance of the different limit states for MS records.

Building story	Probability of exceedance in 50 years (%)	Probabilities of exceedance and non-exceedance of the different limit states				Non-collapse probability	Collapse probability	$P_{NS} = P_C + P_{NC} \times (1 - P_{NE,NE})$
		$P_{NE,NE}$	$P_{E,NE}$	$P_{E,E}$	$P_{NE,E}$	P_{NC}	P_C	
2	10	0.950	0.019	0.029	0.002	0.966	0.034	0.082
	5	0.890	0.002	0.042	0.066	0.794	0.206	0.293
	2	0.856	0.000	0.000	0.144	0.373	0.627	0.681
4	10	0.610	0.339	0.050	0.001	0.979	0.021	0.403
	5	0.494	0.271	0.210	0.025	0.814	0.186	0.598
	2	0.552	0.078	0.169	0.201	0.358	0.642	0.802
8	10	0.917	0.080	0.003	0.000	0.968	0.032	0.112
	5	0.820	0.133	0.043	0.004	0.827	0.173	0.322
	2	0.947	0.006	0.013	0.034	0.473	0.527	0.552
12	10	0.908	0.015	0.008	0.069	0.987	0.013	0.104
	5	0.909	0.019	0.030	0.042	0.885	0.115	0.196
	2	0.964	0.001	0.001	0.034	0.529	0.471	0.490

TABLE 9 | Collapse and non-collapse probabilities and probabilities of exceedance and non-exceedance of the different limit states for MSAS sequence records.

Building story	Probability of exceedance in 50 years (%)	Probabilities of exceedance and non-exceedance of the different limit states				Non-collapse probability	Collapse probability	$P_{NS} = P_C + P_{NC} \times (1 - P_{NE,NE})$
		$P_{NE,NE}$	$P_{E,NE}$	$P_{E,E}$	$P_{NE,E}$	P_{NC}	P_C	
2	10	0.880	0.025	0.089	0.006	0.901	0.099	0.207
	5	0.815	0.002	0.073	0.110	0.656	0.344	0.465
	2	0.820	0.000	0.000	0.180	0.269	0.731	0.779
4	10	0.549	0.388	0.062	0.001	0.978	0.022	0.463
	5	0.449	0.239	0.272	0.040	0.813	0.187	0.635
	2	0.512	0.082	0.157	0.249	0.357	0.643	0.817
8	10	0.916	0.077	0.007	0.000	0.966	0.034	0.115
	5	0.844	0.110	0.041	0.005	0.814	0.186	0.313
	2	0.939	0.005	0.011	0.045	0.445	0.555	0.582
12	10	0.847	0.027	0.018	0.108	0.983	0.017	0.167
	5	0.948	0.014	0.018	0.020	0.870	0.130	0.175
	2	0.923	0.004	0.004	0.069	0.508	0.492	0.531

of different limit states, robust ground motion selection, and multivariate inelastic seismic demands are vitally important in the assessment. In this paper, a robust seismic evaluation tool, within the performance-based earthquake engineering framework, is developed. Two EDPs, MaxISDR and ResISDR, are used to determine the severity of seismic damage and consequences. The joint probability distribution and dependency are modeled using the advanced copula technique. Following SEAOC (1995) and FEMA P-58-1 (2012), the two EDPs reaching different performance limit states are defined. Moreover, the aftershock ground motions are incorporated with the conventional seismic performance evaluation methodology, and furnished a better representation of the prevalent risk. The proposed evaluation tool can indeed be used for existing structures or design of new buildings.

The proposed framework was applied to 2-, 4-, 8-, and 12-story non-ductile RC buildings located in Victoria, BC, Canada. Considering regional seismicity in southwestern BC (i.e., shallow crustal earthquakes, off-shore mega-thrust interface earthquakes from the Cascadia subduction zone, and deep inslab earthquakes), 50 MS records and 50 MSAS sequence records were

selected. Subsequently, IDA was performed and the computed MaxISDR and ResISDR data were used for developing collapse fragility curves and for developing probabilistic inelastic seismic demand models using copulas.

The general conclusions related to the aftershock effects are as follows:

- The MSAS sequence earthquake has significant influence on the 2-story building, where the median collapse capacity is reduced by 13%.
- The MSAS sequence records for the 4-, 8-, and 12-story buildings showed no marked differences in the collapse fragility. This partly may be ascribed to the considered model limitation. The collapse limit states as modeled in this paper are associated with flexure. The model does not consider shear failure, and gravity load collapse.
- The MSAS sequence records, however, have shown marked differences in the non-collapse limit states.

The unconditional probability of unsatisfactory seismic performance P_{NS} integrates the collapse and non-collapse limit states and thus can be used as an overall seismic performance measure of structures. The general conclusions related to the P_{NS} results for the four non-ductile RC frames are as follows:

- With increasing hazard levels (10, 5, and 2% PE in 50 years), the corresponding P_{NS} is increasing.
- Seismic capacities at the LS performance level (i.e., return periods of 500–1000 years) are judged to be satisfactory (i.e., P_{NS} is relatively low).
- Seismic capacities at the CP performance level (i.e., return period of 2500 years) may not be satisfactory (i.e., P_{NS} is high). This highlights the need for undertaking seismic retrofitting to improve the seismic performance.

Finally, the proposed performance-based seismic screening criteria and methods can be used for Canadian buildings. The methodology can be extended to different building types and seismicity (e.g., Eastern Canada). The consideration of MSAS sequences as seismic input was found to be important for the seismic risk assessment of low- to mid-rise buildings, and further investigations are warranted in the future. Furthermore, the aftershock effects should also be integrated in the design of Canadian buildings.

Acknowledgments

Ground motion data for Japanese earthquakes and worldwide crustal earthquakes were obtained from the K-NET/KiK-net/ SK-net databases at http://www.kyoshin.bosai.go.jp/ and http:// www.sknet.eri.u-tokyo.ac.jp/, and the PEER-NGA database at http://peer.berkeley.edu/nga/index.html, respectively. This work was supported by the Natural Science Engineering Research Council Canada (RGPIN-2014-05013) to the first author and the Engineering and Physical Sciences Research Council (EP/ M001067/1) to the second author.

References

Altoontash, A. (2004). *Simulation and Damage Models for Performance Assessment of Reinforced Concrete Beam-Column Joints.* Doctoral Dissertation, Stanford University, Stanford, CA.

ATC. (2002). *Rapid Visual Screening of Buildings for Potential Seismic Hazard: A Handbook,* 2nd Edn. Washington, DC: Applied Technology Council.

Atkinson, G. M. (2004). "An overview of developments in seismic hazard analysis," in *13th World Conference on Earthquake Engineering* Vancouver, BC, Paper No. 5001.

Atkinson, G. M., and Goda, K. (2011). Effects of seismicity models and new ground motion prediction equations on seismic hazard assessment for four Canadian cities. *Bull. Seismol. Soc. Am.* 101, 176–189. doi:10.1785/0120100093

Baker, J. W. (2011). The conditional mean spectrum: a tool for ground motion selection. *J. Struct. Eng.* 137, 322–331. doi:10.1061/(ASCE)ST.1943-541X.0000215

Bruneau, M., and Lamontagne, M. (1994). Damage from 20th century earthquakes in Eastern Canada and seismic vulnerability of unreinforced masonry buildings. *Can. J. Civil Eng.* 21, 643–662. doi:10.1139/l94-065

Christidis, A. A., Dimitroudi, E. G., Hatzigeorgiou, G. D., and Beskos, D. E. (2013). Maximum seismic displacements evaluation of steel frames from their post-earthquake residual deformation. *Bull. Earthq. Eng.* 11, 2233–2248. doi:10.1007/s10518-013-9490-z

Christopoulos, C., Pampanin, S., and Priestley, M. J. N. (2003). Performance-based seismic response of frame structures including residual deformations – part I: single-degree-of-freedom system. *J. Earthq. Eng.* 7, 97–118. doi:10.1080/13632460309350443

Cornell, C. A., and Krawinkler, H. (2000). *Progress and Challenges in Seismic Performance Assessment.* PEER Center News 3. Available at: http://peer.berkeley.edu/news/2000spring/performance.html

DeVall, R. H. (2003). Background information for some of the proposed earthquake design provisions for the 2005 edition of the National Building Code of Canada. *Can. J. Civil Eng.* 30, 279–286. doi:10.1139/l02-048

Elwood, K. J. (2013). Performance of concrete buildings in the 22 February 2011 Christchurch earthquake and implications for Canadian codes. *Can. J. Civil Eng.* 40, 759–776. doi:10.1139/cjce-2011-0564

Erochko, J., Christopoulos, C., Tremblay, R., and Choi, H. (2011). Residual drift response of SMRFs and BRB frames in steel buildings designed according to ASCE 7-05. *J. Struct. Eng.* 137, 589–599. doi:10.1061/(ASCE)ST.1943-541X.0000296

FEMA 356. (2000). *Prestandard and Commentary for the Seismic Rehabilitation of Buildings.* Washington, DC: Federal Emergency Management Agency.

FEMA P-58-1. (2012). *Seismic Performance Assessment of Buildings Volume 1 – Methodology.* Washington, DC: Federal Emergency Management Agency.

Foo, S., and Davenport, A. (2003). Seismic hazard mitigation for buildings. *Nat. Hazards (Dordr.)* 28, 517–535. doi:10.1023/A:1022950629065

Goda, K., and Atkinson, G. M. (2011). Seismic performance of wood-frame houses in South-Western British Columbia. *Earthq. Eng. Struct. Dyn.* 40, 903–924. doi:10.1002/eqe.1068

Goda, K., Pomonis, A., Chian, S. C., Offord, M., Saito, K., Sammonds, P., et al. (2013). Ground motion characteristics and shaking damage of the 11th March 2011 Mw9.0 Great East Japan earthquake. *Bull. Earthq. Eng.* 11, 141–170. doi:10.1007/s10518-012-9371-x

Goda, K., and Taylor, C. A. (2012). Effects of aftershocks on peak ductility demand due to strong ground motion records from shallow crustal earthquakes. *Earthq. Eng. Struct. Dyn.* 41, 2311–2330. doi:10.1002/eqe.2188

Goda, K., and Tesfamariam, S. (2015). Multi-variate seismic demand modelling using Copulas: application to non-ductile reinforced concrete frame in Victoria, Canada. *Struct. Saf.* 56, 39–51. doi:10.1016/j.strusafe.2015.05.004

Goda, K., Wenzel, F., and De Risi, R. (2015). Empirical assessment of nonlinear seismic demand of mainshock-aftershock ground motion sequences for Japanese earthquakes. *Front. Built Environ.* 1:6. doi:10.3389/fbuil.2015.00006

Haselton, C. B., Liel, A. B., Lange, S. T., and Deierlein, G. G. (2008). *Beam-Column Element Model Calibrated for Predicting Flexural Response Leading to Global Collapse of RC Frame Buildings.* PEER Report 2007/03. Berkeley, CA: PEER Center, University of California, Berkeley.

Hatzigeorgiou, G. D., and Beskos, D. E. (2009). Inelastic displacement ratios for SDOF structures subjected to repeated earthquakes. *Eng. Struct.* 31, 2744–2755. doi:10.1016/j.engstruct.2009.07.002

Hatzigeorgiou, G. D., and Liolios, A. A. (2010). Nonlinear behaviour of RC frames under repeated strong ground motions. *Soil Dyn. Earthq. Eng.* 30, 1010–1025. doi:10.1016/j.soildyn.2010.04.013

Hatzigeorgiou, G. D., Papagiannopoulos, G. A., and Beskos, D. E. (2011). Evaluation of maximum seismic displacements of SDOF systems from their residual deformation. *Eng. Struct.* 33, 3422–3431. doi:10.1016/j.engstruct.2011.07.006

Hyndman, R. D., and Rogers, G. C. (2010). Great earthquakes on Canada's west coast: a review. *Can. J. Earth Sci.* 47, 801–820. doi:10.1139/E10-011

Ibarra, L. F., Medina, R. A., and Krawinkler, H. (2005). Hysteretic models that incorporate strength and stiffness deterioration. *Earthq. Eng. Struct. Dyn.* 34, 1489–1511. doi:10.1002/eqe.495

ICBO. (1967). *Uniform Building Code.* Pasadena, CA: International Conference of Building Officials.

Kawashima, K., MacRae, G. A., Hoshikuma, J., and Nagaya, K. (1998). Residual displacement response spectrum. *J. Struct. Eng.* 124, 523–530. doi:10.1061/(ASCE)0733-9445(1998)124:5(523)

Leite, J., Lourenco, P. B., and Ingham, J. M. (2013). Statistical assessment of damage to churches affected by the 2010-2011 Canterbury (New Zealand) earthquake sequence. *J. Earthq. Eng.* 17, 73–97. doi:10.1080/13632469.2012.713562

Liel, A. B., and Deierlein, G. G. (2008). *Assessing the Collapse Risk of California's Existing Reinforced Concrete Frame Structures: Metrics for Seismic Safety Decisions.* Technical Report No. 166. Stanford, CA: John A. Blume Center Earthquake Engineering Center.

Luco, N., and Bazzurro, P. (2007). Does amplitude scaling of ground motion records result in biased nonlinear structural drift responses? *Earthq. Eng. Struct. Dyn.* 36, 1813–1835. doi:10.1002/eqe.695

MacRae, G. A., and Kawashima, K. (1997). Post-earthquake residual displacements of bilinear oscillators. *Earthq. Eng. Struct. Dyn.* 26, 701–716. doi:10.1002/(SICI)1096-9845(199707)26:7<701::AID-EQE671>3.3.CO;2-9

McKenna, F., Fenves, G. L., and Scott, M. H. (2000). *Open System for Earthquake Engineering Simulation.* Berkeley, CA: University of California Berkeley.

McNeil, A. J., Frey, R., and Embrechts, P. (2005). *Quantitative Risk Management: Concepts, Techniques and Tools.* Princeton, NJ: Princeton University Press.

Mitchell, D., Paultre, P., Tinawi, R., Saatciouglu, M., Tremlay, R., Elwood, K., et al. (2010). Evolution of seismic design provisions in the National Building Code of Canada. *Can. J. Civil Eng.* 37, 1157–1170. doi:10.1139/L10-054

NRCC. (1941). *National Building Code of Canada. Associate Committee on the National Building Code.* Ottawa, ON: National Research Council of Canada.

NRCC. (1965). *National Building Code of Canada. Associate Committee on the National Building Code.* Ottawa, ON: National Research Council of Canada.

NRCC. (1970). *National Building Code of Canada. Associate Committee on the National Building Code.* Ottawa, ON: National Research Council of Canada.

Onur, T., Ventura, C. E., and Finn, W. D. L. (2005). Regional seismic risk in British Columbia – damage and loss distribution in Victoria and Vancouver. *Can. J. Civil Eng.* 32, 361–371. doi:10.1139/l04-098

Otani, S. (2004). Earthquake resistant design of reinforced concrete buildings past and future. *J. Adv. Concr. Technol.* 2, 3–24. doi:10.3151/jact.2.3

Pampanin, S., Christopoulos, C., and Priestley, M. J. N. (2003). Performance-based seismic response of frame structures including residual deformations – part II: multi-degree-of-freedom systems. *J. Earthq. Eng.* 7, 119–147. doi:10.1080/13632460309350444

Rainer, J. H. (1992). *Manual for Screening of Buildings for Seismic Investigation.* Institute for Research in Construction. Ottawa: National Research Council of Canada.

Ramirez, C. M., and Miranda, E. (2009). *Building-Specific Loss Estimation Methods & Tools for Simplified Performance-Based Earthquake Engineering.* Technical Report No. 171. Stanford, CA: John A. Blume Center Earthquake Engineering Center.

Ruiz-García, J., and Miranda, E. (2006). Evaluation of residual drift demands in regular multi-storey frames for performance-based seismic assessment. *Earthq. Eng. Struct. Dyn.* 35, 1609–1629. doi:10.1002/eqe.593

Ryu, H., Luco, N., Uma, S. R., and Liel, A. B. (2011). "Developing fragilities for mainshock-damaged structures through incremental dynamic analysis," in *Proceedings, Ninth Pacific Conference on Earthquake Engineering, Building an Earthquake-Resilient Society,* Auckland.

Saatcioglu, M., Shooshtari, M., and Foo, S. (2013). Seismic screening of buildings based on the 2010 National Building Code of Canada. *Can. J. Civil Eng.* 40, 483–498. doi:10.1139/cjce-2012-0055

SEAOC. (1995). *Vision 2000: Performance-Based Seismic Engineering of Buildings.* Sacramento, CA: Structural Engineers Association of California.

Tesfamariam, S., and Goda, K. (2015). Loss estimation for non-ductile reinforced concrete building in Victoria, British Columbia, Canada: effects of mega-thrust M_w9-class subduction earthquakes and aftershocks. *Earthq. Eng. Struct. Dyn.* 44, 2303–2320. doi:10.1002/eqe.2585

Tesfamariam, S., Goda, K., and Mondal, G. (2015). Seismic vulnerability of RC frame with unreinforced masonry infill due to mainshock-aftershock earthquake sequences. *Earthq. Spectra* 31, 1427–1449. doi:10.1193/042313EQS111M

Tesfamariam, S., and Saatcioglu, M. (2008). Risk-based seismic evaluation of reinforced concrete buildings. *Earthq. Spectra* 24, 795–821. doi:10.1193/1.2952767

Uma, S. R., Pampanin, S., and Christopoulos, C. (2010). Development of probabilistic framework for performance-based seismic assessment of structures considering residual deformations. *J. Earthq. Eng.* 14, 1092–1111. doi:10.1080/13632460903556509

Vamvatsikos, D., and Cornell, C. A. (2002). Incremental dynamic analysis. *Earthq. Eng. Struct. Dyn.* 31, 491–514. doi:10.1002/eqe.141

Ventura, C. E., Finn, W. D. L., Onur, T., Blanquera, A., and Rezai, M. (2005). Regional seismic risk in British Columbia – classification of buildings and development of damage probability functions. *Can. J. Civil Eng.* 32, 372–387. doi:10.1139/l04-099

Conflict of Interest Statement: The authors declare that the research was conducted in the absence of any commercial or financial relationships that could be construed as a potential conflict of interest.

14

Three-dimensional energy transmitting boundary in the time domain

Naohiro Nakamura*

Research and Development Institute, Takenaka Corporation, Chiba, Japan

Although the energy transmitting boundary (TB) is accurate and efficient for the finite element method earthquake response analysis, it could be applied in the frequency domain only. In the previous papers, the author proposed an earthquake response analysis method using the time domain energy TB for two-dimensional (2D) problems. In this paper, this technique is expanded for three-dimensional (3D) problems. The inner field is supposed to be a hexahedron shape, and the approximate time domain boundary is explained, first. Next, 2D antiplane time domain boundary is studied for a part of the approximate 3D boundary method. Then, accuracy and efficiency of the proposed method are confirmed by example problems.

Keywords: energy transmitting boundary, FEM, time domain, three-dimensional analysis, soil–structure interaction, viscous boundary

Edited by:
Nikos D. Lagaros,
National Technical University of
Athens, Greece

Reviewed by:
Vagelis Plevris,
School of Pedagogical and
Technological Education, Greece
Sameh Samir F. Mehanny,
Cairo University, Egypt
Anaxagoras Elenas,
Democritus University of Thrace,
Greece

*Correspondence:
Naohiro Nakamura
nakamura.naohiro@takenaka.co.jp

INTRODUCTION

In order to accurately estimate the behavior of buildings during severe earthquakes, both the soil–structure interaction and non-linear effects must be taken into consideration. In addition, three-dimensional (3D) models are needed to express the complex shape of buildings, basements, and piles. In the case where buildings are built close to each other, structure–soil–structure interaction, i.e., Lou et al. (2011) should be considered. Moreover, the collective behavior of buildings during a seismic excitation (city effect), i.e., Ghergu and Ionescu (2009), and the interaction between large group of buildings and the subsoil (site–city interaction), i.e., Guidotti et al. (2012), were also studied. Therefore, in recent years, large scale 3D time history non-linear analyses by the finite element method (FEM) have been carried out.

Although the soil has a semi-infinite extent, the soil model needs to be generated as a finite region model in the FEM analyses. Therefore, artificial wave boundary models are needed especially at the side of the soil model. Currently, simple models, such as the cyclic boundary and the viscous boundary (VB) (Lysmer and Kuhlelameyer, 1969), are often used mainly as the side wave boundary model. They are simple to use, but their accuracy is not high. As a result, the wave boundary cannot be placed close to the analysis object, the analysis modeling domain size and the analysis load are enlarged. For this reason, it is desirable to improve the wave boundary accuracy and reduce the analysis domain size (see **Figure 1**).

Many investigations into this problem have been conducted, i.e., Smith (1973), Kim and Yun (2000), and Wolf and Song (1996). Wolf (2003) proposed a semianalytical method called the scaled boundary FEM, which combines the advantages of the boundary and FEMs by discretizing the boundary spatially without using fundamental solution. It can be applied to unbounded mediums outside of the inner FEM area like a surface finite element. Berenger (1994) introduced the perfectly matched layer

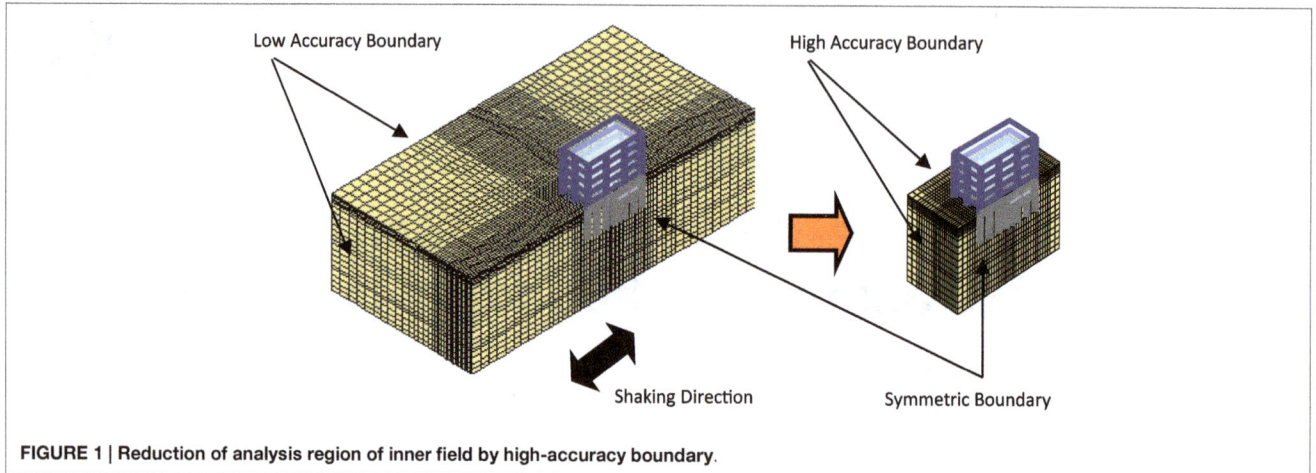

FIGURE 1 | Reduction of analysis region of inner field by high-accuracy boundary.

in the electromagnetic field. That is an absorbing layer model for linear wave equations. Hastings et al. (1996) applied this model to elastic wave propagation, and Basu and Chopra (2004) studied the soil–structure interaction problems using this method. FEM–BEM coupling method was also used to study the soil–pile–structure effect (Millan and Dominguez, 2009) and structure–soil–structure effect (Padron et al., 2011) in the frequency domain. Although there were certain results from these studies, limited application examples are shown at present. Therefore, more practical methods for actual complex problems are needed.

In contrast, the energy transmitting boundary (TB) used in FLUSH (Lysmer et al., 1975a) and ALUSH (Lysmer et al., 1975b) is a highly accurate and efficient side wave boundary. However, TB could only be applied to frequency domain linear analysis and equivalent linear analysis, i.e., Fattah et al. (2012). It is possible to significantly reduce the analysis load for 3D time history FEM analysis by transforming TB to the time domain.

The author has previously proposed the time domain transform methods of strongly frequency-dependent dynamic stiffness and proved that these methods are accurate yet simple (Nakamura, 2006). As an application of the methods, TB for a two-dimensional (2D) in-plane problem that corresponds to FLUSH was transformed to a time domain. It is confirmed that highly accurate analyses in the time domain are also possible as in the frequency domain. Then, non-linear response of an inner field building was calculated and favorable results were obtained (Nakamura, 2009). A study was also conducted to consider the semi-infinity condition at the bottom of TB (Nakamura, 2012b).

In this paper, 3D time history FEM analyses with TB are studied based on these results. The axisymmetric boundary model used in ALUSH is known as a 3D problem TB. However, in many cases, the orthogonal coordinate system is preferred to the axisymmetric coordinate system for actual problems as shown in **Figure 1**. Therefore, in this paper, the orthogonal coordinate system is used for modeling of the inner field (see **Table 1**).

Accordingly, the TB should also be formulated using orthogonal coordinates rather than axisymmetric coordinates, but it is not possible to obtain such a theoretical solution. Therefore, an approximate 3D boundary model (hereinafter referred to as 3D-TB model) from a combination of a 2D in-plane problem TB (hereinafter referred to as SV-TB) and a 2D antiplane problem TB (Lysmer and Waas, 1972) (hereinafter referred to as SH-TB) is used.

At first, the outline of this model is explained. Next, a component of the model, the SH-TB, which has not been studied using time domain transform, is studied.

Then, the characteristics of soil impedance and input motion using 3D-TB are studied. Finally, time history response analysis of the 3D soil–structure interaction system using proposed 3D-TB model is conducted, and the effectiveness of the model is evaluated.

The VB, which is currently thought to be the most practical method for time domain analysis, is used for comparison in this study. Furthermore, it is known that accuracy is improved if the excavation force (EF) is applied to VB. EF is a correction force vector calculated as the product of free field soil displacements and frequency-independent stiffness matrix (refer to Supplementary Material). In order to further clarify the practical applicability of the proposed method, VB with EF is also compared in this study.

OUTLINE OF THE PROPOSED ANALYSIS METHOD

The TB is a highly accurate boundary model located at the outer side of the inner soil model, which is formed by parallel layers on the rigid bedrock. In a horizontal direction, the formulation is theoretical and rigorous. In a vertical direction, the formulation is approximate since it follows the element displacement assumption. The TB is able to almost completely absorb wave

TABLE 1 | Shape of inner field and type of TB.

Shape of inner field	Axisymmetric Shape	Rectangular Solid
Coordinate system	Axisymmetric	Orthogonal
Transmitting boundary	Theoretical method	Approximate method (proposed method)

motion from an arbitrary direction. Even when the bottom of soil is semi-infinite condition, a favorable evaluation is possible by adding a sufficient amount of elements to the soil bottom, in the frequency domain.

In this paper, a time domain 3D-TB model, which corresponds to orthogonal coordinate system and uses SV-TB and SH-TB approximately, is proposed. An outline of this is described hereinafter.

Outline of the 3D-TB

The image of an inner field model is shown in **Figure 2**. In the figure, a vertical nodal group (hereinafter referred to as a "nodal line") is considered on the boundary surface. This is placed as a

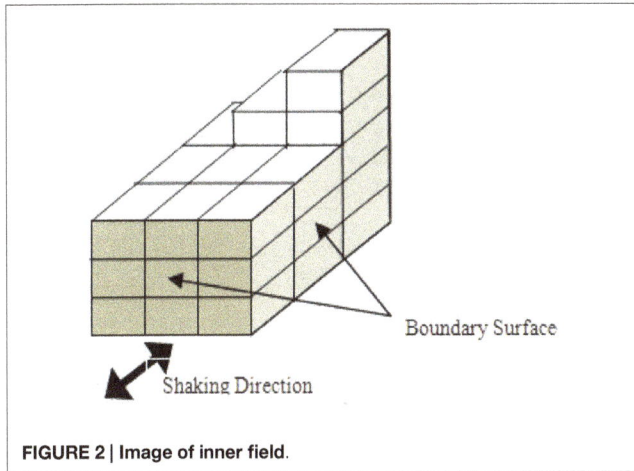

FIGURE 2 | Image of inner field.

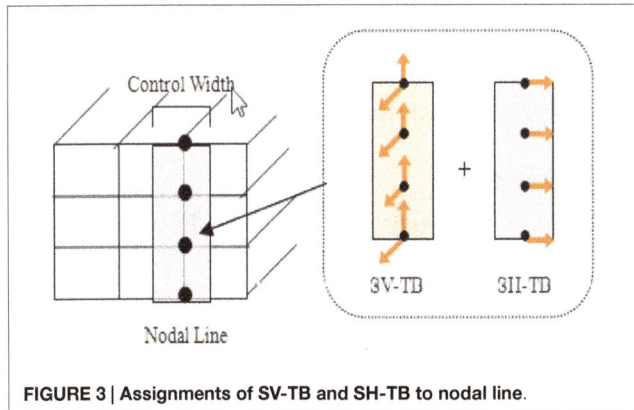

FIGURE 3 | Assignments of SV-TB and SH-TB to nodal line.

basic unit to form the boundary model. The boundary surface is expressed as a collection of these nodal lines.

The control width of one nodal line extends to the center of the adjacent nodal lines. Both SV-TB and SH-TB are assigned in this nodal line (refer to **Figure 3**). Therefore, the degree of freedom within a nodal line is coupled, but the degree of freedom with the other nodal lines is not coupled. Theoretically, all nodal lines should be coupled with each other, but in the proposed model the efficiency of the calculation is improved by disregarding this.

Furthermore, if the soil properties are the same, each nodal line becomes a TB with identical properties, and only the control width is different. For this reason, a TB with a unit width nodal line that corresponds to the type of the soil properties is prepared. This is multiplied by the control width and assigned into the entire boundary surface.

The analysis flow is shown in **Figure 4**. The SH-TB and SV-TB are calculated in the frequency domain. These TB matrices are transformed to the time domains and assigned to the overall equation of motion.

Transform of TB Matrices to Time Domain

The reaction force from TB has to be calculated in the time domain. The calculation is not easy, because the components of the TB matrix are strongly frequency dependent. In this section, the concept of the transform of TB to the time domain and the obtained reaction force in the time domain are briefly explained, using a simple single DOF equation.

Although many methods to transform frequency dependent impedance function to the time domain have been proposed, most of them employed either the past displacement or the past velocity in the formulation of the impulse response. The author proposed transform methods using both the past displacement and velocity, then he confirmed that the accuracy of these methods is high (Nakamura, 2007, 2012a).

In this paper, the following methods were used for the transform. Here, Eq. 1 in the frequency domain is considered. $Y(\omega)$ is the reaction force, $H(\omega)$ is the frequency-dependent function (this corresponds to TB), and $x(\omega)$ is the displacement. The objective is to obtain the reaction force in the time domain $y(t)$. In the proposed methods, $Y(\omega)$ and $H(\omega)$ are approximated by $Y'_B(\omega)$ and $H'_B(\omega)$ as shown in Eq. 2. This equation is expressed as Eq. 3 in the time domain, where $y'_B(t)$ and $x(t)$ are the reaction force and the displacement in the time domain, respectively.

$$Y(\omega) = H(\omega) \cdot x(\omega) \tag{1}$$

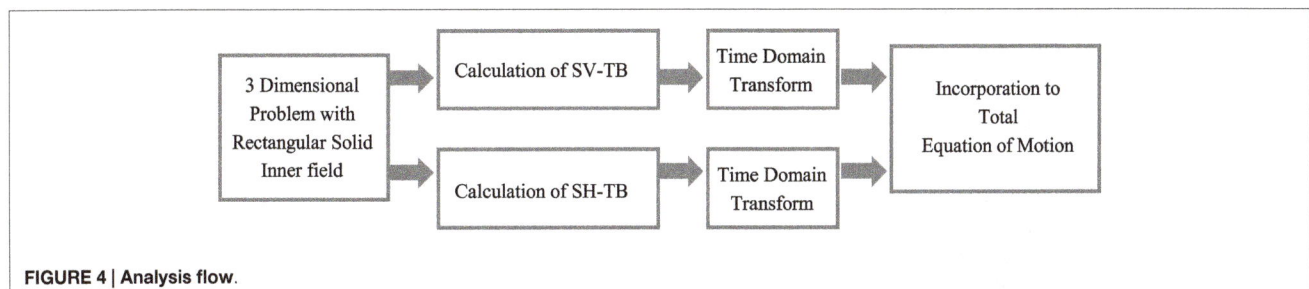

FIGURE 4 | Analysis flow.

$$Y'_B(\omega) = H'_B(\omega) \cdot x(\omega)$$

$$= \left(\sum_{j=0}^{n'} {}_0h_j \cdot e^{-i\omega t_j} + i\omega \cdot \sum_{j=0}^{n'-1} {}_1h_j \cdot e^{-i\omega t_j} - \omega^2 \cdot {}_2h_0 \right) \cdot x(\omega) \quad (2)$$

$$y'_B(t) = \sum_{j=0}^{n'} {}_0h_j \cdot x(t-t_j) + \sum_{j=0}^{n'-1} {}_1h_j \cdot \dot{x}(t-t_j) + {}_2h_0 \cdot \ddot{x}(t) \quad (3)$$

$t_j = j \, \Delta t$ where Δt is the discrete time interval for the transform. ${}_0h_j$, ${}_1h_j$, and ${}_2h_0$ are the coefficients of the impulse response. ${}_0h_0$, ${}_1h_0$, and ${}_2h_0$ are called the simultaneous components, because they correspond to the current time t. ${}_1h_1 \sim {}_1h_{n'}$ and ${}_0h_1 \sim {}_0h_{n'}$ are called the time-delay components, since they correspond to the past time $(t - t_j)$. All of the unknown coefficients of the impulse response are obtained by simultaneous equations with given function data for $H(\omega i)$ ($i = 0, 1, 2, \ldots, N$). This method is called as method B'.

In the case when the hysteretic damping is large, the accuracy of the transform tends to decrease. To improve this problem, the simultaneous components (${}_2h_0$, ${}_1h_0$, and ${}_0h_0$) are corrected with (Δ_2h_0, Δ_1h_0, and Δ_0h_0), where Δ_2h_0, Δ_1h_0, and Δ_0h_0 indicate the modification terms determined by the least square method. The improved reaction force [$Y'_C(t)$ and $y'_C(\omega)$] can be expressed using Eqs 4 and 5. This method is called as method C'.

$$Y'_C(\omega) = (H'_B(\omega) - \omega^2 \cdot \Delta_2h_0 + i\omega \cdot \Delta_1h_0 + \Delta_0h_0) \cdot x(\omega) \quad (4)$$

$$Y'_C(t) = Y'_B(t) + (\Delta_0h_0 \cdot x(t) + \Delta_1h_0 \cdot \dot{x}(t) + \Delta_2h_0 \cdot \ddot{x}(t)) \quad (5)$$

Using Eqs 4 and 5, all the components of $[T_B]$ can be transformed to the time domain. The details of the transform are shown in Nakamura (2007, 2012a).

Remarks for Application

With the method proposed in this paper, the nodal lines are mutually discontinuous as above. Therefore, it is thought that accuracy will decrease when the neighboring free field soil conditions differ greatly. Furthermore, it is necessary to calculate the SV-TB and SH-TB of the nodal line for each type of soil properties. Thus, when there are many types of soil properties, the calculation time increases, and the analysis becomes less efficient. For this reason, it is thought that the proposed method is effective when the types of soil properties are not so many.

STUDY OF SH-TB

At first, the properties and applicability of the time domain SH-TB, which is a component of the 3D-TB model, are verified in preparation for analysis of this model. The applicability and accuracy of SV-TB was already confirmed in Nakamura (2009, 2012a).

Analysis Conditions

The analysis model is shown in **Figure 5**. The soil is multilayered with the shear velocity V_s in the range 200–400 m/s, on the bedrock with $V_s = 500$ m/s. A height difference of 10 m is set at one side of the soil (only left side). The characteristics of the bedrock are evaluated using the bottom VB in the inner field.

The building is represented by a lumped mass model with shear elements. Its width is 20 m, the height of above-ground part is 24 m, and the height of the underground part is 10 m. The causal hysteretic damping model (Nakamura, 2007) is used. The damping ratio is set to be 3% for the building and 2% for the soil. The material properties of the soil and the building are shown in **Tables 2** and **3**, respectively. In this paper, the soil properties and building materials of all analysis models are assumed to be stayed in the linear initial condition, because that can express the differences of the wave boundaries clearly.

Three analysis models, with the boundary at a distance of $L = 5$, 40, and 100 m from the building outer edge, were studied.

For estimating the semi-infinity of the bottom soil, the elements for 100 m height of the material properties of the bedrock were added to the lowest part of the soil model in the calculation

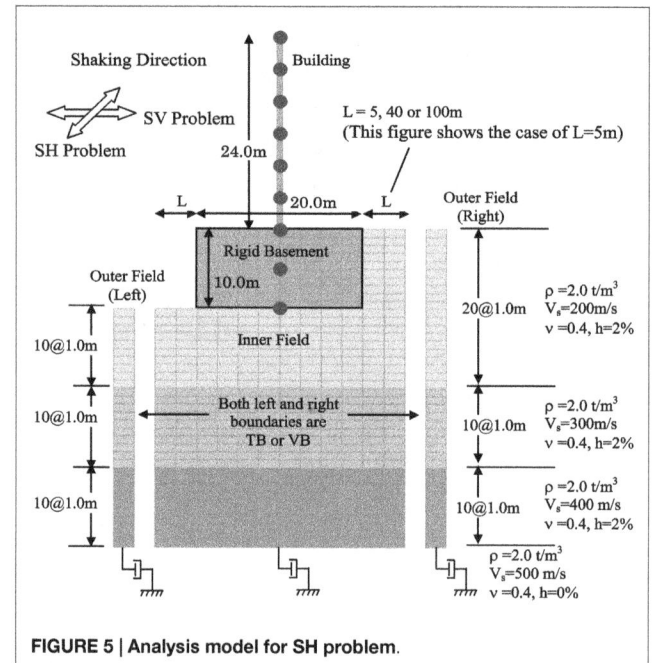

FIGURE 5 | Analysis model for SH problem.

TABLE 2 | Property of soil.

	V_s (m/s)	Poisson ratio ν	Density ρ (t/m³)	Damping ratio h	Thickness (m)
Surface 1	200	0.4	2.0	0.02	20
2	300				10
3	400				10
Bedrock	500			0	–

TABLE 3 | Property of building.

Story	Height (m)	Weight (t)	Rotational inertia (×10⁵ t/m²)	Shear stiffness (×10⁶ kN/m)
6	4.0	480	0	0.4935
5	4.0	480	0	0.9047
4	4.0	480	0	1.234
3	4.0	480	0	1.480
2	4.0	480	0	1.645
1	4.0	480	0	1.727
B1	5.0	720	0	∞
B2	5.0	720	1.68	∞

of the TB matrix. The conditions for time domain transform for the TB matrix are shown in **Table 4**.

The input ground motion was El Centro 1940NS wave (duration of 10 s, time step ΔT of 0.01 s), with the maximum acceleration set to 500 Gal and defined as 2E (double the ascending wave) at the bottom VB. As the time integral method, Newmark-β method ($\beta = 1/4$) was used.

Analysis Results

Figure 6 shows a comparison of the maximum response values (acceleration, displacement, and shear force) for the above-ground part of the building obtained by frequency domain and time domain analyses using the SH-TB. In each figure, the response results for each case are almost identical, and the results vary very slightly in accordance with L. **Figure 7** shows the results of the time domain analysis using a VB. For comparison, the results of the TB for $L = 100$ m are also shown in the figure. The response results for each case are almost identical as are the results for the TB.

In the study of the SV problem (Nakamura, 2009, 2012b), the superiority of the TB is clearly displayed. However, it can be said that the difference between the TB and VB is slight in the case of the SH problem.

Incidentally, when input ground motion from a vertically downward direction is assumed for the SH problem, as it is in this analysis, EF is not required in the calculation.

SOIL IMPEDANCE AND INPUT MOTION OF THE 3D-TB

The characteristics of soil impedance and input motion were evaluated in order to study the efficiency of the proposed 3D-TB. The same study was conducted with a VB as a target of comparison.

TABLE 4 | Conditions for transform into the time domain.

Impedance			Impulse response	
No. of data	Frequencies of data (Hz)	Δt (s)	Simultaneous components	Time delay components
21	0.1, 1.0, 2.0, 3.0, ..., 19.0, 20.0	0.05	k_0, c_0, m_0	$k_1 \sim k_{20}, c_1 \sim c_{19}$

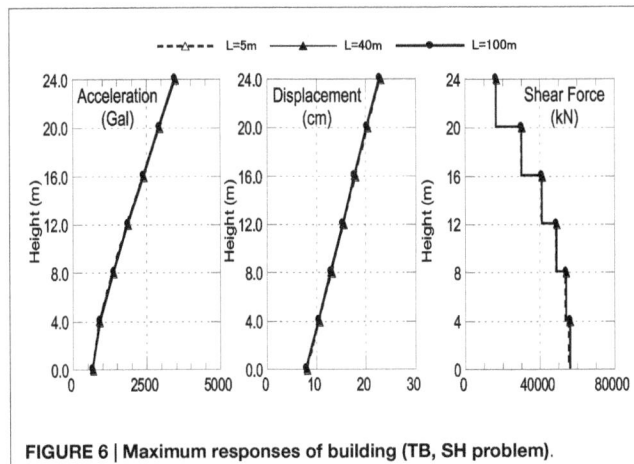

FIGURE 6 | Maximum responses of building (TB, SH problem).

Analysis Conditions

The massless rigid foundation embedded in the multilayered soil is studied. The FEM analysis model is shown in **Figure 8**. This model is the soil model in the previous section transformed to 3D, and each dimension is the same. However, in this model, there is not a height difference at both sides of the soil, and the foundation is entirely embedded. Similarly to the previous section, two types of side boundary, 3D-TB and VB, were studied in the time domain at distances L of 5, 40, and 100 m between the outer edge of the foundation and the boundary.

Study of Soil Impedance

Impulse excitation was performed for the massless rigid foundation in order to calculate the time history wave for the displacement of the foundation. The impulse excitation consists of many different and constant frequency components (it is so-called as the white noise), so it is very convenient to study the frequency

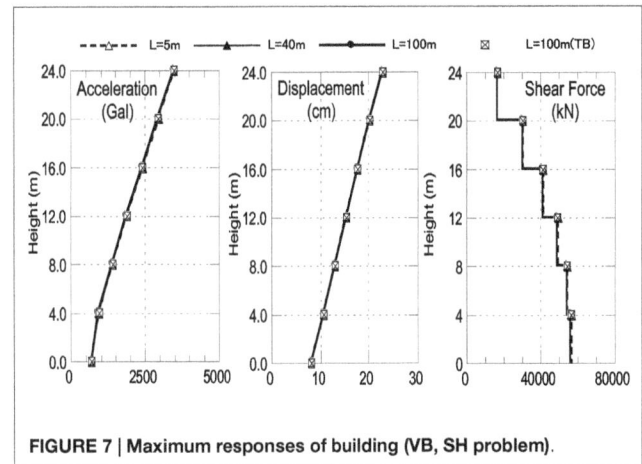

FIGURE 7 | Maximum responses of building (VB, SH problem).

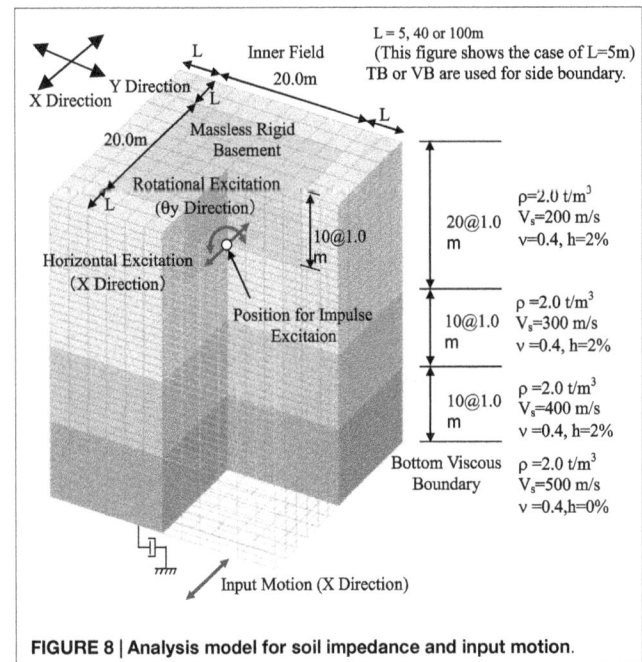

FIGURE 8 | Analysis model for soil impedance and input motion.

dependency of the given function in the time domain. The time integral method was the same as that described in the previous section. The impulse excitation time history wave and the foundation displacement time history wave were transformed using Fourier transformation, and the divisions for frequency domain were performed to calculate impedance. Two components of impedance – horizontal (K_x) and rotational ($K_{\theta y}$) – were studied. The thin-layer element method (TLEM) (Tajimi, 1980) is used as a target for comparison.

Figure 9 shows a comparison between TLEM results and the horizontal components of a soil impedance obtained using 3D-TB and VB. 3D-TB results in **Figures 9A,B** correspond favorably with TLEM totally, while slight fluttering can be seen when $L = 5$ and 40 m in the real part. In contrast, VB results in **Figures 9C,D** indicate that the difference with TLEM increases in the case of $L = 5$ m. The other cases generally correspond with TLEM, but the fluctuation in values for both the real part and imaginary part becomes greater in the vicinity of 0 Hz. This is thought to be because the bottom of the model is also VB, and therefore reaction force for excitation similar to static loading cannot be obtained.

For the rotational component in **Figure 10**, the tendency is almost the same. 3D-TB corresponds favorably with TLEM in all cases. The difference between VB and TLEM is large when $L = 5$ m, but in all other cases, VB generally corresponds favorably to TLEM.

Study of Input Motion

An impulse wave was applied as the input ground motion from the bottom of the model, and time history response analysis is conducted. The acceleration response wave is calculated at the center of the massless rigid foundation at soil surface level. The acceleration response and the time history wave of the impulse input motion are transformed to the frequency domain by Fourier transform, and divisions are performed to calculate the transfer function of the input motion. Two studies were conducted for VB, one when EF is applied and one when EF is not considered.

The analysis results are shown in **Figure 11**. For the 3D-TB in **Figure 11A**, the results for all cases of L are almost identical. In the case of VB without EF in **Figure 11B**, on the other hand, differences exist between each L case. In the results for $L = 100$ m, fluttering that was not apparent in **Figure 11A** can also be seen in the frequency range of 3–8 Hz. **Figure 11C** shows the results of the case of VB with FE. In all cases except for $L = 5$ m, the results of VB became almost identical with 3D-TB results. Thus, it is considered that the accuracy of VB is improved by EF.

Summary of Impedance and Input Motion

The accuracy of soil impedance and input motion was studied for the massless rigid foundation in the multilayered soil. The results when using 3D-TB were as follows.

-   For soil impedance, the results corresponded favorably with the analytical solution (TLEM) generally. Although there was slight fluttering in the case of $L = 5$ and 40 m, the results were favorable totally.
-   All cases are almost identical for input motion.

In contrast, the results when using VB were as follows.

-   For soil impedance, there was a large difference in the case of $L = 5$ m. But for all other cases, the results generally corresponded favorably with the analytical solution.
-   In the case of input motion, the disparity in each case was large when EF was not applied, and in all cases, the results

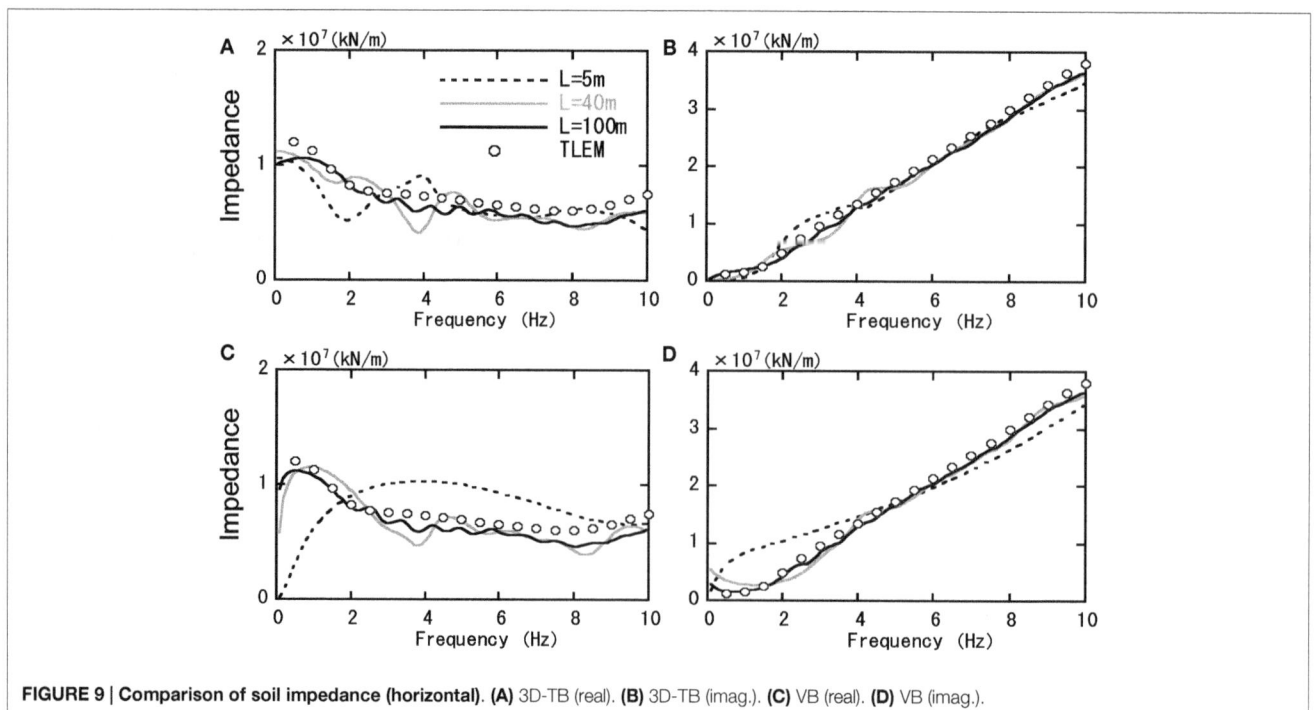

FIGURE 9 | Comparison of soil impedance (horizontal). (A) 3D-TB (real). **(B)** 3D-TB (imag.). **(C)** VB (real). **(D)** VB (imag.).

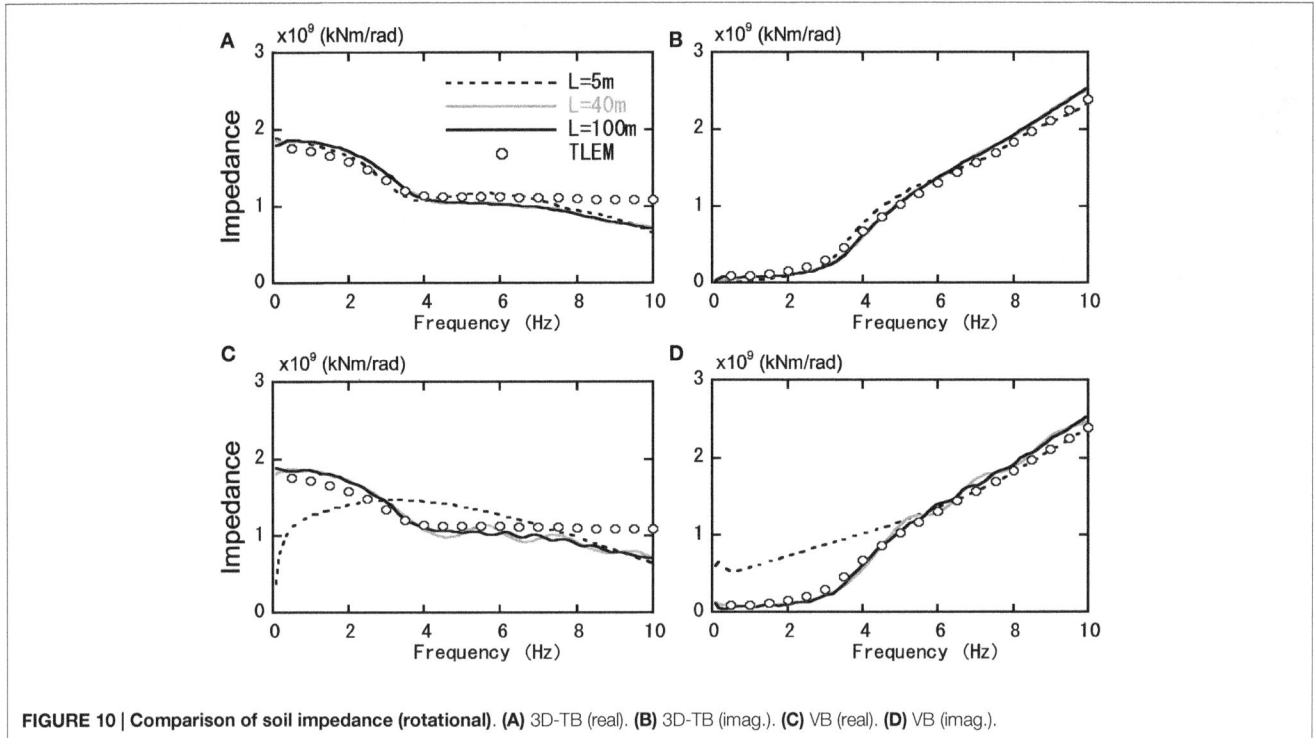

FIGURE 10 | Comparison of soil impedance (rotational). (A) 3D-TB (real). (B) 3D-TB (imag.). (C) VB (real). (D) VB (imag.).

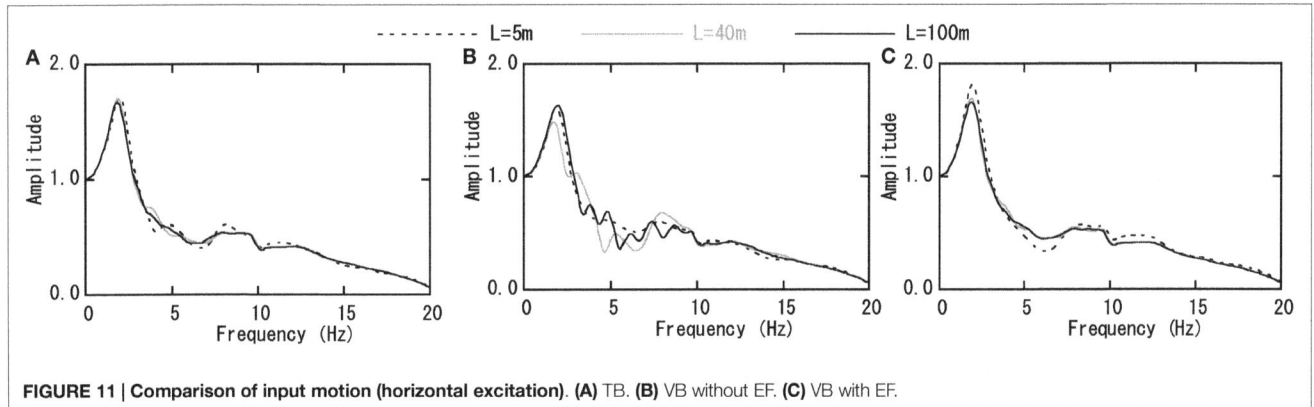

FIGURE 11 | Comparison of input motion (horizontal excitation). (A) TB. (B) VB without EF. (C) VB with EF.

differed from the results for the TB. Accuracy improved when EF was applied, and the results corresponded with the results for the TB in all cases except for $L = 5$ m.

STUDY OF TIME HISTORY ANALYSIS USING 3D-TB

Time history seismic analysis of the soil and structure interaction system is conducted using the proposed 3D-TB, and the accuracy and the efficiency of the method are studied.

Analysis Conditions

The analysis model is shown in **Figure 12**. This model is the model from **Figure 4** transformed to 3D. Therefore, the soil and building properties are the same as in Section "Study of SH-TB." The input ground motion conditions and the time integration method are the same as in Section "Study of SH-TB," but in this Section "Outline of the Proposed Analysis Method," types of excitation are studied, excitation in the X and Y direction (hereafter referred as "X excitation" and "Y excitation", respectively). The distance L from the outer edge of the building to the boundary is set as 5, 10, 20, 40, 60, 80, or 100 m, and the study conducted.

Comparison of Responses for the Soil Near the Building

The maximum response values (acceleration and displacement) for soil near the building when 3D-TB is used are shown in **Figure 13A**. Responses were compared for three cases, $L = 5$, 40, and 100 m. Although the maximum acceleration values of $L = 5$ m are slightly different to the other cases in Y excitation, the all values generally are almost identical in all cases for both X and Y excitations.

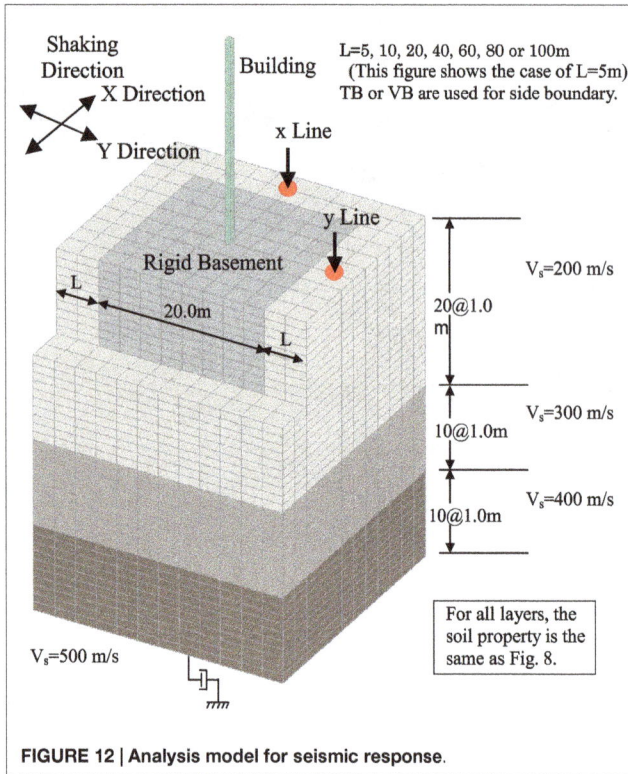

FIGURE 12 | Analysis model for seismic response.

The maximum response values using VB without EF are shown in **Figure 13B**. The results for the 3D-TB at $L = 100$ m, which are thought to be the most accurate among all cases (hereafter referred as "the high-accuracy values"), are also shown in this figure.

The results for VB at $L = 100$ m almost correspond with the high-accuracy values. It can be ascertained from this that even with a VB, good accuracy can be achieved if a sufficiently large L is applied. On the other hand, the difference from the high-accuracy values becomes greater in the results for $L = 5$ and 40 m.

The results when VB with EF is used are shown in **Figure 13C**. Overall, the accuracy is improved compared to **Figure 13B**, and the values in the case of $L = 40$ m correspond favorably with the high-accuracy values. In a contrast, there is a large disparity in the case of $L = 5$ m. This is thought to be due to the effect of the difference in soil impedance in the previous section.

Comparison of Horizontal Response Values of the Building

Figure 14A shows the horizontal maximum response values (acceleration, displacement, and shear force) for the above-ground part of the building when 3D-TB is used. The same three cases as in the previous section, $L = 5$, 40, and 100 m, were compared. Although there are slight differences in some parts of acceleration values and shear force values between the case when $L = 5$ m, and the other cases for Y excitation, generally the results for all cases correspond favorably for both X and Y excitations.

Table 5a shows the ratios of these maximum values corresponding to the high-accuracy values. Black field in the table indicates that the maximum difference exceeds 20%, and gray

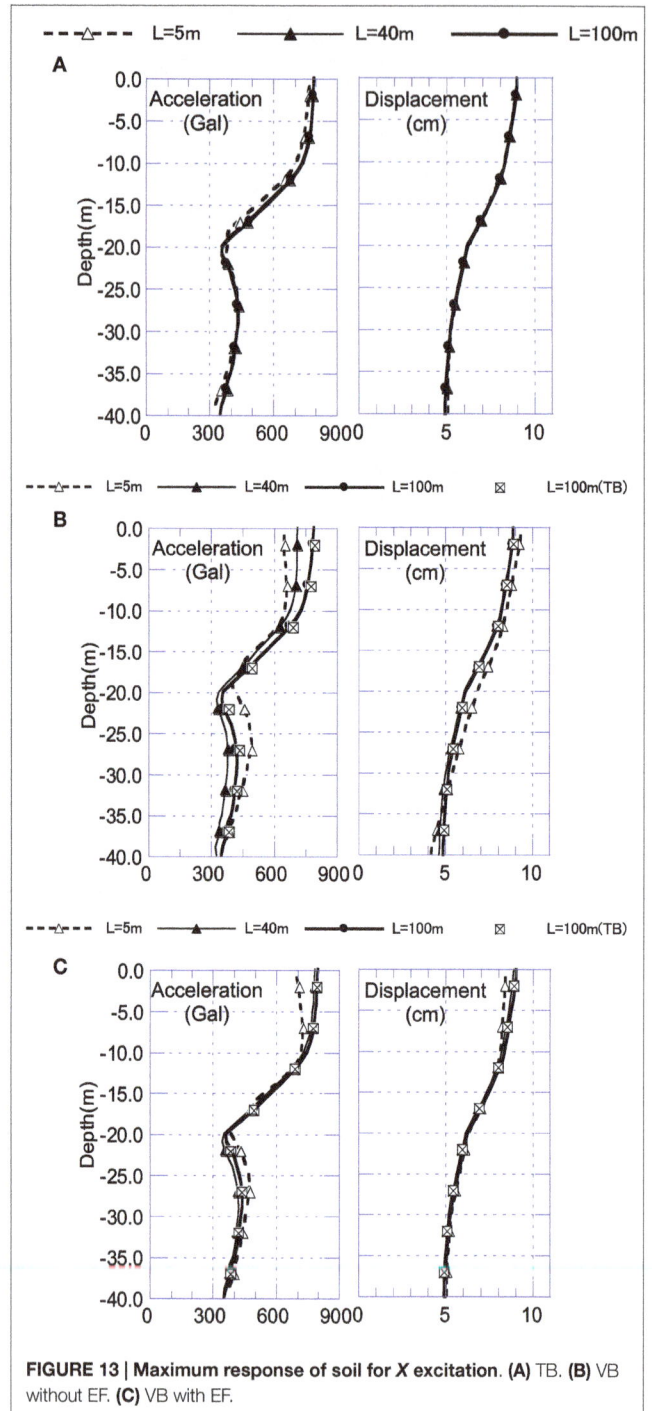

FIGURE 13 | Maximum response of soil for *X* excitation. (A) TB. (B) VB without EF. (C) VB with EF.

field indicates that the maximum difference is from 10 to 20%. When the 3D-TB was used for analysis, the maximum difference for all response values was <10%, and favorable response results could be obtained even in the case of $L = 5$ m.

The maximum response values for the above-ground part of the building when VB without EF is used are shown in **Figure 14B**. The results for VB at $L = 100$ m almost correspond with the high-accuracy values, but the results at $L = 5$ and 40 m are different significantly from the high-accuracy values. In

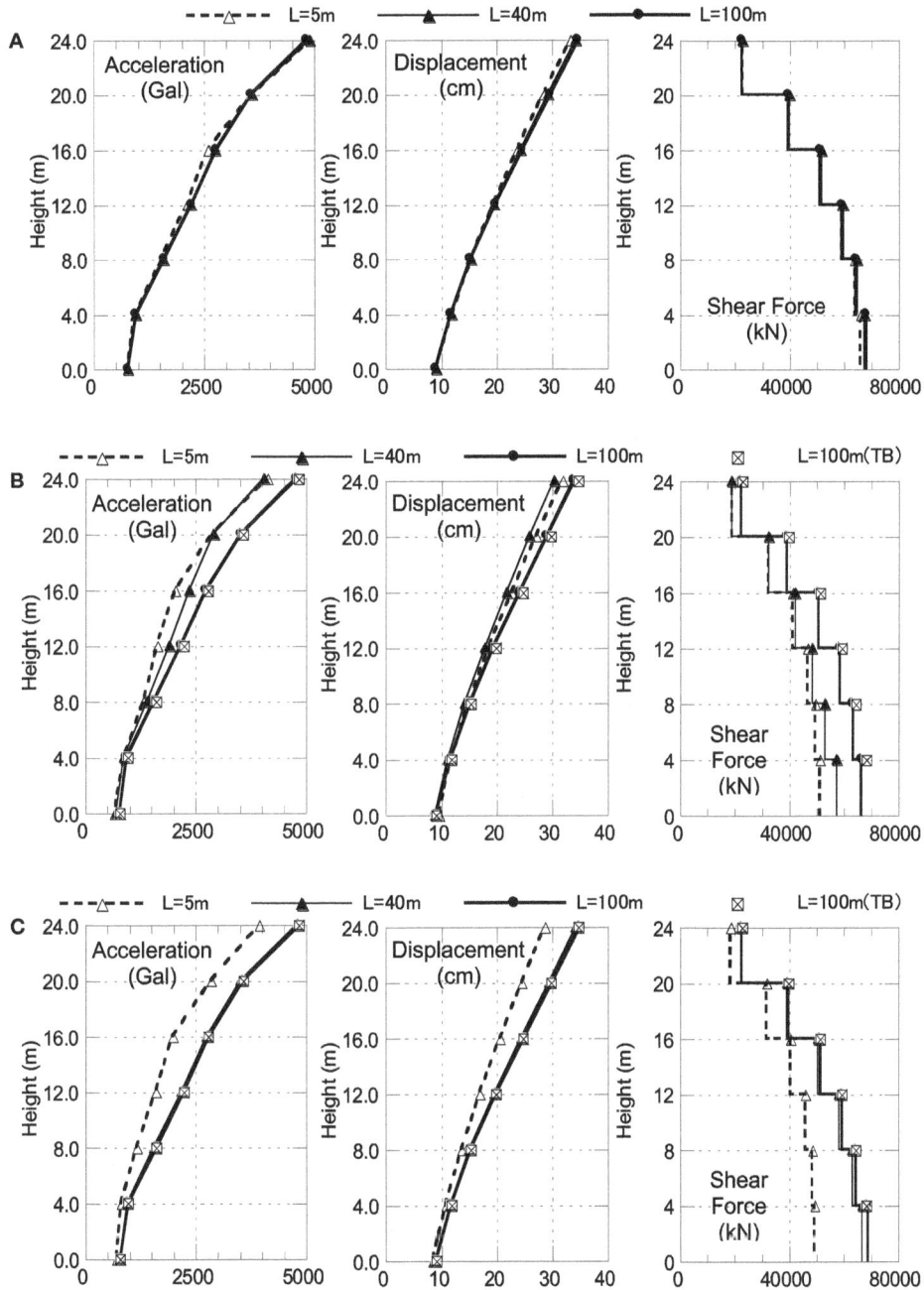

FIGURE 14 | Maximum response of building for *X* excitation. (A) TB. **(B)** VB without EF. **(C)** VB with EF.

Table 5b, the cases of $L = 60$ and 80 m are also included for comparison, in addition to the above cases. When using VB, the differences exceed 20% in some fields in the case of $L = 5$ and 40 m. Even at $L = 60$ m, the differences in half the fields exceeded 10%. The cases of $L = \geq 80$ m, differences of all values are <10%. The results when using VB with EF are shown in **Figure 14C**. Although the differences are large at $L = 5$ m, the accuracy is favorable at $L = 40$ m. In **Table 5c**, the cases of $L = 10$ and 20 m are also included for comparison. In the cases of $L = \geq 20$ m, all differences are <10%.

Figure 15 shows the transfer function of the response acceleration at the top node of the building for the input ground motion. When 3D-TB is used, shown in **Figure 15A**, the results for the values of both $L = 5$ and 40 m corresponded favorably with those of $L = 100$ m. When VB without EF was used, there is a significant difference in terms of the peak height between the cases of $L = 5$ and 100 m, as shown in **Figure 15B**. Between the cases of $L = 40$ and 100 m, the peak height and positions corresponded to each other, but a difference can be seen at 1.9–2.5 Hz. When the EF was applied to VB, as shown in **Figure 15C**, the accuracy for $L = 40$ m improved.

However, the accuracy for $L = 5$ m remained poor, as shown in **Figure 15B**. These results correspond to the tendency shown in **Figures 14A–C**.

Summary of Response Behavior

The above tendency is consistent with the results in Section "Soil Impedance and Input Motion of the 3D-TB." Thus, the following can be concluded.

- When the 3D-TB is used, response accuracy is favorable even when L is small. This is thought to be because the accuracy for both soil impedance and input motion is high. The horizontal response accuracy was favorable (the difference is <10%) at $L = 5$ m (1/4 of the building width).
- When VB without EF is used, the accuracy of the response results is low at $L = 5$ and 40 m. This is thought to be because the accuracy of both soil impedance and input motion are low at $L = 5$ m, and the accuracy of input motion is low at $L = 40$ m. The horizontal response accuracy was favorable at $L = 80$ m (four times the building width).
- When VB with EF is used, the response values at $L = 40$ m become favorable. This is thought to be because the accuracy of the input motion is improved due to the application of EF. On the other hand, the accuracy at $L = 5$ m remained low. This is thought to be because of the low accuracy of the soil impedance. The horizontal response accuracy was favorable at $L = 20$ m (one time the building width).

Study of Analysis Load

Table 6 provides a comparison of the analysis loads for the cases that provided favorable horizontal response results for the building in **Table 5**, the case of $L = 5$ m of 3D-TB, the case of $L = 80$ m of VB without FE, and the case of $L = 20$ m of VB with EF. Model shapes of $L = 5$, 20, and 80 m are shown in **Figure 16**.

As for the analysis load, the required memory size and the analysis time during the calculation were counted using a single core Xeon7560 (2.26 GHz) processor. This processing unit has 256 GB of main memory space, and the calculations for all cases were conducted within the main memory. Furthermore, the 3D-TB calculation time in the frequency domain and the time domain transform time (total for both for the SV problem and the SH problem is 1.2 min) are included in the 3D-TB analysis time.

Compared to VB (without EF) case, the 3D-TB case has around 1/30 of the number of inner field nodal points and elements. It is also ~1/13 of the memory and analysis time. Furthermore, this required memory and analysis time is reduced to approximately half that required in the case of VB with EF applied.

CONCLUSION

In this paper, an approximate time domain TB that can be used with a 3D orthogonal coordinate system was studied. First, 3D-TB with high calculation efficiency that can be applied in a

TABLE 5 | Comparison of maximum response of building.

Case	Excitation	Acceleration	Displacement	Shear force
(a) TB				
$L = 5$ m	X	0.93–1.01	0.95–1.01	0.97–1.01
	Y	0.90–0.95	0.94–0.99	0.91–0.95
$L = 40$ m	X	0.99–1.01	0.99–1.01	0.99–1.01
	Y	0.96–0.99	0.97–1.00	0.97–0.97
(b) VB without EF				
$L = 5$ m	X	0.73–0.91	0.91–1.06	0.75–0.84
	Y	0.73–0.89	0.97–1.07	0.84–0.88
$L = 40$ m	X	0.82–0.91	0.87–0.99	0.82–0.84
	Y	0.79–0.92	0.87–0.98	0.79–0.84
$L = 60$ m	X	0.89–0.92	0.91–0.99	0.90–0.92
	Y	0.89–0.91	0.91–0.98	0.88–0.90
$L = 80$ m	X	0.95–0.99	0.95–0.99	0.96–0.99
	Y	0.96–0.98	0.96–0.98	0.96–0.97
(c) VB with EF				
$L = 5$ m	X	0.71–0.90	0.82–0.94	0.73–0.82
	Y	0.74–0.93	0.87–0.96	0.81–0.86
$L = 10$ m	X	0.83–0.96	0.89–0.96	0.86–0.93
	Y	0.83–0.96	0.93–0.98	0.92–0.95
$L = 20$ m	X	0.94–0.98	0.94–0.99	0.95–0.98
	Y	0.95–0.98	0.95–0.99	0.96–0.98
$L = 40$ m	X	0.98–0.99	0.99–0.99	0.98–0.99
	Y	0.96–1.00	0.98–0.99	0.96–0.98

Values in this table show the range of maximum responses (ratios to the response of TB, L = 100 m). The color of each field shows the maximum difference (black: >20%, gray: between 10 and 20%, and white: ≤10%).

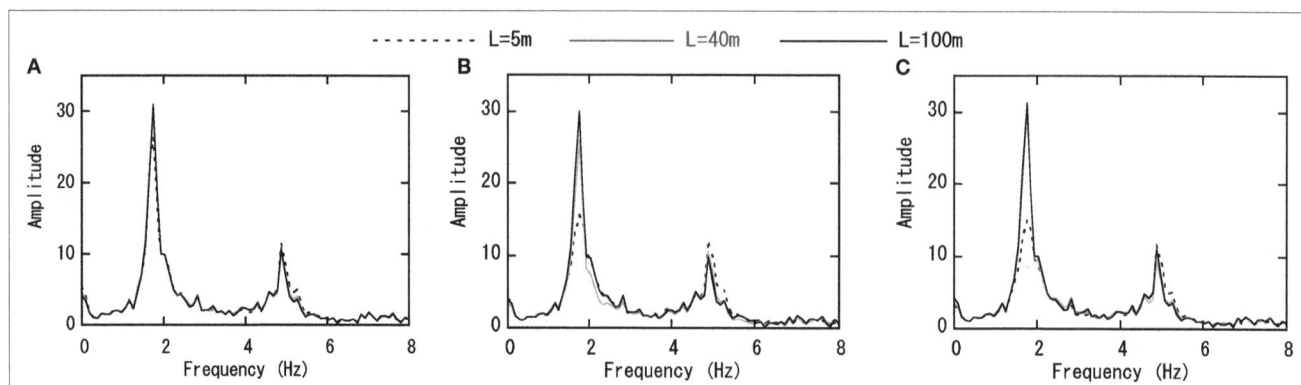

FIGURE 15 | Transfer function (building top node/input motion) for X excitation. (A) TB. **(B)** VB without EF. **(C)** VB with EF.

TABLE 6 | Comparison of analysis load (cases whose differences of building response are <10%).

	L (m)	No. of node	No. of elem.	Required memory (GB)	Analysis time (min)
TB	5	6,675	5,520	1.3	39
VB without EF	80 (16.0)	195,135 (29.2)	184,336 (33.4)	18.0 (13.8)	516 (13.2)
VB with EF	20 (4.0)	23,631 (3.5)	21,136 (3.8)	2.4 (1.8)	71 (1.8)

The values in the parenthesis mean the magnification to TB.

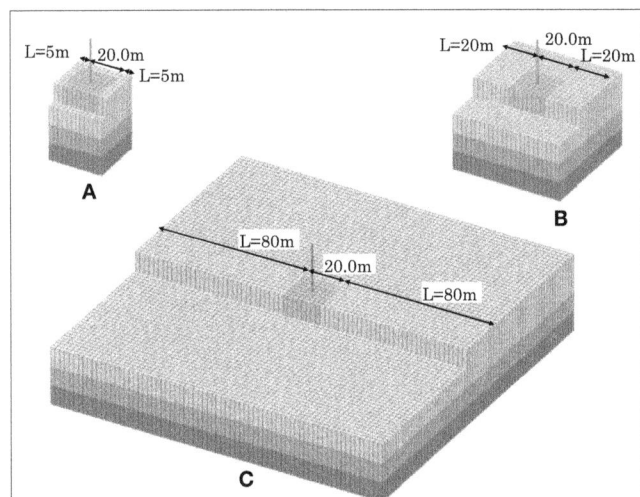

FIGURE 16 | Comparison of analysis model [(A) L = 5 m, (B) L = 20 m, and (C) L = 80 m].

rectangular analysis domain was explained. A nodal line on the boundary surface is considered to be a single unit, and the SV-TB and the SH-TB are assigned to it.

Next, the properties of the component, the SH-TB, were studied and verified for favorable accuracy. Then the impedance and input motion of the rigid foundation embedded in the multilayered soil were calculated using the 3D-TB, and favorable correspondence with the analysis solution was obtained.

Furthermore, seismic response analysis of the 3D problem was conducted using the proposed 3D-TB. From the aspect of the accuracy of horizontal response values, improvement effects were obtained at ~1/13 of the required memory and analysis time compared to VB without EF and approximately half of the required memory and analysis time compared to VB with EF. It can be concluded from this that the effectiveness of the proposed 3D-TB has been confirmed.

REFERENCES

Basu, U., and Chopra, A. K. (2004). Perfectly matched layers for transient elasto-dynamics of unbounded domains. *Int. J. Numer. Methods Eng.* 59, 1039–1074. doi:10.1002/nme.896

Berenger, J. P. (1994). A perfectly matched layer for the absorption of electromagnetic wave. *J. Comput. Phys.* 114, 185–200. doi:10.1006/jcph.1994.1159

Fattah, M. Y., Schanz, T., and Dawood, S. H. (2012). The role of transmitting boundaries in modeling dynamic soil-structure interaction problems. *Int. J. Eng. Technol.* 2, 236–258.

Ghergu, M., and Ionescu, I. R. (2009). Structure-soil-structure coupling in seismic excitation and city effect. *Int. J. Eng. Sci.* 47, 342–354. doi:10.1016/j.ijengsci.2008.11.005

Guidotti, R., Mazzieri, I., Stupazzini, M., and Dagna, P. (2012). "3D numerical simulation of the site-city interaction during the 22 February 2011 MW 6.2 Christchurch earthquake," in *15th World Conference of Earthquake Engineering*. Lisbon.

Hastings, F. D., Schneider, J. B., and Broschat, S. L. (1996). Application of the perfectly matched layer (PML) absorbing boundary condition to elastic wave propagation. *J. Acoust. Soc. Am.* 100, 3061–3069. doi:10.1121/1.417118

Kim, D. K., and Yun, C. B. (2000). Time domain soil-structure interaction analysis in two – dimensional medium based on analytical frequency – dependent infinite elements. *Eng. Struct.* 22, 258–271. doi:10.1016/S0141-0296(98)00070-4

Lou, M., Wang, H., Chen, X., and Zhai, Y. (2011). Structure-soil-structure interaction: literature review. *Soil Dyn. Earthquake Eng.* 31, 1724–1731. doi:10.1016/j.soildyn.2011.07.008

Lysmer, J., and Kuhlelameyer, R. L. (1969). Finite dynamic model for infinite area. *J. Eng. Mech. Div.* 95, 859–877.

Lysmer, J., Udaka, T., Seed, H. B., and Hwang, R.N. (1975a). *FLUSH A Computer Program for Approximate 3-D Analysis of Soil-Structure Interaction Problems.* Report No.EERC75-30. Berkeley, CA: University of California.

Lysmer, J., Udaka, T., Tsai, C.-F., and Seed, H. B. (1975b). *ALUSH A Computer Program for Seismic Response Analysis of Axisymmetric Soil-Structure Systems.* Report No.EERC75-31. Berkeley, CA: University of California.

Lysmer, J., and Waas, G. (1972). Shear wave in plane infinite structures. *J. Eng. Mech. Div.* 98, 85–105.

Millan, M. A., and Dominguez, J. (2009). Simplified BEM/FEM model for dynamic analysis of structures on piles and pile group in viscoelastic and poroelastic soils. *Eng. Anal. Bound. Elem.* 33, 25–34. doi:10.1016/j.enganabound.2008.04.003

Nakamura, N. (2006). Improved methods to transform frequency dependent complex stiffness to time domain. *Earthquake Eng. Struct. Dyn.* 35, 1037–1050. doi:10.1002/eqe.520

Nakamura, N. (2007). Practical causal hysteretic damping. *Earthquake Eng. Struct. Dyn.* 36, 597–617. doi:10.1002/eqe.644

Nakamura, N. (2009). Nonlinear response analysis of soil-structure interaction system using transformed energy transmitting boundary in the time domain. *Soil Dyn. Earthquake Eng.* 29, 799–808. doi:10.1016/j.soildyn.2008.08.004

Nakamura, N. (2012a). A basic study on the transform method of frequency dependent functions into time domain – relation to Duhamel's integral and time domain transfer function. *J. Eng. Mech.* 138, 276–285. doi:10.1061/(ASCE)EM.1943-7889.0000330

Nakamura, N. (2012b). Two-dimensional energy transmitting boundary in the time domain. *Int. J. Earthquakes Struct.* 3, 97–115. doi:10.12989/eas.2012.3.2.097

Padron, L. A., Aznarez, J. J., and Maeso, O. (2011). 3-D boundary element – finite element method for the dynamic analysis of piled buildings. *Eng. Anal. Bound. Elem.* 35, 465–477. doi:10.1016/j.enganabound.2010.09.006

Smith, W. (1973). A non-reflecting plane boundary for wave propagation problems. *J. Comput. Phys.* 15, 492–503. doi:10.1016/0021-9991(74)90075-8

Tajimi, H. (1980). "A contribution to theoretical prediction of dynamic stiffness of surface foundations," in *Proceeding of 7th World Conference on Earthquake Engineering* (Istanbul), 105–112.

Wolf, J. P. (2003). *The Scaled Boundary Finite Element Method*. West Sussex: John Wiley & Sons Ltd.

Wolf, J. P., and Song, C. (1996). *Finite-Element Modelling of Unbounded Media*. West Sussex: John Wiley & Sons Ltd.

Conflict of Interest Statement: The author declares that the research was conducted in the absence of any commercial or financial relationships that could be construed as a potential conflict of interest.

15

A computer-based environment for processing and selection of seismic ground motion records: OPENSIGNAL

Gian Paolo Cimellaro[1,2]* and Sebastiano Marasco[2]

[1] Department of Civil and Environmental Engineering, University of California Berkeley, Berkeley, CA, USA, [2] Department of Structural, Geotechnical and Building Engineering (DISEG), Politecnico di Torino, Turin, Italy

A new computer-based platform has been proposed whose novelty consists in modeling the local site effects of the ground motion propagation using a hybrid approach based on an *equivalent linear model*. The soil behavior is modeled assuming that both the shear modulus and the damping ratio vary with the shear strain amplitude. So, the hysteretic behavior of the soil is described using the shear modulus degradation and damping ratio curves. In addition, another originality of the proposed system architecture consists in the evaluation of the conditional mean spectrum on the entire Italian territory automatically, knowing the geographical coordinates. The computer-based platform based on signal processing has been developed using a modular programing approach, to enable the selection and the processing of earthquake ground motion records. The proposed computer-based platform combines in unified environment different features, such as (i) selection of ground motion records using both spectral and waveform matching, (ii) signal processing, (iii) response spectra analysis, and (iv) soil response analysis. The computer-based platform OPENSIGNAL is freely available for the general public at http://areeweb.polito.it/ricerca/ICRED/Software/OpenSignal.php.

Keywords: soil response analysis, ground motion selection, spectral matching, filtering, conditional mean spectrum

Edited by:
Solomon Tesfamariam,
The University of British Columbia,
Canada

Reviewed by:
Iolanda-Gabriela Craifaleanu,
Technical University of Civil
Engineering Bucharest, Romania
Flavia De Luca,
University of Bristol, UK

*Correspondence:
Gian Paolo Cimellaro,
Department of Civil and
Environmental Engineering,
University of California Berkeley,
Davis Hall, Berkeley,
CA 94720-1710, USA
gianpaolo.cimellaro@polito.it

Introduction

Nowadays, the state-of-practice in earthquake engineering design has progressively moved toward the use of dynamic non-linear time history analysis with respect to response spectrum analysis, because of the exponential increment of computational power. All these methods need as prerequisite the selection of a proper suite of earthquake ground motions to be reliable. In fact, among all possible sources of uncertainty (e.g., structural material properties, modeling approximations, design and analysis assumptions, etc.), the selection of earthquake ground motion has the highest effect on the variability of the structural response (Padgett and Desroches, 2007). The selection of earthquake records on most of seismic design codes are based on parameters obtained by disaggregated seismic hazard maps at a specific site, such as the magnitude, M, and the source-to-site distance, R, but other parameters can also be used, such as the soil type, the source mechanism, and the duration. Other parameters can also be used based on intensity measures, such as the peak ground acceleration, pga, the spectral acceleration at the fundamental period of the structure $S_a(T_1)$. Other selection criteria are based on spectral matching to a specific target spectrum, such as (1) a design code spectrum, (2) a seismic scenario determined from a ground motion prediction relationship (Campbell and Bozorgnia, 2008), (3) a uniform hazard spectrum (UHS), and (4) a conditional mean spectrum (CMS) (Baker, 2011). Using design code spectrum and UHS might bring to over-softening and over-damping during the analysis (Baker, 2011); therefore, a matching procedure based on the CMS has been developed and presented for the Italian territory in this paper. The local seismic response has been modeled using an *equivalent linear model*, assuming that both the shear modulus and the damping ratio vary with the shear strain amplitude. So, the hysteretic behavior of the soil is described using the shear modulus degradation and damping ratio curves.

A large number of computer programs, public and commercial, are available at the Observatories and Research Facilities for European Seismology (ORFEUS) data center[1], such as SMARTS 2.0, Shake-91, DIMAS, and PickEv 2000, and at the Pacific Earthquake Engineering research center[2], such as SIMQKE-I and SIMQKE-II. Most of existing public signal processing software are developed to analyze a single seismic earthquake record at a time (e.g., Seismosignal – available at http://seismosoft.com/ or view wave available at http://iisee.kenken.go.jp/staff/kashima/viewwave.html). For multiple records analysis, commercial software are needed, such as Bispec (Hachem, 2003), but they have the inconvenient that they are not freely available in the market. Other examples of data processing software are USDP (Akkar, 2008) and TSPP (Boore, 2009). However, most of these programs can be used after the earthquake records are selected, but they are not able to guide you through the ground motion selection process from a given database, for example, so they do not provide users with the ability to perform all of these functions in an integrated fashion. Iervolino et al. (2010) developed a Matlab-based software called REXEL, which allows ground motion selection using the Italian database, Itaca (Luzi et al., 2008); however, the proposed software is not able to perform multiple signal processing and it is not able to build the CMS on the Italian territory automatically, by using as input the GIS coordinates. Similar software have also been developed by Corigliano et al. (2012) who implement a software called ASCONA (Corigliano et al., 2012) for the selection of compatible natural ground motions. Recently, also Katsanos and Sextos (2013) developed a Matlab-based software environment integrating finite element analysis with earthquake records selection which works with the PEER database. However, signal processing, soil response analysis, and the possibility to use new target spectra, such as the Conditional Mean Spectra, is not included in the programs mentioned above.

Selecting Accelerograms in Engineering Analysis

The process which is usually followed by each designer in order to select "reliable" earthquake records is shown in **Figure 1** that also describes the structure of the computer platform which has

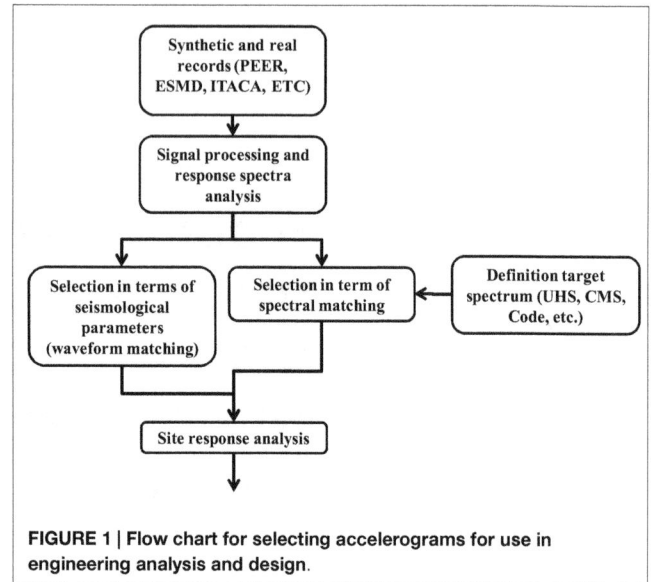

FIGURE 1 | Flow chart for selecting accelerograms for use in engineering analysis and design.

been developed. Each part of the computer environment has been implemented in MATLAB (2012) and has a graphical user interface that is simple and intuitive to be used.

Description of the Computer Platform

The advantage of the proposed platform is combining all the steps that are described in **Figure 1** together. It can read data a large variety of file formats from the most common ground motion databases, such as the PEER-NGA strong motion database (PEER)[3], the European Strong-Motion database (ESMD)[4], Chilean Database (UCHILE) and from ITalian ACcelerometric Archive (ITACA)[5], but it allows also reading manually seismic records selecting the *free format*. It is composed of several interactive graphical interfaces that integrate the most common signal processing and selection criteria techniques used in earthquake engineering. In the next paragraphs, each part of the platform is described in detail, where more attention is given to the spectral matching procedure and at the site response analysis.

Signal Processing and Filtering

Many ground motion parameters, such as peak displacements and velocities, are used often in different field of earthquake and geotechnical engineering (Boore and Bommer, 2005); however, their values are affected by the noise of the earthquake ground motion. The influence of noise in ground motion records is evident at low and high frequencies where the signal-to-noise ratio is usually lower compared to the mid spectrum. In particular, the effect of low frequencies noise (<1 Hz) on strong motion intensity parameters, such as ground velocities, displacements, and response spectra ordinates, is evident. Then, filtering operations

---

[1]http://www.orfeus-eu.org/software.html
[2]http://nisee.berkeley.edu/software/

[3]http://ngawest2.berkeley.edu/
[4]http://www.isesd.hi.is/ESD_Local/frameset.htm
[5]http://itaca.mi.ingv.it

became the primary tool for correcting the ground motion records and consequentially it has become standard practice to cut low and high frequencies by looking at the spectra of the Fourier amplitude spectra and the signal-to-noise ratio. The Butterworth filter is the most used in seismic applications and it is designed to have a flat frequency response in the pass-band range, while it is equal to 0 in the stop-band range. Analytically, the frequency response amplitude of the Butterworth filter is given by

$$|H(\omega)| = \frac{1}{\sqrt{1+\left(\dfrac{\omega}{\omega_c}\right)^{2N}}} \qquad (1)$$

where ω is the generic angular frequency, ω_c represents the cutoff frequency, and N is the order of the filter. This type of filter is often used for processing ground motion records, because, for example, the low frequencies can generate unrealistic soil permanent deformations. The source of this error as mentioned above is generated by the high and low frequency noise which contaminate the signal.

Baseline corrections can be applied using different techniques, in the *time domain* to remove unwanted trends and in the *frequency domain* to remove unwanted frequencies.

In the frequency domain, the noise is most easily removed by the use of a bandpass filter, like the *low-pass filter* which is set up to values <0.1. In addition, the effect of aliasing can be eliminated by filtering the original ground motion beyond the Nyquist frequency (12.5 Hz). In most of the databases, ground motion records are already filtered, but there might be cases in which some records are unfiltered and in that case, the filtering option in the environmental platform can be used.

Figure 2 shows the main user dialog window of the computer environment. The number identifies the steps to be followed in order to obtain a correct processing procedure and they are summarized below:

1. Import the record with "*Open*" or insert the name.dat of the file.
2. Select the correct database.

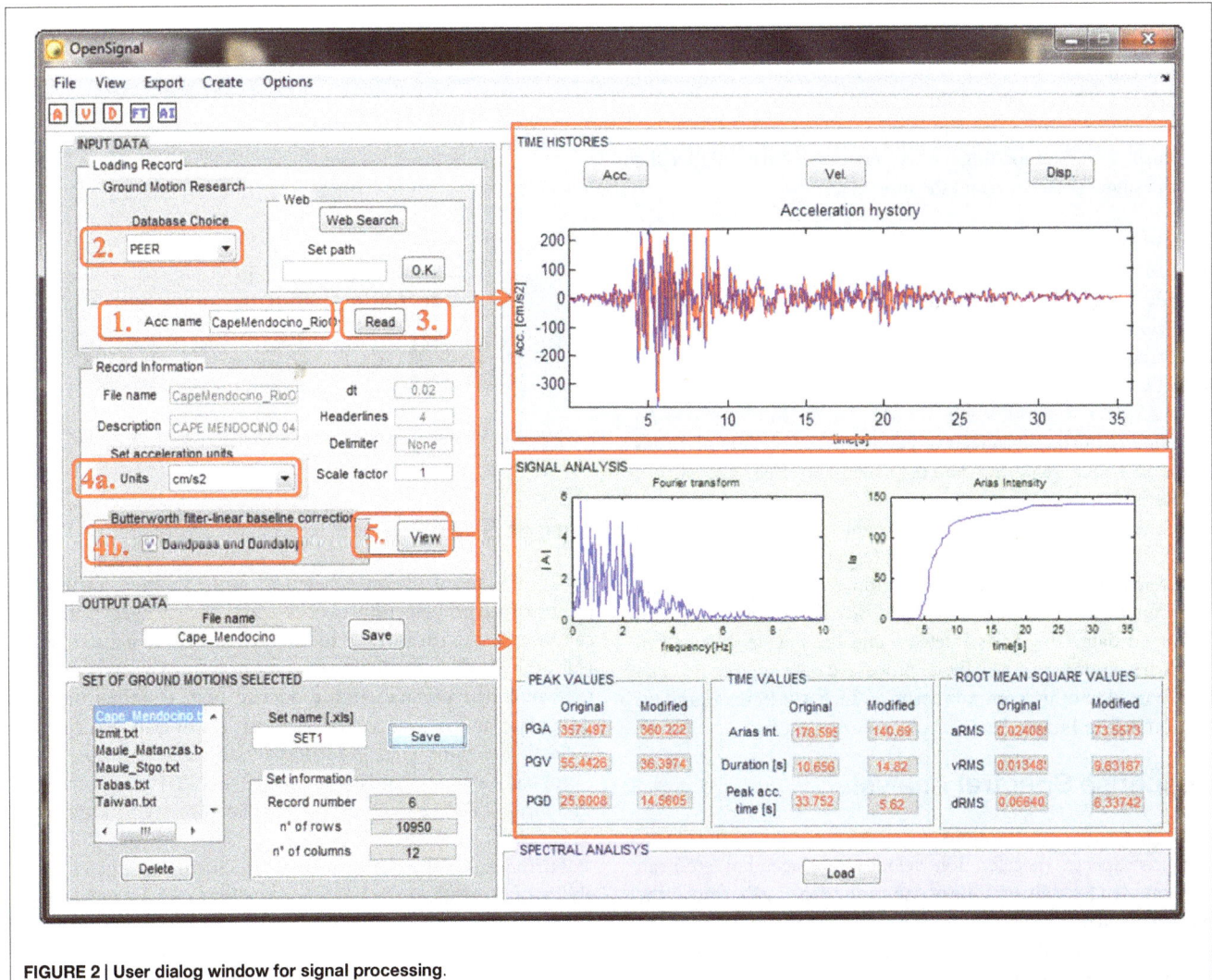

FIGURE 2 | User dialog window for signal processing.

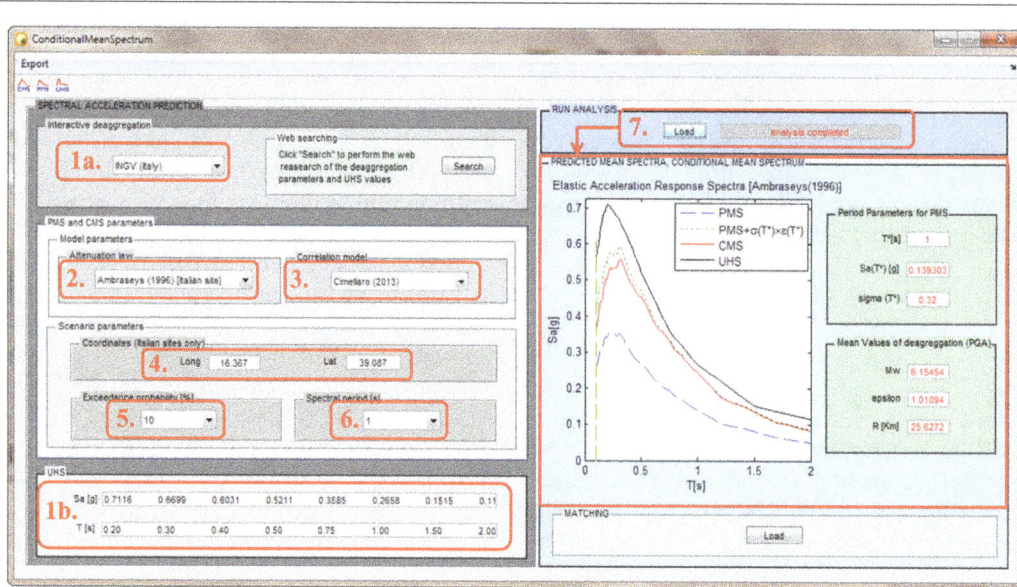

FIGURE 3 | User dialog window for defining the CMS.

3. Push "*Read*" for reading the file and extracting of the main information.
4a. Select the measure units for the accelerations.
4b. Tip the "*Bandpass and Bandstop*" for filtering the signal.
5. Push "*View*" for loading the acceleration history and plotting the signal properties and the time histories.

In the *Input Data* (upper left), the records are uploaded and read automatically for the selected ground motion databases (PEER, ESMD, UCHILE and ITACA) or using the option "*free format.*" The *signal processing module* (**Figure 2**) allows to correct the ground motion records with the Butterworth filter by modifying the default set up values ($f_{min} = 0.25$ Hz, $f_{max} = 25$ Hz, $n = 4$) if needed. The effect of the filter is shown in the *Time Histories* visualization panel where the time histories of accelerations, velocities, and displacements, both filtered and unfiltered are displayed.

In the *Signal Analysis* block (low right in **Figure 2**), the main parameters of the earthquake records (e.g., peak ground acceleration, velocity and displacement, duration, etc.) both *peak* and *root mean square values* are calculated and saved both for filtered and unfiltered data. The Arias Intensity and the Fourier Transform graph are plotted, too. Finally, all processed data and records can be exported in other common formats, such as MS Excel and txt (bottom left in **Figure 2**).

Response Spectral Analysis

Once the set of ground motions are selected and filtered, the Elastic Response Spectra (acceleration, velocity, displacement, etc.) can be computed for a given value of damping ratio. Furthermore, the *mean* and *median* acceleration response spectra of the uploaded set of records with the associated range of dispersion ($\pm\sigma$) can be also evaluated and plotted using also log and semi-log scales.

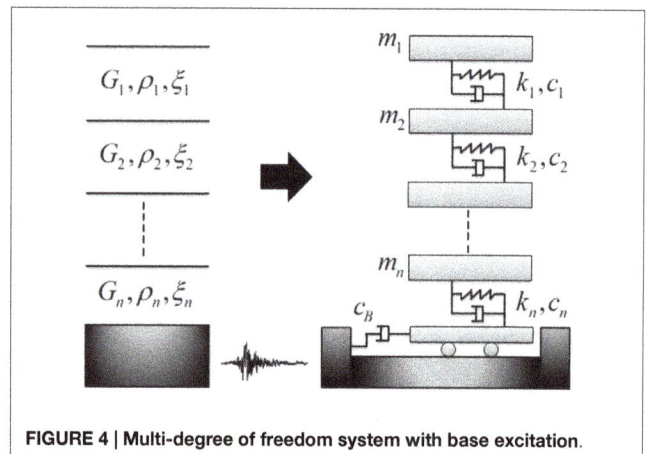

FIGURE 4 | Multi-degree of freedom system with base excitation.

Target Spectrum

Different types of target spectrum can be defined during the ground motion selection process. The design spectrum (DS) can be evaluated according to the Italian seismic standards, the NTC 2008 (NTC-08, 2008) for any point in the Italian territory, once the parameters are defined (e.g., nominal life, soil category, damping ratio, over strength factor q to describe the inelastic behavior, etc.). Additionally, the DS according to the European seismic standard, EC8 (CEN, 2004), and to the US standards (FEMA, 2009) can be evaluated inserting the proper parameters.

Furthermore, the platform allows evaluating for a given probability of exceedance the *UHS*, and the *Predicted Mean Spectrum (PMS)* using different ground motion prediction equations (GMPE), which are currently available: Ambraseys et al. (1996), Campbell and Bozorgnia (2008), Boore and Atkinson (2008), and

Chiou and Youngs (2008). Two additional attenuation laws have been recently inserted to define the CMS for the Chilean sites (Contreras and Boroschek, 2012) and the Regional Indian attenuation equations for Indian sites (Iyengar et al., 2010). However, the novelty of the proposed system architecture is that it allows evaluating the *CMS* (Baker and Cornell, 2006) for the first time on the entire Italian territory automatically knowing the GIS coordinates and in any other site worldwide knowing the proper parameters.

Non-linear response history analyses are used to generate sets of demands, which are predictive of the buildings performance. FEMA P-58-1 (FEMA, 2012) identifies three different types of performance assessment: *intensity-based assessment, scenario-based assessment,* and *time-based assessment*. Each of these methodologies includes the development of an appropriate target acceleration response spectrum, the selection of an appropriate suite of earthquake ground motions, and the scaling of motions for consistency with the target spectrum. Evaluation of a response spectrum as target spectrum is the first step to apply the procedure above mentioned and for this purpose *Seismic Performance Assessment of Buildings* rules assert that the spectral shape should be consistent with the geologic characteristics of the site. The two most used spectra are the *UHS* and *CMS*. The first one is created with referring to a given hazard level and probability of exceedance by enveloping the results of the probabilistic seismic hazard analysis (PSHA) for each period. Furthermore, this is an efficient way of representing seismic hazards for probabilistic performance evaluation of structures, but, at the same time, the spectral values at each period cannot occur in a single ground motion. In other words, the amplitude of a single ground motion is not equally spaced from the UHS at all period. Thus, the UHS is not very representative as target spectrum for any individual ground motion. This limitation has led to focus on the CMS, which is obtained conditioning on a spectral acceleration at only one period. The deaggregation parameters (M, R, and ϵ), obtained from the PSHA as mean

values, depending on the period of interest, must be determined to calculate the predicted mean and SD of log spectral acceleration values at all periods using an adequate GMPE. Once the GMPE is selected, the CMS can be defined as the sum of two contributions and is given by

$$\log\left(S_a(T_i)\right)_{/\log(S_a(T_{ref}))} = \log\left(S_a(T_{ref})\right) + \rho\left(T_i, T_{ref}\right)\epsilon\left(T_{ref}\right)\sigma_{\log(S_a)}(T_i)$$
(2)

where the first term is the logarithmic spectral acceleration $\log(S_a(T_{ref}))$; the second term is the product between the conditional mean ϵ value, for the period of interest T_{ref}, the SD of log distribution $\sigma_{\log(Sa)}$, and the correlation coefficient ($\rho(T_i, T_{ref})$).

The parameter ϵ is a measure of the difference between the log spectral acceleration of a record and the mean log spectral demand predicted, while the correlation coefficient ρ defines the linear correlation between a pair of ϵ values associated to two different periods. While specific correlation equations exist for the California sites (Chiou and Youngs, 2008; Baker and Jayaram, 2008), a new correlation equation (Cimellaro, 2013) has been developed for the European sites analyzing 595 strong motion records and considering the Ambraseys GMPE which is given by

$$\rho_{\epsilon(T_1)\epsilon(T_2)} = 1 - \left(\frac{A_0 + A_2\log(T_{min}) + A_4(\log(T_{max}))^2}{1 + A_1\log(T_{max}) + A_3(\log(T_{min}))^2}\right)\ln\left(\frac{T_{min}}{T_{max}}\right)$$
(3)

where $T_{min} = \min(T_1, T_2)$, $T_{max} = \max(T_1, T_2)$, while A_0, A_1, A_2, A_3, A_4 are the model parameters. This correlation equation has been implemented in the platform together with the correlation models proposed by Chiou and Youngs (2008) for the Californian sites.

Figure 3 shows the main user dialog window for defining the CMS. The number identifies the steps to be followed in order to obtain the CMS and the PMS at a given site and they are summarized below:

1a. Select the geological institute (**INGV(Italy)**) providing the deaggregation parameters.
1b. Insert the spectral acceleration of the UHS to be found in the geological institute internet site. This step is not necessary for defining the CMS and PMS.
2. Select the attenuation model (Ambraseys et al., 1996).
3. Select the correlation coefficient model.
4. Insert the geographic coordinates.
5. Select the exceedance probability.
6. Select the referring period.
7. Run the analysis with "**Load.**"

TABLE 1 | OPENSIGNAL soil response analysis vs. EERA.

Feature	OPENSIGNAL	EERA
Discretization	Lumped mass	Continuous layers
Type of solution	Time domain	Frequency domain
Type of analysis	Step-by-step integration	Transfer function
Soil model	$G(\gamma)$ and $D(\gamma)$ curves	$G(\gamma)$ and $D(\gamma)$ curves
Damping model	Rayleigh formulation (RF)	Kelvin–Voigt model
Non-linearity	Solution with parameters uploading at every step	Iterative approximation of equivalent linear response

TABLE 2 | Six ground motion records from the three databases.

Database	Station ID	Event ID	Earthquake name	Date	PGA [cm/s²]	M_L	R [km]
ITACA	TLM1	IT-1976-0001	Friuli	06/05/1976	95.7	4.5	28.8
ITACA	FMC	IT-1981-0002	Basilicata	16/01/1981	103.9	4.6	16.4
ESMD	291	424	Sicilia Orientale	13/12/1990	103.0	5.4	29.0
ESMD	1353	692	Levkas island	15/03/1994	94.2	4.0	13.0
PEER	5038 Sun.	P0539	N. Palm Springs	08/07/1986	91.2	5.9	44.4
PEER	1117 G.G.P.	P0024	San Francisco	22/03/1957	93.5	5.3	–

When the analysis is completed the mean values of the deaggregation function of referring period, the mean values of deaggregation function of PGA, the spectra and the main information about the CMS are plotted in the spaces with the blue contour.

Earthquake Records Selection Criteria

Ground motion *selection* and *scaling* procedures are applied in order to obtain a set of motions that are usually used in dynamic elastic and even non-linear response history analysis.

The proposed framework retrieves records from the PEER-NGA strong motion database (PEER - available at http://ngawest2.berkeley.edu/), the European strong motion database (ESMD - available at http://www.isesd.hi.is/ESD_Local/frameset.htm). The current PEER database includes 21,336 three-component records from 600 shallow crustal events with small-to-moderate magnitude located in California. It covers a magnitude range of 3–7.9, and a rupture distance range of 0.05–1533 km. The estimated or measured time-averaged shear-wave velocity in the top 30 m at the recording sites (Vs30) ranges from 94 to 2100 m/s. The European strong motion database includes around 3000 uniformly processed and formatted European strong-motion records and associated earthquake, station, and waveform parameters (Ambraseys et al., 1996). In details, it includes 462 triaxial strong-motion records from 110 earthquakes and 261 stations in Europe and the Middle East.

Two selection criteria can be used which are based on *Waveform Matching* and *Spectral Matching*.

The *Waveform Matching* can be obtained selecting some specific seismological parameters obtained by the disaggregated seismic hazard maps at a specific site, such as the moment magnitude, M_w, the fault distance or Joyner–Boor distance (R or R_{JB}, expressed in kilometer), the *fault mechanism*, the *soil type* according to EC8, and the *waveform parameters* (e.g.,

peak ground acceleration, peak ground velocity, peak ground displacement).

The *Spectral Matching* instead requires an additional step, which is the definition of the target spectrum and the type of matching to be carried out. The current available options in the platform are three: (i) *Single period*, (ii) *Multi periods* (*up to three values*), and (iii) *Mean Deviation*. A selected percentage error is defined in all cases to vary the number of earthquakes selected. The second step is the selection of the Target Spectrum among the CMS, the DS, the UHS, the PMS, or any User Defined (UDS) response spectrum.

After the selection of the target spectrum, the search of the records between the ground motion databases available in the computer environment is performed. Both horizontal and vertical components of ground motion can be considered in both search methods. All the records that are spectrum compatible are identified. Then, the records found can be preselected manually in a table and visually inspected comparing both response spectra and other data (e.g., location) and only after this further check the records can be downloaded and saved.

Site Response Analysis

In the Italian territory, the epsilon values, which are necessary to evaluate the CMS, have been evaluated assuming a perfectly rigid soil with a flat topography. Therefore, if the target spectrum is the CMS, the selection procedure does not lead to a representative set of acceleration time histories at the site. Therefore, it is necessary to implement site response analysis and project the earthquake record from the bedrock to the soil surface. In reality, the soil parameters affect the seismic response of a geotechnical system, as the soil filters the seismic input and specific frequencies may be amplified while others might not. So, in order to take into account the local site effects of the ground motion propagation, a system of dynamic equations is solved by direct integration using

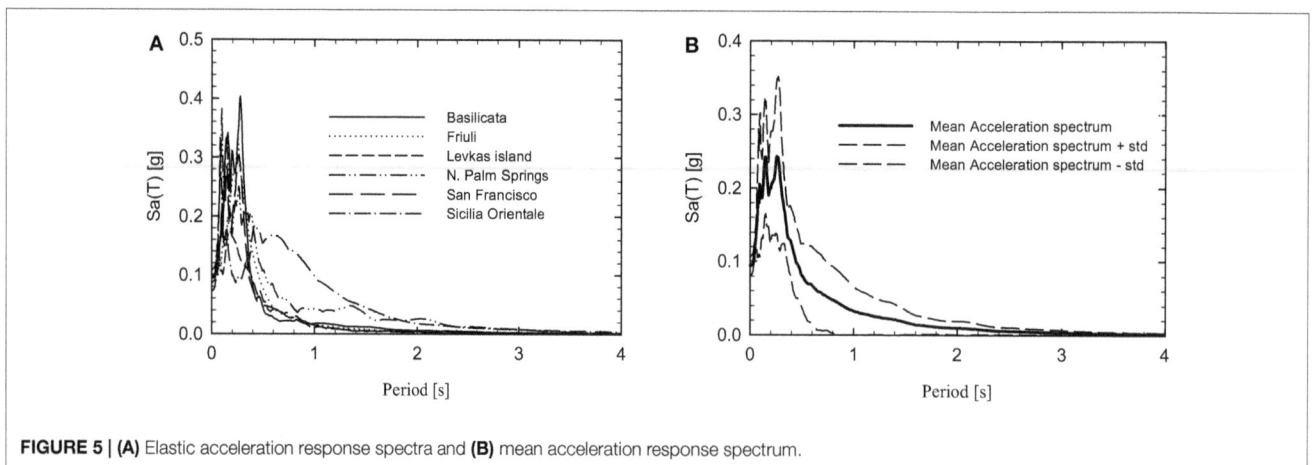

FIGURE 5 | (A) Elastic acceleration response spectra and **(B)** mean acceleration response spectrum.

TABLE 3 | Signal properties of the unfiltered vs. filtered records (Friuli earthquake).

	PGA [cm/s²]	PGV [cm/s]	PGD [cm]	Ia	Duration [s]	Peak acc. time [s]	a_{RMS} [cm/s²]	v_{RMS} [cm/s]	d_{RMS} [cm]
Unfiltered	95.67	4.39	0.43	3.17	1.85	0.50	8.77E-03	2.53E-03	2.44E-03
Filtered	101.72	4.77	0.32	3.17	1.85	0.50	2.47E-03	9.52E-03	3.40E-03

the implicit Newmark method, which is given by the following equations

$$\dot{u}_{i+1} = \dot{u}_i + \left[(1-\gamma)\Delta t \right]\ddot{u}_i + \left(\gamma \Delta t\right)\ddot{u}_{i+1} \qquad (4)$$

$$u_{i+1} = u_i + (\Delta t)\dot{u}_i + \left[\left(\frac{1}{2} - \beta\right)\Delta t^2\right]\ddot{u}_i + \left(\beta \Delta t^2\right)\ddot{u}_{i+1} \qquad (5)$$

where Δt defines the time step, u, $\dot{u}$, and $\ddot{u}$ represent the displacement, velocity, and acceleration of the system, respectively. The parameters are determined at the time i + 1 starting from the known values at time i. The parameters β and γ define the variation of the acceleration over the time step and in the present paper they have been assumed equal to $\beta = 1/4$ and $\gamma = 1/4$ (average acceleration method). The layered soil column above the bedrock is modeled as a multi-degree of freedom system with lumped parameters (spring-dashpot system) and the seismic excitation is imposed at the base of the physical model (bedrock) as an acceleration history (**Figure 4**).

The equations of motion of the system can be expressed in the following matrix format as

$$[M]\{\ddot{u}\} + [C]\{\dot{u}\} + [K]\{u\} = -[M]\{I\}\ddot{u}_g \qquad (6)$$

where $[M]$, $[C]$, and $[K]$ are the mass, damping, and stiffness matrices, respectively, while $\{\ddot{u}\}$, $\{\dot{u}\}$, and $\{u\}$ are the vectors of the absolute nodal accelerations, velocities, and displacements, respectively. The term $\{I\}\ddot{u}_g$ represents the earthquake load, where each component of the vector $\{I\}$ is equal to a unit value. In a non-linear formulation, the energy of the system is dissipated through the hysteretic loading–unloading cycles, thus, an equivalent viscous damping matrix may be defined in order to simulate the process of dissipation. In addition, in the time domain analyses, the damping depends on the frequencies. The damping matrix is determined using the Rayleigh damping formulation where the damping matrix [C] is defined as follows

$$[C] = a_0[M] + a_1[K] \qquad (7)$$

with

$$a_0 = \xi\frac{4\pi(f_0 f_1)}{f_0 + f_1} \quad a_1 = \xi\frac{1}{\pi(f_0 + f_1)} \qquad (8)$$

where ξ is the damping ratio of the soil system and f_0 and f_1 are the two control frequencies. The main approximation of this procedure consists in the underestimation of the damping at frequencies (Hashash and Park, 2002) between f_0 and f_1, and the overestimation of the damping at frequencies lower than f_0 and higher than f_1. Thus, the selection of the two control frequencies is very important in order to obtain reliable results. The proposed procedure evaluates the damping matrix according to Hudson et al. (1994) in which f_0 is the fundamental frequency of the soil column, while f_1 represents the predominant frequency of the ground motion. In addition, the variation of damping ratio among the soil layers is taken into account in the computer environment calculating the damping matrix [C] by assembling the elements of the damping matrices

$$[C] = \frac{4\pi(f_0 f_1)}{f_0 + f_1}\begin{bmatrix} \xi_1 m_1 & 0 & \ldots & 0 \\ 0 & \xi_2 m_2 & 0 & 0 \\ \ldots & 0 & \ldots & \ldots \\ 0 & 0 & \ldots & \xi_n m_n \end{bmatrix}$$

$$+ \frac{1}{\pi(f_0 + f_1)}\begin{bmatrix} \xi_1 k_1 & -\xi_1 k_1 & \ldots & 0 \\ -\xi_1 k_1 & \xi_1 k_1 + \xi_2 k_2 & -\xi_2 k_2 & \ldots \\ \ldots & -\xi_2 k_2 & \ldots & -\xi_{n-1} k_{n-1} \\ 0 & \ldots & -\xi_{n-1} k_{n-1} & \xi_{n-1} k_{n-1} + \xi_n m_n \end{bmatrix} \qquad (9)$$

The proposed method uses an *equivalent linear model* (hybrid approach), which describes the soil behavior assuming that both the shear modulus and the damping ratio vary with the shear strain amplitude. So, the hysteretic behavior of the soil is described using the shear modulus degradation curve ($G - \gamma$) and the damping ratio curve ($\xi - \gamma$). The use of this curve in combination with Eq. 6 is explained in the work of Bardet et al. (2000). In the proposed platform, clay, sand, and rock degradation curves are available by default (Bardet et al., 2000). The dynamic equations of the system given in Eq. 6 are solved using Newmark method evaluating the nodal displacements and the

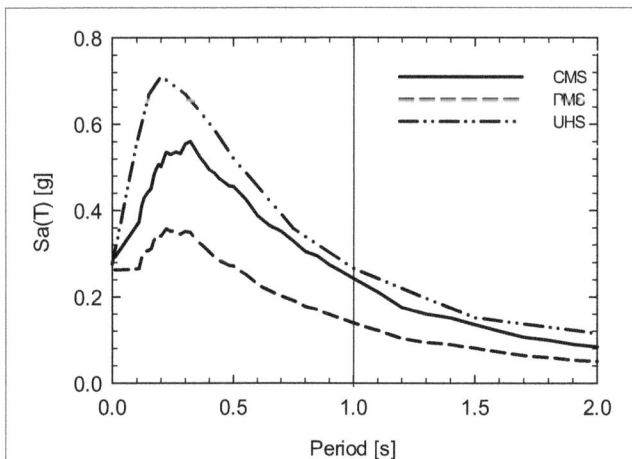

FIGURE 6 | Conditional mean spectrum (CMS), predicted mean spectrum (PMS), and uniform hazard spectrum (UHS) in Soveria Mannelli, Italy.

TABLE 4 | Spectral matching parameters.

Database	Matching criteria	Component	% Error/mean deviation	$T_i/T_{min} - T_{max}$ [s]
ESMD	Single period	Y-component	10	1
PEER	Single period	Y-component	10	1
ESMD	Multi period	Y-component	30	0.2 and 1
PEER	Multi period	Y-component	30	0.2 and 1
ESMD	Mean deviation	Y-component	0.5	0.2–1
PEER	Mean deviation	Y-component	0.9	0.2–1

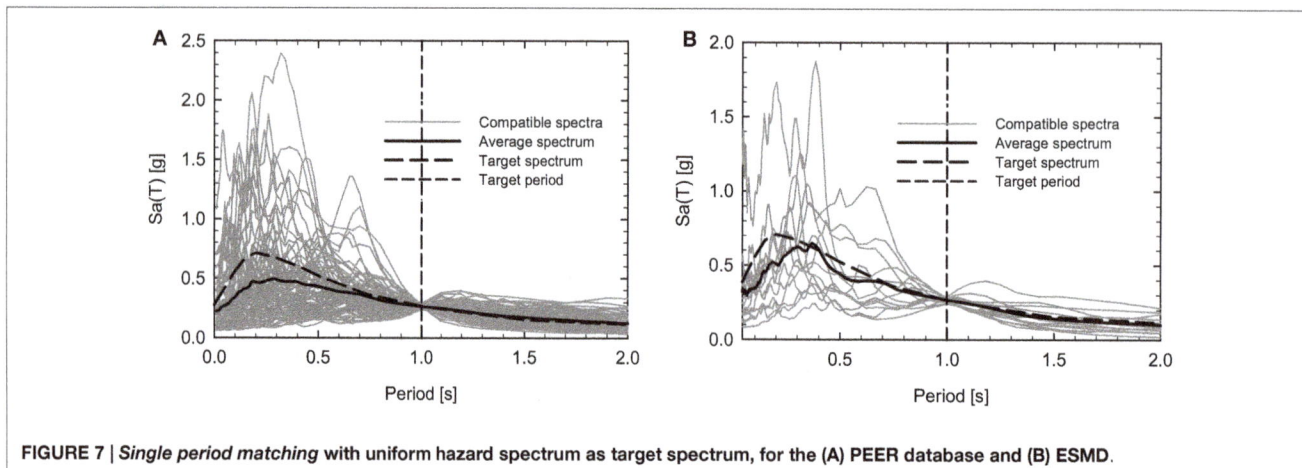

FIGURE 7 | *Single period matching* with uniform hazard spectrum as target spectrum, for the (A) PEER database and (B) ESMD.

corresponding shear deformations γ at a given time instant *t*. The γ values are inserted in the curves to update the shear modulus *G* and the damping ratio ξ, which are used to define the new stiffness and damping (Eq. 9) matrices defined at the same time instant *t*. The proposed method has some limitation at large shear strain deformations, because in that case, the soil presents a non-linear behavior and both stiffness and damping depend on the number of loading–unloading cycles. Nevertheless, it was observed that at medium deformations, the non-linear behavior of the soil is not significantly influenced by the load path (Jardine et al., 1986). Thus, the proposed hybrid approach can lead to reliable results for the range of medium deformations. The comparison between the proposed approach and the method implemented in EERA (Bardet et al., 2000) is shown in **Table 1**.

It necessary to mention that the site response analysis tool of OPENSIGNAL is independent from the other tools (matching, filtering, etc.). However, if you decide to use the tools in sequence you should be aware that it can be applied only to records of the database, which have been recorded on rock site.

Case Study

As illustrative example to show the applicability of the proposed methodology and the capabilities of the computer-based platform environment, six ground motion records (two records from each database) have been chosen from three different databases to test the record processing tool. The list of records selected is given in **Table 2**, where are shown two different bin sets of six records taken from each database.

The six uncorrected records have values of PGA between 90 and 105 cm/s² and local magnitude between 4.0 and 6.0. Every uncorrected record in **Table 2** has been filtered with a Butterworth filter having $f_{min} = 0.25$ Hz, $f_{max} = 25$ Hz, and $n = 4$. The filtered set of records is then used for the response spectral analysis. **Figure 5** shows the elastic response spectra of the ground motion set in **Table 2** with a damping ratio equal to 5%.

TABLE 5 | Ground motion characteristics.

Station ID	Event name	Date	M_L
TLM1	Friuli	06/05/1976	4.5
CSC	Val Nerina	19/09/1979	5.5

After uploading the records, the computer platform allows computing the main signal parameters divided in three main categories:

- Peak values: *PGA*, *PGV*, and *PGD*;
- Time values: arias intensity, duration, peak acceleration time;
- Root mean square values: a_{RMS}, v_{RMS}, d_{RMS}.

The parameters are calculated for both the unfiltered (*Original*) and filtered (*Modified*) records. As example, the signal properties of "*Friuli*" earthquake are summarized in **Table 3**.

Spectral Matching

Wide varieties of techniques have been developed for selecting a reliable set of earthquake records to be used in the dynamic structural analysis (Cimellaro et al., 2011). One selection criteria is based on spectral matching to a specific target spectrum. Thus, the definition of a target spectrum represents the preliminary phase of the spectral matching, and for this purpose, OPENSIGNAL allows choosing between five different spectra:

- Design spectrum (DS) according to NTC-08, 2008, EC8, and FEMA 302;
- Uniform hazard spectrum (UHS);
- Predicted mean spectrum (PMS) according to Ambraseys et al. (1996), Campbell and Bozorgnia (2008), and Boore and Atkinson (2008), Contreras and Boroschek (2012) and Iyengar et al. (2010) GMPE;
- Conditional mean spectrum (CMS);
- User defined spectrum (UDS).

TABLE 6 | Geotechnical soil characteristics.

Layer number	Soil	Thickness [m]	Shear-wave velocity [m/s]	Initial shear modulus [MPa]	Initial damping ratio [%]	Unit weight [kg/m³]
1	Sand	6.5	136.21	37.18	0.24	2004.08
2	Sand	5.0	176.15	62.19	0.24	2004.08
3	Clay	9.0	404.46	348.35	0.24	2129.46
4	Sand	8.0	225.52	101.93	0.24	2004.08
5	Clay	6.0	275.84	162.03	0.24	2129.46
6	Sand	8.0	207.46	86.26	0.24	2004.08

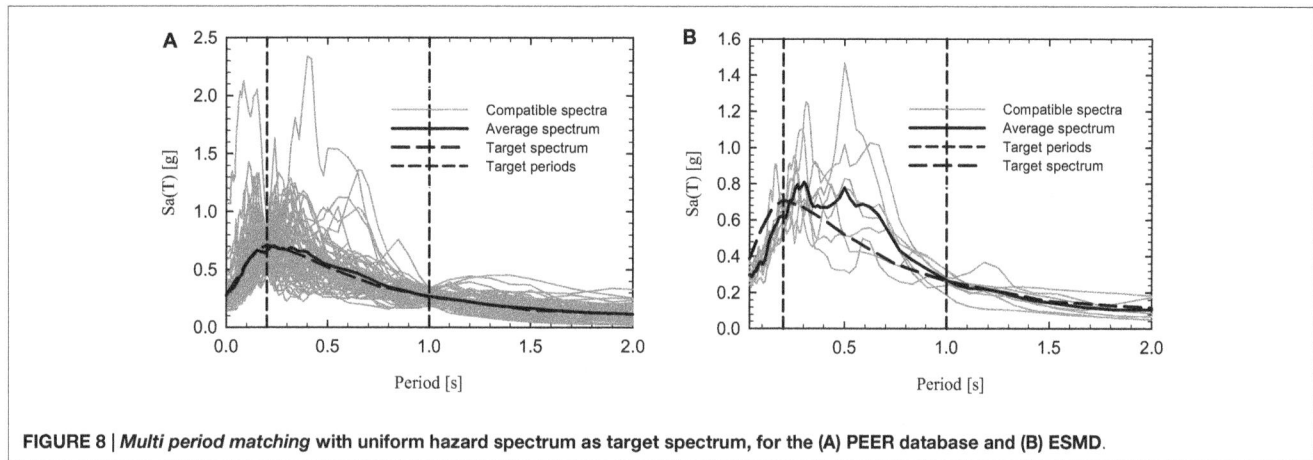

FIGURE 8 | *Multi period matching* with uniform hazard spectrum as target spectrum, for the (A) PEER database and (B) ESMD.

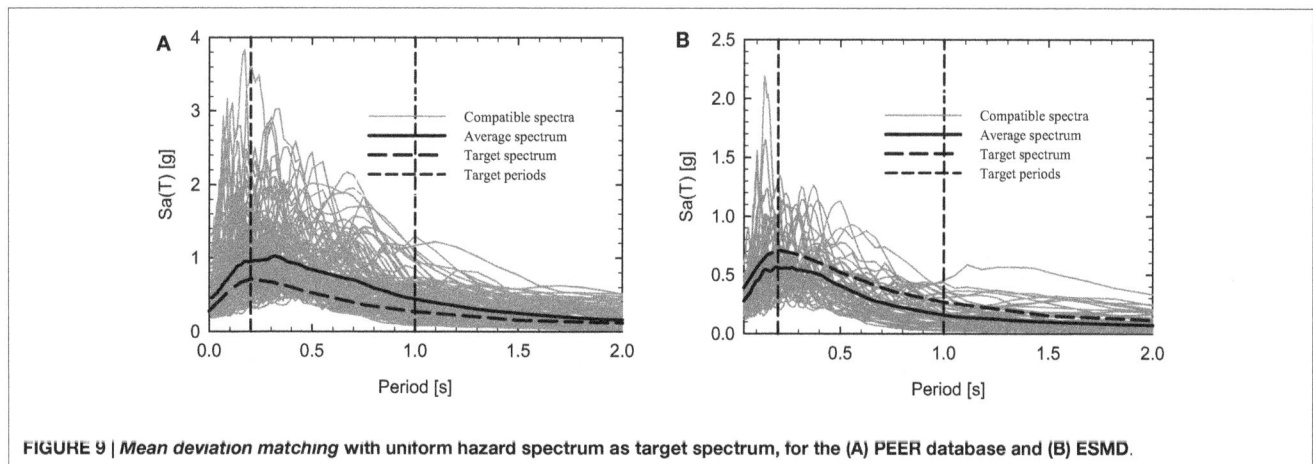

FIGURE 9 | *Mean deviation matching* with uniform hazard spectrum as target spectrum, for the (A) PEER database and (B) ESMD.

As example, the site of Soveria Mannelli (16.3859° longitude, 39.0969° latitude, close to Lamezia Terme) in southern Italy has been selected. The Spectral matching is carried out considering the UHS spectrum and the CMS spectrum as target spectrum. The CMS has been defined for the period of 1 s and for a probability of exceedance of 10% in 50 years. The CMS has been built taking into account the deaggregation values associated to the referring period (Barani et al., 2009; Cimellaro, 2013). The information related to the UHS is taken from the INGV internet website[6] for the Italian sites and from the USGS internet website[7] for the US

sites. **Figure 6** is shown together the three target response spectra (CMS, UHS, and PMS) at the specific site, which are going to be used for selecting the earthquake records.

OPENSIGNAL is available in three different approaches for spectral matching called:

- Single period approach,
- Multi period approach, and
- Mean deviation approach.

The "single period approach" does matching using a single control point, while the multi period approach uses more points for the matching. The mean deviation approach defines the mean error during the matching.

---

[6]http://esse1-gis.mi.ingv.it/
[7]http://geohazards.usgs.gov/deaggint/2008/

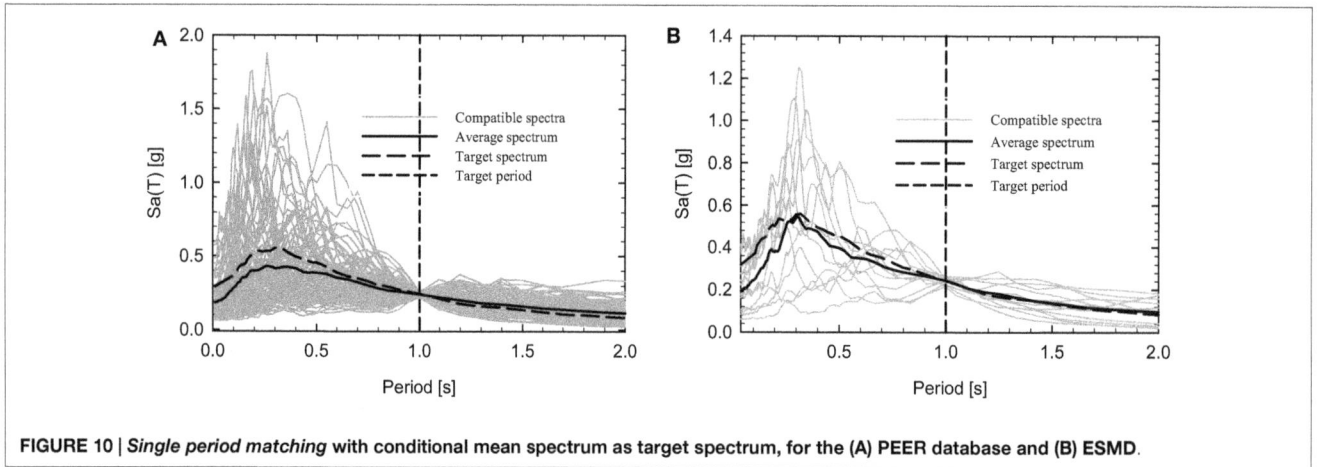

FIGURE 10 | *Single period matching* with conditional mean spectrum as target spectrum, for the (A) PEER database and (B) ESMD.

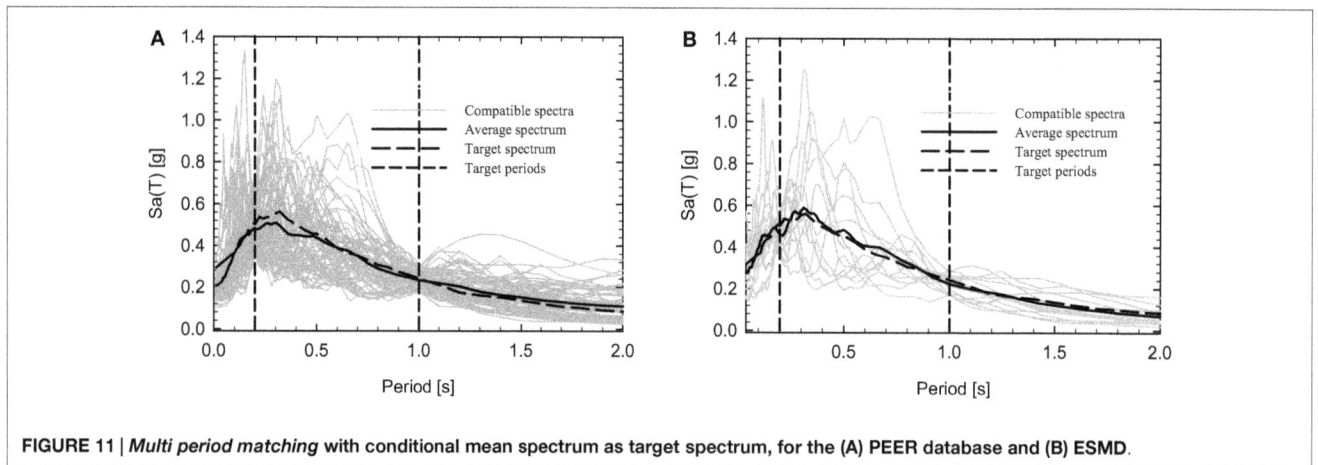

FIGURE 11 | *Multi period matching* with conditional mean spectrum as target spectrum, for the (A) PEER database and (B) ESMD.

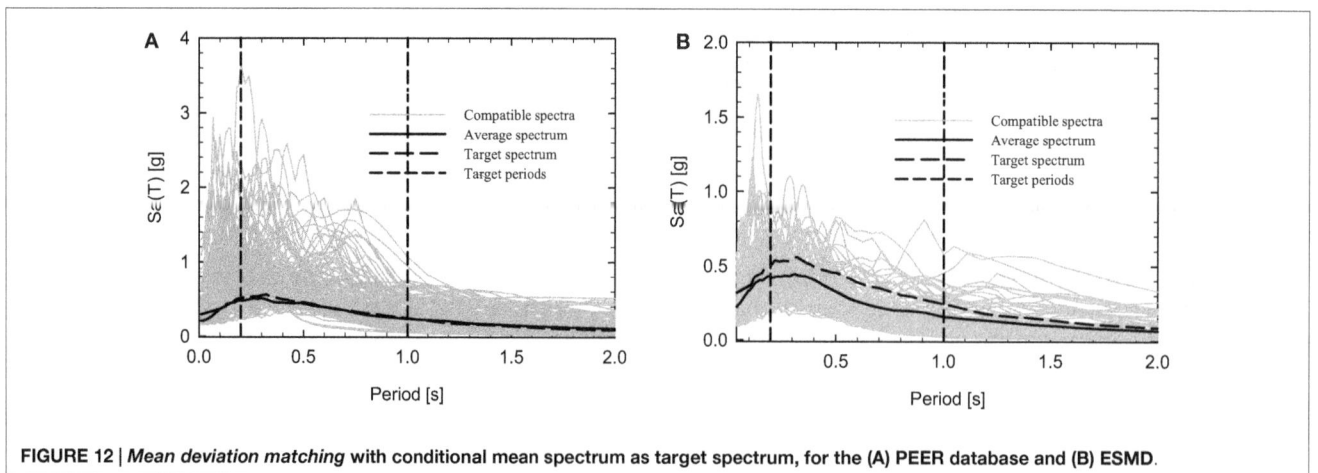

FIGURE 12 | *Mean deviation matching* with conditional mean spectrum as target spectrum, for the (A) PEER database and (B) ESMD.

The computer environment allows using any of the matching procedures above mentioned, selecting the tolerance in term of percentage error for the first two approaches or in term of mean deviation for the latter one. The search can be performed for the

X, Y, or Z components for both the ESMD and PEER records. In **Table 4**, the parameters used for the matching criteria for both the ESMD and PEER database are summarized, while the results of the three spectral matching procedures for ESMD and

PEER database are shown in **Figures 7–12** for the case when the fundamental period of the structure is assumed equal to $T = 1$ s.

Site Response Analysis

In order to show the applicability of the proposed site response analysis method, two different ground motions records, Val Nerina and Friuli earthquake, have been considered (**Table 5**).

The soil stratigraphy selected for the application is summarized in **Table 6** in which layer 1 is closest to the soil surface, while layer 6 is closest to the bedrock.

The accuracy of the numerical solution depends on the number of sub-layers, or rather on the degrees of freedom. Since the generic earthquake half wave length should be described by three to four points at least for each layer, the thickness to be assigned to each layer should not be greater than the ratio between the shear-wave velocity (V_s) and six-height times the predominant seismic frequency (f_{max}). In order to satisfy this requirement, the entire soil stratigraphy has been divided in equal parts using the following equation

$$h_{max} \simeq \frac{V_s}{7 f_{max}} \tag{10}$$

The amplification effect due to the soil stratigraphy is shown in **Figure 13** in which the *response spectrum on bedrock* is compared with the *response spectrum on the soil surface*, for both Val Nerina (**Figure 13A**) and Friuli (**Figure 13B**) earthquake.

Finally, the results obtained with OPENSIGNAL and EERA are compared in **Figure 14** for the same earthquake records.

It is important to mention that the proposed hybrid method for soil response analysis presents some limitations. In fact, it can lead to inaccurate results for high amplitude records, because they produce large shear deformations in the soil column and in the large strain range, it is necessary to consider the real $\tau - \gamma$ curve in order to appreciate the non-linear behavior of the soil. In these cases, therefore, it is better to use specific non-linear site response analysis software (e.g., EERA). In addition, since the proposed method uses the Rayleigh formulation to define the damping at each step, the solution is strictly dependent on the f_0/f_1 ratio. At this purpose, for high value of the frequency ratio,

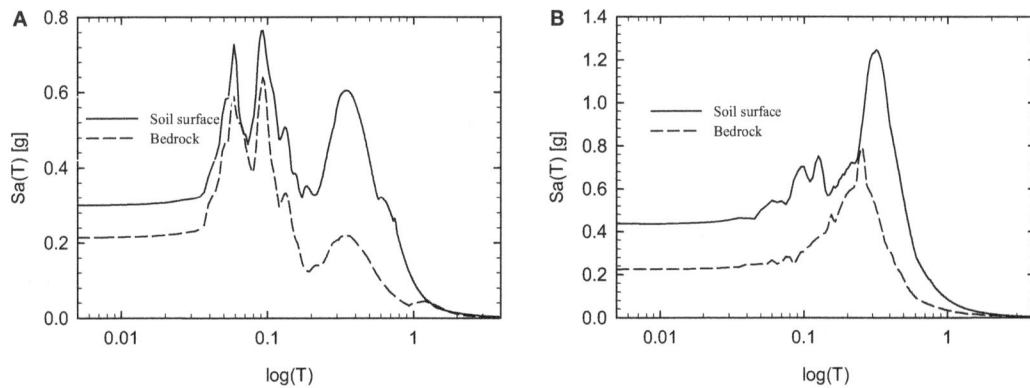

FIGURE 13 | Val Nerina earthquake (A) and Friuli earthquake (B) comparisons between bedrock and soil surface results.

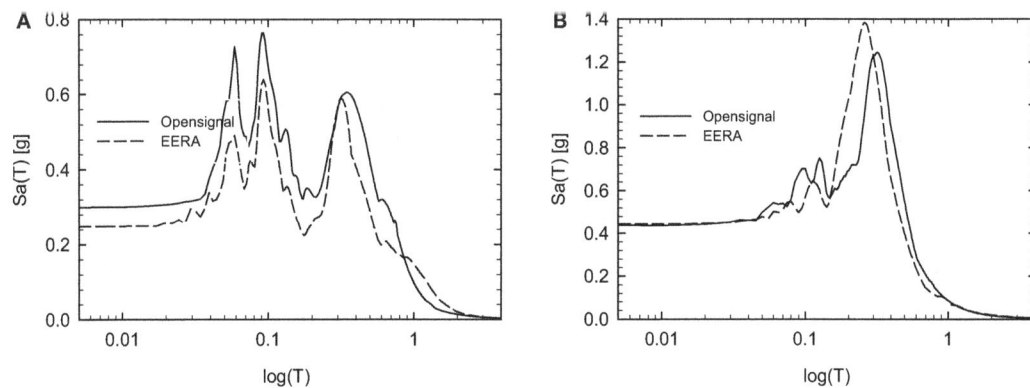

FIGURE 14 | Val Nerina earthquake (A) and Friuli earthquake (B) comparisons between EERA and OPENSIGNAL results.

the Rayleigh damping approach might lead to underestimated damping values, so the final record might be amplified.

Concluding Remarks

The use of ground motion data is growing worldwide due to the increasing availability of records and increased interest from the earthquake engineering community in using non-linear response history analysis in seismic analysis and design. In particular, the selection and processing of earthquake records plays a key role in seismic risk assessment of buildings and structures in general.

The paper presents a new software platform for processing and selection of seismic records, called "OPENSIGNAL," that are freely available for the general public. The platform consists of a number of modules, integrated in a unified environment and aimed for: selection of ground motion records, signal processing, response spectra analysis, soil spectra analysis, etc.

The main novelties of the platform are (i) the capacity of modeling the local site effects of the ground motion propagation, using a hybrid approach based on an equivalent linear model and (ii) the evaluation of the CMS according to seven different

attenuation models, using the geographical coordinates (for Italian sites) or the seismological characteristics (for any site). The platform provides the possibility of using various ground motion record formats (PEER, EMSD, ITACA, and UCHILE), as well as free format records.

OPENSIGNAL allows the automatic reading of the ground motion records from the mentioned databases, reducing the processing time. The possibility to choose the filtering parameters and to modify the time history by scaling, provide the user a useful and flexible tool.

All the above mentioned modules can also work independently other than in sequence allowing more flexibility in the utilization of the software.

Acknowledgments

The research leading to these results has received funding by the European Community's Seventh Framework Programme – the Marie Curie International Outgoing Fellowship (IOF) Actions-FP7/2007-2013 under the Grant Agreement no. PIOF-GA-2012-329871 of the project IRUSAT – Improving Resilience of Urban Societies through Advanced Technologies.

References

Akkar, S. (2008). "An introduction to utility software for data processing (USDP)," in *4th BSHAP Project Workshop 16-17 December 2008* (Budva).

Ambraseys, N. N., Simpson, K. A., and Bommer, J. (1996). Prediction of horizontal response spectra in Europe. *Earthq. Eng. Struct. Dyn.* 25, 371–400. doi:10.1002/(SICI)1096-9845(199604)25:4<401::AID-EQE551>3.0.CO;2-B

Baker, J. W. (2011). Conditional mean spectrum: tool for ground-motion selection. *J. Struct. Eng.* 137, 322–331. doi:10.1061/(ASCE)ST.1943-541X.0000215

Baker, J. W., and Cornell, C. A. (2006). Spectral shape, epsilon and record selection. *Earthq. Eng. Struct. Dyn.* 35, 1077–1095. doi:10.1002/eqe.571

Baker, J. W., and Jayaram, N. (2008). Correlation of spectral acceleration values from NGA ground motion models. *Earthq. Spectra* 24, 299–317.

Barani, S., Spallarossa, D., and Bazzurro, P. (2009). Disaggregation of probabilistic ground-motion hazard in Italy. *Bull. Seismol. Soc. Am.* 99, 2638–2661. doi:10.1785/0120080348

Bardet, J., Ichii, K., and Lin, C. (2000). *EERA: A Computer Program for Equivalent-Linear Earthquake Site Response Analyses of Layered Soil Deposits.* Los Angeles, CA: Department of Civil Engineering; University of Southern California.

Boore, D. M. (2009). *"TSPP – A Collection of FORTRAN Programs for Processing and Manipulating Time Series."* Reston, VA: U.S. Geological Survey, Open-File, Report 2008-1111, v. 2.0, 52.

Boore, D. M., and Atkinson, G. M. (2008). Ground-motion prediction equations for the average horizontal component of PGA, PGV, and 5%-damped PSA at spectral periods between 0.01 s and 10.0 s. *Earthq. Spectra.* 24, 99–138. doi:10.1193/1.2830434

Boore, D. M., and Bommer, J. J. (2005). Processing of strong-motion accelerograms: needs, options and consequences. *Soil Dyn. Earthq. Eng.* 25, 93–115. doi:10.1016/j.soildyn.2004.10.007

Campbell, K. W., and Bozorgnia, Y. (2008). NGA ground motion model for the geometric mean horizontal component of PGA, PGV, PGD and 5% damped linear elastic response spectra for periods ranging from 0.01 to 10 s. *Earthq. Spectra.* 24, 139–171. doi:10.1193/1.2857546

CEN. (2004). *Eurocode 8: Design of Structures for Earthquake Resistance Part 1: General Rules, Seismic Actions and Rules for Buildings.* Brussels: European Committee for Standardization.

Chiou, B. S.-J., and Youngs, R. R. (2008). An NGA model for the average horizontal component of peak ground motion and response spectra. *Earthq. Spectra.* 24, 173–215. doi:10.1193/1.2894832

Cimellaro, G. P. (2013). Correlation in spectral accelerations for earthquakes in Europe. *Earthq. Eng. Struct. Dyn.* 42, 623–633. doi:10.1002/eqe.2248

Cimellaro, G. P., Reinhorn, A. M., D'Ambrisi, A., and De Stefano, M. (2011). Fragility analysis and seismic record selection. *J. Struct. Eng.* 137, 379–390. doi:10.1061/(ASCE)ST.1943-541X.0000115

Contreras, V., and Boroschek, R. (2012). "Strong ground motion attenuation relations for Chilean subduction zone interface earthquakes," in *15 WCEE* (Lisbon).

Corigliano, M., Lai, C., Rota, M., and Strobbia, C. (2012). ASCONA: automated selection of compatible natural accelerograms. *Earthq. Spectra* 28, 965–987. doi:10.1193/1.4000072

FEMA. (2009). *NEHRP Recommended Seismic Provisions for New Buildings and Other Structures. Federal Emergency Management Agency (FEMA P-750).* Washington, DC: Federal Emergency Management Agency.

FEMA. (2012). *FEMA P-58-1: Seismic Performance Assessment of Buildings. Volume 1–Methodology.* Washington, DC: Federal Emergency Management Agency.

Hachem, M. (2003). *BISPEC Program Description and User's Manual.* Available at: http://www.ce.memphis.edu/7137/PDFs/BispecHelpManual.pdf

Hashash, Y. M. A., and Park, D. (2002). Viscous damping formulation and high frequency motion propagation in non-linear site response analysis. *Soil Dyn. Earthq. Eng.* 22, 611–624. doi:10.1016/S0267-7261(02)00042-8

Hudson, M., Idriss, M. I., and Beikae, M. (1994). *QUAD4M – A Computer Program to Evaluate the Seismic Response of Soil Structures Using Finite Element Procedures and Incorporating a Compliant Base.* Davis, CA: Center for Geotechnical Modeling, Department of Civil and Environmental Engineering, University of California; Washington, DC: The National Science Foundation.

Iervolino, I., Galasso, C., and Cosenza, E. (2010). REXEL: computer aided record selection for code-based seismic structural analysis. *Bull. Earthq. Eng.* 8, 339–362. doi:10.1007/s10518-009-9146-1

Iyengar, R. N., Chadha, R. K., Rao, K. B., and Kanth, S. T. G. R. (2010). "Development of probabilistic seismic hazard map of India," in *The National Disaster Management Authority*, ed. R. N. Iyengar (New Delhi: Government of India), 86 p.

Jardine, R., Potts, D., Fourie, A., and Burland, J. (1986). Studies of the influence of nonlinear stress-strain characteristics in soil structure interaction. *Geotechnique* 36, 377–396. doi:10.1680/geot.1986.36.3.377

Katsanos, E. I., and Sextos, A. G. (2013). ISSARS: an integrated software environment for structure-specific earthquake ground motion selection. *Adv. Eng. Software* 58, 70–85. doi:10.1016/j.advengsoft.2013.01.003

Luzi, L., Hailemikael, S., Bindi, D., Pacor, F., Mele, F., and Sabetta, F. (2008). ITACA (ITalian ACcelerometric Archive): a web portal for the dissemination of Italian strong-motion data. *Seismol. Res. Lett.* 79, 716–722. doi:10.1785/gssrl.79.5.716

MATLAB. (2012). *MATLAB Version 2012b*, Version 2012b Edition. Natick, MA: The MathWorks Inc.

NTC-08. (2008). *Nuove Norme Tecniche per le Costruzioni (NTC08) (in Italian). Gazzetta Ufficiale Della Repubblica Italiana*. Rome: Consiglio Superiore dei Lavori Pubblici, Ministero delle Infrastrutture, 29.

Padgett, J., and Desroches, R. (2007). Sensitivity of seismic response and fragility to parameter responses. *J. Struct. Eng.* 133, 1710–1718. doi:10.1061/(ASCE)0733-9445(2007)133:12(1710)

Conflict of Interest Statement: The authors declare that the research was conducted in the absence of any commercial or financial relationships that could be construed as a potential conflict of interest.

16

Critical double impulse input and bound of earthquake input energy to building structure

Kotaro Kojima, Kohei Fujita and Izuru Takewaki*

Department of Architecture and Architectural Engineering, Graduate School of Engineering, Kyoto University, Kyoto, Japan

A theory of earthquake input energy to building structures under single impulse is useful for disclosing the property of energy transfer function. This property shows that the area of the energy transfer function is constant irrespective of natural period and damping of building structures. However, single impulse may be unrealistic from a certain viewpoint because the frequency characteristic of input cannot be expressed by this input. In order to resolve such issue, a double impulse is introduced in this paper. The frequency characteristic of the Fourier amplitude of the double impulse is found in an explicit manner and a critical excitation problem is formulated with an interval of two impulses as a variable. The solution to that critical excitation problem is derived. An upper bound of the earthquake input energy is then derived by taking full advantage of the property of the energy transfer function that the area of the energy transfer function is constant. The relation of the double impulse to the corresponding one-cycle sinusoidal wave as a representative of near-fault pulse-type waves is also investigated.

Keywords: earthquake input energy, double impulse, critical excitation method, energy transfer function, upper bound of input energy

Edited by:
Nikos D. Lagaros,
National Technical University of
Athens, Greece

Reviewed by:
Chara C. Mitropoulou,
National Technical University of
Athens, Greece
Alfredo Camara,
City University London, UK

*Correspondence:
Izuru Takewaki,
Department of Architecture and
Architectural Engineering, Graduate
School of Engineering, Kyoto
University, Kyotodaigaku-Katsura,
Nishikyo, Kyoto 615-8540, Japan
takewaki@archi.kyoto-u.ac.jp

Introduction

In the history of seismic resistant design of building structures, the earthquake input energy has played an important role together with deformation and acceleration (for example, Housner, 1959, 1975; Berg and Thomaides, 1960; Housner and Jennings, 1975; Zahrah and Hall, 1984; Akiyama, 1985; Ohi et al., 1985; Uang and Bertero, 1990; Leger and Dussault, 1992; Fajfar and Vidic, 1994; Kuwamura et al., 1994; Riddell and Garcia, 2001; Trifunac et al., 2001; Takewaki, 2004a,b; Trifunac, 2008). While deformation and acceleration can predict and evaluate the performance of a building structure mainly for serviceability, the energy can evaluate the performance of a building structure mainly for safety. Especially, energy is appropriate for describing the performance of building structures of different sizes in a unified manner because energy is a global index different from deformation and acceleration as local indices.

Compared with most of the previous works dealing with time histories, the earthquake input energy is formulated here in the frequency domain (Lyon, 1975; Ordaz et al., 2003; Takewaki, 2004a,b, 2005a,b; Takewaki and Fujita, 2009; Kojima et al., 2015) to enable the derivation of bound of earthquake input energy, which is useful for the design of building structures under uncertain conditions. Another advantageous feature to introduce the upper bound of input energy is to avoid the infinite numerical integration required in the frequency-domain formulation (Kojima et al., 2015). When the structure becomes stiffer, the contribution from higher excitation frequencies is not negligible (a smaller time increment is required in the time domain). In such a case, the avoidance of the infinite numerical integration in the frequency domain may be useful.

A theory of earthquake input energy to building structures under single impulse has been shown to be useful for disclosing the property of energy transfer function (Takewaki, 2004a). This property means that the area of the energy transfer function is constant. The property of the energy transfer function similar to the case of a simple single-degree-of-freedom (SDOF) model has also been clarified for a swaying-rocking model. By using this property, the mechanism of earthquake input energy to the swaying-rocking model including the soil amplification has been made clear under the input of single impulse (Kojima et al., 2015). However, single impulse may be unrealistic because the frequency characteristic of input cannot be expressed by this input. In order to resolve such issue, double impulse is introduced in this paper.

The double impulse represents a simplified version of near-fault pulse-type waves. For this class of ground motions, many useful research works have been conducted. Mavroeidis and Papageorgiou (2003) investigated the characteristics of this class of ground motions in detail and proposed some simple models (for example, Gabor wavelet and Berlage wavelet). Xu et al. (2007) employed a kind of Berlage wavelet and applied it to the performance evaluation of passive energy dissipation systems. Takewaki and Tsujimoto (2011) used the Xu's approach and proposed a method for scaling ground motions from the viewpoints of drift and input energy demand. Takewaki et al. (2012) employed a sinusoidal wave for pulse-type waves. In this paper, a one-cycle sinusoidal wave is employed as a representative of near-fault pulse-type waves and is compared with the double impulse.

The frequency characteristic of the Fourier amplitude of the double impulse is found in an explicit manner and a critical excitation problem is formulated with an interval of two impulses as a variable. The solution to that critical excitation problem is then derived. An upper bound and a narrower upper bound of the earthquake input energy are derived by taking full advantage of the property of the energy transfer function that the area of the energy transfer function is constant. The narrower upper bound enables the evaluation of the upper bound of the earthquake input energy without infinite integration. Only a linear elastic response of an SDOF model is considered here for the introduction of the frequency-domain approach.

Earthquake Input Energy in Frequency Domain

Consider a damped linear SDOF system of mass m, stiffness k, and damping coefficient c as shown in **Figure 1**. Let $\Omega = \sqrt{k/m}, h = c/(2\Omega m)$, and x denote the undamped natural circular frequency, the damping ratio, and the displacement of the mass relative to the ground, respectively. The time derivative is denoted by an over-dot. The input energy to this SDOF system by a unidirectional ground acceleration $\ddot{u}_g(t)$ from $t = 0$ to $t = t_0$ (end of input) can be defined by the work made by the ground on the structural system and is expressed by

$$E_I = \int_0^{t_0} m(\ddot{u}_g + \ddot{x})\dot{u}_g dt \qquad (1)$$

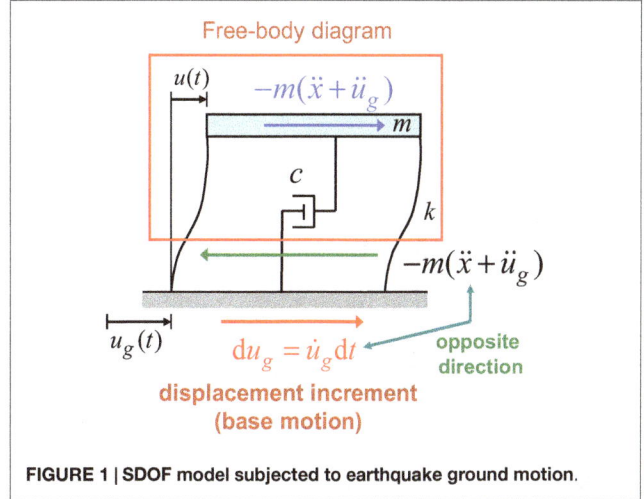

FIGURE 1 | SDOF model subjected to earthquake ground motion.

The term $m(\ddot{u}_g + \ddot{x})$ indicates the inertial force with minus sign and is equal to the sum of the restoring force kx of the spring and the damping force $c\dot{x}$ of the dashpot in the system. Integration by parts of Eq. (1) provides

$$E_I = \int_0^{t_0} m(\ddot{x} + \ddot{u}_g)\dot{u}_g dt = \int_0^{t_0} m\ddot{x}\dot{u}_g dt + \left[(1/2)m\dot{u}_g^2\right]_0^{t_0}$$
$$= [m\ddot{x}\dot{u}_g]_0^{t_0} - \int_0^{t_0} m\dot{x}\ddot{u}_g dt + \left[(1/2)m\dot{u}_g^2\right]_0^{t_0} \qquad (2)$$

If the initial and terminal conditions are expressed by $\dot{x} = 0$ at $t = 0$ and $\dot{u}_g = 0$ at $t = 0$ and $t = t_0$, the input energy can be reduced to the following form:

$$E_I = -\int_0^{t_0} m\ddot{u}_g\dot{x}dt \qquad (3)$$

It is known (Page, 1952; Lyon, 1975; Ordaz et al., 2003; Takewaki, 2004a,b, 2005a,b; Takewaki and Fujita, 2009; Kojima et al., 2015) that the input energy per unit mass can also be expressed in the frequency domain by use of Fourier and inverse Fourier transformations.

$$E_I/m = -\int_{-\infty}^{\infty} \dot{x}\ddot{u}_g dt = -\int_{-\infty}^{\infty}\left[(1/2\pi)\int_{-\infty}^{\infty}\dot{X}e^{i\omega t}d\omega\right]\ddot{u}_g dt$$
$$= -(1/2\pi)\int_{-\infty}^{\infty}\ddot{U}_g(-\omega)\left\{H_V(\omega; \Omega, h)\ddot{U}_g(\omega)\right\}d\omega$$
$$= \int_0^{\infty}\left|\ddot{U}_g(\omega)\right|^2\left\{-\text{Re}\left[H_V(\omega; \Omega, h)\right]/\pi\right\}d\omega$$
$$\equiv \int_0^{\infty}\left|\ddot{U}_g(\omega)\right|^2 F(\omega)d\omega \qquad (4)$$

where $H_V(\omega; \Omega, h)$ is the velocity transfer function defined by $\dot{X}(\omega) = H_V(\omega; \Omega, h)\ddot{U}_g(\omega)$ and $F(\omega) = -\text{Re}[H_V(\omega; \Omega, h)]/\pi$. $\dot{X}$ and $\ddot{U}_g(\omega)$ are the Fourier transforms of $\dot{x}$ and $\ddot{u}_g(t)$, respectively. The function $F(\omega)$ is called the "energy transfer function" from the expression of Eq. (4). The symbol i denotes the imaginary unit. The velocity transfer function $H_V(\omega; \Omega, h)$ can be expressed explicitly by

$$H_V(\omega; \Omega, h) = -i\omega/(\Omega^2 - \omega^2 + 2ih\Omega\omega) \qquad (5)$$

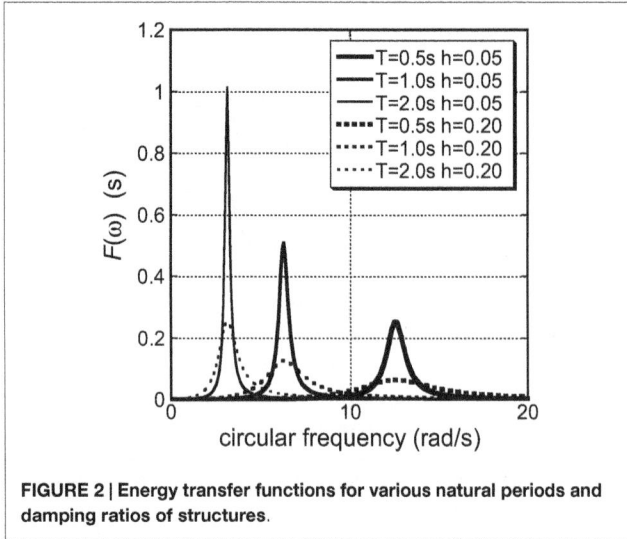

FIGURE 2 | Energy transfer functions for various natural periods and damping ratios of structures.

The energy transfer function $F(\omega)$ can then be expressed by

$$F(\omega) = \frac{2h\Omega\omega^2}{\pi\{(\Omega^2 - \omega^2)^2 + (2h\Omega\omega)^2\}} \quad (6)$$

Equation (4) indicates that the earthquake input energy to a damped linear elastic SDOF system does not depend on the phase property of input motions and this fact is well known (Page, 1952; Lyon, 1975; Kuwamura et al., 1994; Ordaz et al., 2003; Takewaki, 2004a,b, 2005a,b; Takewaki and Fujita, 2009; Kojima et al., 2015). It can also be understood from Eq. (4) that the function $F(\omega)$ defined in Eq. (6) plays an important role in the evaluation of the earthquake input energy and may have some influence on the investigation of constancy property of the earthquake input energy for structures with various model parameters (natural period and damping ratio). The functions $F(\omega)$ for various natural periods $T = 0.5, 1.0, 2.0$ s and damping ratios $h = 0.05, 0.20$ are plotted in **Figure 2**. It is noteworthy that the area of $F(\omega)$ can be proved to be constant regardless of Ω and h. This fact for any damping ratio has already been pointed out by Ordaz et al. (2003). Its proof has been presented by Takewaki (2004a) and the property is shown in the following section.

Property of Energy Transfer Function

Consider the earthquake input energy to the SDOF model subjected to single impulse $\ddot{u}_g(t) = V\delta(t)$ with $|\ddot{U}_g(\omega)| = V$, where $\delta(t)$ is the Dirac delta function. From Eq. (4), this earthquake input energy in a normalized form can be evaluated by

$$E_I/(mV^2) = \int_0^\infty F(\omega)\,d\omega = \frac{1}{2} \quad (7)$$

Equation (7) can be proved by taking into account that single impulse $\ddot{u}_g(t) = V\delta(t)$ with $|\ddot{U}_g(\omega)| = V$ is equivalent to the impulsive loading with the initial velocity of V in time domain (Takewaki, 2004a). Another proof may be possible with the help of the residue theorem (Ordaz et al., 2003; Takewaki, 2004a). This property of Eq. (7) will be used effectively in deriving the upper bounds of input energy subjected to double impulse in the Section "Bounds of Earthquake Input Energy to SDOF System Subjected to Double Impulse."

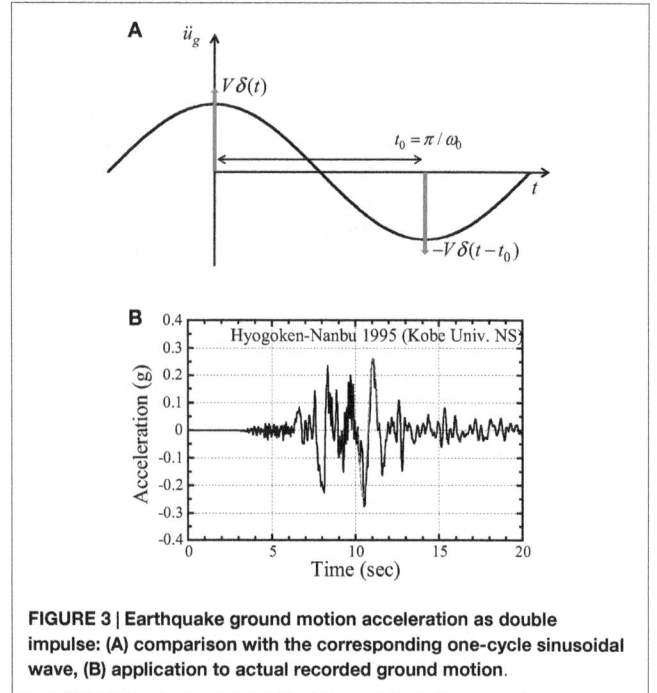

FIGURE 3 | Earthquake ground motion acceleration as double impulse: (A) comparison with the corresponding one-cycle sinusoidal wave, (B) application to actual recorded ground motion.

Double Impulse Input

Consider a ground motion acceleration $\ddot{u}_g(t)$ as double impulse, as shown in **Figure 3A**, expressed by

$$\ddot{u}_g(t) = V\delta(t) - V\delta(t - t_0) \quad (8)$$

where V is the given initial velocity and $t_0 = \pi/\omega_0$ (ω_0: interval circular frequency) is the time interval between two impulses. The comparison with the corresponding one-cycle sinusoidal wave is also plotted in **Figure 3A**. The application of the double impulse and one-cycle sinusoidal wave to an actual recorded ground motion (NS-component at Kobe University during 1995 Hyogoken-Nanbu earthquake) is shown in **Figure 3B**. It can be observed that a one-cycle sinusoidal wave of a predominant period 1.2 s can be a good substitute of a part of this recorded ground motion. The corresponding velocity and displacement of such double impulse are plotted in **Figure 4**. The Fourier transform of $\ddot{u}_g(t)$ can be derived as

$$\begin{aligned}\ddot{U}_g(\omega) &= \int_{-\infty}^\infty \{V\delta(t) - V\delta(t - t_0)\}e^{-i\omega t}dt \\ &= \int_{-\infty}^\infty \{V\delta(t)e^{-i\omega t} - V\delta(t-t_0)e^{-i\omega t_0}e^{-i\omega(t-t_0)}\}dt \\ &= V(1 - e^{-i\omega t_0})\end{aligned}$$

$$(9)$$

The squared Fourier amplitude of the double impulse can then be computed as

$$|\ddot{U}_g(\omega)|^2 = V^2(2 - 2\cos\omega t_0) \quad (10)$$

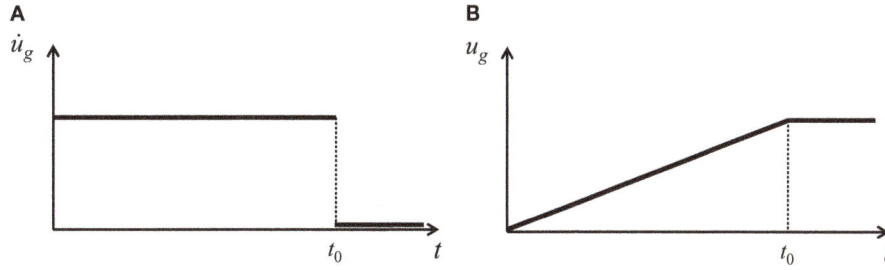

FIGURE 4 | Velocity and displacement of double impulse: (A) velocity, (B) displacement.

Earthquake Input Energy by Double Impulse and the Corresponding Critical Excitation Problem

The substitution of Eq. (10) into Eq. (4) leads to

$$E_I/(mV^2) = \int_0^\infty F(\omega)(2 - 2\cos\omega t_0)\mathrm{d}\omega \qquad (11)$$

The critical excitation problem (Drenick, 1970; Takewaki, 2001a, 2013; Abbas and Manohar, 2002; Moustafa et al., 2010) can be formulated as

[*Critical Excitation Problem*]: Find the double impulse interval t_0 for the fixed double impulse velocity V so as to maximize the earthquake input energy $E_I/(mV^2)$.

By using $F(\omega) = 0$ at $\omega = 0$, $\omega \to \infty$ and the integration by parts,

$$\int_0^\infty F(\omega)\cos\omega t_0 \mathrm{d}\omega = \left[F(\omega)\frac{\sin\omega t_0}{t_0}\right]_0^\infty$$

$$- \int_0^\infty \left\{\frac{\mathrm{d}}{\mathrm{d}\omega}F(\omega)\right\}\frac{\sin\omega t_0}{t_0}\mathrm{d}\omega, \qquad (12a)$$

it can be shown that

$$\lim_{t_0 \to \infty} \int_0^\infty F(\omega)\cos\omega t_0 \mathrm{d}\omega = 0 \qquad (12b)$$

$$\left(\therefore \lim_{t_0 \to \infty} \int_0^\infty F(\omega)(2 - 2\cos\omega t_0)\mathrm{d}\omega = \int_0^\infty 2F(\omega)\mathrm{d}\omega\right)$$

$$\lim_{t_0 \to 0} \int_0^\infty F(\omega)(2 - 2\cos\omega t_0)\mathrm{d}\omega = 0 \qquad (12c)$$

Then, the normalized earthquake input energy $E_I/(mV^2)$ with respect to t_0 can be sketched as shown in **Figure 5**. The wavy property comes from the timing of the correspondence of peaks of the energy transfer function $F(\omega)$ and the normalized squared Fourier amplitude $(2 - 2\cos\omega t_0)$ of the double impulse shown in **Figure 6**.

The condition to characterize the critical value t_0 maximizing the earthquake input energy can be described as

$$\frac{\partial}{\partial t_0}\{E_I/(mV^2)\} = \frac{\partial}{\partial t_0}\left(\int_0^\infty F(\omega)(2 - 2\cos\omega t_0)\mathrm{d}\omega\right)$$

$$= \frac{\partial}{\partial t_0}\left(\int_0^\infty F(\omega)\cos\omega t_0 \mathrm{d}\omega\right) = 0 \qquad (13)$$

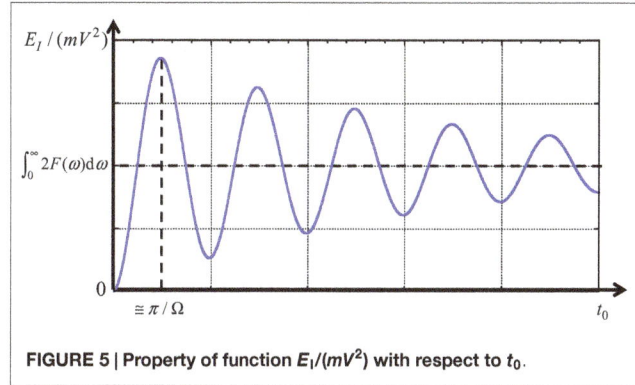

FIGURE 5 | Property of function $E_I/(mV^2)$ with respect to t_0.

This is the stationarity condition and the condition is expressed more explicitly by expanding the manipulation in Eq. (13) as follows.

$$\int_0^\infty F(\omega)\omega\sin\omega t_0 \mathrm{d}\omega = 0 \qquad (14)$$

The solution to the present critical excitation problem can be obtained as the first peak of $E_I/(mV^2)$ as shown in **Figure 5**.

Figures 6A–C show two examples of the relation of energy transfer function with the normalized squared Fourier amplitude of ground motion (double impulse: $t_0 = 1.0$ s). The normalized squared Fourier amplitude of the corresponding one-cycle sinusoidal wave is also plotted in **Figure 6C**. The normalization has been done for the square V^2 of velocity amplitude. It can be understood that the double impulse is a good substitute of a one-cycle sinusoidal wave except the amplitude within a certain range. Since the first peak plays an important role as shown later (see "Numerical Example"), this limited correspondence is sufficient for the present formulation. It is further observed that the one-cycle sinusoidal wave exhibits a frequency characteristic slightly shorter than that for the double impulse. This is because, zero initial conditions of velocity and displacement are used for the one-cycle sinusoidal wave and the period of velocity and displacement waves become slightly shorter than that of the acceleration.

Bounds of Earthquake Input Energy to SDOF System Subjected to Double Impulse

Uncertainties exist in the Fourier amplitude of the double impulse. For example, the value V in Eq. (8) may be uncertain. If the value V becomes smaller, its Fourier amplitude can be bounded by the

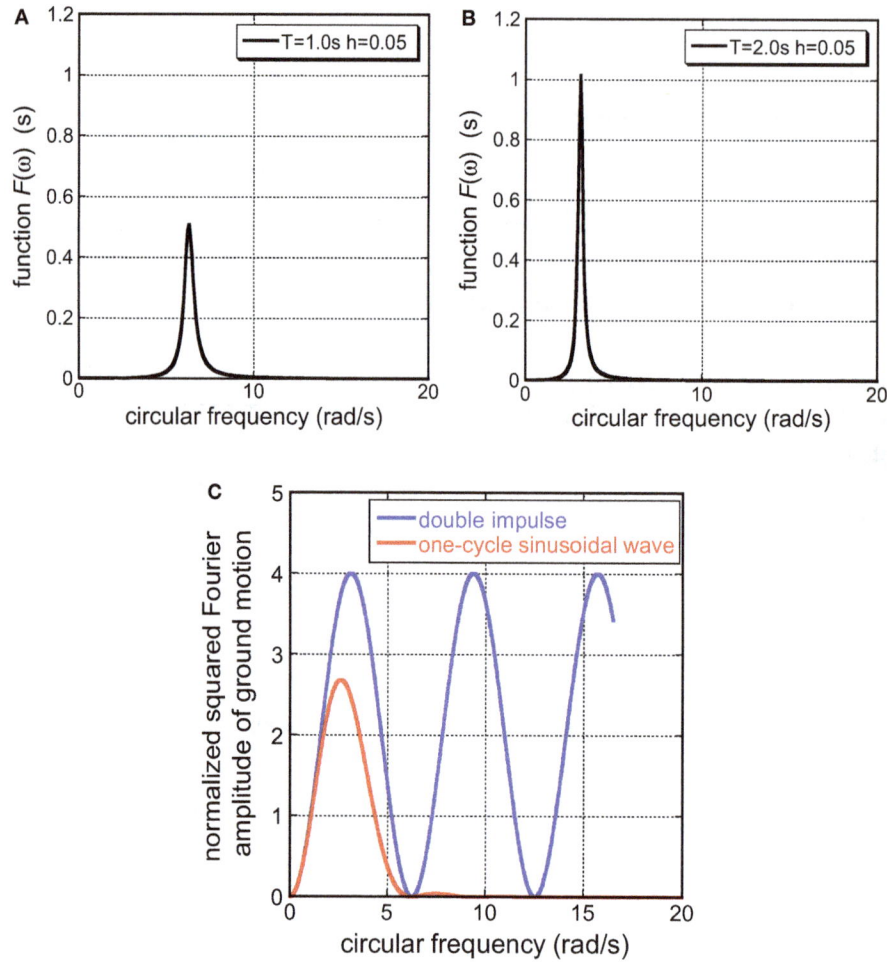

FIGURE 6 | Relation of energy transfer function with normalized squared Fourier amplitude of double impulse: **(A)** energy transfer function with $T = 1.0$ s, $h = 0.05$, **(B)** energy transfer function with

$T = 2.0$ s, $h = 0.05$, **(C)** normalized squared Fourier amplitude of double impulse [$t_0 = 1.0$ s] and the corresponding one-cycle sinusoidal wave.

original one. It is therefore meaningful to discuss the upper bound of the earthquake input energy to the SDOF system.

Consider the bounds of the scaled earthquake input energy defined by Eq. (11). Since the energy transfer function $F(\omega)$ in Eq. (6) is usually positive, it is sufficient to discuss the envelope function of $\left| \ddot{U}_g(\omega) \right|^2$.

Let E_I^U and $\hat{E}_I$ denote the upper bound and the proposed narrower upper bound of the earthquake input energy using a narrower bound of Fourier amplitude (see **Figure 7**). E_I^U and $\hat{E}_I$ can then be derived as follows:

$$E_I^U/(mV^2) = 2 \qquad (15a)$$

$$E_I/(mV^2) = \int_0^{\omega_U} F(\omega)\left[4 - \{4 - (2 - 2\cos\omega t_0)\}\right] d\omega$$
$$+ \int_{\omega_U}^{\infty} F(\omega)(2 - 2\cos\omega t_0) d\omega$$
$$\leq \int_0^{\omega_U} F(\omega)\left[4 - \{4 - (2 - 2\cos\omega t_0)\}\right] d\omega$$

$$+ \int_{\omega_U}^{\infty} 4F(\omega) d\omega$$
$$= \int_0^{\infty} 4F(\omega) d\omega$$
$$- \int_0^{\omega_U} F(\omega)\{4 - (2 - 2\cos\omega t_0)\} d\omega$$
$$= 2 - \int_0^{\omega_U} F(\omega)(2 + 2\cos\omega t_0) d\omega = \hat{E}_I/(mV^2)$$

$$(15b)$$

In Eq. (15b), ω_U denotes the upper limit of circular frequency for computation of integration shown in **Figure 7**. The term $\{4 - (2 - 2\cos\omega t_0)\}$ in Eq. (15b) indicates the shaded portion in **Figure 7**. The validity of inequality in Eq. (15b) can be proven by the property of $F(\omega)$ as a positive function and the relation $0 \leq 2 - 2\cos\omega t_0 \leq 4$. The positivity of $F(\omega)$ can be shown from the fact that, if some parts of $F(\omega)$ are negative, it contradicts the positivity of the energy consumption (total input energy) in the SDOF model subjected to an infinitely long sinusoidal ground motion expressed by a Dirac delta function at the corresponding

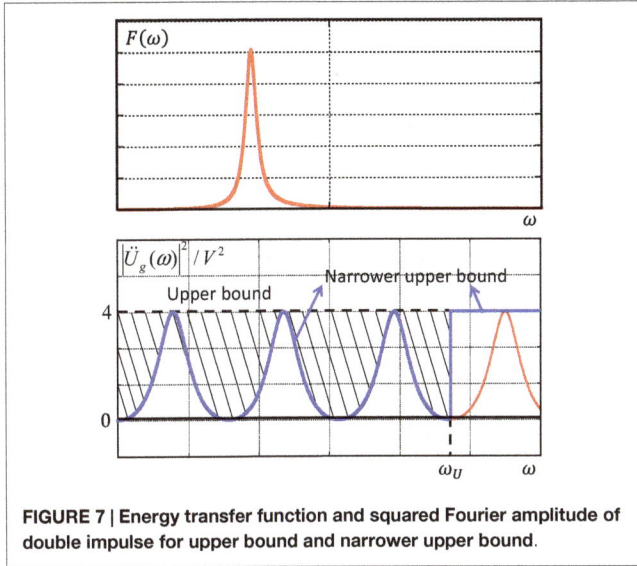

FIGURE 7 | Energy transfer function and squared Fourier amplitude of double impulse for upper bound and narrower upper bound.

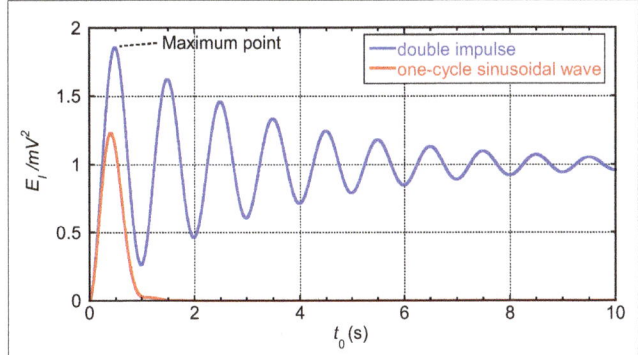

FIGURE 8 | Normalized input energy $E_I/(mV^2)$ with respect to t_0 for $\Omega = 2\pi$ rad/s and $h = 0.05$ under double impulse and the corresponding one-cycle sinusoidal wave.

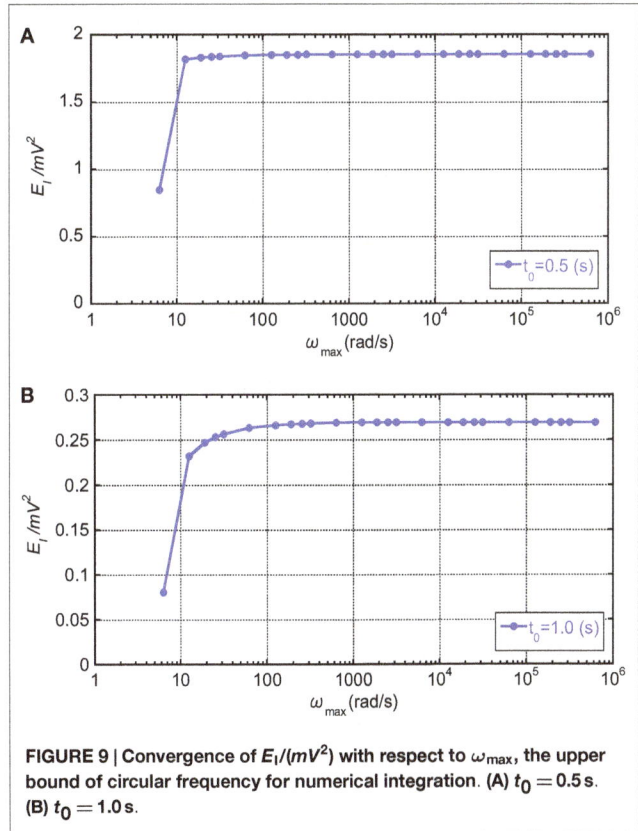

FIGURE 9 | Convergence of $E_I/(mV^2)$ with respect to ω_{max}, the upper bound of circular frequency for numerical integration. **(A)** $t_0 = 0.5$ s. **(B)** $t_0 = 1.0$ s.

frequency. Equation (15b) enables the evaluation of the upper bound of the scaled earthquake input energy without infinite integration by taking full advantage of Eq. (7).

Numerical Example

The validity of the solution to the critical excitation problem shown in the Section "Double Impulse Input" and the accuracy of the upper bound derived in the Section "Earthquake Input Energy by Double Impulse and the Corresponding Critical Excitation Problem" are demonstrated here. The accuracy of the frequency-domain formulation in the computation of earthquake input energy was demonstrated by Ordaz et al. (2003) and Takewaki (2004b) through the comparison with the result by the time-domain formulation.

Figure 8 shows the normalized input energy $E_I/(mV^2)$ with respect to t_0 for the SDOF model of $\Omega = 2\pi$ rad/s and $h = 0.05$ under the double impulse and the corresponding one-cycle sinusoidal wave. The principal property for the one-cycle sinusoidal wave can be captured by the double impulse except the amplitude (i.e., the critical period of the one-cycle sinusoidal wave can be obtained within a good approximation). The amplitudes depend on the normalization of both inputs and their difference does not cause any difficulty because the principal objective is to obtain the critical period of the one-cycle sinusoidal wave and the critical interval of the double impulse. In addition, a slightly shorter period characteristic can be observed for the one-cycle sinusoidal wave. This phenomenon results from the fact explained in **Figure 6**. It should be noted that the SDOF structural model is fixed as explained just before, and the interval of the double impulse is varied for finding the critical interval. During the variation of the interval, the velocity amplitude V is kept constant. Such a treatment of variation of the interval may be difficult for recorded ground motions because the amplitude of acceleration has to be changed depending on the interval for the constant velocity amplitude. It can be observed that $E_I/(mV^2)$ exhibits the maximum value approximately at $t_0 = 0.5$ s because

$\pi/\Omega = 0.5$ ($\Omega = 2\pi$ rad/s) in this model and certainly converges to $\int_0^\infty 2F(\omega)d\omega$, i.e., 1. From the practical point of view, t_0 should be set between 0.25 and 2 s in order to express the characteristic period between 0.5 and 4 s of pulse-type waves.

Figure 9 illustrates the convergence of $E_I/(mV^2)$ with respect to ω_{max}, the upper bound of circular frequency for numerical integration. It can be understood that $\omega_{max} = 20$ rad/s is almost sufficient for the estimation of the maximum value, which occurs approximately at $t_0 = 0.5$ s in this case. On the other hand, $\omega_{max} = 100$ rad/s is necessary for the estimation of the local minimum value, which occurs approximately at $t_0 = 1.0$ s.

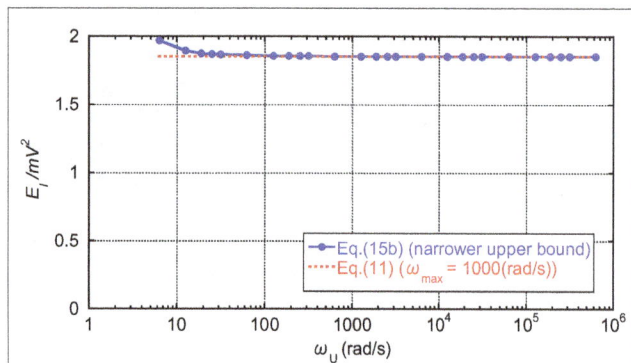

FIGURE 10 | Comparison of narrower upper bound given by Eq. (15b) with the exact one by Eq. (11).

This difference may result from the fact that, while the error is relatively large at the local minimum point, the error is relatively small at the maximum point. Since only the maximum point is meaningful, the large setting of ω_{max} for the local minimum value of earthquake input energy does not cause any problem.

Figure 10 presents the comparison of the narrower upper bound given by Eq. (15b) with the exact one by Eq. (11). The parameters are $\Omega = 2\pi$ rad/s, $h = 0.05$, and $\omega_{max} = 1000$ rad/s. The upper bound given by Eq. (15a) is 2.0. It can be observed that, as ω_U (upper limit circular frequency for numerical integration) becomes larger than 20 rad/s, the narrower upper bound given by Eq. (15b) converges rapidly to the exact one by Eq. (11). About the triple of the fundamental natural circular frequency [$\Omega = 2\pi$ rad/s in this case] seems to be sufficient for practical computation.

Conclusion

The conclusions may be summarized as follows:

(1) When the ground motion is white-like (constant Fourier amplitude spectrum), the input energy to the structure is constant regardless of the natural period and damping ratio of the structure, i.e., input energy constant property. This input corresponds to single pulse. The input energy constant property can be proved by considering the physical meaning of the constant Fourier spectrum of the input ground motion

in the time domain, i.e., the input of initial velocity at zero time.

(2) Double impulse is more realistic because the input frequency characteristic can be introduced. A critical excitation problem with an interval of two impulses as a variable can be formulated in the frequency domain and the solution to that critical excitation problem can be derived by drawing the graph of the normalized earthquake input energy with respect to the interval of two impulses. The solution to the present critical excitation problem can be obtained as the first peak of the normalized earthquake input energy.

(3) An upper bound and a narrower upper bound of the earthquake input energy to a SDOF model under double impulse input can be derived by taking full advantage of the property of the energy transfer function that the area of the energy transfer function is constant (input energy constant property) and introducing the envelope function in the Fourier transform of the double impulse input. The narrower upper bound enables the evaluation of the upper bound of the normalized earthquake input energy without infinite integration.

(4) Numerical examples demonstrate that the double impulse represents a one-cycle sinusoidal wave as an approximation of near-fault ground motions and can capture the critical property of near-fault ground motions (i.e., the critical period of the one-cycle sinusoidal wave can be obtained within a good approximation). Furthermore, the proposed upper bound of earthquake input energy can converge to an exact value as the upper limit of frequency ω_U for numerical integration becomes larger.

Only elastic structures have been treated for simple presentation of the theory and the present method takes advantage of the energy transfer function approach, which can be used for elastic structures. An equivalent linearization technique (Caughey, 1960; Roberts and Spanos, 1990; Takewaki, 2001b) may be promising for inelastic structures. This formulation will be presented in the future.

Acknowledgments

Part of the present work is supported by the Grant-in-Aid for Scientific Research of Japan Society for the Promotion of Science (No. 24246095, No. 15H04079). This support is greatly appreciated.

References

Abbas, A. M., and Manohar, C. S. (2002). Investigations into critical earthquake load models within deterministic and probabilistic frameworks. *Earthquake Eng. Struct. Dyn.* 31, 813–832. doi:10.1002/eqe.124.abs

Akiyama, H. (1985). *Earthquake Resistant Limit-State Design for Buildings.* Tokyo: University of Tokyo Press.

Berg, G. V., and Thomaides, T. T. (1960). "Energy consumption by structures in strong-motion earthquakes," in *Proceedings of 2nd World Conference on Earthquake Engineering*, Tokyo, 681–696.

Caughey, T. K. (1960). Random excitation of a system with bilinear hysteresis. *J. Appl. Mech.* 27, 649–652. doi:10.1115/1.3644077

Drenick, R. F. (1970). Model-free design of aseismic structures. *J. Eng. Mech. Div.* 96, 483–493.

Fajfar, P., and Vidic, T. (1994). Consistent inelastic design spectra: hysteretic and input energy. *Earthquake Eng. Struct. Dyn.* 23, 523–537. doi:10.1002/eqe. 4290230505

Housner, G. W. (1959). Behavior of structures during earthquakes. *J. Eng. Mech. Div.* 85, 109–129.

Housner, G. W. (1975). "Measures of severity of earthquake ground shaking," in *Proc. of the US National Conf. on Earthquake Engineering*, Ann Arbor, MI, 25–33.

Housner, G. W., and Jennings, P. C. (1975). "The capacity of extreme earthquake motions to damage structures," in *Structural and Geotechnical Mechanics: A Volume Honoring N.M.Newmark*, ed. W. J. Hall (Englewood Cliff, NJ: Prentice-Hall), 102–116.

Kojima, K., Sakaguchi, K., and Takewaki, I. (2015). Mechanism and bounding of earthquake energy input to building structure on surface ground subjected to engineering bedrock motion. *Soil Dyn. Earthquake Eng.* 70, 93–103. doi:10.1016/j.soildyn.2014.12.010

Kuwamura, H., Kirino, Y., and Akiyama, H. (1994). Prediction of earthquake energy input from smoothed Fourier amplitude spectrum. *Earthquake Eng. Struct. Dyn.* 1994, 1125–1137. doi:10.1002/eqe.4290231007

Leger, P., and Dussault, S. (1992). Seismic-energy dissipation in MDOF structures. *J. Struct. Eng.* 118, 1251–1269. doi:10.1061/(ASCE)0733-9445(1992)118:5(1251)

Lyon, R. H. (1975). *Statistical Energy Analysis of Dynamical Systems.* Cambridge, MA: The MIT Press.

Mavroeidis, G. P., and Papageorgiou, A. S. (2003). A mathematical representation of near-fault ground motions. *Bull. Seismol. Soc. Am.* 93, 1099–1131. doi:10.1785/0120020100

Moustafa, A., Ueno, K., and Takewaki, I. (2010). Critical earthquake loads for SDOF inelastic structures considering evolution of seismic waves. *Earthquake Struct.* 1, 147–162. doi:10.12989/eas.2010.1.2.147

Ohi, K., Takanashi, K., and Tanaka, H. (1985). A simple method to estimate the statistical parameters of energy input to structures during earthquakes. *J. Struct. Construct. Eng. Archi. Inst. Jpn.* 347, 47–55.

Ordaz, M., Huerta, B., and Reinoso, E. (2003). Exact computation of input-energy spectra from Fourier amplitude spectra. *Earthquake Eng. Struct. Dyn.* 32, 597–605. doi:10.1002/eqe.240

Page, C. H. (1952). Instantaneous power spectra. *J. Appl. Phys.* 23, 103–106. doi:10.1063/1.1701949

Riddell, R., and Garcia, J. E. (2001). Hysteretic energy spectrum and damage control. *Earthquake Eng. Struct. Dyn.* 30, 1791–1816. doi:10.1002/eqe.93

Roberts, J. B., and Spanos, P. D. (1990). *Random Vibration and Statistical Linearization.* New York, NY: Wiley.

Takewaki, I. (2001a). A new method for nonstationary random critical excitation. *Earthquake Eng. Struct. Dyn.* 30, 519–535. doi:10.1002/eqe.21

Takewaki, I. (2001b). Probabilistic critical excitation for MDOF elastic-plastic structures on compliant ground. *Earthquake Eng. Struct. Dyn.* 30, 1345–1360. doi:10.1002/eqe.66

Takewaki, I. (2004a). Bound of earthquake input energy. *J. Struct. Eng.* 130, 1289–1297. doi:10.1061/(ASCE)0733-9445(2004)130:9(1289)

Takewaki, I. (2004b). Frequency domain modal analysis of earthquake input energy to highly damped passive control structures. *Earthquake Eng. Struct. Dyn.* 33, 575–590. doi:10.1002/eqe.361

Takewaki, I. (2005a). Bound of earthquake input energy to soil-structure interaction systems. *Soil Dyn. Earthquake Eng.* 25, 741–752. doi:10.1016/j.soildyn.2004.11.017

Takewaki, I. (2005b). Frequency domain analysis of earthquake input energy to structure-pile systems. *Eng. Struct.* 27, 549–563. doi:10.1016/j.engstruct.2004.11.014

Takewaki, I. (2013). *Critical Excitation Methods in Earthquake Engineering*, 2nd Edn. Oxford: Elsevier.

Takewaki, I., and Fujita, K. (2009). Earthquake input energy to tall and base-isolated buildings in time and frequency dual domains. *J. Struct. Des. Tall Spec. Build.* 18, 589–606. doi:10.1002/tal.497

Takewaki, I., Moustafa, A., and Fujita, K. (2012). *Improving the Earthquake Resilience of Buildings: The Worst Case Approach.* London: Springer.

Takewaki, I., and Tsujimoto, H. (2011). Scaling of design earthquake ground motions for tall buildings based on drift and input energy demands. *Earthquake Struct.* 2, 171–187. doi:10.12989/eas.2011.2.2.171

Trifunac, M. D. (2008). Energy of strong motion at earthquake source. *Soil Dyn. Earthquake Eng.* 28, 1–6. doi:10.1039/c4sm00280f

Trifunac, M. D., Hao, T. Y., and Todorovska, M. I. (2001). *On Energy Flow in Earthquake Response.* Report CE 01-03. University of Southern California, Los Angeles.

Uang, C. M., and Bertero, V. V. (1990). Evaluation of seismic energy in structures. *Earthquake Eng. Struct. Dyn.* 19, 77–90. doi:10.1002/eqe.4290190108

Xu, Z., Agrawal, A. K., He, W.-L., and Tan, P. (2007). Performance of passive energy dissipation systems during near-field ground motion type pulses. *Eng. Struct.* 29, 224–236. doi:10.1016/j.engstruct.2006.04.020

Zahrah, T. F., and Hall, W. J. (1984). Earthquake energy absorption in SDOF structures. *J. Struct. Eng.* 110, 1757–1772. doi:10.1061/(ASCE)0733-9445(1984)110:8(1757)

Conflict of Interest Statement: The authors declare that the research was conducted in the absence of any commercial or financial relationships that could be construed as a potential conflict of interest.

17

Ground Motion Characteristics of the 2015 Gorkha Earthquake, Survey of Damage to Stone Masonry Structures and Structural Field Tests

Rishi Ram Parajuli[1]* and Junji Kiyono[2]

[1]*Department of Urban Management, Graduate School of Engineering, Kyoto University, Kyoto, Japan,* [2]*Graduate School of Global Environmental Studies, Kyoto University, Kyoto, Japan*

On April 25, 2015, a M7.8 earthquake rattled central Nepal; ground motion recorded in Kantipath, Kathmandu, 76.86 km east of the epicenter suggested that the low-frequency component was dominant. We consider data from eight aftershocks following the Gorkha earthquake and analyze ground motion characteristics; we found that most of the ground motion records are dominated by low frequencies for events with a moment magnitude >6. The Gorkha earthquake devastated hundreds of thousands of structures. In the countryside, and especially in rural mountainous areas, most of the buildings that collapsed were stone masonry constructions. Detailed damage assessments of stone masonry buildings in Harmi Gorkha was done, with an epicentral distance of about 17 km. Structures were categorized as large, medium, and small depending on their plinth area size and number of stories. Most of the structures in the area were damaged; interestingly, all ridge-line structures were heavily damaged. Moreover, Schmidt hammer tests were undertaken to determine the compressive strength of stone masonry and brick masonry with mud mortar for normal buildings and historical monuments. The compressive strengths of stone masonry and brick masonry were found to be 12.38 and 18.75 MPa, respectively. Historical structures constructed with special bricks had a compressive strength of 20.50 MPa. Pullout tests were also conducted to determine the stone masonry-mud mortar bond strength. The cohesive strength of mud mortar and the coefficient of friction were determined.

Keywords: Gorkha earthquake, ground motion characteristics, damage survey, stone masonry, field test, Schmidt hammer test

Edited by:
Katsuichiro Goda,
University of Bristol, UK

Reviewed by:
Siau Chen Chian,
National University of Singapore,
Singapore
Rama Mohan Pokhrel,
University of Tokyo, Japan

*Correspondence:
Rishi Ram Parajuli
parajuli.ram.27z@st.kyoto-u.ac.jp

INTRODUCTION

Nepal lies in an active seismic zone in the Himalayan belt within the boundary between the Eurasian and Indian plates. Records of large earthquakes that have devastated Nepal, claiming a significant number of lives, have been kept for more than seven centuries. On June 7, AD 1255, a mega earthquake was the first ever documented earthquake in the region; it was likely to have had an intensity of MMI X and killed about one-third of the people in the current capital Kathmandu, including King Abhaya Malla of the Malla era [BECA World International (New Zealand) et al.,

1993]. Other major historical earthquakes occurred in 1408, destroying the Machhendra Nath temple in Patan, 1681 and 1810. Bilham (1995) stated that the major earthquake event of August 26, 1833 had a moment magnitude of 7.5–7.9 with a possible rupture length of 70 km and an epicenter located 50 km North or North-East of Kathmandu, and was preceded by two large foreshocks that took place 5 h and 15 min prior to the main shock. This alarmed people and caused them to stay outside their houses, thereby probably saving many lives. Another well-known devastating earthquake prior to the Gorkha earthquake was the Nepal–Bihar earthquake of 1934 with a Richter magnitude of 8.4. Bramha Smasher JBR stated in his book (Rana, 1935) that the 1934 mega-earthquake claimed 8,591 lives in total with 4,296 in Kathmandu valley, and destroyed 56,231 structures, including 492 temples and schools. A Richter magnitude 6.6 earthquake in August 1988 was another earthquake that devastated the eastern part of Nepal, having its epicenter in Udayapur. This earthquake claimed 721 lives in eastern Nepal, along with injuries to 6,213 people. A total of 14,965 dwellings were completely destroyed, most of which were constructed with mud-stone or clay brick masonry (Sato et al., 1989).

The Gorkha earthquake that struck on April 25 at 11:56 a.m. (NST) had an epicenter in Barpak, Gorkha. It ruptured to the east of the epicenter for a length of about 100 km at a strike angle of 295° (USGS, 2015). The size of this earthquake is 7.8 in moment magnitude and is 7.6 in local magnitude, as measured by Nepal's seismological center (NSC). The recent Gorkha earthquake claimed a total of 8,857 lives (as of August 8) (Government of Nepal, 2015). The greatest death toll was in the Sindhupalchok district, in the eastern part of Nepal, near to the estimated end point of the rupture. In this region, a total of 3,532 people lost their lives, whereas just 1,573 were seriously injured due to the quake. Most of the structures in this district are stone masonry buildings with mud mortar, reinforced with concrete frame structures exist only in few small towns (Central Bureau of Statistics, 2012). The district with the next highest death toll was the capital, Kathmandu, where 1,226 deaths were recorded, along with injuries to 7,952 people. Considering the three districts in the Kathmandu valley, the total death toll rises to 1,739, significantly more than that in Gorkha, the district where the epicenter was located, where the death toll was 449.

The death toll was affected by the timing of event, as it happened at noon when most of people in the hardest hit areas were out of their houses at work in the fields. Another factor that lowered the death toll and damage was the low-frequency dominant component of ground motion. The main shock of the earthquake had dominant frequencies of roughly 0.23, 0.23, and 0.27 Hz corresponding to the East-West (EW), North-South (NS), and Up-Down (UD) components recorded in Kathmandu. Recorded ground acceleration of the Gorkha earthquake in Kathmandu shows the peak value of <200 cm/s^2, where probabilistic seismic hazard analysis of Nepal suggested that PGA is around 100 cm/s^2 considering return period of 98 years and 450 cm/s^2 for return period of 475 years in soft soil areas (Parajuli et al., 2008). In this study, we analyze the characteristics of ground motions for nine earthquake events, including the "main shock." Ground

motion data recorded by the USGS in Kathmandu (station code KATNP) have been downloaded from the strong motion archive (CESMD, 2015). Nepal has a total population of nearly 26.5 million, with about 17% of the people in urban areas and the rest in rural areas. Almost half of the population of Nepal lives in the relatively flat Terai region, with hilly areas retaining 43% of the population, and only 7% in the mountainous region. Building structure types used throughout Nepal are shown in **Figure 1**; most of the structures are of stone/brick masonry with mud mortar (SBMM); in the Terai region, stone/brick masonry with cement mortar (SBCM) is also common. Reinforced cement concrete (RCC) structures have only a small share, whereas wooden frame structures (WFS) are widely used in the Terai region. Structural types that cannot be characterized as above are specified as other (OTH), along with structures not specified (NS) during data collection.

The structures built to provide shelter for half the populations of the country in the hilly and mountainous region are mostly of stone masonry with mud mortar. Specifically, SBMM constructions account for 50% of buildings in hilly regions and 47% of buildings in mountainous regions. The use of SBMM for outer wall construction in rural areas is nearly 83%. In mountainous and hilly areas, 93 and 65% use SBMM for foundation, and 89 and 62% use SBMM for the outer walls, respectively. Outside the Kathmandu valley, with 19.63% of the fatalities, the death toll is much higher in mountainous and hilly areas, such as Sindhupalchok, Nuwakot, Dhading, Rasuwa, and Gorkha with 3,532, 1,109, 679, 660, and 449 deaths, respectively, and accounting for 73% of the total (Government of Nepal, 2015). A map of these five districts and the Kathmandu valley with locations of epicenter of the main shock and aftershocks are shown in **Figure 2**. The Gorkha earthquake most greatly affected areas with a greater share of SBMM constructions. Sindhupalchok district (92% of buildings), Nuwakot (93%), Dhading (87%), Rasuwa (90%), and Gorkha (88%) are all dominated by structures with such foundation. In those five districts, 90% of structures were built with mud mortar (Central Bureau of Statistics, 2012). In hilly areas, stone is commonly locally available, so more of the structures are built with it. Studying the damage patterns for such structures, and developing corresponding countermeasures for those, has to be in focus to increase the resiliency of such structures in rural areas.

Some damage surveys have been already conducted since the Gorkha earthquake. Goda et al. (2015) revealed that the damage scenario is not widespread, but localized in the Kathmandu valley. The damage assessments in the small towns of Melamchi, Trishuli, and Baluwa found that majority of stone and brick masonry buildings were severely damaged. We conducted a detailed damage survey in Harmi, a rural village in the Gorkha district, where all of the structures are made of stone masonry with mud mortar.

Local building materials in rural Nepal are spatially variable, even within a few kilometers. However, the general construction methods in rural Nepal consist of a foundation of stone masonry with mud mortar that rises up to a ridge supporting the outer walls. Timber columns and beams are commonly used to support

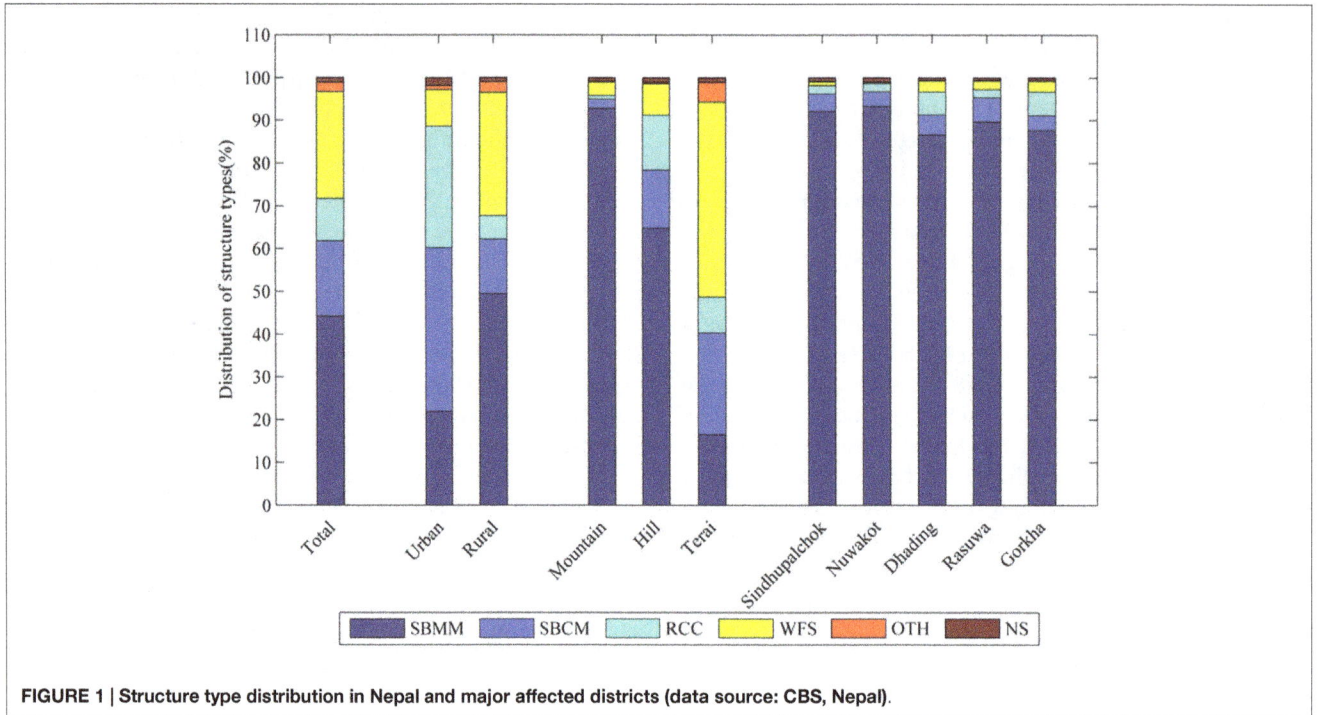

FIGURE 1 | Structure type distribution in Nepal and major affected districts (data source: CBS, Nepal).

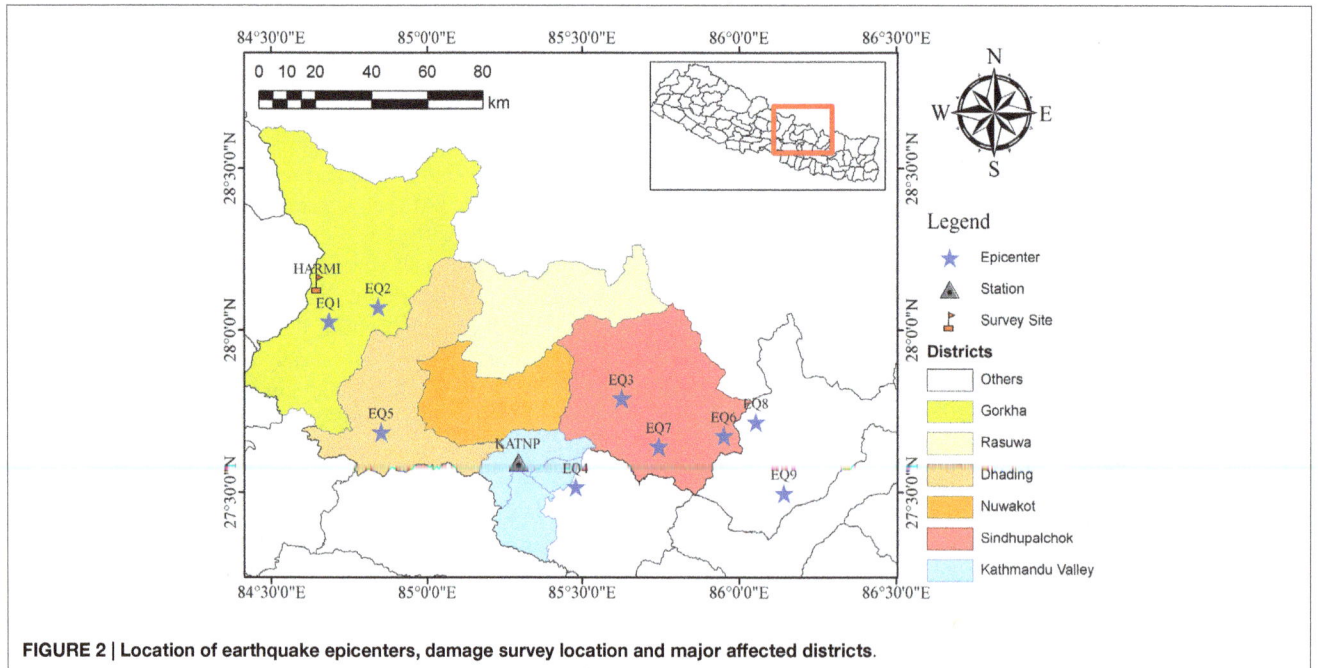

FIGURE 2 | Location of earthquake epicenters, damage survey location and major affected districts.

extended roofs and slabs as intermediate support. The material properties of such structures are not commonly studied, so we have undertaken field pullout tests to assess the strength of mortar. Similarly, Schmidt hammer tests have also been used for stone masonry structures and brick masonry structures, even though they are not well defined for use with stone masonry. In comparison to typical buildings in the region, historical monuments usually have special types of materials used in construction; mostly they consist of special brick masonry in three layers (inner, outer, and infill layers) with mud (Ranjitkar, 2000), and occasionally with lime-surkhi mortar. We have also tested the strength of such walls in the Gorkha palace using the Schmidt hammer.

GROUND MOTION CHARACTERISTICS

Nepal does not have a dense network of accelerometers; however, the USGS has established a station (KATNP) that records earthquakes in the capital, Kathmandu, and data from that station are analyzed in this paper. In total, nine independent datasets available from strongmotioncenter.org are analyzed and discussed here.

Table 1 presents detailed information regarding trigger dates and times, moment magnitudes, the locations of epicenters, and the epicentral distances from the recording station KATNP (27.7120°N, 85.3155°E). Earthquakes are numbered 1–9, with EQ1 representing the main shock, and EQ8 the major aftershock to the east of the fault plane. Earthquake events range from moment magnitude 5.2–7.8, with epicentral distances as far as 83.90 km and as near as 18.5 km.

The spatial distribution of the earthquakes extends to the east and west of the recording station, which help evaluate the effect of directivity of the seismic waves. **Figure 2** shows the location of the earthquakes relative to the recording station (KATNP) in Kathmandu and the damage survey site Harmi. Data are sampled at an interval of 0.005 s, and the length of recorded data varies for each event. For analysis, we have chosen a record length of 81.92 s (16,384 samples). This data selection of 2^{14} samples facilitates using fast Fourier transforms, which require a power of 2 for calculation. Records that are shorter than the required length were extended with null values for the remaining duration.

Ground motion, Fourier spectra and response spectra of the EW components of all earthquake events are shown in **Figure 3**, respectively, from left to right. All of the events are stacked into a single figure where base line accelerations for EQ1, EQ2, EQ3, EQ4, EQ5, E6, EQ7, EQ8, and EQ9 are 0, 300, 400, 500, 600, 700, 800, 900, and 1000 cm/s², respectively, as shown by dotted lines in the figure. The main shock of the Gorkha earthquake had an epicentral distance of 76.86 km NW from KATNP; maximum recorded accelerations were 155, 162, and 184 cm/s² for the EW, NS, and UD components, respectively. Fourier transforms to the frequency domain showed that all three components were dominated by low frequencies. **Figure 3** clearly shows that the dominant frequencies of large aftershocks (EQ2, EQ6, and EQ8) are low: even the small ones are in a higher range. In contrast to the Fourier spectra, spectral accelerations (**Figure 3**) show aftershock ground motions that are greater and in a higher frequency

range, even though the main shock has a higher value over a lower range of frequencies (0.22 Hz).

Figure 4 shows the dominant frequencies of all earthquakes in all three directions. In four of the events [EQ1 (M7.8), EQ2 (M6.6) EQ6 (M6.7), and EQ8 (M7.3)], all of the components are dominated by low frequencies ≤ 1 Hz. Three of the events [EQ3 (M5.5), EQ4 (M5.3), and EQ5 (M5.2)] have dominant frequencies in all three components ≥ 1 Hz. EQ7 (M5.3) is low-frequency dominant in the EW and NS components, while the UD component had a slightly higher value of 1.26 Hz. The final event, EQ9 (M6.3), has variable frequency content, with peak Fourier amplitudes for the EW component at 0.28 Hz, the NS component at 2.43 Hz, and the UD component at 1.17 Hz.

The response of a structure to earthquake ground motion with a single degree of freedom is represented by response spectra for various natural frequency ranges for the structure. A damping ratio of 5% is assumed in the calculation of response spectra. **Figure 5** shows the tripartite plot of pseudo velocity spectra (centimeter per second) with axes for displacement (centimeter) and pseudo acceleration (square centimeter). Four earthquake events (EQ1, EQ2, EQ6, and EQ8) exceeded a velocity of 10 cm/s with peak values in range of 0.2–0.5 Hz. Despite EQ1, the main shock, other earthquake events had a small peak in the higher frequency range of 0.8–3 Hz but the main shock surges only at a lower frequency range with crossing value of 100 cm/s in range of 0.08–0.2 Hz. EQ9 also has the same trend as the other three stated above, but the value peaks at slightly <10 cm/s. Apart from EQ3, EQ7, and EQ8, the other events crossed the spectral acceleration value of 100 cm/s² in the range of 2.5–10 Hz; EQ4 and EQ5 have a peak value only in this range.

The response acceleration of the Gorkha earthquake (EQ1) has an almost flat shape in the range of 0.3–10 Hz. Maximum displacement during the main shock was nearly 300 cm for the structure with a frequency of nearly 0.25 Hz at a velocity of 380 cm/s and 500 cm/s² as acceleration. The phenomenon of such spectral parameters will be discussed briefly later in the discussion.

The characteristics of ground motion have an impact on damage scenarios all over the affected area. Low-rise masonry and reinforced concrete buildings in the Kathmandu valley have high natural frequency. Super high-rise, base isolated buildings could have suffered severe damage if they had been built in the affected area. The natural frequencies of various structures are shown in

TABLE 1 | Earthquake data.

SN	Description	Time (UTC)	Magnitude (M_w)	Location		Epicentral distance (km)
				Latitude	Longitude	
1	EQ1	25-04-015 06:11	7.8	28.1473	84.7079	76.86
2	EQ2	25-04-015 06:45	6.6	28.1927	84.8645	69.30
3	EQ3	25-04-015 06:56	5.5	27.9100	85.6501	33.00
4	EQ4	25-04-015 08:55	5.3	27.6364	85.5029	18.50
5	EQ5	25-04-015 23:16	5.2	27.8052	84.8744	43.60
6	EQ6	26-04-015 07:09	6.7	27.7945	85.9739	67.20
7	EQ7	26-04-015 16:26	5.3	27.7612	85.7704	44.80
8	EQ8	12-05-015 07:05	7.3	27.8368	86.0772	75.10
9	EQ9	12-05-015 07:36	6.3	27.6180	86.1659	83.90

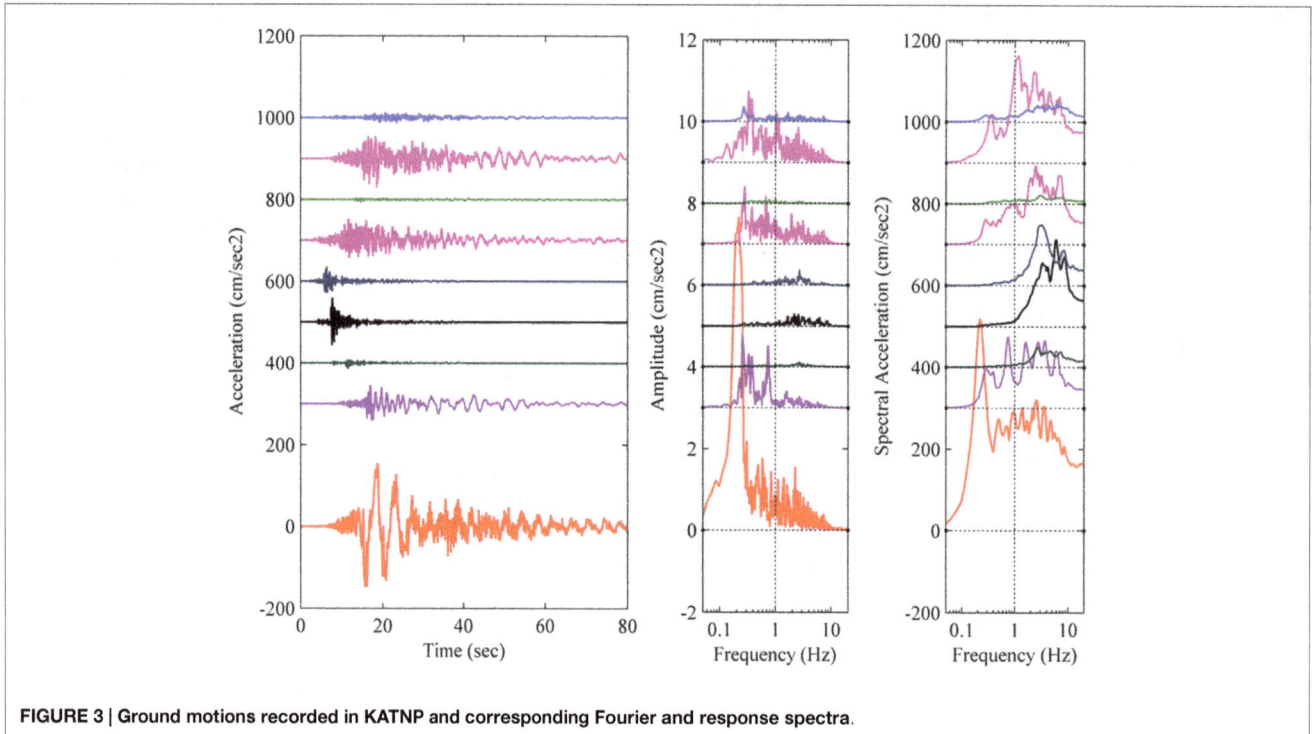

FIGURE 3 | Ground motions recorded in KATNP and corresponding Fourier and response spectra.

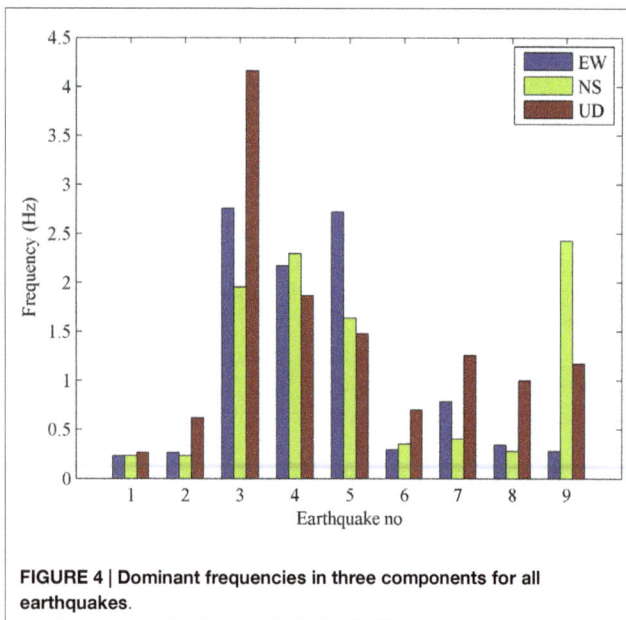

FIGURE 4 | Dominant frequencies in three components for all earthquakes.

Figure 6. Damage of any structure during earthquake directly relates to the strength itself and the amount of earthquake force that pushed it. Strength of the structures relies on materials used and the technique of construction. We found that in most affected areas, people live in stone masonry buildings with mud mortar, which is vulnerable for lateral loads. Even though earthquake ground motion records outside the Kathmandu valley are unavailable, we attempt to evaluate damage scenarios in rural areas. Most of the structures are two stories and some are three stories.

The natural frequency of such structures is not so low to resonate with earthquake ground motion frequency.

DAMAGE SURVEY

Most settlements in the mountainous region of Nepal are in rural areas that are dominated by shelters constructed with stone masonry. Brick masonry structures and reinforced concrete structures are found in a few areas, mainly newly developed towns and areas accessible by road. The epicenter of the earthquake was in Barpak, Gorkha, which is a rural mountainous area where all of the structures are stone masonry with mud mortar with an exception of a few reinforced concrete buildings.

We chose a cluster of 149 structures in Harmi, Gorkha. The location is 165 km from Kathmandu by road. It is reached by following the Prithvi highway to the west up to Dumre, then along the Dumre–Beshisahar–Chame highway to Turture, and from there along the Turture–Palungtar road to Harmi. This area is about 17 km SW of the epicenter of the main shock. The topography of the area was selected as it starts from the ridge of a mountain, at an altitude of 1162 m extending down to 600 m at the bottom of a hill (**Figure 7**). We found that the damage scenario in these rural areas was localized with topography, so we plotted the locations of surveyed structures on a contour map of the area. To construct the contours, we used a free-source digital elevation model (Aster Gdem, 2009), with an accuracy of 30 m. Image tile "N28E084" was used as a base and the data were extracted for the study area. Contour lines were drawn at interval of 20 m. The north facing slope of the study area has a small local ridge at a level between 880 and 960 m. Another main ridge of that hill is found above 1160 m.

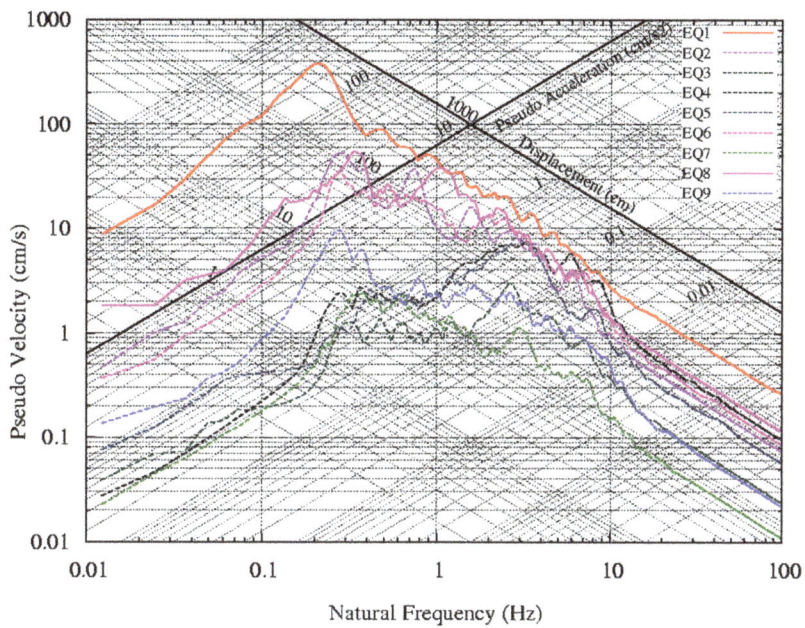

FIGURE 5 | Combined velocity, displacement, and acceleration response spectra of the earthquakes.

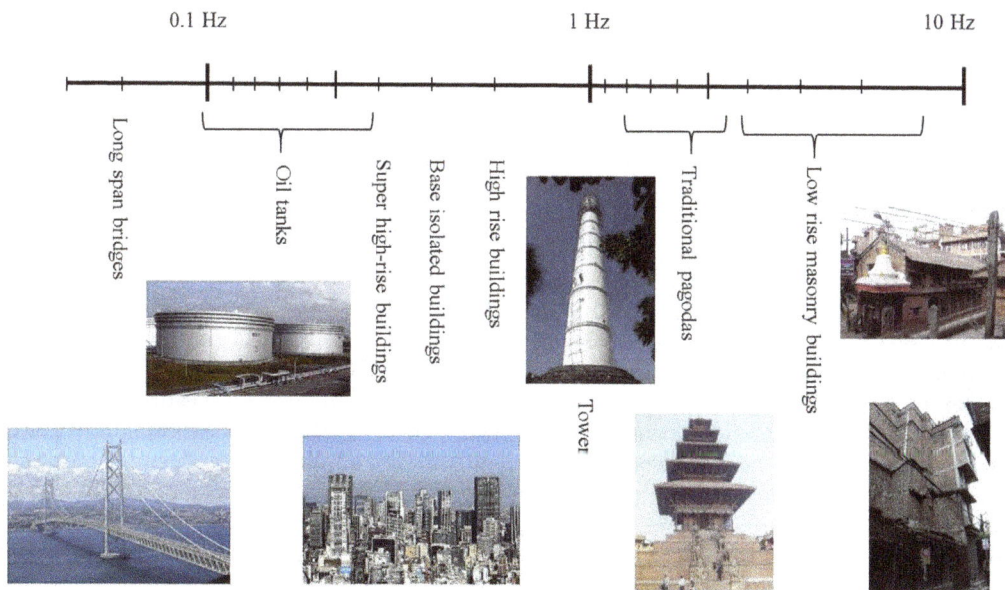

FIGURE 6 | Natural frequency of various structures.

The structures were categorized into three groups by size, where all of the structures are of stone masonry with mud mortar; a few of them have cement pointing on their outer faces. Large-sized structures (L) are of two to four stories and are larger in plinth area (around 75 m²). Medium-sized structures (M) are single or double storied, having plinth area in the range of 45–75 m². The rest of the structures fall under the small (S)

category. The damage grades used in the study lie in the range from 0 and 5, where 0 denotes no damage and 5 represents totally collapsed in all sides. A damage grade of 4 represents severely damaged structures where only cracked ground floor walls still stand, and the roof and upper floors have been brought down to the ground. Structures with severe damage but with building shape preserved, albeit with major cracks in the walls or partial

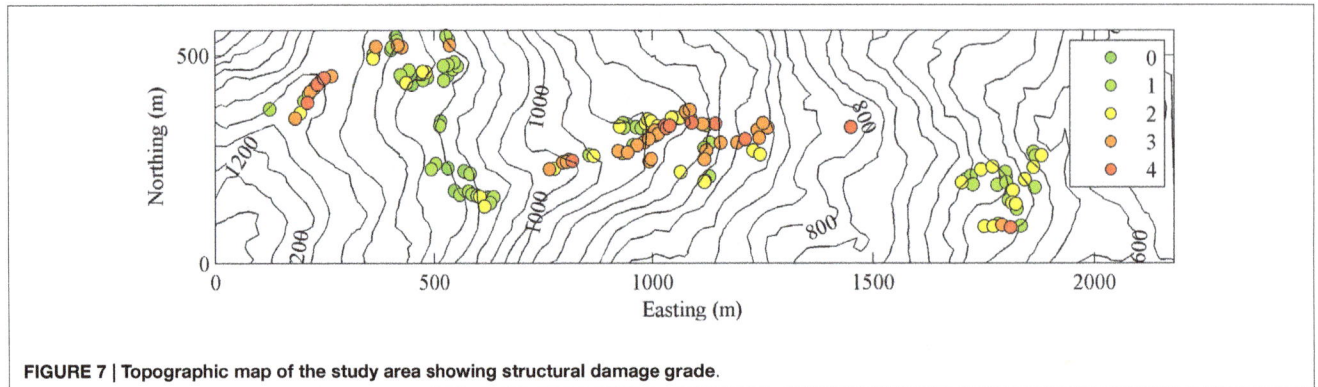

FIGURE 7 | Topographic map of the study area showing structural damage grade.

collapses, are categorized as grade 3. These structures are accessible with special precautions taken. Structures having a few major cracks in walls, but accessible even though they are not habitable without intensive maintenance work, fall under grade 2. The structural category for grade 1 corresponds to excessive minor cracks throughout the walls; these structures are habitable with little maintenance work. Intact structures with no damage or only a few minor cracks, which are habitable with little or no maintenance work, are categorized in grade 0.

In the study cluster, there are 149 structures consisting of 58 large, 68 medium, and 23 small-sized buildings with 39, 46, and 15% of weightage, respectively. Damage grade and location were recorded using GPS at the site. **Figure 7** shows the damaged structures on a topographic map. Green colored dots represent grade 0 structures, whereas red dots represent the location of grade 4 structures (as we do not have any grade 5 structures). **Figure 8** shows the damage grade of structures with percentages of structures that include categories of structure sizes. We found that 8% of the structures had a damage grade of 0; 38% were in grade 1; and grade 2 and grade 3 structures were 24% each. The remaining 6% of the structures were damaged severely, at grade 4.

Here, we can see most of the structures fall under damage grades 1, 2, and 3, with less coming from grades 0 and 4. From the survey, we found that most of the buildings on ridge lines suffered heavy damage but those on side-slopes were not damaged as much. The study area comprises an area that includes a mountain ridge along with a local ridge line formed on the middle of the slope. Hence, we categorized the structures as ridge-line structures, those that are located on the ridge line. There are a total of 52 structures located on the ridge line, including the main and local ridge lines. Damage grade details of the structures on the ridge line are shown in **Figure 9**. There are no structures that fall under grade 0; in fact only 15% of the structures graded as 1 with 19% in grade 2. More than half of the structures, i.e., 52%, were grade 3 and the remaining 14% fell under grade 4. A small structure that was graded as 3 on the ridge line is shown in Image S1 in Supplementary Material.

The failure mechanisms of structures constructed with stone masonry with mud mortar are mainly seen in two categories. Delamination of the wall is the major failure mechanism and shear failure is secondary. The methods for constructing stone masonry with mud mortar are based around building two wall layers: an inner and an outer; however, this layered single wall

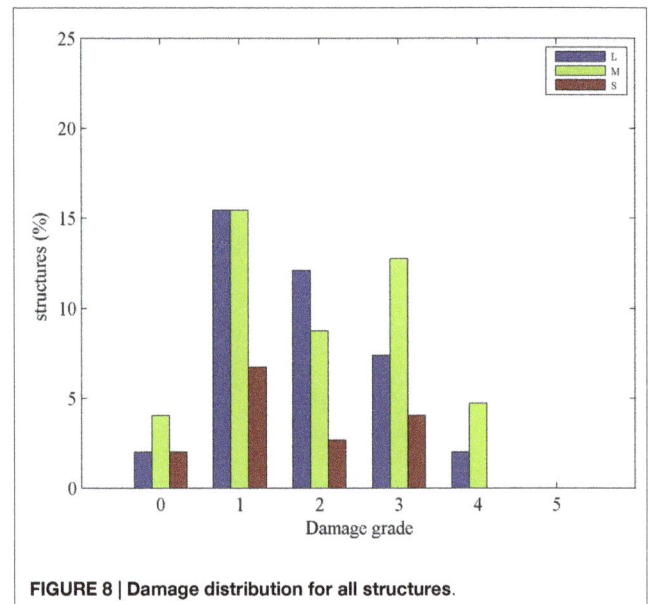

FIGURE 8 | Damage distribution for all structures.

can be a main cause for delamination. Bonding of the inner and outer walls does not exist, which causes the wall to act as two independent walls during an earthquake, thereby causing severe damage. There are many structures with vertical cracks appearing in association with the shear failure of the wall. Structures with horizontal bands of chiseled stone have a few cracks compared to those without the horizontal bans.

FIELD TEST

Pullout Test to Assess Bonding Strength of Mud Mortar

Materials used in local constructions are not of any specific standard. Most of the stone masonry structures in rural Nepal are constructed using local stone and mud. The properties of such materials are not well known. After the Bam earthquake in Iran, adobe and masonry structures were investigated to further characterize the bonding strength of mortar (Kiyono and Kalantari, 2004). We have done similar simple field tests here to determine material properties.

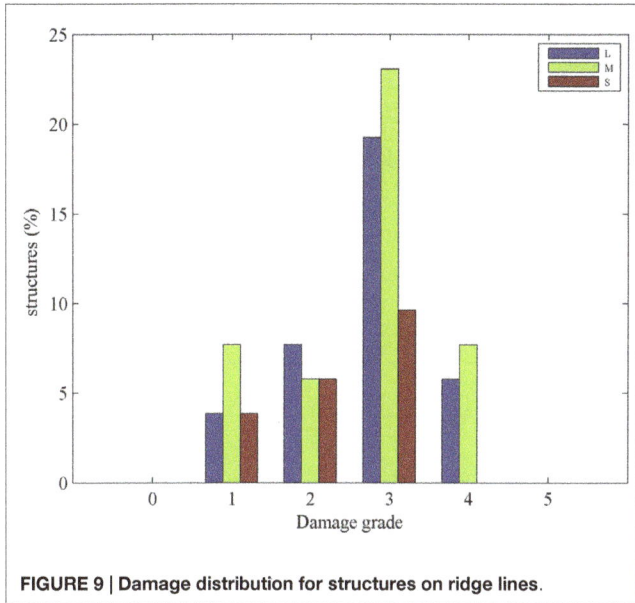

FIGURE 9 | Damage distribution for structures on ridge lines.

We conducted pullout tests in the field to determine the bonding strength of stone and mud mortar joints. Damaged buildings were chosen for sampling, selecting the most undisturbed sample from the remaining parts of a structure. Sample stone was carefully freed on three sides so that only the bottom remained bonded with mud mortar. A simple weighting gage and a rope to connect with sample and weighing gage were used in test. Weighing gage consisted of the spring type gage that shows the pulling force in kilogram, which can be adjusted in some range to make it 0. Weighing gage was tightened with a rope that bound the stone from the sides. We set the force applied during the stretching of rope to 0 from adjustable screw. Force was applied gradually to pull the stone out and the reading in the weighing gage (S) was recorded. After pulling out the stone, we measured the mortar joint area (A) that exactly bonded with the stone, ignoring voids at the joint surface. The weight of the sample stone (W) was also measured to facilitate the calculation of the normal stress acting on this surface.

Three samples were taken to calculate normal stress ($\sigma = W/A$) and shear stress ($\tau = S/A$) (shown in **Table 2**). Equation 1 shows the theoretical relationship of shear and normal stress with bonding stress (c) and coefficient of friction (µ), considering the equilibrium of forces in the horizontal direction.

$$\tau = c + \mu\sigma \qquad (1)$$

In fitting the data from the test result, we found the value of cohesive strength (c) and coefficient of friction (µ) of stone masonry with mud mortar joints to be 0.001137 MPa, and 0.6, respectively.

Samples of the test are not enough to conclude the material strength; hence, we compare these values with the test results from the 2003 Iran Bam earthquake damage survey (Kiyono and Kalantari, 2004). Results of the test conducted in Iran and Nepal are shown in **Figure 10**. In Iran, tests were conducted for sun-dried and baked brick masonry structures, where the shear

strengths of mortar bonding were estimated to be 0.0029 and 0.0097 MPa, respectively. The frictional coefficient for the joint was found to be 0.62 and 0.54, respectively, for sun dried and baked brick masonry. Test results from Nepal show that the frictional coefficient lies between the values of sun-dried and baked masonry structures in Iran, but shear bonding strength is much lower than that of both brick masonry structures.

Schmidt Hammer Test

A non-destructive test device, the Schmidt hammer, is often used to determine the surface hardness and penetration resistance of concrete or rock. Even though the device is designed for concrete structures, we have successfully used it for stone and brick masonry structures. To use a Schmidt hammer for stone and brick masonry structures, we must assume that the masonry components themselves stand as uniform blocks with mortar forming the matrix between hard elements. Rebounds of a hammer depend on the strength of the mortar too and, therefore, represent the overall strength of the masonry structure. There are some drawbacks in this assumption, but we anticipate that these measurements might be used as a reference for future studies. We conducted the test at several points on the surface of the structure, with a minimum distance between test points set to 30 mm. Conversion of rebound numbers to the probable strength of the structure is done using a chart based on the pressure resistance on a 15 cm cube of concrete, as provided by the manufacturer (Proceq, 2006). Categorically, we discuss three types of structures, i.e., stone masonry, brick masonry with mud mortar, and historical monument structure.

Stone Masonry with Mud Mortar

Stone masonry structures were tested at two sites in Harmi, Gorkha. One was a large structure constructed 38 years ago that had collapsed up to the first floor, but with intact ground floor walls (Image S3 in Supplementary Material). The other one was a small structure. In the large structure, we conducted the test at 26 points where we found large variations in rebound numbers. Some locations in joint areas could not show the data (i.e., they were below the lowest range value for hammer 10) and in some locations there were relatively large stone blocks that caused high rebound values and led to an overestimation of strength. Hence, we disregard data below the lower range and above rebound number 30; which corresponds to 26 MPa. A total of four data points from each of the lower and higher ranges were omitted and the remaining 18 data were taken into consideration to calculate the strength. Average rebound numbers range between 15 and 30 with an average of 21.47 and a standard deviation of 4.4. From the conversion chart, we found that the compressive strength of stone masonry is 12.0 MPa.

Similarly, we conducted the test on a small structure where a total of eight points were sampled. This structure was built only two and half years ago. In this structure, we did not find lower values, as there was a band of relatively large stone blocks. Ignoring two points having rebound number values >30, the remaining six data points had an average of 21.5 and a standard deviation of 5.12. Using the conversion chart, the probable strength of the stone masonry with mud mortar was found to be 12.75 MPa. From

TABLE 2 | Pullout test for bonding strength.

Sample no.	Weight (N)	Joint area (mm²)	Normal stress σ (MPa)	Pullout force (N)	Shear stress τ (MPa)
1	129.49	24,000.00	0.0053955	107.91	0.00449625
2	103.01	28,000.00	0.00367875	85.35	0.00304811
3	56.90	19,500.00	0.00291785	60.82	0.00311908

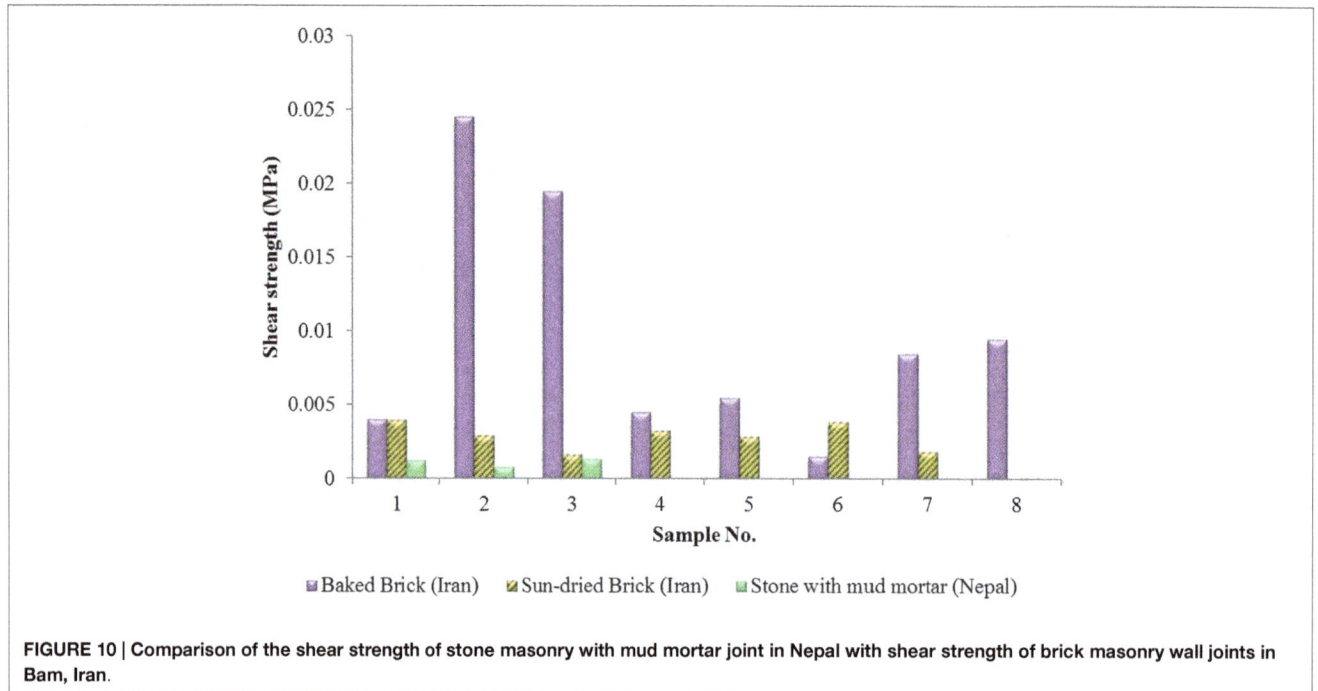

FIGURE 10 | Comparison of the shear strength of stone masonry with mud mortar joint in Nepal with shear strength of brick masonry wall joints in Bam, Iran.

these two tests of stone masonry with mud mortar structures, the probable compressive strength was determined to be 12.38 MPa, roughly the average of the sampled structures.

Brick Masonry with Mud Mortar

Brick masonry with mud mortar structures are common in newly developed towns and cities in Nepal. For testing, we chose a small two-story building that had some cracks in the walls due to the earthquake. This structure is located in Palungtar municipality, Gorkha. The load-bearing main wall of the structure had dimensions of 3750 mm × 5700 mm with a thickness of 350 mm and a height of 3900 mm. Four sampling points were selected in the short wall side of the structure, maintaining 35 mm for the edge distance. As the walls of the structure were cracked, we can make measurements in just four locations. Rebound numbers recorded in those points are 22, 27, 25, and 28. Hence, the average rebound value is 25.5 with a SD of 2.65, corresponding to a probable compressive strength of 18.75 MPa.

Historical Brick Masonry Structure with Mud Mortar

We chose the historical monument of the Gorkha durbar for structural testing (Image S4 in Supplementary Material). This structure was originally built in AD 1640 and was made of brick masonry and timber. This monument stands on a ridge of the same hill that hosts the Gorkha bazar on its southern slope. The

structure experienced severe damage during the main earthquake at an epicentral distance of 27 km. The Gorkha durbar is a three-story building with a tile roof. We selected sampling points on the ground floor wall along two basal lines: one 380 mm from plinth level and another 350 mm above the first. Horizontal pitches of the sampling points were taken at 500 mm. A total of 42 blows were made on the wall, with the highest and lowest rebound numbers being 50 and 11, respectively. During the test we found, in some places, a brick element that was not intact and caused lower rebound values. Hence, we neglect such sampling points during the calculations. By not using two sampling points, we end up with 40 samples to evaluate the strength of the masonry wall in the historical structure. Rebound numbers ranged from 20 to 50, with an average of 32.4 and a SD of 6.92. From the conversion chart, the corresponding probable compressive strength of the wall is 29.5 MPa.

DISCUSSION

The earthquake ground motion observed during the Gorkha earthquake was dissimilar from previous earthquakes in the region. Many researchers expect that the triggering of such an earthquake would damage lots of structures in Kathmandu and claim tens of thousands of lives (Dixit et al., 2000; Wyss, 2005), which overestimates the actual toll by at least an order of magnitude. One of the main reasons behind less damage is that the low-frequency ground motion reduced vulnerability in high-frequency structures. Most residential housing in the affected area does not have a natural frequency low enough to be in resonance with the ground motion recorded in Kathmandu.

The characteristics of ground motion alone, as recorded in KATNP, cannot adequately define the phenomena of such acceleration time history. The likelihood of amplification of the low-frequency component by soil strata is high, but is this the only reason for slow ground motions in Kathmandu? People who were surveyed in Gorkha concerning the shaking pattern and described the scene as buildings moving to and fro and trees behaving like swings. Considering these observations, we can argue that the source of the earthquake had an effective rupture mechanism that radiated low-frequency dominant ground motions. This was not only so for the main shock but also for the aftershocks, which had similar low-frequency component characteristics recorded at KATNP. This supports the evidence for low-frequency amplifying behavior in the soils of the Kathmandu basin. We should also consider the non-linearity of soil behavior; excitations with higher acceleration cause soil layers to act as filters for the high-frequency components while amplifying low frequencies with the resonance effect. Epicentral distance also has a key role in components of frequency range; events with spectra with higher frequencies correspond to nearer events, and those having low frequencies are generally distant events. Smaller events of less than moment magnitude 6 have higher frequency dominant acceleration, including the M6.3 event EQ9 aftershock on 12 May, which had high-frequency dominance. The dominant frequencies of all components for all earthquake events are shown in **Figure 11**, as related to epicentral distance and moment magnitude. These data recorded with high-frequency dominance focus the issues back on the characteristics of the source of earthquake mechanism not only in the local site condition that are responsible for the generation of ground motion events with different dominant frequencies.

In hilly areas, where most of the structures are built of stone masonry with mud mortar, damage along ridge lines is particularly notable. Structures located on slopes, with foundations lying over some layers of soil, generally had very low levels of damage even at short epicentral distances. The conventional thought of building safe houses on ridge lines, over hard rock foundations now becomes suspect. Local site effects of ground motion tended to amplify high-frequency components along ridge lines where bedrock is shallower. Previously, we showed a figure of a damage scenario in **Figure 9**. Now, considering the ratio of total structures to ridge structures, damage scenarios of higher grades are mostly concentrated along ridges. **Table 3** shows the percentage of structures damaged on a ridge line in the study area.

Large structures on ridge lines constituted 100% of the grade 4 damage. Damage at a grade 3 level also has a higher contribution from ridge-line structures. There are few structures having damage at grade 2 or even grade 1 level that exist due to special attention during construction. Horizontal bands of chiseled stone were used for a more esthetic appearance and also had external cement pointing on the walls.

During the 2011 Mw6.9 Nepal–Sikkim earthquake damage in stone masonry structures was reported widely, where delamination of walls is the major failure pattern (Shakya et al., 2013). They mentioned about the severe damages in Taplejung, Ilam, and Panchthar districts of Nepal, up to about 90 km (distance to Ilam bazar) as epicentral distance. We have similar topography in mountainous area so we can compare the scenario in eastern part (affected by the Nepal–Sikkim earthquake) and mid and western part (affected by the Gorkha earthquake). Spreading of damage due to Gorkha earthquake is not that high as compared to that of the smaller Nepal–Sikkim earthquake.

The pullout test conducted in damaged structures to find out the joint properties. Here, we compare the data with the test conducted in Iran after the Bam earthquake where the test was done similarly on the damaged structures. Shear strength of the joint from the test is very low that can be neglected for the modeling but frictional coefficient of mortar joint found significant.

The Schmidt hammer test for stone masonry with mud mortar was performed on walls of two, large and small structures having damage grade of 4 and 3, respectively. The wall itself in the area of hammer blow was intact (only with some minor cracks), which reflected on low bouncing values. We had neglected such values during the analysis; hence, we can generalize the result for all cases. The structure built up of brick masonry with mud mortar had some cracks in other sides but tested wall was intact during the time of test. The historical Gorkha durbar had also suffered from some damages on other sides but the front wall, where test was conducted had minor cracks with loosening of cladding bricks, which also appeared in result that we excluded for analysis. Hence, all the test results are not affected significantly by damage state of the structure.

Material properties for old masonry structures in Kathmandu studied previously (Parajuli et al., 2011) proposed the compressive strength of brick to be 11 MPa, where the same for mortar and wall are 1.6 and 1.8 MPa, respectively. Results from our tests in comparison with the previous study are almost ten times higher for wall strength. If we consider only the brick element, the resulting value from this test is almost 50% more than those experiments conducted previously. Brick quality for the experiment used in tested structures is different; hence, the results we obtained are able to take into account the compressive strength of stone and brick element itself rather than the integrated wall with mortar.

CONCLUSION

The ground motion characteristics of the Gorkha earthquake seem unique. The reasons for such characteristics require high priority research in the field of seismology. Source mechanisms, directivity, wave paths, and local site conditions should be investigated intensively. The western part of Nepal has a large seismic gap. Earthquakes with the same or even stronger shaking may occur in near future. The Gorkha earthquake had low-frequency ground motion with accelerations of <200 cm/s^2, but the velocity was relatively high which caused damage. One of the reasons behind the collapse of many historical structures, including Dharahara,

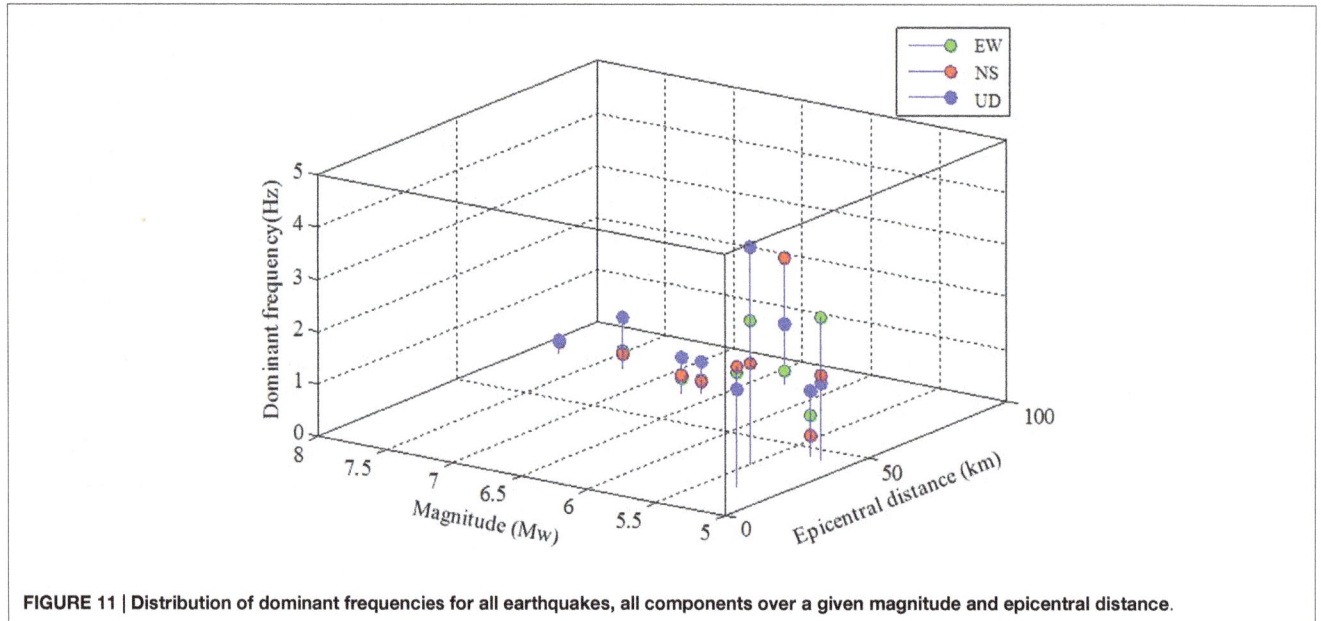

FIGURE 11 | Distribution of dominant frequencies for all earthquakes, all components over a given magnitude and epicentral distance.

TABLE 3 | Percentage of ridge-line structures damaged.

SN	Grade	Structure size		
		L (%)	M (%)	S (%)
1	0	0	0	0
2	1	9	17	20
3	2	22	23	75
4	3	91	63	83
5	4	100	57	NA
6	5	NA	NA	NA

a tower structure monument, in comparison with general buildings, is likely to be lower frequency dominant ground motion. We should consider the epicentral distance and rupture line during the interpretation of ground motion frequency components.

Rural areas in Nepal have a large stock of stone masonry structures used for shelter and other purposes. These need to be reinforced using locally available materials to make them more resilient. Ridge structures are at a higher risk of earthquake damage relative to structures on slopes. Local construction methods should be improved technically, by providing longitudinal and transverse bonding during construction.

The study of material properties used locally should be advanced in order to analyze the structural behaviors of various materials during an earthquake. Even though accuracy could not be assured for the Schmidt hammer tests (designed for reinforced concrete), we have shown test results that provide a probable strength for the stone/brick masonry structures. Stone used in masonry with mud mortar has a probable compressive strength of 12.38 MPa, where local bricks used in masonry with mud mortar have at strength of 18.75 MPa and bricks used in masonry with mud mortar for historical structures are at 29.5 MPa. Note that these results are based only on the surface hardness; masonry structures are not as homogeneous as concrete structures. Also

the strength of the mortar is not well represented in such tests, even though loosening and degradation of mortar result in a drop in rebound number. Hence, these values should be used with caution. The bonding strength of stone masonry with mud mortar was investigated using a pullout test on site, which results in a cohesive strength of mud mortar of 0.001137 MPa, with a coefficient of friction of 0.6. Therefore, to study stone masonry with mud mortar, we can use mortar strength combined with the compressive strength of the stone.

ACKNOWLEDGMENTS

The authors thank Prof. Masakatsu Miyazima, Prof. Prem Nath Maskey, and Dr. Hari Ram Parajuli for providing insight and expertise. We are grateful to the inhabitants of Harmi, Gorkha for their kind support and cooperation during the damage survey in Nepal. We also show our gratitude to the USGS, http://www.strongmotioncenter.org/, for sharing data with us, and we thank the reviewers for their insight.

FUNDING

A part of the research was conducted under the support by JST J-RAPID program and JSPS KAKENHI Grant Number 26249067.

REFERENCES

Aster Gdem. (2009). *Database – Aster Gdem.* Available at: http://gdem.ersdac.jspacesystems.or.jp/

BECA World International (New Zealand), SILT Consultants Pvt. Ltd. (Nepal), TAEC Consult Pvt. Ltd. (Nepal), Golder Associates (Canada), and Urban Regional Research (USA). (1993). *Seismic Hazard Mapping and Risk Assessment for Nepal.* UNDP/HMGN/UNCHS (Habitat) Subproject NEP/88/054/21.03.

Bilham, R. (1995). Location and magnitude of the 1833 Nepal earthquake and its relation to the rupture zones of contiguous great Himalayan earthquakes. *Curr. Sci.* 69, 155–187.

Central Bureau of Statistics. (2012). *National Population and Housing Census 2011 (National Report) Government of Nepal.* Vol. 01. Kathmandu: Government of Nepal.

CESMD. (2015). *Internet Data Reports for Earthquakes of 2015.* Available at: http://strongmotioncenter.org/

Dixit, A. M., Dwelley-Samant, L. R., Nakami, M., and Pradhanang, S. B. (2000). "The Kathmandu valley earthquake risk management project," in *12th World Conference on Earthquake Engineering.* (New Zealand: 12th WCEE held in Auckland), 1–8.

Goda, K., Kiyota, T., Pokhrel, R. M., Chiaro, G., Katagiri, T., Sharma, K., et al. (2015). The 2015 Gorkha Nepal earthquake: Insights from earthquake damage survey. *Front. Built Environ.* 1:8. doi:10.3389/fbuil.2015.00008

Government of Nepal. (2015). *Nepal Disaster Risk Reduction Portal.* Available at: http://drrportal.gov.np/

Kiyono, J., and Kalantari, A. (2004). Collapse mechanism of adobe and masonry structures during the 2003 Iran Bam earthquake. *Bull. Earthquake Res. Inst.* 79, 157–161.

Parajuli, H., Kiyono, J., Ono, Y., and Tsutsumiuchi, T. (2008). Design earthquake ground motions from probabilistic response spectra : case study of Nepal. *J. Jpn. Assoc. Earthquake Eng.* 8, 16–28. doi:10.5610/jaee.8.4_16

Parajuli, H. R., Kiyono, J., and Taniguchi, H. (2011). "Structural assessment of the Kathmandu world heritage buildings," in *Proceedings of the 31 St Conference on Earthquake Engineering* (Tokyo: JSCE), 1–5.

Proceq. (2006). *Proceq Concrete Test Hammer Manual.* Available at: http://www.proceq.com/fileadmin/documents/proceq/products/Concrete/Original_Schmidt/English/Proceq_Operating_Instructions_Original_Schmidt_E.pdf

Rana, B. S. J. R. (1935). *Great Nepal Earthquake 1934 (In Nepali).* Kathmandu: Jorganesh Publishers.

Ranjitkar, R. (2000). *Seismic Strengthening of the Nepalese Pagoda.* Kathmandu: Kathmandu Valley Preservation Trust.

Sato, T., Fujiwara, T., Murakami, H. O., and Kubo, T. (1989). *Reconnaissance Report on the 21 August 1988 Earthquake in the Nepal-India Border Region.* Research Report on natural disasters. Japanese Group for the study of Natural Disaster science.

Shakya, K., Pant, D. R., Maharjan, M., Bhagat, S., Wijeyewickrema, A. C., and Maskey, P. N. (2013). Lessons learned from performance of buildings during the September 18, 2011 earthquake in Nepal. *Asian J. Civ. Eng.* 14, 719–733.

USGS. (2015). *M7.8 Nepal Earthquake of 25 April 2015.* Available at: http://earthquake.usgs.gov/earthquakes/eqarchives/poster/2015/NepalSummary.pdf

Wyss, M. (2005). Human losses expected in Himalayan earthquakes. *Nat. Hazards* 34, 305–314. doi:10.1007/s11069-004-2073-1

Conflict of Interest Statement: The authors declare that the research was conducted in the absence of any commercial or financial relationships that could be construed as a potential conflict of interest.

18

Effect of Non-linearity of Connecting Dampers on Vibration Control of Connected Building Structures

Masatoshi Kasagi, Kohei Fujita, Masaaki Tsuji and Izuru Takewaki*

Department of Architecture and Architectural Engineering, Graduate School of Engineering, Kyoto University, Kyoto, Japan

The connection of two building structures with dampers is one of the effective vibration control systems. In this vibration control system, both buildings have to possess different vibration properties in order to provide a higher vibration reduction performance. In addition to such condition of different vibration properties of both buildings, the connecting dampers also play an important role in the vibration control mechanism. In this paper, the effect of non-linearity of connecting dampers on the vibration control of connected building structures is investigated in detail. A high-damping rubber damper and an oil damper with and without relief mechanism are treated. It is shown that while the high-damping rubber damper is effective in a rather small deformation level, the linear oil damper is effective in a relatively large deformation level. It is further shown that while the oil dampers reduce the response in the same phase as the case without dampers, the high-damping rubber dampers change the phase. The merit is that the high-damping rubber can reduce the damper deformation and keep the sufficient space between both buildings. This can mitigate the risk of building pounding.

Keywords: building connection, passive damper, non-linearity, structural control, high-damping rubber damper, oil damper, relief mechanism, smart structure

Edited by:
Nikos D. Lagaros,
National Technical University of
Athens, Greece

Reviewed by:
Ehsan Noroozinejad Farsangi,
Kerman Graduate University of
Advanced Technology, Iran
Ali Koçak,
Yildiz Technical University, Turkey

***Correspondence:**
Izuru Takewaki
takewaki@archi.kyoto-u.ac.jp*

INTRODUCTION

The connection of multiple building structures with dampers is one of the effective vibration control systems (for example, Iwanami et al., 1996; Luco and de Barros, 1998). In the case where these building structures have different natural frequencies, each building disturbs the vibration of other building structures. Furthermore, it can be assumed that the properties of connecting dampers affect strongly the performance of vibration reduction of such connected building structures. It appears therefore useful to investigate the effect of the non-linear properties of connecting dampers on the vibration reduction mechanisms of connected building structures.

A high-damping rubber damper (Tani et al., 2009) and an oil damper with and without relief mechanism (Soong and Dargush, 1997; Hanson and Soong, 2001) are treated in this paper. The high-damping rubber dampers have a high performance of energy absorption for a cyclic loading and possess not only the large elastic–plastic deformation capacity like metallic hysteretic dampers but also the sufficient amount of viscous damping capacity (see Tani et al., 2009). Especially the large deformation capacity (almost 300% shear deformation) and the extremely large performance for accumulated plastic deformation are two major advantages over other dampers. On the other hand, the relief mechanism is usually adopted in the oil dampers in order to decrease the force applied to surrounding structural members.

As for the vibration control of buildings with connecting dampers, Iwanami et al. (1996) investigated the optimal quantity of connecting elements (stiffness and damping). This theory is well known as "Fixed-Point Theory." In this method, the stiffness of the connecting element is used for adjusting the heights of the fixed points, and the minimum transmissibility has been achieved by the damping of the connecting element. Luco and de Barros (1998) derived an optimal interconnecting element location in the connected building system. Takewaki (2007, 2015) and Fukumoto and Takewaki (2015) introduced the energy approach in the design of buildings with the connecting damper system in which the energy transfer function plays a key role for assessing the effectiveness of the connecting dampers. Cimellaro and Lopez-Garcia (2011) studied the optimal damper distribution in the buildings with a connecting damper system. Patel and Jangid (2011) investigated the response of two buildings connected by friction dampers. Richardson et al. (2013a,b) developed a closed-form expression of the optimal connecting dampers, which minimize the absolute displacement transmissibility.

It is shown in this paper that while the high-damping rubber damper is effective in a rather small deformation level, the oil damper is effective in a relatively large deformation level. Furthermore, it will be remarked that the response velocity in the oil dampers is too large compared to the limit value specifically in low-rise buildings and careful attention should be paid in their installation to low-rise buildings.

MODELING OF NON-LINEAR DAMPERS AND CONNECTED BUILDING STRUCTURES

Modeling of Non-Linear Dampers

The oil damper used in this paper obeys the following damper force f_V – relative velocity $\dot{u}$ relation as shown in **Figure 1**.

$$f_V = \begin{cases} c_V\dot{u} & \text{for } |\dot{u}| \leq |\dot{u}_r| \\ \kappa c_V\dot{u} + \eta_{V,c} & \text{for } |\dot{u}| > |\dot{u}_r| \end{cases} \quad (1)$$

where κ denotes the ratio of the post-relief damping coefficient to the initial value c_V and $\eta_{V,c}$ is the damping force at 0 velocity in the post-relief relation. When the velocity attains the specific value

$\dot{u}_r$, the relief mechanism works and goes into the second branch. An example of the damping force–relative displacement relation of an oil damper without and with the relief mechanism is shown in **Figure 2A**.

The high-damping rubber damper (called high-damping rubber later) used in this paper possesses the hybrid characteristics of hysteretic one like metallic dampers and viscoelastic one (Tani et al., 2009). The damper is used as a shear deformation type and the area and thickness of the high-damping rubber damper are the characteristic parameters. An example of the shear stress–shear strain relation is presented in **Figure 2B**.

Modeling of Connected Building Structures

In the usual connecting damper systems, the fundamental natural frequency of the main structure is smaller than that of the substructure as shown in **Figure 3**.

Consider first the single-degree-of-freedom (SDOF) model, subjected to the base acceleration $\ddot{x}_g$, of the main structure (building A) connected with the substructure (building B) using a connecting damper as shown in **Figure 4**. Let m_A, k_A, and c_A denote the mass, stiffness, and damping coefficient of the main structure and let m_B, k_B, and c_B denote those of the substructure. The damping coefficient of the oil damper is denoted by c_D and the area of the high-damping rubber damper is by A_D. The thickness of the high-damping rubber damper is to be given. The model parameters of this system are shown in **Table 1** and the properties of the dampers are presented in **Table 2**. The quantities of oil dampers and high-damping rubbers have been determined so that the pre-relief damping ratio in the lowest mode due to the added oil damper is about 0.07 and the response reduction ratio by the high-damping rubber is almost equivalent to that by the oil damper.

NON-LINEAR RESPONSE OF CONNECTED BUILDINGS WITH NON-LINEAR DAMPERS: HARMONIC EXCITATION

Input Base Acceleration

Figure 5 shows the input base acceleration. The input frequency has been determined so that the frequency is resonant to the natural

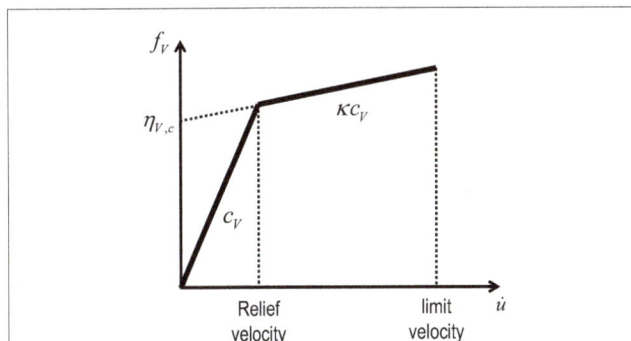

FIGURE 1 | Force–velocity relation of oil damper with relief mechanism.

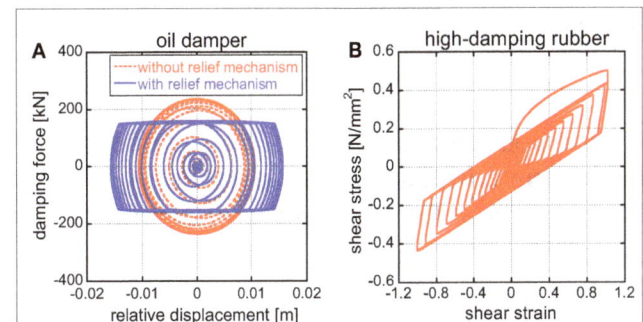

FIGURE 2 | Connecting damper properties: (A) oil damper and (B) high-damping rubber damper.

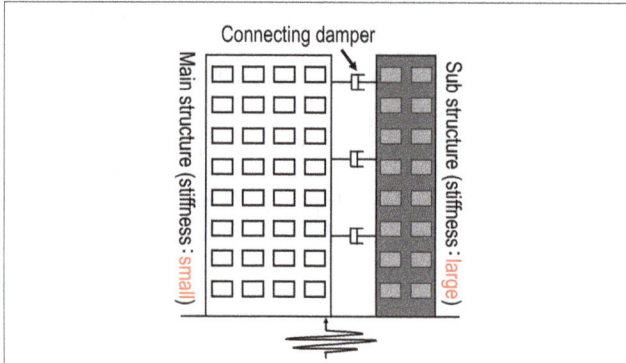

FIGURE 3 | Connecting damper systems in which the fundamental natural frequency of the main structure is smaller than that of the substructure.

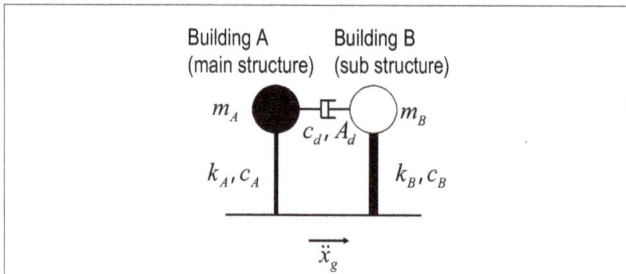

FIGURE 4 | SDOF model, subjected to the base acceleration $\ddot{x}_g$, of the main structure (building A) connected with the substructure (building B) using a connecting damper.

TABLE 1 | Model parameters.

Building A (main structure)	Mass m_A	3.0×10^5 (kg)
	Stiffness k_A	147 (kN/mm)
	Damping ratio h_A	0.02
	Natural period T_A	0.28 (s)
Building B (substructure)	Mass m_B	3.0×10^5 (kg)
	Stiffness k_B	294 (kN/mm)
	Damping ratio h_B	0.02
	Natural period T_B	0.20 (s)

frequency of the main structure (Takewaki, 2006; Takewaki and Tsujimoto, 2011; Murase et al., 2013; Fukumoto and Takewaki, 2015). After a transition process, the input goes into a steady state.

Response Characteristics of Connected Buildings with Different Connecting Dampers and Response Characteristics with Respect to Input Level

Figure 6 shows the relative building displacements without damper and with high-damping rubbers or oil dampers ($T_B = 0.20$ s, oil dampers with relief mechanism). In order to investigate the effect of the separation of building natural periods between buildings A and B, two cases ($T_B = 0.16, 0.24$ s) are

TABLE 2 | Comparison of high-damping rubber and oil damper.

High-damping rubber	Area A_D	980,000 (mm²)
	Thickness	0.015 (m) [limit deformation: 0.045 (m)]
Oil damper with relief mechanism	Damping coefficient c_D	1,200 [kN/(m/s)]
	Relief force f_r	250 (kN)
	Damping coefficient ratio κ	0.05
	Limit damping force ratio α (limit damping force/relief load)	1.1 [limit damping force 275 (kN)]

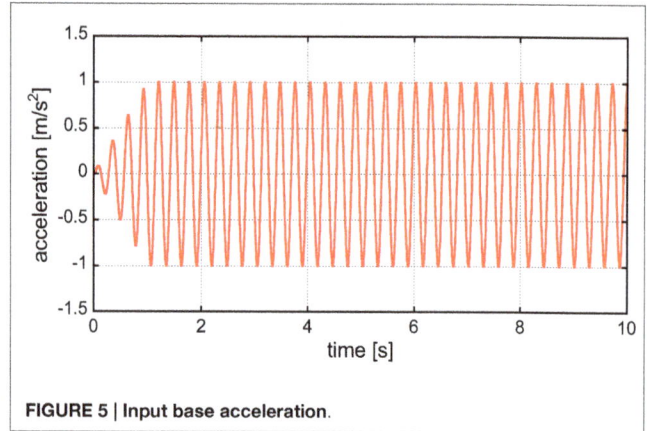

FIGURE 5 | Input base acceleration.

added to the model shown in **Table 1**. However, since the input frequency is resonant to the natural period of building A, the modification of the natural period of building B does not affect so much the relative building displacements. Therefore, the figures for the cases ($T_B = 0.16, 0.24$ s) are not shown here. It can be observed from **Figure 6** that while the oil damper reduces the vibration amplitude without the phase change, the high-damping rubber modified the vibration phase slightly. **Figure 7** illustrates the horizontal displacement of building A and building B with and without high-damping rubbers. On the other hand, **Figure 8** presents the horizontal displacements of building A and building B with and without oil dampers. It can also be seen that while the oil damper reduces the vibration amplitude without the phase change, the high-damping rubber changes the vibration phase. Furthermore, as the separation of building natural periods between buildings A and B becomes large, the effect of dampers on vibration reduction becomes remarkable.

In order to investigate the effect of building damping ratios on the response, the additional models with the damping ratios $h_A = h_B = 0.03, 0.04$, and 0.05 have been treated. **Table 3** shows the maximum displacements of buildings A and B. It can be observed that as the structural damping increases, the maximum displacement of building A resonant to the input decreases remarkably. Furthermore, it can also be found that the vibration reduction effect by dampers is retained even for the increased structural damping, although the reduction rate is slightly decreased.

Figure 9 shows the maximum displacement and acceleration of building A with and without dampers with respect to the input acceleration amplitude. The input frequency has been given so as

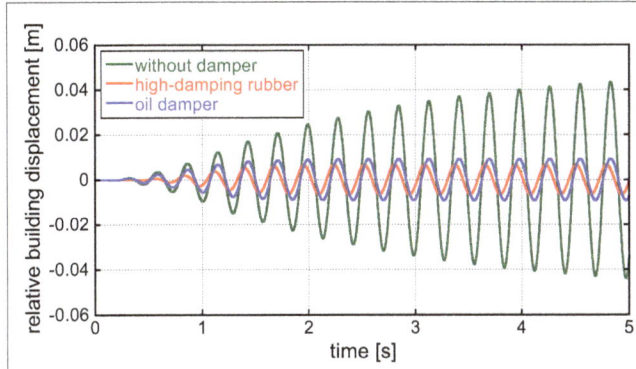

FIGURE 6 | Relative building displacement without damper and with high-damping rubber or oil damper.

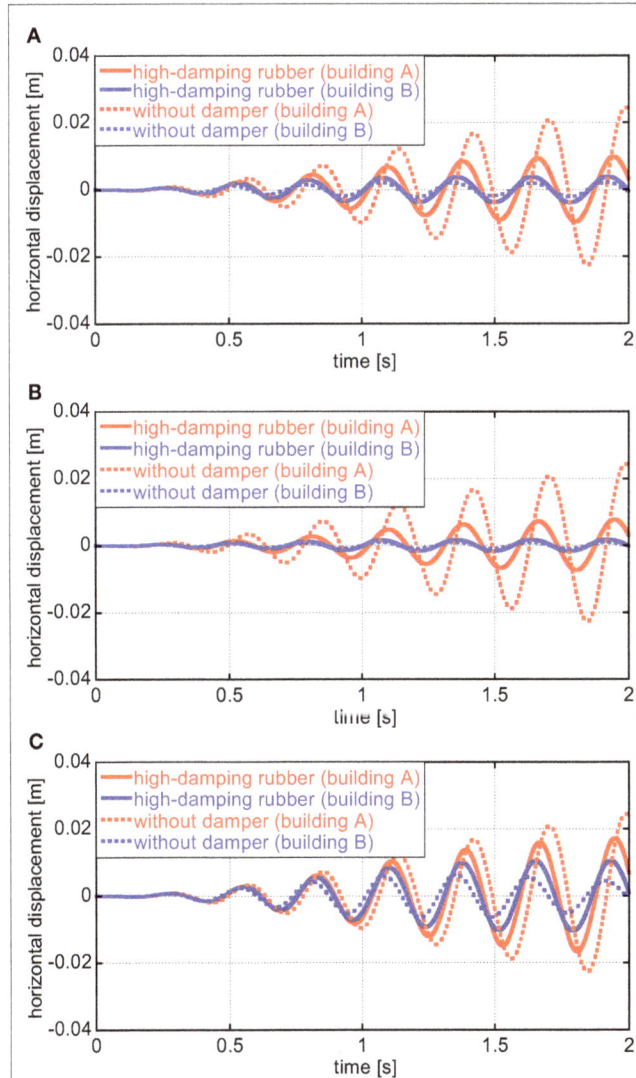

FIGURE 7 | Horizontal displacement of building A and building B with and without high-damping rubbers: (A) $T_B = 0.2$ s, (B) $T_B = 0.16$ s, and (C) $T_B = 0.24$ s.

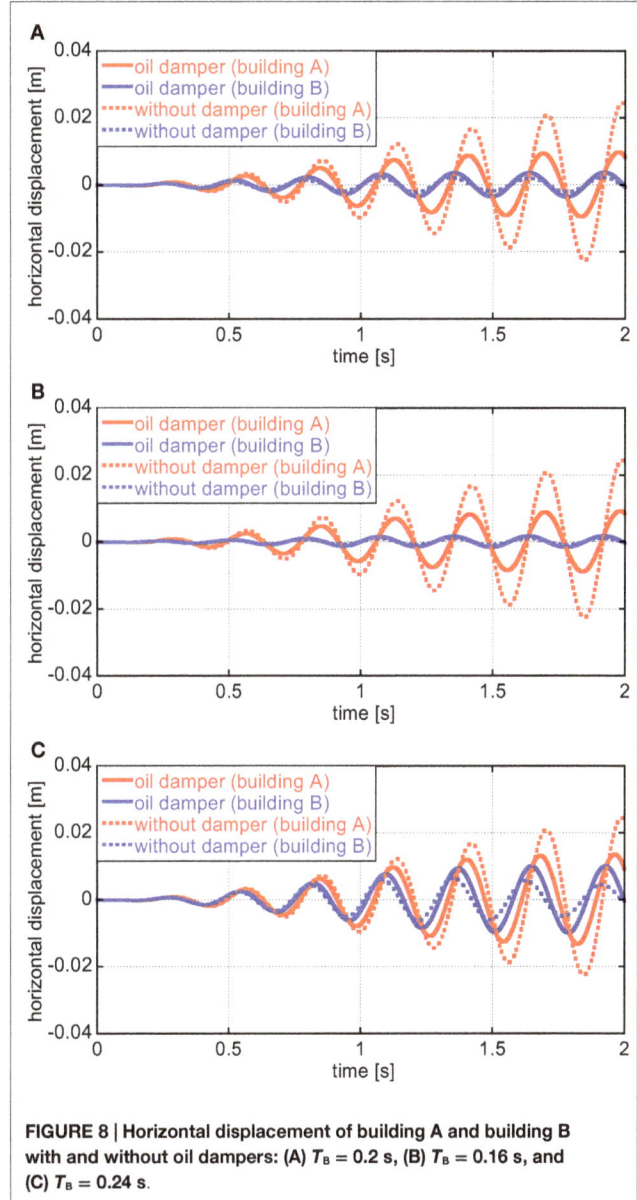

FIGURE 8 | Horizontal displacement of building A and building B with and without oil dampers: (A) $T_B = 0.2$ s, (B) $T_B = 0.16$ s, and (C) $T_B = 0.24$ s.

to be resonant to the natural frequency of the main structure. It can be found that while linear oil dampers without relief mechanism can reduce the vibration of the building A effectively for the increasing input level, the performance of the oil dampers with relief mechanism and the high-damping rubber deteriorates for the increasing input level. In particular, while the deterioration rate of the high-damping rubber is gradual, that of the oil dampers with relief mechanism is rapid after a certain limit. **Figure 10** presents the maximum relative displacement and damping force with and without dampers with respect to input acceleration amplitude. It can be observed that the high-damping rubber can reduce the relative displacement clearly compared to the oil dampers with relief mechanism.

The following conclusion may be drawn from **Figures 9** and **10**. Since the high-damping rubber has stiffness (especially large stiffness in the small deformation range), the resonant phenomenon may be avoided in the small input level. On the other hand, the damping force in the high-damping rubber becomes large.

NON-LINEAR RESPONSE OF CONNECTED BUILDINGS WITH NON-LINEAR OIL DAMPERS: SIMULATED EARTHQUAKE GROUND MOTIONS

In order to investigate the response characteristics of the connected buildings with the linear and non-linear oil dampers for earthquake ground motions, some numerical examples are shown in this section.

Simulated Earthquake Ground Motions

In this paper, general design ground motions compatible with a specific code-specified design response spectrum in Japan is used. Two representative phase properties are employed to represent the two types of ground motions, i.e., El Centro NS 1940 for the near-field ground motion and Hachinohe NS 1968 for the far-field ground motion. These simulated ground motions have been generated following the method by Gasparini and Vanmarcke (1976). **Figure 11** shows the acceleration time history and the acceleration response spectrum with the code-specified design acceleration response spectrum in Japan.

Effect of Relief Mechanism of Oil Dampers on Non-Linear Response

In this section, the effect of relief mechanism of oil dampers on non-linear response is investigated. Since it has been reported

(Adachi et al., 2013a,b) that if the ratio of the relief load to the maximum damping force in linear case is approximately equal or larger than 0.5, the displacement response of the structure is not affected so much by the relief mechanism. Based on this fact, the relief load shown in **Table 4** is used here.

Figure 12A shows the horizontal displacements of building A with and without relief mechanism and energy consumptions by oil dampers (phase: El Centro). On the other hand, **Figure 12B** illustrates those for the simulated ground motion with the phase of Hachinohe. As pointed out just before, the horizontal displacements of building A is not affected so much by the introduction of relief mechanism. However, it can be seen that the energy dissipation performance of oil dampers with relief mechanism is slightly low compared to the linear oil damper. This performance deterioration occurs early in the El Centro-phase motion and at the intermediate stage in the Hachinohe-phase motion. This timing seems to correspond to the stage at which the large response displacement occurs.

Table 5 presents the response amplification due to relief mechanism. It can be observed that a remarkable response increase is seen for the simulated ground motion of Hachinohe.

MULTI-STORY BUILDING MODEL

Consider a three-story connected building system as shown in **Figure 13**. Although many researches on optimal damper locations have been proposed for linear dampers (for example, Tsuji and Nakamura, 1996; Takewaki, 1997, 2009; Trombetti and Silvestri, 2004; Lavan and Levy, 2006; Aydin et al., 2007; Cimellaro, 2007; Cimellaro and Retamales, 2007), the research on non-linear dampers is very limited (Adachi et al., 2013a). The properties of the main structure and the substructure are shown in **Tables 5** and **6**. In case of using the same dampers, those are located uniformly at every story.

TABLE 3 | Maximum displacements of buildings A and B for various levels of structural damping (meter).

Damper	Damping ratio	0.02	0.03	0.04	0.05
Without damper	Building A	0.0510	0.0340	0.0255	0.0204
	Building B	0.0023	0.0022	0.0022	0.0022
High-damping rubber	Building A	0.0100	0.0091	0.0083	0.0077
	Building B	0.0039	0.0038	0.0037	0.0036
Oil damper	Building A	0.0100	0.0092	0.0085	0.0080
	Building B	0.0037	0.0035	0.0034	0.0033

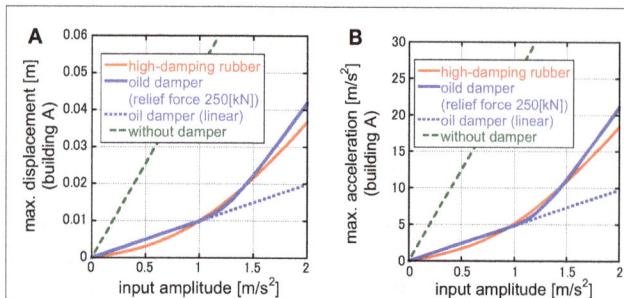

FIGURE 10 | Maximum relative displacement and damping force with and without dampers with respect to input acceleration amplitude: (A) maximum relative displacement and (B) maximum damping force.

TABLE 4 | Relief load for each input acceleration.

Phase	Maximum damping force in linear case (kN)	Relief load (kN)
El Centro	530	265
Hachinohe	529	264.5

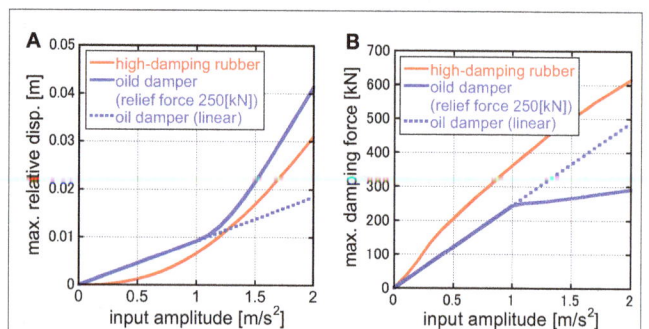

FIGURE 9 | Maximum displacement and acceleration of building A with and without dampers with respect to input acceleration amplitude: (A) maximum displacement and (B) maximum acceleration.

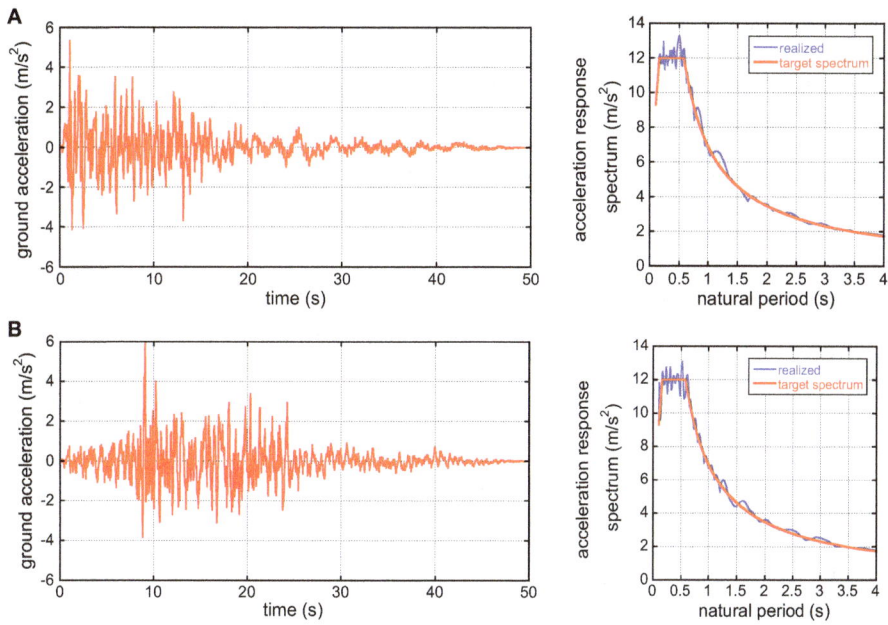

FIGURE 11 | Design earthquake ground motions compatible with the design response spectrum in Japan revised in 2000: (A) phase of El Centro NS 1940 and (B) phase of Hachinohe NS 1968.

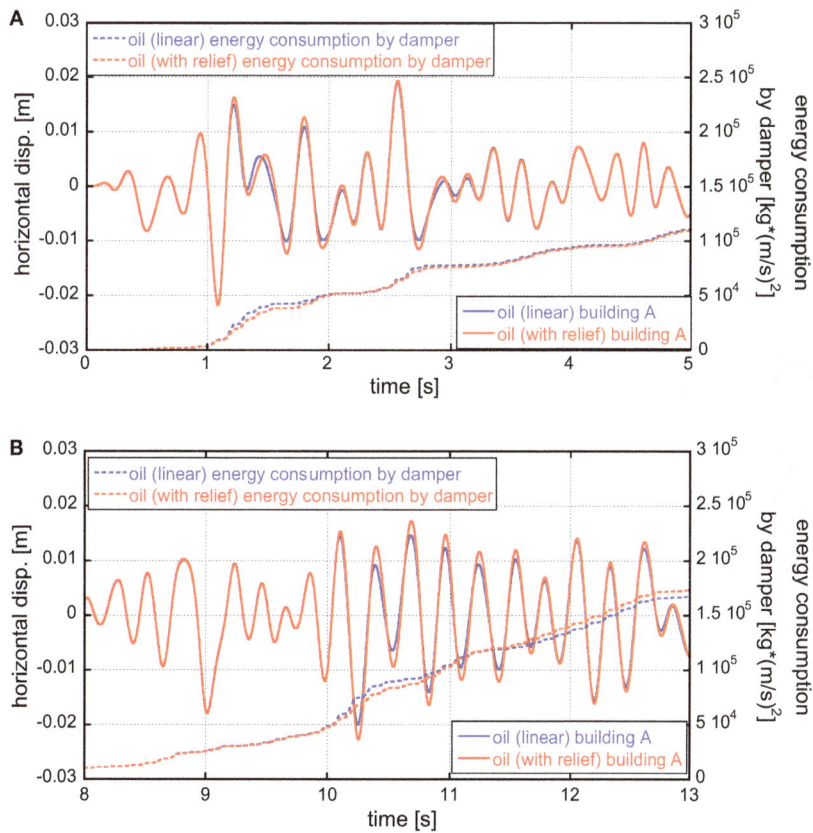

FIGURE 12 | Horizontal displacement and energy consumption by oil dampers: (A) phase: El Centro and (B) phase: Hachinohe.

Figure 14 shows the input base acceleration. As in the previous case for an SDOF model, the input frequency has been determined so that the frequency is resonant to the fundamental natural frequency of the main structure (Murase et al., 2013). After a transition process, the input goes into a steady state.

Table 6 presents the fundamental natural period and damping ratio of the main structure and the substructure, and **Table 7** shows story mass and stiffness of both structures.

In the present section, various damper combinations are considered. **Table 8** indicates the relationship between the damper combinations and top-mass displacement. The quantities of the dampers are the same as for the SDOF model in the previous section. The oil damper is with the relief mechanism. In this case, the model with the oil dampers in all the stories exhibits the smallest top displacement. However, since this property depends on the input amplitude and other parameters, e.g., the quantity of dampers, a careful attention should be paid. Especially the response velocity in the oil dampers is too large compared to the limit value in low-rise buildings. As for the high-damping rubber dampers, there is no limitation on the response velocity.

Figure 15 shows the damping force deformation relation in the first, second, and third stories for the model with oil dampers in all stories and the model with high-damping rubber in the first story and oil dampers in other stories. **Table 9** presents

the relative building displacement and damping force. It can be observed that the high-damping rubber can reduce the relative building displacement in the first story compared to the model of oil dampers in all stories. Since the high-damping rubber has an issue of damper stroke, it should be used in lower stories.

LIMIT VALUE OF MAXIMUM VELOCITY OF OIL DAMPERS

It is important to investigate the feasibility of oil dampers. An example of the limit value of the maximum velocity of oil dampers with the limit stroke ± 100 (mm) is 150–300 (mm/s).

Table 10 shows the maximum response velocities of the oil dampers computed in Section "Effect of Relief Mechanism of Oil Dampers on Non-Linear Response." It should be remarked that while the phases of the adopted earthquake ground motions are different, the target acceleration response spectrum is the same. This condition may lead to a similar maximum velocity response of dampers. However, these values are larger than the limit value stated above. It may be concluded that the response velocity in the oil dampers is too large compared to the limit value specifically in low-rise buildings and a special attention should be paid in its use.

CONCLUSION

The following conclusions have been derived:

(1) The oil damper and the high-damping rubber damper are both effective dampers for the building connecting system.
(2) While a high-damping rubber damper is effective in a rather small deformation level, an oil damper without relief mechanism is effective in a relatively large deformation level.

TABLE 5 | Response amplification due to relief mechanism.

Phase	Maximum disp. in linear case (m)	Maximum disp. with relief mechanism (m)	Response increase (%)
Hachinohe	0.0200	0.0226	13
El Centro	0.0216	0.0218	0.9

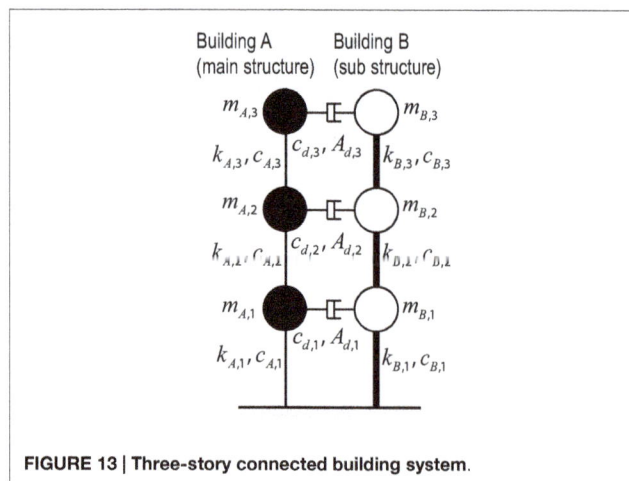

FIGURE 13 | Three-story connected building system.

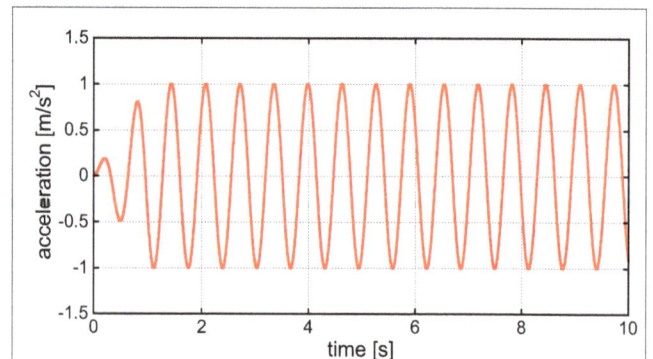

FIGURE 14 | Input base acceleration.

TABLE 6 | Fundamental natural period and damping ratio of main structure and substructure.

	Main structure	Substructure
Fundamental natural period (s)	0.638	0.451
Structural damping ratio (stiffness-proportional)	0.02	0.02

TABLE 7 | Story mass and stiffness.

		Main structure	Substructure
Mass at each node (kg)		3.0×10^5	3.0×10^5
Story stiffness (kN/mm)	Third story	147	294
	Second story	147	294
	First story	147	294

TABLE 8 | Damper location and top-mass displacement.

Story	Small ← response → large								
3	Oil	Oil	Oil	Rubber	Oil	Rubber	Rubber	Rubber	Without
2	Oil	Oil	Rubber	Oil	Rubber	Oil	Rubber	Rubber	Without
1	Oil	Rubber	Oil	Oil	Rubber	Rubber	Rubber	Oil	Without
Top disp. (m)	0.0409	0.0446	0.0548	0.0553	0.0567	0.0569	0.0599	0.0607	0.300

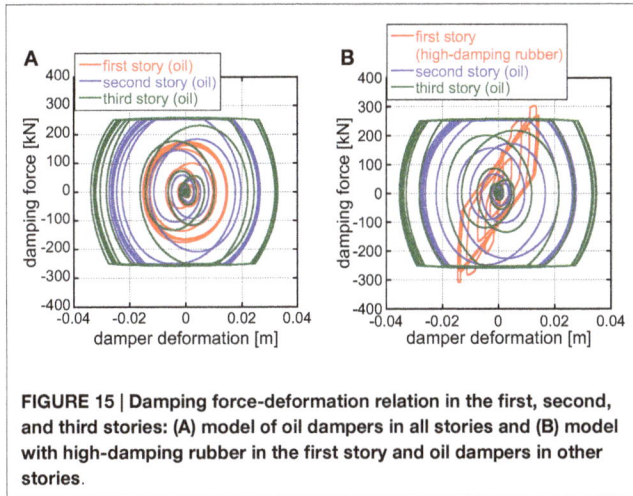

FIGURE 15 | Damping force-deformation relation in the first, second, and third stories: (A) model of oil dampers in all stories and (B) model with high-damping rubber in the first story and oil dampers in other stories.

(3) While the oil dampers reduce the response in the same phase as the case without dampers, the high-damping rubber dampers change the phase. The merit is that the high-damping rubber can reduce the damper deformation and keep the sufficient space between both buildings. This can mitigate the risk of building pounding.

(4) The oil dampers are effective for the reduction of acceleration. However, the response velocity in the oil dampers is too large compared to the limit value specifically in low-rise buildings as for the high-damping rubber dampers, there is no limitation on the response velocity.

(5) The vibration amplification in the oil damper with relief mechanism should be discussed carefully.

REFERENCES

Adachi, F., Yoshitomi, S., Tsuji, M., and Takewaki, I. (2013a). Nonlinear optimal oil damper design in seismically controlled multi-story building frame. *Soil Dyn. Earthq. Eng.* 44, 1–13. doi:10.1016/j.soildyn.2012.08.010

Adachi, F., Fujita, K., Tsuji, M., and Takewaki, I. (2013b). Importance of interstory velocity on optimal along-height allocation of viscous oil dampers in super high-rise buildings. *Eng. Struct.* 56, 489–500. doi:10.1016/j.engstruct.2013.05.036

Aydin, E., Boduroglub, M. H., and Guney, D. (2007). Optimal damper distribution for seismic rehabilitation of planar building structures. *Eng. Struct.* 29, 176–185. doi:10.1016/j.engstruct.2006.04.016

Cimellaro, G. P. (2007). Simultaneous stiffness-damping optimization of structures with respect to acceleration, displacement and base shear. *Eng. Struct.* 29, 2853–2870. doi:10.1016/j.engstruct.2007.01.001

Cimellaro, G. P., and Lopez-Garcia, D. (2011). Algorithm for design of controlled motion of adjacent structures. *Struct. Control Health Monit.* 18, 140–148. doi:10.1002/stc.357

TABLE 9 | Relative building displacement and damping force.

Story	Oil damper (all stories)		High-damping rubber (1st story); oil damper (2nd and 3rd stories)	
	Relative building disp. (m)	Damping force (kN)	Relative building disp. (m)	Damping force (kN)
3	0.0338	258	0.0353	259
2	0.0272	254	0.0280	254
1	0.0152	176	0.0145	309

TABLE 10 | Maximum response velocities of oil dampers computed in Section "Effect of Relief Mechanism of Oil Dampers on Non-Linear Response."

	Without relief mechanism (mm/s)	With relief mechanism (mm/s)
El Centro-phase	442.5	490.7
Hachinohe-phase	440.7	532.3

AUTHOR CONTRIBUTIONS

MK formulated the problem, conducted the computation, and wrote the paper. KF helped the computation. MT discussed the results. IT supervised the research and wrote the paper.

FUNDING

Part of the present work is supported by the Grant-in-Aid for Scientific Research of Japan Society for the Promotion of Science (No. 15H04079) and the joint research with Sumitomo Rubber Industry, Co., in Japan. These supports are greatly appreciated.

Cimellaro, G. P., and Retamales, R. (2007). Optimal softening and damping design for buildings. *Struct. Control Health Monit.* 14, 831–857. doi:10.1002/stc.181

Fukumuto, Y., and Takewaki, I. (2015). Critical demand of earthquake input energy to connected building structures. *Earthq. Struct.* 9, 1133–1152.

Gasparini, D. A., and Vanmarcke, E. H. (1976). *Simulated Earthquake Motions Compatible with Prescribed Response Spectra – SIMQKE, A Computer Program Distributed by NISEE/Computer Applications*. Berkeley, CA.

Hanson, R. D., and Soong, T. T. (2001). *Seismic Design with Supplemental Energy Dissipation Devices*. Oakland, CA: EERI.

Iwanami, K., Suzuki, K., and Seto, K. (1996). Vibration control method for parallel structures connected by damper and spring. *JSME Int. J.* 39, 714–720.

Lavan, O., and Levy, R. (2006). Optimal design of supplemental viscous dampers for linear framed structures. *Earthq. Eng. Struct. Dyn.* 35, 337–356. doi:10.1002/eqe.524

Luco, J. E., and de Barros, F. C. P. (1998). Control of seismic response of a composite tall building modeled by two interconnected shear beams. *Earthq.*

Eng. Struct. Dyn. 17, 205–242. doi:10.1002/(SICI)1096-9845(199803)27:3<205:: AID-EQE712>3.0.CO;2-X

Murase, M., Tsuji, M., and Takewaki, I. (2013). Smart passive control of buildings with higher redundancy and robustness using base-isolation and inter-connection. *Earthq. Struct.* 4, 649–670. doi:10.12989/eas.2013.4.6.649

Patel, C. C., and Jangid, R. S. (2011). Dynamic response of adjacent structures connected by friction dampers. *Earthq. Struct.* 2, 149–169. doi:10.12989/eas.2011.2.2.149

Richardson, A., Walsh, K. K., and Abdullah, M. M. (2013a). Closed-form design equations for controlling vibrations in connected structures. *J. Earthq. Eng.* 17, 699–719. doi:10.1080/13632469.2013.771590

Richardson, A., Walsh, K. K., and Abdullah, M. M. (2013b). Closed-form design equations for coupling linear structures using stiffness and damping elements. *Struct. Control Health Monit.* 20, 259–281. doi:10.1002/stc.490

Soong, T. T., and Dargush, G. F. (1997). *Passive Energy Dissipation Systems in Structural Engineering*. Chichester: John Wiley & Sons.

Takewaki, I. (1997). Optimal damper placement for minimum transfer functions. *Earthq. Eng. Struct. Dyn.* 26, 1113–1124. doi:10.1002/(SICI)1096-9845(199711)26:11<1113::AID-EQE696>3.0.CO;2-X

Takewaki, I. (2006). *Critical Excitation Methods in Earthquake Engineering*. Amsterdam: Elsevier Science.

Takewaki, I. (2007). Earthquake input energy to two buildings connected by viscous dampers. *J. Struct. Eng.* 133, 620–628. doi:10.1061/(ASCE)0733-9445(2007)133:5(620)

Takewaki, I. (2009). *Building Control with Passive Dampers: Performance-Based Design for Earthquakes*. John Wiley & Sons.

Takewaki, I. (2015). "Fundamental properties of earthquake input energy on single and connected building structures, Chapter 1," in *New Trends in Seismic Design of Structures*, eds. Lagaros N. D., Tsompanakis Y., and Papadrakakis M. (Stirlingshire: Saxe-Coburg Publisher), 1–28.

Takewaki, I., and Tsujimoto, H. (2011). Scaling of design earthquake ground motions for tall buildings based on drift and input energy demands. *Earthq. Struct.* 2, 171–187. doi:10.12989/eas.2011.2.2.171

Tani, T., Yoshitomi, S., Tsuji, M., and Takewaki, I. (2009). High-performance control of wind-induced vibration of high-rise building via innovative high-hardness rubber damper. *J. Struct. Des. Tall Spec. Build.* 18, 705–728. doi:10.1002/tal.457

Trombetti, T., and Silvestri, S. (2004). Added viscous dampers in shear-type structures: the effectiveness of mass proportional damping. *J. Earthq. Eng.* 8, 275–313. doi:10.1080/13632460409350490

Tsuji, M., and Nakamura, T. (1996). Optimum viscous dampers for stiffness design of shear buildings. *J. Struct. Des. Tall Build.* 5, 217–234. doi:10.1002/(SICI)1099-1794(199609)5:3<217::AID-TAL70>3.0.CO;2-R

Conflict of Interest Statement: The authors declare that the research was conducted in the absence of any commercial or financial relationships that could be construed as a potential conflict of interest.

19

Beyond uncertainties in earthquake structural engineering

Izuru Takewaki *

Department of Architecture and Architectural Engineering, Kyoto University, Kyoto, Japan

Keywords: earthquake engineering, uncertainty, ground motion, structural parameter, interval analysis, robustness, redundancy

Uncertainties

Ground Motion Definition

Earthquake events and realized earthquake ground motions are extremely uncertain even with the present knowledge, and it is not easy to predict forthcoming events precisely both in time and frequency (Anderson and Bertero, 1987; Takewaki et al., 1991; 2013; 2011a; Conte et al., 1992; Ariga et al., 2006; Minami et al., 2013; Çelebi et al., 2014). For example, recently reported near-field ground motions (Northridge 1994, Kobe 1995, Turkey 1999, and Chi–Chi, Taiwan 1999), the Mexico Michoacan motion 1985, and the Tohoku motion 2011, had some peculiar characteristics that could not have been predicted. It is also true that civil, mechanical, and aerospace engineering structures are often subjected to disturbances with inherent uncertainties due mainly to their "low rate of occurrence." Worst-case analysis (Drenick, 1970; Shinozuka, 1970; Takewaki, 2006/2013; Elishakoff and Ohsaki, 2010), combined with proper information based on reliable physical data, is expected to play an important role in avoiding difficulties caused by such uncertainties. Approaches based on the concept of "critical excitation" seem promising.

There are various buildings in a city (**Figure 1A**). Each building has its own natural period and its idiosyncratic structural properties. Earthquakes trigger various kinds of ground motions in the city. The relation of the building's natural period with the predominant period of the induced ground motion may lead to disastrous phenomena, as many observations from past historical earthquakes have demonstrated. Once a large earthquake occurs, some building codes are typically upgraded, but such makeshift efforts never resolve all issues and new damage problems have occurred even recently. In order to overcome this problem, a new paradigm has to be posed. In my view, the concept of "critical excitation," and structural design based upon it, could become a powerful new paradigm. Critical excitation methods were pioneered by Drenick (1970) and Shinozuka (1970). Just as the investigation of limit states of structures plays an important role in the specification of allowable response and performance levels of structures during disturbances, the clarification of critical excitations for a given (group of) structure(s) can provide structural designers with useful information for determining excitation parameters.

After Drenick and Shinozuka's pioneering work (1970), versatile researches have been developed (Iyengar and Manohar, 1985; 1987; Pirasteh et al., 1988; Srinivasan et al., 1992; Manohar and Sarkar, 1995; Pantelides and Tzan, 1996; Tzan and Pantelides, 1996; Takewaki, 2000; 2001a;b; 2008a; Abbas and Manohar, 2002; Fujita et al., 2010a; Moustafa and Takewaki, 2010a;b; Moustafa et al., 2010; Moustafa, 2011; Takewaki et al., 2012). Details of critical excitation methods are given in Takewaki (2006/2013).

In the case where influential active faults are known during the design stage of a structure (especially an important structure), the effects of these active faults should be taken into account through the concept of critical excitation. While influential active faults are not necessarily known in advance, virtual or scenario faults and their *energy* can be predefined, especially for the design of important

Edited by:
Nikos D. Lagaros,
National Technical
University of Athens, Greece

Reviewed by:
Sameh Samir F. Mehanny,
Cairo University, Egypt

*Correspondence:
takewaki@archi.kyoto-u.ac.jp

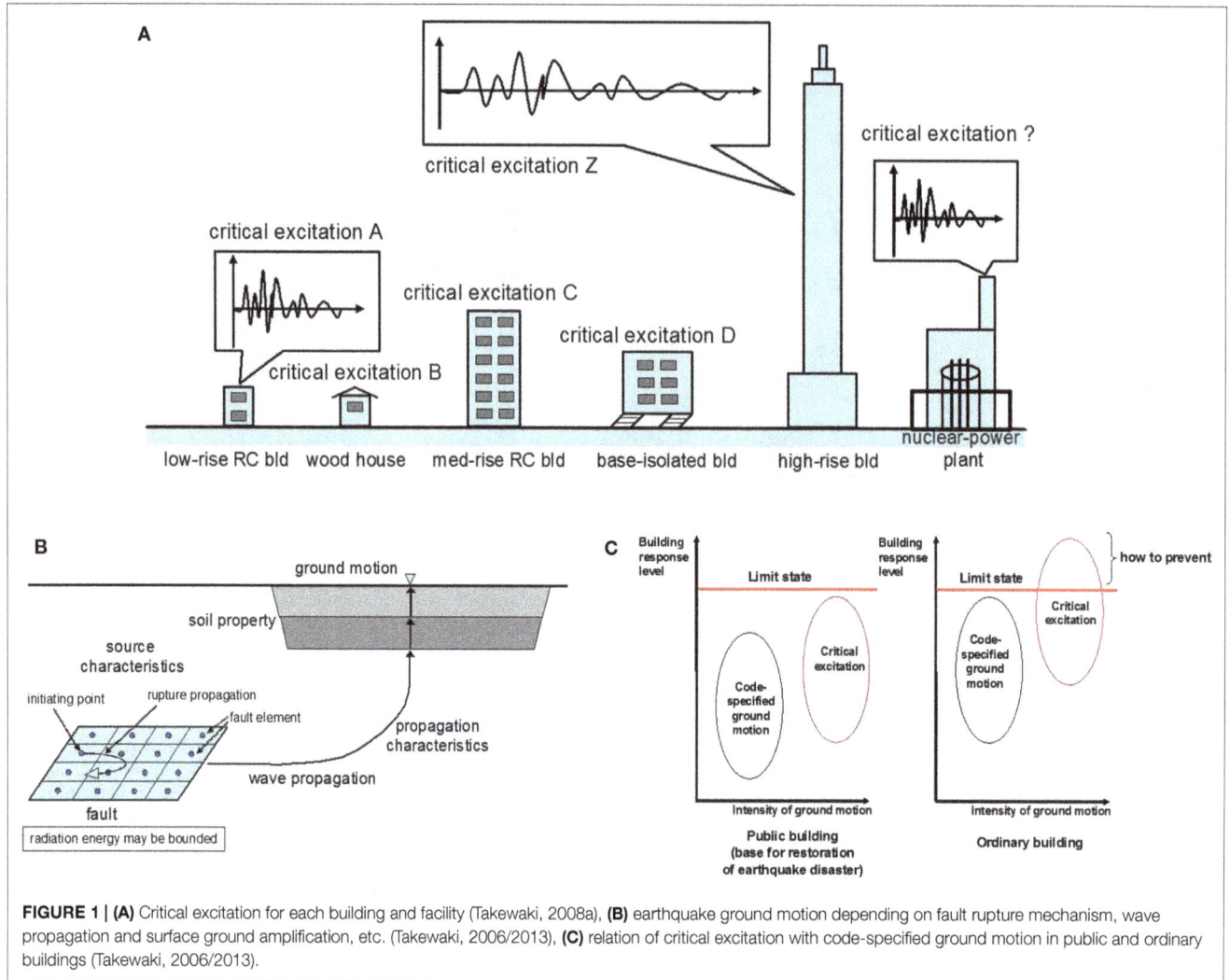

FIGURE 1 | (A) Critical excitation for each building and facility (Takewaki, 2008a), **(B)** earthquake ground motion depending on fault rupture mechanism, wave propagation and surface ground amplification, etc. (Takewaki, 2006/2013), **(C)** relation of critical excitation with code-specified ground motion in public and ordinary buildings (Takewaki, 2006/2013).

and socially influential structures. It is believed that earthquakes have an upper bound on magnitude (Strasser and Bommer, 2009). The combination of worst-case analysis (Takewaki, 2004; 2005) with appropriate specification of energy levels (Boore, 1983) derived from the analysis of various factors, e.g., the fault rupture mechanism and earthquake occurrence probability, enables more robust and reliable seismic-resistant design methods (**Figure 1B**). The appropriate setting of energy levels or information used in the worst-case analysis is important, and more research should be conducted on this subject.

In other words, the earthquake energy radiated from the fault has an upper bound (Trifunac, 2008). The problem is to find the most unfavorable ground motion for a (group of) building(s) (**Figure 1A**) (Takewaki et al., 2013). A ground motion displacement spectrum or acceleration spectrum has been proposed at the rock surface depending on the seismic moment, distance from the fault, etc. (**Figure 1B**) (Boore, 1983). Such spectra may have uncertainties. One possibility or approach is to specify the acceleration or velocity power while allowing variability of the spectrum (Takewaki and Tsujimoto, 2011; Takewaki et al., 2013).

The problem of ground motion variability is very important and difficult. Code-specified design ground motions are usually constructed by taking past observations and probabilistic insights into account. However, as stated above, uncertainties in the occurrence of earthquakes (or ground motions), fault rupture mechanisms, wave propagation mechanisms, ground properties, etc., cause much difficulty in defining reasonable design ground motions, especially for important buildings in which damage or collapse has to be avoided absolutely (see **Figure 1C**) (Anderson and Bertero, 1987; Geller et al., 1997; Stein, 2003; Takewaki, 2006/2013).

A long-period ground motion has been observed in Japan recently (Takewaki et al., 2011a). This type of ground motion is reported to require large seismic demands on such structures as high-rise buildings, base-isolated buildings, oil tanks, etc., which have a long natural period. This large seismic demand is induced by the resonance between the long-period ground motion and the long natural period of these constructed facilities. To the best of the author's knowledge, a promising approach is to shift the natural period of the building and add damping to the building by taking full advantage of technologies via seismic control (Takewaki, 2009). However, it is also understood that seismic control is still under development, while sufficient time is necessary to respond to uncertain ground motions. It is hoped that the approach of critical excitation methods (Takewaki et al., 2013) will help the development of new seismic-resistant design methods of buildings for such unpredicted or unpredictable ground motions. Critical excitation problems for fully non-stationary excitation models [see, for examples, Conte and Peng (1997), Fang and Sun (1997)] and critical excitation problems for elastic–plastic responses subjected to those excitations seem to be challenging problems.

As for response combination by multiple actions, Menun and Der Kiureghian (2000a;b) discussed the evaluation methods of envelopes for seismic response vectors. The normal stress in a structural member under combined loading of axial and bending actions may be one example. This problem is related to interval analysis (Fujita and Takewaki, 2011a;b) and its further development is warranted.

Structural Parameter Specification

Structural control with passive dampers has a successful history in mechanical and aerospace engineering, probably because these fields usually deal with predictable external loading and environments with little uncertainty. This technique is also supported by various methods of structural health monitoring (Takewaki et al., 2011b). However, in civil engineering, the situation is different (Housner, 1997; Soong and Dargush, 1997; Srinivasan and McFarland, 2000; Cheng et al., 2008; Takewaki, 2009). Building and civil structures are often subjected to severe earthquake ground motions, wind disturbances, and other external loading with large uncertainties (Takewaki, 2006/2013). It is therefore inevitable to take these uncertainties into account in structural design and its application to actual structures.

Interval analysis [see, for example, Moore (1966), Alefeld and Herzberger (1983), Qiu (2003), Chen and Wu (2004), Chen et al. (2009)] in terms of uncertain structural parameters is an effective tool for evaluating the sustainability of buildings in earthquake-prone countries. The number of combinations of uncertain structural parameters increases exponentially, but this difficulty can be overcome by introducing a sensitivity analysis or Taylor series expansion.

The critical combination of interval parameters is found by introducing an assumption of "inclusion monotony" as well as sensitivity information from Taylor series expansion. It has been demonstrated that the proposed method is useful for the development of the concept of sustainable building design under uncertain structural-parameter environments.

The concept of sustainable building design under uncertain structural-parameter environments is illustrated in **Figure 2A**. The member stiffness and strength of buildings are uncertain due to various factors resulting from randomness, material deterioration, temperature dependence, etc. The damping coefficients of structural members and/or passive dampers may also be uncertain (Takewaki and Ben-Haim, 2005). Several kinds of methods have been proposed to describe this uncertainty (Ben-Haim and Elishakoff, 1990; Ben-Haim et al., 1996; Ben-Haim, 2001/2006).

The time variation of Young's modulus and damping coefficients are shown in **Figure 2A** as representative examples. Karbhari and Lee (2009) discuss the service life estimation and extension of civil engineering structures from the viewpoints of material deterioration. These member and/or damper uncertainties lead to response variability of buildings under earthquake ground motions. Efficient and reliable methods are desired for predicting the upper bound of such building response.

As stated above, interval analysis in terms of uncertain structural parameters is an effective tool for evaluating the response variability and the sustainability of buildings in earthquake-prone countries. The number of combinations of end-points of uncertain structural parameters increases exponentially, while the evaluation of the upper and lower bounds of the objective function requires elaborate manipulation. It has been shown that this difficulty can be overcome by introducing the sensitivity or Taylor series expansion analysis.

Recently, various kinds of problems with uncertain parameters have been dealt with (Kanno and Takewaki, 2006; Takewaki, 2008b; Fujita and Takewaki, 2011a;b; 2012; Takewaki and Fujita, 2014).

Unpredicted Phenomena

In recent years, unexpected phenomena in earthquake engineering have proved to be possible in a real world: for example, resonance of building vibration to ground motion in Mexico (1985) and Tohoku (2011), pulse-type ground motion in Northridge (1994) and Kobe (1995), large fault displacement in Chi–Chi (1999), giant tsunami, long-period long-duration ground motion, and soil liquefaction under smaller vibration level with longer duration in Tohoku (2011). The effects by torsional response of buildings with eccentricity and soil-structure interaction under rather soft ground may cause further unpredicted phenomena.

Because super high-rise buildings in megacities in Japan had never been shaken intensively by so-called long-period ground motions before March 11, 2011, the response of high-rise buildings to such long-period ground motions is one of the most controversial subjects and issues in the field of earthquake-resistant design in Japan (Takewaki et al., 2012; 2011a). Ground motions with large levels of velocity response spectrum in a broad frequency range including the period of 5–10 s is called the long-period ground motion. It is worth noting that most of high-rise and super-high-rise buildings in Japan have never been designed based on careful recognition of such issue. The analysis of possibility of occurrence of long-period ground motions should be investigated in more detail. The inspection of deep ground profiles may be absolutely necessary (Takewaki et al., 2013; 2012; 2011a).

Smart prediction and preparedness are extremely important and suitable for responding to such unexpected phenomena and for addressing issues related to earthquake engineering.

Strategy for Uncertainties

Worst-Case Analysis

A significance of critical excitation is supported by its broad perspective. In general, there are two classes of buildings in a

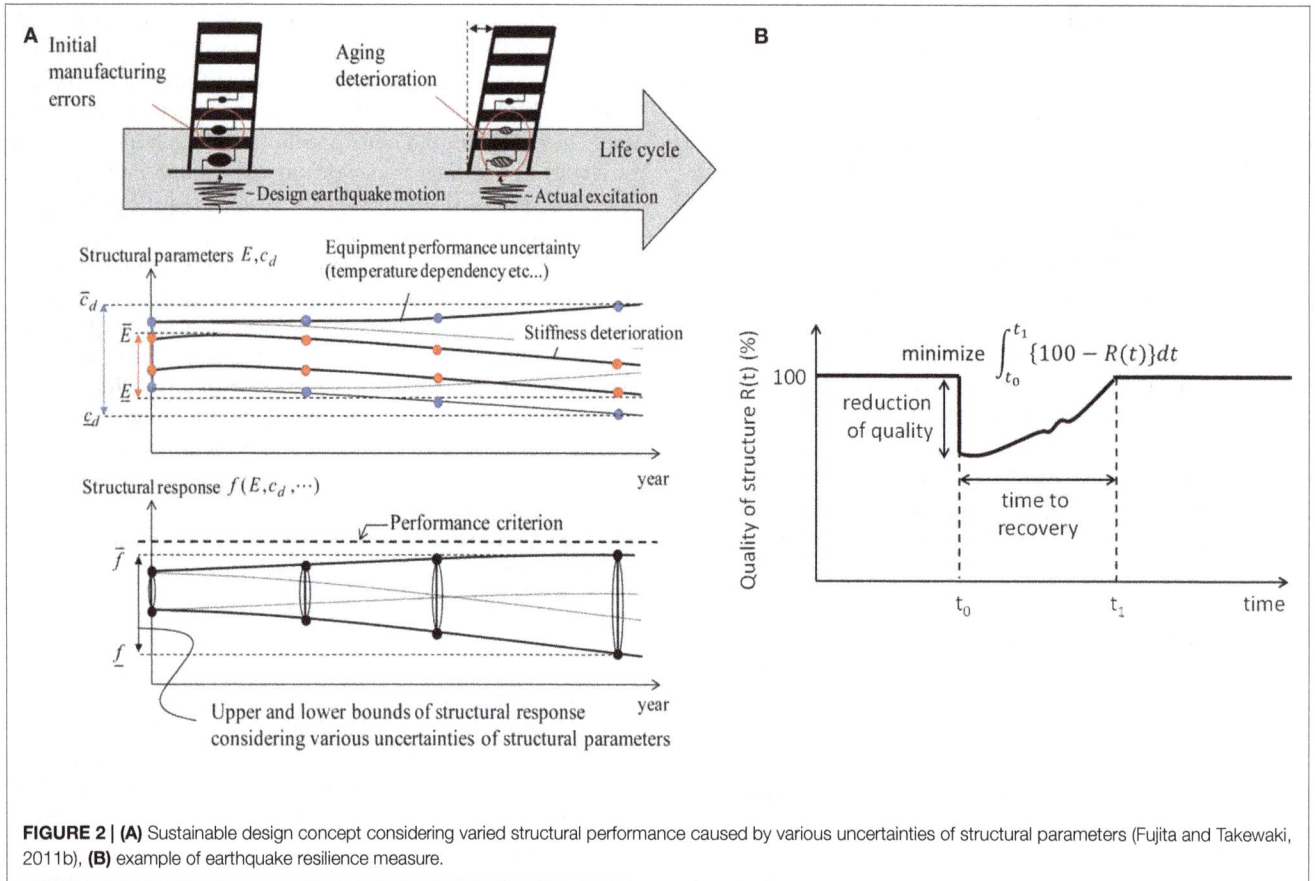

FIGURE 2 | (A) Sustainable design concept considering varied structural performance caused by various uncertainties of structural parameters (Fujita and Takewaki, 2011b), **(B)** example of earthquake resilience measure.

city. One is the important buildings, which play an important role during disastrous earthquakes. The other is ordinary buildings. The former should not be damaged during earthquakes, while some partial damage is acceptable for the latter, especially under critical excitation that is larger than code-specified design earthquakes (see **Figure 1C**). The concept of critical excitation may enable structural designers to make ordinary buildings more seismic-resistant (Takewaki, 2006/2013; Takewaki et al., 2012). The worst-case analysis is also characterized by the word of "Anti-optimization" (Elishakoff and Ohsaki, 2010). While the design of minimum cost corresponds to the design of minimum response for limited materials and a specified input, the design obtained by the anti-optimization means the design of maximum response for variable inputs.

Structural Control and Health Monitoring

While structural control is a promising and smart tool for sustainable building design (Fujita et al., 2010b), it is also true that a lot of uncertainties should be quantified for reliable implementation of these techniques (Takewaki and Ben-Haim, 2005). The sustainable building design under uncertain structural-parameter environment may be one of the most challenging issues in the building structural engineering. Even if all the design constraints are satisfied at the initial construction stage, some responses to external loadings (earthquakes, strong winds, etc.) may ultimately come to violate them over service life due to randomness, material

deterioration, temperature dependence, etc. To overcome such difficulties, response evaluation methods for uncertain structural-parameter environments are needed. By predicting the response variability accurately, the elongation of service life of buildings may be possible.

Enhancement of Earthquake Resilience

Bruneau and Reinhorn (2006) discussed the earthquake resilience of building structures and infrastructures. They defined "the resilient structures" as (1) those with small collapse probability, (2) those with reduced consequences from failures in terms of lives lost, damage, and negative economic and social consequences, (3) reduced time to recovery. **Figure 2B** shows the temporal variation of performance and functionality of a structure after an earthquake. The requirements of (2) and (3) may be understood so that the minimization of the time integral of the reduction of quality, (100-Quality), corresponds to the upgrade of the resilience of the structure. They proposed four resilience measures; (1) robustness, (2) redundancy, (3) resourcefulness, (4) rapidity.

Concluding Remarks

There exist aleatory and epistemic uncertainties in the seismic structural design under earthquake ground motions. The aleatory uncertainty represents the uncertainty related to inherent randomness of a phenomenon, which cannot be reduced by the advancement of research and the epistemic uncertainty means the uncertainty concerned with knowledge which can be reduced by the development of research. While uncertainties in modeling earthquake ground motions seem to include both aleatory and epistemic uncertainties because of their extremely small probability of occurrence, uncertainties in modeling structural properties of buildings seem to contain mostly aleatory uncertainties based

on the rapid advance of research in this field (although compared to the nature and level of input uncertainties). It is desired to narrow the region of epistemic uncertainties both in modeling earthquake ground motions and structural properties. Worst-case analysis, structural control and health monitoring, and introduction of the concept of earthquake resilience may be promising strategies for overcoming such unavoidable uncertainties.

References

Abbas, M., and Manohar, C. S. (2002). Investigations into critical earthquake excitations within deterministic and probabilistic frameworks. *Earthq. Eng. Struct. Dyn.* 31, 813–832. doi:10.1002/eqe.124.abs

Alefeld, G., and Herzberger, J. (1983). *Introduction to Interval Computations*. New York: Academic Press.

Anderson, J. C., and Bertero, V. V. (1987). Uncertainties in establishing design earthquakes. *J. Struct. Eng. ASCE* 113, 1709–1724. doi:10.1061/(ASCE)0733-9445(1987)113:8(1709)

Ariga, T., Kanno, Y., and Takewaki, I. (2006). Resonant behavior of base-isolated high-rise buildings under long-period ground motions. *Struct. Des. Tall Spec. Build.* 15, 325–338. doi:10.1002/tal.298

Ben-Haim, Y. (2001/2006). *Infomation-Gap Decision Theory: Decisions Under Severe Uncertainty*. London: Academic Press.

Ben-Haim, Y., Chen, G., and Soong, T. T. (1996). Maximum structural response using convex models. *J. Eng. Mech. ASCE* 122, 325–333. doi:10.1061/(ASCE)0733-9399(1996)122:4(325)

Ben-Haim, Y., and Elishakoff, I. (1990). *Convex Models of Uncertainty in Applied Mechanics*. Amsterdam: Elsevier.

Boore, D. M. (1983). Stochastic simulation of high-frequency ground motions based on seismological models of the radiated spectra. *BSSA* 73, 1865–1894.

Bruneau, M., and Reinhorn, A. (2006). "Overview of the resilience concept," in *Proc. 8th US Nat. Conf. on Earthq. Eng.*, San Francisco.

Çelebi, M., Okawa, I., Kashima, T., Koyama, S., and Iiba, M. (2014). Response of a tall building far from the epicenter of the 11 March 2011 M9.0 Great East Japan earthquake and aftershocks. *Struct. Des. Tall Spec. Build.* 23, 427–441. doi:10.1002/tal.1047

Chen, S. H., Ma, L., Meng, G. W., and Guo, R. (2009). An efficient method for evaluating the natural frequency of structures with uncertain-but-bounded parameters. *Comput. Struct.* 87, 582–590. doi:10.1016/j.compstruc.2009.02.009

Chen, S. H., and Wu, J. (2004). Interval optimization of dynamic response for structures with interval parameters. *Comput. Struct.* 82, 1–11. doi:10.1016/j.compstruc.2003.09.001

Cheng, F. Y., Jiang, H., and Lou, K. (2008). *Smart Structures: Innovative Systems for Seismic Response Control*. CRC Press.

Conte, J. P., and Peng, B. F. (1997). Fully nonstationary analytical earthquake ground motion model. *J. Eng. Mech. ASCE* 123, 15–24. doi:10.1061/(ASCE)0733-9399(1997)123:1(15)

Conte, J. P., Pister, K. S., and Mahin, S. A. (1992). Nonstationary ARMA modeling of seismic motions. *Soil Dyn. Earthq. Eng.* 11, 411–426. doi:10.1016/0267-7261(92)90005-X

Drenick, R. F. (1970). Model-free design of aseismic structures. *J. Eng. Mech. Div. ASCE* 96, 483–493.

Elishakoff, I., and Ohsaki, M. (2010). *Optimization and Anti-Optimization of Structures Under Uncertainty*. London: Imperial College Press.

Fang, T., and Sun, M. (1997). A unified approach to two types of evolutionary random response problems in engineering. *Arch. Appl. Mech.* 67, 496–506. doi:10.1007/s004190050134

Fujita, K., Moustafa, A., and Takewaki, I. (2010a). Optimal placement of viscoelastic dampers and supporting members under variable critical excitations. *Earthq. Struct.* 1, 43–67. doi:10.12989/eas.2010.1.1.043

Fujita, K., Yamamoto, K., and Takewaki, I. (2010b). An evolutionary algorithm for optimal damper placement to minimize intestorey-drift transfer function. *Earthq. Struct.* 1, 289–306. doi:10.12989/eas.2010.1.3.289

Fujita, K., and Takewaki, I. (2011a). An efficient methodology for robustness evaluation by advanced interval analysis using updated second-order Taylor series expansion. *Eng. Struct.* 33, 3299–3310. doi:10.1016/j.engstruct.2011.08.029

Fujita, K., and Takewaki, I. (2011b). Sustainable building design under uncertain structural-parameter environment in seismic-prone countries. *Sustain. Cities Soc.* 1, 142–151. doi:10.1016/j.scs.2011.07.001

Fujita, K., and Takewaki, I. (2012). Robust passive damper design for building structures under uncertain structural parameter environments. *Earthq. Struct.* 3, 805–820. doi:10.12989/eas.2012.3.6.805

Geller, R. J., Jackson, D. D., Kagan, Y. Y., and Mulargia, F. (1997). Earthquakes cannot be predicted. *Science* 275, 1616. doi:10.1126/science.275.5306.1616

Housner, G., Bergman, L. A., Caughey, T. K., Chassiakos, A. G., Claus, R. O., Masri, S. F., et al. (1997). Special issue, structural control: past, present, and future. *J. Eng. Mech. ASCE* 123(9), 897–971.

Iyengar, R. N., and Manohar, C. S. (1985). "System dependent critical stochastic seismic excitations," in *M15/6, Proc. of the 8th Int. Conf. on SMiRT*, Brussels.

Iyengar, R. N., and Manohar, C. S. (1987). Nonstationary random critical seismic excitations. *J. Eng. Mech. ASCE* 113, 529–541. doi:10.1061/(ASCE)0733-9399(1987)113:4(529)

Kanno, Y., and Takewaki, I. (2006). Sequential semidefinite program for maximum robustness design of structures under load uncertainties. *J. Optim. Theory Appl.* 130, 265–287. doi:10.1007/s10957-006-9102-z

Karbhari, V. M., and Lee, L. S.-W. (2009). "Vibration based damage detection techniques for structural health monitoring of civil infrastructure systems," in *Chapter 6 in Structural Health Monitoring of Civil Infrastructure Systems*, ed. V. M. Karbhari and F. Ansari (Cambridge: CRC Press/Woodhead Publishing), 177–212.

Manohar, C. S., and Sarkar, A. (1995). Critical earthquake input power spectral density function models for engineering structures. *Earthq. Eng. Struct. Dyn.* 24, 1549–1566. doi:10.1002/eqe.4290241202

Menun, C., and Der Kiureghian, A. (2000a). Envelopes for seismic response vectors: I theory. *J. Struct. Eng. ASCE* 126, 467–473. doi:10.1061/(ASCE)0733-9445(2000)126:4(474)

Menun, C., and Der Kiureghian, A. (2000b). Envelopes for seismic response vectors: II application. *J. Struct. Eng. ASCE* 126, 474–481. doi:10.1061/(ASCE)0733-9445(2000)126:4(474)

Minami, Y., Yoshitomi, S., and Takewaki, I. (2013). System identification of super high-rise buildings using limited vibration data during the 2011 Tohoku (Japan) earthquake. *Struct. Control Health Monit.* 20, 1317–1338. doi:10.1002/stc.1537

Moore, R. E. (1966). *Interval Analysis, Englewood Cliffs*. New Jersey: Prentice-Hall.

Moustafa, A. (2011). Damage-based design earthquake loads for single degree-of-freedom inelastic structures. *J. Struct. Eng. ASCE* 137, 456–467. doi:10.1061/(ASCE)ST.1943-541X.0000074

Moustafa, A., and Takewaki, I. (2010a). Critical characterization and modeling of pulse-like near-fault strong ground motion. *Struct. Eng. Mech.* 34, 755–778. doi:10.12989/sem.2010.34.6.755

Moustafa, A., and Takewaki, I. (2010b). Deterministic and probabilistic representation of near-field pulse-like ground motion. *Soil Dyn. Earthq. Eng.* 30, 412–422. doi:10.1016/j.soildyn.2009.12.013

Moustafa, A., Ueno, K., and Takewaki, I. (2010). Critical earthquake loads for SDOF inelastic structures considering evolution of seismic waves. *Earthq. Struct.* 1, 147–162. doi:10.12989/eas.2010.1.2.147

Pantelides, C. P., and Tzan, S. R. (1996). Convex model for seismic design of structures: I analysis. *Earthq. Eng. Struct. Dyn.* 25, 927–944. doi:10.1002/(SICI)1096-9845(199609)25:9<927::AID-EQE594>3.0.CO;2-H

Pirasteh, A. A., Cherry, J. L., and Balling, R. J. (1988). The use of optimization to construct critical accelerograms for given structures and sites. *Earthq. Eng. Struct. Dyn.* 16, 597–613. doi:10.1002/eqe.4290160410

Qiu, Z. P. (2003). Comparison of static response of structures using convex models and interval analysis method. *Int. J. Numer. Methods Eng.* 56, 1735–1753. doi:10.1002/nme.636

Shinozuka, M. (1970). Maximum structural response to seismic excitations. *J. Eng. Mech. Div. ASCE* 96, 729–738.

Soong, T. T., and Dargush, G. F. (1997). *Passive Energy Dissipation Systems in Structural Engineering*. Chichester: John Wiley & Sons.

Srinivasan, A. V., and McFarland, D. M. (2000). *Smart Structures: Analysis and Design*. Cambridge: Cambridge University Press.

Srinivasan, M., Corotis, R., and Ellingwood, B. (1992). Generation of critical stochastic earthquakes. *Earthq. Eng. Struct. Dyn.* 21, 275–288. doi:10.1002/eqe.4290210401

Stein, R. S. (2003). Earthquake conversations. *Sci. Am.* 288, 72–79. doi:10.1038/scientificamerican0103-72

Strasser, F. O., and Bommer, J. J. (2009). Large-amplitude ground-motion recordings and their interpretations. *Soil Dyn. Earthq. Eng.* 29, 1305–1329. doi:10.1016/j.soildyn.2009.04.001

Takewaki, I. (2000). Optimal damper placement for critical excitation. *Probab. Eng. Mech.* 15, 317–325. doi:10.1016/S0266-8920(99)00033-8

Takewaki, I. (2001a). A new method for nonstationary random critical excitation. *Earthq. Eng. Struct. Dyn.* 30, 519–535. doi:10.1002/eqe.21

Takewaki, I. (2001b). Critical excitation for elastic-plastic structures via statistical equivalent linearization. *Probab. Eng. Mech.* 17, 73–84. doi:10.1016/S0266-8920(01)00030-3

Takewaki, I. (2004). Bound of earthquake input energy. *J. Struct. Eng. ASCE* 130, 1289–1297. doi:10.1061/(ASCE)0733-9445(2004)130:9(1289)

Takewaki, I. (2005). Bound of earthquake input energy to soil-structure interaction systems. *Soil Dyn. Earthq. Eng.* 25, 741–752. doi:10.1016/j.soildyn.2004.11.017

Takewaki, I. (2006/2013). *Critical Excitation Methods in Earthquake Engineering*, 1st and 2nd Edn. London: Elsevier.

Takewaki, I. (2008a). *"Critical Excitation Methods for Important Structures", Invited as a Semi-Plenary Speaker*. Southampton: EURODYN.

Takewaki, I. (2008b). Robustness of base-isolated high-rise buildings under code-specified ground motions. *Struct. Des. Tall Spec. Build.* 17, 257–271. doi:10.1002/tal.350

Takewaki, I. (2009). *Building Control with Passive Dampers: Optimal Performance-Based Design for Earthquakes*. Singapore: John Wiley & Sons Ltd.

Takewaki, I., and Ben-Haim, Y. (2005). Info-gap robust design with load and model uncertainties. *J. Sound Vib.* 288, 551–570. doi:10.1016/j.jsv.2005.07.005

Takewaki, I., Conte, J. P., Mahin, S. A., and Pister, K. S. (1991). A unified earthquake-resistant design method for steel frames using ARMA models. *Earthq. Eng. Struct. Dyn.* 20, 483–501. doi:10.1002/eqe.4290200508

Takewaki, I., and Fujita, K. (2014). "Robust control of building structures under uncertain conditions," in *Encyclopedia of Earthquake Engineering*, eds M. Beer, E. Patelli, I. Kougioumtzoglou, and I. Siu-Kui (Berlin, Heidelberg: Springer-Verlag).

Takewaki, I., Fujita, K., and Yoshitomi, S. (2013). Uncertainties in long-period ground motion and its impact on building structural design: case study of the 2011 Tohoku (Japan) earthquake. *Eng. Struct.* 49, 119–134. doi:10.1016/j.engstruct.2012.10.038

Takewaki, I., Moustafa, A., and Fujita, K. (2012). *Improving the Earthquake Resilience of Buildings: The Worst Case Approach*. London: Springer.

Takewaki, I., Murakami, S., Fujita, K., Yoshitomi, S., and Tsuji, M. (2011a). The 2011 off the Pacific coast of Tohoku earthquake and response of high-rise buildings under long-period ground motions. *Soil Dyn. Earthq. Eng.* 31, 1511–1528. doi:10.1016/j.soildyn.2011.06.001

Takewaki, I., Nakamura, M., and Yoshitomi, S. (2011b). *System Identification for Structural Health Monitoring*. Southampton: WIT Press.

Takewaki, I., and Tsujimoto, H. (2011). Scaling of design earthquake ground motions for tall buildings based on drift and input energy demands. *Earthq. Struct.* 2, 171–187. doi:10.12989/eas.2011.2.2.171

Trifunac, M. D. (2008). Energy of strong motion at earthquake source. *Soil Dyn. Earthq. Eng.* 28, 1–6. doi:10.1039/c4sm00280f

Tzan, S. R., and Pantelides, C. P. (1996). Convex models for impulsive response of structures. *J. Eng. Mech. ASCE* 122, 521–529. doi:10.1061/(ASCE)0733-9399(1996)122:6(521)

Conflict of Interest Statement: The author declares that the research was conducted in the absence of any commercial or financial relationships that could be construed as a potential conflict of interest.

Permissions

The contributors of this book come from diverse backgrounds, making this book a truly international effort. This book will bring forth new frontiers with its revolutionizing research information and detailed analysis of the nascent developments around the world.

We would like to thank all the contributing authors for lending their expertise to make the book truly unique. They have played a crucial role in the development of this book. Without their invaluable contributions this book wouldn't have been possible. They have made vital efforts to compile up to date information on the varied aspects of this subject to make this book a valuable addition to the collection of many professionals and students.

This book was conceptualized with the vision of imparting up-to-date information and advanced data in this field. To ensure the same, a matchless editorial board was set up. Every individual on the board went through rigorous rounds of assessment to prove their worth. After which they invested a large part of their time researching and compiling the most relevant data for our readers.

The editorial board has been involved in producing this book since its inception. They have spent rigorous hours researching and exploring the diverse topics which have resulted in the successful publishing of this book. They have passed on their knowledge of decades through this book. To expedite this challenging task, the publisher supported the team at every step. A small team of assistant editors was also appointed to further simplify the editing procedure and attain best results for the readers.

Apart from the editorial board, the designing team has also invested a significant amount of their time in understanding the subject and creating the most relevant covers. They scrutinized every image to scout for the most suitable representation of the subject and create an appropriate cover for the book.

The publishing team has been an ardent support to the editorial, designing and production team. Their endless efforts to recruit the best for this project, has resulted in the accomplishment of this book. They are a veteran in the field of academics and their pool of knowledge is as vast as their experience in printing. Their expertise and guidance has proved useful at every step. Their uncompromising quality standards have made this book an exceptional effort. Their encouragement from time to time has been an inspiration for everyone.

The publisher and the editorial board hope that this book will prove to be a valuable piece of knowledge for researchers, students, practitioners and scholars across the globe.

List of Contributors

Kohei Fujita, Ayumi Ikeda and Izuru Takewaki
Department of Architecture and Architectural Engineering, Kyoto University, Kyoto, Japan

Kotaro Kojima and Izuru Takewaki
Department of Architecture and Architectural Engineering, Graduate School of Engineering, Kyoto University, Kyoto, Japan

Kotaro Kojima and Izuru Takewaki
Department of Architecture and Architectural Engineering, Graduate School of Engineering, Kyoto University, Kyoto, Japan

Katsuichiro Goda and Raffaele De Risi
Department of Civil Engineering, University of Bristol, Bristol, UK,

Friedemann Wenzel
Geophysical Institute, Karlsruhe Institute of Technology, Karlsruhe, Germany

Ayumi Ikeda, Kohei Fujita and Izuru Takewaki
Department of Architecture and Architectural Engineering, Kyoto University, Kyoto, Japan

Tomaso Trombetti, Michele Palermo, Antoine Dib, Giada Gasparini, Stefano Silvestri and Luca Landi
Department of Civil, Chemical, Environmental, and Materials Engineering, University of Bologna, Bologna, Italy

Kotaro Kojima
Department of Architecture and Architectural Engineering, Graduate School of Engineering, Kyoto University, Kyoto, Japan

Philippe Guéguen and Ismael Riedel
Institute of Earth Science, Université Grenoble Alpes/ CNRS/IFSTTAR, Grenoble, France

Hugo Yepes
Institute of Earth Science, Université Grenoble Alpes/ CNRS/IFSTTAR, Grenoble, France
Instituto Geofísico, Escuela Politécnica Nacional, Quito, Ecuador

Oren Lavan and David Abecassis
Faculty of Civil and Environmental Engineering, Technion – Israel Institute of Technology, Haifa, Israel

Katsuichiro Goda
Department of Civil Engineering, University of Bristol, Bristol, UK,

Takashi Kiyota, Rama Mohan Pokhrel, Gabriele Chiaro and Toshihiko Katagiri
Institute of Industrial Science, University of Tokyo, Tokyo, Japan,

Keshab Sharma
Department of Civil and Environmental Engineering, University of Alberta, Edmonton, AB, Canada,

Sean Wilkinson
School of Civil Engineering and Geosciences, Newcastle University, Newcastle upon Tyne, UK

Panagiotis Mergos
Research Centre for Civil Engineering Structures, Department of Civil Engineering, City University London, London, UK

Katrin Beyer
Earthquake Engineering and Structural Dynamics Laboratory, Department of Civil Engineering, École Polytechnique Fédérale de Lausanne, Lausanne, Switzerland

Angelos Tsatsis
Laboratory of Soil Mechanics, School of Civil Engineering, National Technical University of Athens, Athens, Greece

Ioannis Anastasopoulos
Division of Civil Engineering, School of Engineering Physics and Mathematics, University of Dundee, Dundee, UK

Solomon Tesfamariam
School of Engineering, The University of British Columbia, Kelowna, BC, Canada

Katsuichiro Goda
Department of Civil Engineering, University of Bristol, Bristol, UK

Naohiro Nakamura
Research and Development Institute, Takenaka Corporation, Chiba, Japan

Gian Paolo Cimellaro
Department of Civil and Environmental Engineering, University of California Berkeley, Berkeley, CA, USA, Department of Structural, Geotechnical and Building Engineering (DISEG), Politecnico di Torino, Turin, Italy

Sebastiano Marasco
Department of Structural, Geotechnical and Building Engineering (DISEG), Politecnico di Torino, Turin, Italy

Kotaro Kojima, Kohei Fujita and Izuru Takewaki
Department of Architecture and Architectural Engineering, Graduate School of Engineering, Kyoto University, Kyoto, Japan

Rishi Ram Parajuli
Department of Urban Management, Graduate School of Engineering, Kyoto University, Kyoto, Japan,

Junji Kiyono
Graduate School of Global Environmental Studies, Kyoto University, Kyoto, Japan

Masatoshi Kasagi, Kohei Fujita, Masaaki Tsuji and Izuru Takewaki
Department of Architecture and Architectural Engineering, Graduate School of Engineering, Kyoto University, Kyoto, Japan

Izuru Takewaki
Department of Architecture and Architectural Engineering, Kyoto University, Kyoto, Japan